Advances in High-speed Rail Technology

More information about this series at http://www.springer.com/series/13506

Limin Jia · Li Wang · Yong Qin

High-Speed Railway Operation Under Emergent Conditions

Theoretical Methodology and Applications

BJTUP

Springer

Limin Jia
Beijing Jiaotong University
Beijing, China

Li Wang
Beijing Jiaotong University
Beijing, China

Yong Qin
Beijing Jiaotong University
Beijing, China

ISSN 2363-5010 ISSN 2363-5029 (electronic)
Advances in High-speed Rail Technology
ISBN 978-3-662-63031-0 ISBN 978-3-662-63033-4 (eBook)
https://doi.org/10.1007/978-3-662-63033-4

Jointly published with Beijing Jiaotong University Press
The print edition is not for sale in China Mainland. Customers from China Mainland please order the print book from: Beijing Jiaotong University Press.

This Springer imprint is published by the registered company Springer-Verlag GmbH, DE part of Springer Nature.
The registered company address is: Heidelberger Platz 3, 14197 Berlin, Germany

Preface

The development of high-speed railway in the world has gone through three construction upsurges. The first upsurge can be traced back to the 1960s to the early 1990s. The Tokaido Shinkansen, which was put into operation in 1964 in Japan, is the world's first high-speed railway. Subsequently, France and Germany also started the construction of high-speed railways. This period is the beginning stage of high-speed railway construction. The second construction upsurge was from the early 1990s to the early twenty-first century. During this period, more and more developed countries recognized the importance of high-speed railways, especially in Europe, such as France, Germany, Spain, the Netherlands, and the UK, which have begun to build high-speed railways on a large scale. In this period, Europe has initially established a high-speed railway network. The third upsurge is since the beginning of the twenty-first century. The rapid development of China's high-speed railway has led to the revival of high-speed railway construction worldwide. In the meantime, China completed six major speed-ups of the railway and passed the Medium and Long-Term Railway Network Plan in 2004, becoming the country with the longest mileage of high-speed railways (operating mileage accounts for more than half of the world's total mileage).

With the rapid growth of the mileage and scale of high-speed railway construction, the high-speed railway has been networked in some countries (such as China), and there are mutual influence and complex relationship between different lines. The dynamics and spatial and temporal correlation of each line are also stronger. In this complex operating environment, the space affected by emergencies is more extensive, and the handling of incidents has also become much more difficult. Therefore, transportation organization management under emergencies conditions is particularly important.

At the beginning of the construction of high-speed railways, there have been several major accidents. For example, on June 3, 1998, the German high-speed rail ICE 884 train derailed when it was driving to a bridge in Escher, 101 people killed and 88 people injured. On April 25, 2005, the high-speed railway derailed in Amagasaki, Japan, resulting in 107 deaths and 549 people injured. On July 23, 2011, the Chinese Ningbo-Wenzhou railway line was damaged due to bad weather, resulting in a rear-end collision between D301 and D3115 trains. In the accident, 40 people were killed and 172 were injured. These incidents have caused huge losses to the people's lives and property, which has greatly affected the normal production and living order, and the social response is extremely bad. At the same time, these incidents pose serious challenges to railway safety and emergency management. Especially after the high-speed railway networked operation, the impact of emergencies will spread along the network, and the operational manager's work object is no longer an isolated line, but a dynamic network of railways affected by events. Therefore, studying the relevant theories, methods, and techniques of transport organization under the conditions of emergencies is an indispensable part of the rapid development of high-speed railways in the future.

At present, there are two kinds of researches on the emergency disposal of high-speed railway under abnormal conditions: One is to follow the traditional railway idea, and to modify the existing theoretical system and technical methods, but the research emphasis is still the traditional transportation organization. The research does not form a transportation organization strategy specifically for emergencies. Although this method can solve the theoretical problem to a certain extent, it lacks applicability to the complex and varied working environment of high-speed railway. The rapid development of high-speed railway has made the concept of operation and emergency management begin to change: The traditional emergency response is from the operator's point of view, with the primary goal of restoring railway operation as soon as possible, while the new operation and emergency management work aims to reduce overall losses (including social impact, passenger replacement service, minimum impact on the network, and so forth) , reflecting the "people-oriented" thinking. These changes in concept will inevitably lead to major changes in transportation organization strategies, methods, and technologies. The other research follows the general emergency management approach and focuses on the emergency management of railway emergencies. Its main contents include emergency plan system, emergency risk assessment, emergency platform and system construction, and so forth. Although it can provide research ideas for emergency disposal of high-speed railways, it does not involve transportation organization content, and it is not able to provide railway operation organization plans under emergency conditions, and cannot truly direct on-site train operations. There are certain limitations in theoretical systems and methods. In view of the current research status and needs in the field, the author continued to explore and practice the relevant theoretical systems, technical methods, and system platform construction of high-speed railway transportation organizations under the conditions of emergencies, aiming at perfecting, supplementing, and forming China's high-speed railway emergency management system, with more than 20 years of experience and scientific research accumulation. These constitute an important foundation for the formation of this book.

The original intention of this book is to provide a framework and related theories and methods for high-speed rail transportation organization and emergency management under emergency conditions. At the same time, it is also hoped to provide a systematic method for railway managers. It is expected to help improve and reconstruct the actual operation and management of the railway site. In order to balance the needs of the above two aspects, this book focuses on the scientific issues and theoretical methods related to railway emergencies and transportation organizations, and tries to achieve the theory with practice. All the methods, algorithms, and problem-solving processes in the book are based on the actual scene of railway operation, and the results can be used as a reference for actual railway operation or related operational support systems. Based on the theoretical methods and model algorithms proposed in this book, the "National Railway Emergency Plans Management System" and the "Railway Transport Organization Emergency Response Decision Support System (High-speed Railway)" have been developed scaled up and applied on a large scale in China Railways. The content of this book not only adapts to the characteristics of China's high-speed railway, such as long operating mileage, complex space-time relationship, diverse transport organization modes and difficult emergency disposal, but also takes the general needs of other countries and regions in high-speed railway operation organization and management into account, which has universal applicability.

This book explores the organization of high-speed railway transportation under the conditions of emergencies, which is explored and analyzed from three levels: basic theoretical exploration, method and technology research, and system platform construction. The book is divided into 10 chapters. The main contents are as follows.

Chapter 1 introduces the current situation of high-speed rail transportation organizations. From the analysis of the development of the world high-speed railway, the construction and operation of high-speed railways in various countries of the world are introduced, and the connotation and extension of the railway transportation system and transportation organization are expounded from a system perspective. Then, based on the analysis of the technical characteristics of China's high-speed railway operation, the relationship between railway transportation organization and emergency management is discussed. It is pointed out that the railway transportation organization process is an important part of railway emergency management under the emergency conditions. The technical documents and plans from the macro- to the micro-level in the process of railway transportation organization constitute the important management methods, scheduling strategies, and adjustment measures in the four stages of prevention, preparation, response, and recovery of railway emergency management under the emergency conditions.

Chapter 2 introduces some basic concepts and theories related to high-speed rail transportation organization under emergency conditions. These contents laid the foundation for a deep understanding of the essence of high-speed rail transportation organizations under emergencies. Firstly, it expounds the related concepts of emergencies and railway emergencies, then analyzes the differences of railway transportation organization problems under emergencies, and focuses on the definition, organization principles, organization strategy, optimization objectives and constraints, and transportation organization processes under different types of emergencies; finally, the complexity of transportation organization under severe emergencies is analyzed.

Chapter 3 analyzes the difference between the calculation for high-speed railway section carrying capacity under the conditions of emergency and the normal conditions, and points out that the section carrying capacity at this time has the characteristics of ambiguity and randomness. Based on the traditional train subtraction coefficient method, this chapter proposes a calculation method for section carrying capacity based on fuzzy Markov chain for the uncertainty of section carrying capacity under emergency conditions. This method is more fault-tolerant of various factors and their uncertainty. The effectiveness of the algorithm under emergencies is verified by the Beijing-Shanghai high-speed railway.

Chapter 4 analyzes the influencing factors of train route search problems under emergency conditions and establishes a route search model based on sufficient capacity. In addition, for its unique runtime sensitivity, interval passing ability as fuzzy random variables and spatial attributes, a definition of time density of line arc capability based on decision-maker preference is proposed, and the capability-time-density based route searching model is established. The route search model allows the reader to select the appropriate route search method based on actual site conditions.

Chapter 5 introduces the method for service plan compiling of high-speed railway trains under emergency conditions. Under the condition that the scope of the emergency has a wide range and long duration, the transportation organization of the high-speed railway will not only adjust the train timetable plan, but also adjust the train service plan. Firstly, the bi-level programming model of train service plan is introduced. The upper-level plan is used to optimize the stop station scheme, the lower-level plan is used to optimize the passenger flow distribution, and the loop feedback overall optimization between the stop station scheme optimization and passenger flow distribution is realized. Finally, an example analysis is carried out on the background of Taiwan high-speed rail and Beijing-Shanghai high-speed railway. It proves that the new method can generate multiple train service plans for different emergency conditions and increase the practicality of transportation organization strategy.

Chapter 6 is the main content of this book. Firstly, it analyzes the train running rules of high-speed railway under the conditions of emergencies. Then, based on the characteristics of emergencies, it establishes a bi-level programming model for train dispatching optimization and train operation adjustment. The top layer aims to optimize the strategies of detour the train, cancle the train and merge the train. After the top layer dispatching strategies are given, the lower-layer adopts strategies such as changing train interval running time, train stop time and train overtaking mode, and provides adjusted timetable for the lower layer. So as to realize the bi-level iterative overall optimization. Finally, according to the characteristics of emergencies, different objective functions are considered, and two solving algorithms are designed to better meet the actual situation on site.

Chapters 7 and 8 introduce the relevant models and methods for static and dynamic configuration and scheduling optimization of high-speed railway emergency resources. Aiming at the characteristics of high-speed railway emergencies, starting from the simplest single-point emergency point-single resource scheduling problem, and finally implementing multi-emergency point-based multi-resource emergency resource scheduling based on utility function.

Chapter 9 introduces the decision support system for high-speed railway train operation under emergency conditions. Firstly, it analyzes the system architecture, introduces three types of databases that support the operation of the system. Then three subsystems are described in detail: the acquisition and generation subsystem of special operating conditions for high-speed railway, high-speed railway transportation organization emergency plan preparation and decision support subsystem, high-speed train operation plan preparation subsystem under special operation conditions. Finally introduced the national railway emergency plan management system and railway transportation organization emergency response decision support system (high-speed railway). The application of the two demonstration projects laid the foundation for the promotion of relevant theoretical methods, technical systems, and system platforms proposed in this book.

Chapter 10 introduces the open questions that need to be addressed in the field for the majority of readers. These problems include high-speed railway operation rules, passenger flow forecasting under emergencies, coordinated organization and optimization of multiple modes of transportation under emergencies, and comprehensive information push for passengers under emergencies. They involve four stages of high-speed railway emergency managment, including prevention, preparation, response, and recovery. Some of these problems are the core contents of transportation organizations under emergencies, and some are important supplements and conditions. Therefore, these issues are organized into a chapter for scholars to study together.

This book covers both the relevant theoretical methods of transportation organization under emergencies, and specific system implementation and demonstration projects. Different readers can focus on different chapters. Railway field workers can focus on Chaps. 4–6 and Chap. 9 to help them form a more rational transportation organization strategy in their daily management work. At the same time, these contents have practical reference significance for the construction and deployment of the dispatching command system under emergency conditions. Researchers in the field of railway transport organization and emergency management can pay attention to Chaps. 1–3 and Chaps. 7, 8 and 10 to help them understand railway transportation organization and emergency management under emergencies from a new perspective. Thus they can get the research ideas and problem-solving paths aim at the integration of transportation organization and emergency management.

The content of this book integrates the achievement of various research projects undertaken by the author and the team over the years. These projects have been supported by the National Natural Science Foundation of China (Research on fuzzy stochastic hybrid intelligent optimization theory and method for high-speed railway train operation adjustment, No. 61074151; Research on intelligent integration model and theory of railway transportation system security monitoring information, No. 60074002; Research on Theory and Methodology of Train Operation Adjustment in Urban Regional Railway Network under Special Conditions, No.71701010), the National Key Research and Development Programme of China (Safety and Security Technology of High-speed Railway System, No. 2016YFB1200401), the "Eleventh Five-Year" National Science and Technology Support Program (Key technologies and systems for optimal design of high-speed train operation organization plan, No. 2009BAG12A10), Ministry of Railways Science and Technology Development Plan (Research on the comprehensive safety monitoring system of Qinghai-Tibet Railway and research on railway emergency plan management, No. 2008F002), Qinghai-Tibet Railway Corporation Science and Technology Research and Development Plan (Research and development of the Qinghai-Tibet Railway emergency rescue system and emergency rescue command information system, No. QZ2007-Z13), Huo Ying-dong Fund Young Teachers Award Program (Research on high-speed railway transportation organization optimization technology, No. 114023), and an international cooperation program with IBM (Project of Railway Asset Condition Analysis and Scheduling Optimization for Emergency Responses, No: JSA201002001). These projects also received strong funding and support from the State Key Laboratory of Rail Traffic Control and Safety and the School of Transportation and Transportation of Beijing Jiaotong University, as well as Beijing Urban Traffic Information Intelligent Perception and Service Engineering Technology Research Center.

Some of the contents of this book also include the work of the following collaborators, including Dr. Xu Jie, Associate Professor of Beijing Jiaotong University; Dr. Meng Xuelei, Associate Professor of Lanzhou Jiaotong University; Dr. Zhou Huijuan, Associate Professor of North China University of Technology; Dr. Zhu Tao, Senior Engineer of China Railway Information Technology Center and Dr. Guo Jianyuan, Associate Professor of Beijing Jiaotong University and several other experts and scholars. Meng Xuelei and Zhou Huijuan's dissertation are also important sources of relevant content in this book. The authors express their heartfelt thanks to their contributions.

In the process of writing this book, graduate students of Beijing Jiaotong University Zhang Huiru, Wang Mingming, Dou Fei, Gao Jianghua, Zhang Lanxia, and Guo Dan, also participated in the discussion and correction of some contents; Chen yueqin, the responsible editor of this book from Beijing Jiaotong University Press, gave great help to the publication of this book. The authors also express their heartfelt thanks to them.

With the increasing scale of the high-speed railway network, the relationship between transportation efficiency and security is becoming closer, and the impact of emergencies on the overall network is increasing. Transportation organizations under emergencies will inevitably play an increasingly important role in the management of high-speed railway operations.

The author hopes that this book will play a role in the integration of transportation organization and emergency management, and hopes that more colleagues will participate in the research and application of this field.

Due to the author's low level and the limitations of work, the omissions and even inadequacies in the book are unavoidable. I sincerely request readers to criticize and correct.

Beijing, China — Limin Jia
January 2016 — Li Wang

Contents

Chapter 1
High-Speed Railway Transportation Organization Status

High-speed railway refers to an existing railway with a speed of over 200 km/h or a new railway with a speed of over 250 km/h [1]. With the rapid development of the economy and the improvement of people's living standards, people have more and more choices of travel modes, and more and more attention is paid to the improvement of travel quality. High-speed railways have developed rapidly around the world with low energy consumption, low environmental pollution, large transportation capacity, good safety, fast speed, saving land resources, and high punctuality [2]. Since Japan built the first high-speed railway, the Tokaido Shinkansen, the high-speed railway has gradually become the trend of railway development, and it also reflects the comprehensive strength of various countries. Although China's research on high-speed railway-related technologies is carried out relatively late, with the Beijing-Tianjin intercity railway, Wuhan-Guangzhou high-speed railway, Zhengzhou-Xi'an high-speed railway, and Beijing-Shanghai high-speed railway successively opened, China's high-speed railway is leading the development of world's high-speed railway.

At present, high-speed railways have been built and operated in China, Spain, Japan, Germany, France, Belgium, Austria, Sweden, Britain, Italy, Russia, Turkey, South Korea, the Netherlands, Switzerland, Brazil and many other countries and regions. According to the statistics of the International Railway Union, as of September 10, 2016, the total operating mileage of high-speed railways in various countries and regions of the world is 35,834 km, the scale of high-speed railways under construction is 44,648 km, and the planned construction of high-speed railways is 41,880 km [3].

L. Jia et al., *High-Speed Railway Operation Under Emergent Conditions*,
Advances in High-speed Rail Technology,
https://doi.org/10.1007/978-3-662-63033-4_1

1.1 Overview of High-Speed Railway Development in the World

As a safe, reliable, fast and comfortable, large-capacity, low-carbon, and environmentally friendly transportation mode, high-speed railway has become an important trend in the development of the world's railways. Throughout the development of high-speed railway in the world, it can be divided into three stages, forming three construction upsurge.

The first was from the 1960s to the early 1990s. On October 1, 1964, Japan opened the world's first high-speed rail system, the Tokaido Shinkansen, with an operating speed of 270 km/h. Later, the JR East, and Hayabusa were built, which basically formed the main structure of the high-speed railway network nationwide in Japan. Subsequently, France, Italy, Germany also began to build high-speed railways, including the French TGV southeast line and TGV Atlantic line, Italy's Rome to Florence line, and Germany's Hanover to Würzburg line. During this period, the total length of the world high-speed railway reached 3327 km, which was the beginning stage of high-speed railway construction.

The second was from the early 1990s to the beginning of the twenty-first century. After the first high-speed railway construction upsurge, more and more countries are aware of the importance of high-speed railways. Among them, France, Germany, Spain, Netherlands, the UK, and other European countries began to build their own railways on a large scale. Especially in 1994, Britain and France built the first international high-speed railway, which connected London and Paris through the English Undersea Tunnel. In 1997, the "Eurostar" train from Paris connected France, Belgium, Netherlands, and Germany together to form the European high-speed rail network [4]. During this period, the total length of the world high-speed railway reached 6474 km, and the construction of high-speed railway entered a period of rapid development. In just 10 years, the mileage of the world's high-speed railway has doubled.

The third time is the beginning of the 21st century. This phase has embarked on a worldwide wave of high-speed rail construction, involving Asia, North America, South America, Europe, and Oceania. In the meantime, the Chinese railway has speeded up several times. In 2004 China passed the "medium and long-term railway network planning", which marked the beginning of the period of large-scale construction of China's high-speed railway. China's Beijing-Tianjin intercity railway, Beijing-Shanghai high-speed railway, Zhengzhou-Xi'an high-speed railway, and Japan's Sanyo Shinkansen have designed and operated at speeds of more than 300 km per hour. The USA, South Korea, Australia, Belgium, Spain, Switzerland, Canada, India, Turkey, and other countries have also joined the army of high-speed railway construction. In addition, Sweden, Morocco, Russia, the Czech Republic, Brazil, and other countries are also actively planning and preparing for the construction of high-speed railways. Table 1.1 describes the history of high-speed railway construction in countries around the world. As of September 10, 2016, the four countries with the longest operating distances in the world are China, Spain, Japan, and France, with 20,000 km,

Table 1.1 Construction of high-speed railways in countries around the world (As of September 10, 2016)

Serial number	Country	First operating time	Maximum operating speed/(km/h)	Operating mileage/km	Planned construction mileage/km
1	Japan	1964	320	2664	958
2	France	1983	320	2036	3164
3	Italy	1981	300	923	346
4	Germany	1991	300	1352	790
5	Spain	1992	310	3100	3010
6	Belgium	1997	300	209	–
7	China	2003	350	20,000	25,000
8	United Kingdom	2003	300	1377	–
9	Korea	2004	300	819	613
10	Switzerland	2007	250	80	57
11	Netherlands	2009	300	120	
12	Turkey	2009	250	1420	1506
13	Austria	2012	250	292	201
14	United States	2012	300	362	777
15	Saudi Arabia				550
16	Morocco				680
17	Russia	2009	250	645	770
18	Poland	2014	85		627
19	Portugal				1006
20	Sweden				750
21	India				495
22	Brazil				511
23	Denmark	2000	200	5	60

3100 km, 2664 km, and 2036 km [3], respectively. During this period, the mileage of world high-speed railway operations nearly tripled.

1.1.1 High-Speed Railway in Japan

On October 1, 1964, the world's first high-speed railway, the Tokaido Shinkansen in Japan, was put into operation, marking the transition of the world's high-speed railway from experimental stage to the commercial operation stage. The Tokaido Shinkansen was built in 1959 with a total length of 515.4 km and a maximum operating speed of 210 km/h. The operation of the Tokaido Shinkansen has shortened the travel time from Tokyo to Osaka by nearly half, from the original 6 hours 30 minutes to 3 hours 10 minutes. The Tokaido Shinkansen has won the support and welcome of the general public with its advantages of safety, speed, punctuality, and comfort, and quickly occupied the transportation market. The daily average traffic volume reached 360,000 passengers, equivalent to the passenger volume of 10 expressways. The construction of the Tokaido Shinkansen has alleviates the tense passenger transportation demand in the Tokaido area, making traffic smooth and safe in major cities such as Tokyo, Yokohama, Nagoya, and Osaka, and has achieved good economic benefits. The construction of Tokaido Shinkansen makes the railway, which was once dismissed as a "sunset industry", show a strong vitality and bring about a new era of railway development.

Since the completion of the Tokaido Shinkansen in 1964, the Japanese Shinkansen has been developed from the original one to the present six [5]: Tokaido Shinkansen, Sanyo Shinkansen, Tohoku Shinkansen, Joetsu Shinkansen, Hokuriku Shinkansen, Kyushu Shinkansen, operating mileage of 2664 km. In 2004, the Shinkansen transported 294.258 million passengers and completed passenger turnover of 74.67 billion person-km. In 2005, passenger turnover was about 316 million passengers. On March 5, 2011, the fastest train of the Shinkansen in Japan, the bullet train "The Hayabusa," was put into operation.

Since the operation of the Shinkansen in Japan in 1964, no major casualties have occurred, and the average delay time is 1 min. It can be said that high safety and high accuracy have become important features of the Japanese high-speed railway.

1.1.2 High-Speed Railway in French

France has long been working on improving the technology related to train speed. In October 1976, France began construction of the TGV southeast line. In September 1983, the entire line was completed and opened, with a maximum operating speed of 270 km. After the line was opened, the passenger traffic volume increased rapidly. The travel time from Paris to Lyon was shortened from 3 hours and 50 minutes to 2 hours, achieving the expected economic benefits. The TGV southeast line is the first high-speed rail line in Europe. In 1990 and 2007, the test speed set a world record for wheel and rail speed of 515.3 km and 574.8 km, respectively [6].

In order to reduce the number of passenger transfers, the French high-speed railway adopts the transportation mode of high-speed trains running at high-speed lines. By extending the running distance of TGV high-speed trains, it expands its access range and thus expands passenger flow.

1.1.3 High-Speed Railway in German

With the development of high-speed railways, Germany formally developed a high-speed railway system in 1986 to implement the intercity express (ICE) program. A total of 107 km of the Mannheim to Stuttgart line was built in June 1991, and a 327 km Hannover-Würzburg line was built in 1992. The maximum operating speed reached 280 km/h and set an experimental speed record of 406.9 km/h at the time. With the development of ICE, the German capital has been incorporated into the ICE high-speed transportation system. Not only that, ICE also realized international direct transportation from Zurich to Frankfurt.

In Germany's passenger flow organization of high-speed railways, although passengers transfer conditions are very good, Germany has adopted a large number of ICE trains and IC trains off the high-speed line to reduce passenger transfer. In addition, Germany also uses a variety of speed trains on high-speed lines to increase transport capacity.

1.1.4 High-Speed Railway in China

As of September 10, 2016, 70 high-speed railway lines have been completed and put into operation in China. The total mileage exceeded 20,000 km, accounting for more than half of the world's total high-speed railway mileage. According to the plan, the total mileage will reach nearly 45,000 km by 2050.

On December 22, 1994, China's first 160-km-long Guangzhou-Shenzhen high-speed railway was completed and opened to traffic. Its operation provides valuable experience for China's high-speed railway development and the speed up of existing railway lines.

China Railways carried out five large-scale accelerations from 1997 to 2004. In 2007, the sixth major speed increase was successfully implemented by transforming the range radius, bridge tunnel, and switch of 18 existing trunk lines, such as Beijing-Harbin, Beijing-Shanghai, Beijing-Guangzhou, Beijing-Kowloon, Lanzhou-Lianyungang, ShangHai-Kunming, Lanzhou-Xinjiang, Guangzhou-Shenzhen, and Qiangdao-Jinan, as well as overhauling and updating the equipment of train speed raising system, passenger transport facilities, cross-line facilities, and other related facilities. In the early stage of the sixth speed increase. China carried out a series of transformation and innovation on railway vehicles, and the China Railway High-speed Train (CRHT) of 250 km per hour was put into use in batches.

1. "Eight vertical and Eight horizontal" high-speed railway network planning

In 2004, the executive meeting of China State Council discussed and adopted the "Medium- and Long-Term Railway Network Plan", which is the first industry plan approved by the State Council of China and the blueprint for China's railway construction up to 2020. In 2008, the plan was adjusted by the state, and it was determined that by 2020, the scale of China's railway network will reach 150,000 km.

The main busy trunk lines will be operated by passenger and cargo lines respectively. The passenger dedicated line will reach 16,000 km, and the double line rate will reach 50%. The electrification rate is 60%, the transportation capacity meets the needs of national economic and social development, and the main technical equipment reaches or approaches the international advanced level. China's high-speed railway development is "eight vertical" (coastal channel, Beijing-Shanghai channel, Beijing-Hong Kong (Taiwan) channel, Beijing-Harbinand Beijing-Hong Kong-Macao channel, Huhhot-Nanjing channel, Beijng-kunming channel, Baotou-Haikou channel, Lanzhou-Guangzhou channel) eight horizontal (Suifen River-Manzhouli channel, Beijing-Lanzhou channel, Qingdao-Yinchuan channel, land bridge channel, along the river channel, Shanghai-Kunming channel, Xiamen-Chongqing channel, Guangzhou-Kunming channel) is the main skeleton, and the main skeleton of the rapid passenger transportation network is constructed. The following is an introduction to the development of China's high-speed railways, taking the Beijing-Tianjin intercity railway, Zhengzhou_Xi'an high-speed railway, Wuhan-Guangzhou high-speed railway, and Beijing-Shanghai high-speed railway that have been opened to traffic.

2. Beijing-Tianjin intercity railway

The Beijing-Tianjin Intercity Railway is the earliest high-standard railway passenger line built in China. It started construction on July 4, 2005 and was completed on December 15, 2007. This line is operated on August 1, 2008. The total length of Beijing-Tianjin intercity railway is about 120 km, along which there are 4 stations including Beijing South, Yizhuang, Wu Qing and Tianjin.Yongle Station is reserved. About 85% of the line are elevated lines. The railway was designed for a top speed of 350 km/h, and the CRH_3 train hit a speed of 394.3 km/h during the test. The train running time from Beijing to Tianjin is 30 minutes. The line adopts the transportation organization mode of the mixed intercity train and the interline train. The minimum tracking interval of the train is 3 minutes. The trains are usually grouped with 8 vehicles and a capacity of 600 person. During the peak hours, some trains are grouped with 16 vehicles and 1200 person. The Beijing-Tianjin intercity railway train adopts China's CTCS—3D train operation control system with independent intellectual property rights, which marks a qualitative leap in China's railway modernization construction.

3. Zhengzhou-Xi'an high-speed railway

The Zhengzhou-Xi'an high-speed railway is the first high-speed railway in the central and western China. On June 28, 2009, the Zhengzhou-Xi'an Passenger Dedicated Line was completed. On February 6, 2010, the line was officially put into operation. The line starts from Xinzhengzhou Station in the east and Xi'an North Station in the west. The main line is 456.639 km long, and another line extends from Xi'an North Station to the west of Xianyang West Station, with a length of 27.879 km. The Zhengzhou-Xi'an high-speed railway has a total length of 484.518 km, bridges and tunnels accounting for 59.75%, and a two-line design with 13 stations along the way. The design speed is 350 km, and the highest speed is 394.2 km. The construction of the Zhengzhou-Xi'an high-speed railway shortened the travel time from Xi'an to Beijing to 4 hours from the original 11 hours, and the running time from Xi'an to Shanghai was shortened from 15 to 5 h.

The Zhengzhou-Xi'an high-speed railway adopts the world's advanced level of dispatch control technology. Its power dispatching system uses traction power supply and power integration. The centralized dispatching and control of the whole line depends on the CTC transportation dispatch command system. The running train adopts the CRHT with completely independent intellectual property rights. The trains are grouped with 8 vehicles and a capacity of 600 persons, and the maximum annual transportation capacity is 83.4 million.

4. Wuhan-Guangzhou high-speed railway

The Wuhan-Guangzhou high-speed railway is located in the three provinces of Hunan, Hubei, and Guangdong. It starts from Wuhan Station and passes through Changsha, Hengyang, and other stations to reach Guangzhou South Railway Station. There are 15 stations with a total length of about 1068.8 km. It started construction in Changsha on June 23, 2005, and officially opened on December 26, 2009. According to the characteristics of China's high-speed railway and passenger dedicated line built into the network and compatible operation, the Wuhan-Guangzhou high-speed railway implements a common line operation plan for the 350 km high-speed train and the over-speed 250 km/h speed train. The transportation organization mode is divided into daily and weekend. It created a new mode for the world's high-speed rail transport organization. The Wuhan-Guangzhou high-speed railway has a design speed of 350 km and an average speed of 341 km. The total operating time is only 188 min. The maximum speed of the test operation is 394.2 km, which creates the highest speed record of the high-speed railway in the case of two-vehicle reconnection.

The Wuhan-Guangzhou high-speed railway train adopts the CRHT developed by China, and through the innovation of the car body, wheel rail, and braking technology, the train has better operational capability. The minimum tracking interval of the whole line is 3 minutes. Usually, 70 pairs of trains are driven every day, including 60 pairs of CRHT with a speed of 300 km per hour and 10 pairs of CRHT with a speed of 250 km per hour. At the peak period, 80 pairs of CRHT are driven every day.

5. Beijing-Shanghai high-speed railway

The world-famous Beijing-Shanghai high-speed railway started construction on April 18, 2008, and officially opened on June 30, 2011. The line starts at Beijing South Railway Station and ends at Shanghai Hongqiao Station. It covers seven provinces and municipalities including Beijing, Tianjin, Shanghai, Hebei, Anhui, Shandong, and Jiangsu. It is about 1318 km long and is the longest, the most expensive and the highest standard high-speed railway in the history of China. It has a design speed of 350 km and an operating speed of 300 and 250 km, with a local top speed of 380 km. The whole running time is 5–7 h, and the minimum headway time of the train is designed to be 3.5 min. The existing Beijing-Shanghai railway will be diverted by passengers and goods, and the passengers can be transported by more than 80 million in one direction annually.

The CRH380A, CRH380B, CRH380AL, CRH380BL, CRH380C, and CRH380CL are used in the Beijing-Shanghai high-speed railway. In future, the CRH380D and CRH380DL will be put into use. During the train commissioning phase, the CRH380A set a world record speed for railways operating at 486.1 km / h.

1.2 Overview of Railway Transport Organization Research

The previous section described the development of the world's high-speed railways. This section will elaborate on the rail transport organization from a systematic perspective.

1.2.1 Railway Transportation System

The system exists in different forms, and various classifications can be performed according to the reasons generated by the system and the attributes reflected. According to the origin of the system, it is divided into natural systems and artificial systems. According to the scale and structure of the system, it is divided into simple system and large system. According to the time characteristics of the system, it is divided into dynamic system and static system. According to the relationship between the system and the external environment, it is divided into an open system and a closed system.

The Railway Train Operation Organization System is an open artificial system [7]. The railway transportation system under natural and emergency conditions is also an open artificial system. In order to adapt to the constantly changing social environment, people must constantly modify and improve the various components of the railway transportation system, which requires the railway transportation system to have sufficient flexibility and adjustability.

The railway transport system is also a multi-target system. The overall goal of the railway transport system is to achieve macro-social benefits and the benefits of the railway operation department. But the specific goals are diverse, including efficient, fast, economical, comfortable, safe, and environmentally friendly, and it is difficult to meet the above requirements. This is because there is a very strong "alternative profit and loss" or "trade off" phenomenon [8] between the functional elements of the railway transportation system. That is, while the optimization of one function and the interest occur, there will be one or several functional benefits losses, such as the direct train rate and the cost of rail transport, have significant adverse effects. Such multi-objective conflicts are common in railway transportation systems, especially under emergencies. Therefore, it is necessary to coordinate various elemental objectives under the overall goal in order to obtain the overall optimal effect of the railway transportation system. Especially in the case of emergencies, this kind of coordinated operation and the choice or bias of the target is particularly important.

1. The composition of the system

The structure of the system is the way in which the various elements of the system are related to each other and interact, that is, the form of the specific connection between the various elements, which is the inherent basis for the system to maintain integrity and have certain functions.

The elements of the rail transport system are equipment and staff. The equipment is further divided into fixed equipment and mobile equipment. The fixed equipment includes lines and stations, and the mobile equipment includes trains and bullet trains. The staff includes production staff and management staff.

In the system structure of the railway transportation system, the time–space relationship between trains and lines and stations is the most important system relationship. The core work of the railway transportation organization is to arrange the relationship between trains and lines and stations. The specific forms of transportation include transportation organization mode, passenger train service plan, train operation chart, and bullet train operation plan, etc.

Divide the railway transportation system under emergencies with large granularity, that is, divide the system into workers, station lines, equipment. There are different complexities between workers, station lines and equipment between different types of railways. So the internal structure of the railway transportation system under the emergency conditions is very complicated.

The railway transportation system is a large system. China's railway transportation system not only includes the general railway station line, equipment, and staff, but also includes the intercity railway, high-speed railway. It is a system with many factors and a very large scale. Under the conditions of emergencies, the elements of the railway transportation system, that is, various types of station lines, equipment, and workers, need to be uniformly deployed and coordinated. And the relationship between these elements becomes extremely complicated. The structure of the system has also become complicated. Moreover, under the emergency conditions, the elements in the railway transportation system, have a strong randomness in the state attribute, which makes the structure of the railway transportation system also uncertain. Therefore, it can be concluded that the railway transportation system under the emergency conditions has all the characteristics of large system discrimination conditions, and the railway transportation system under the emergency conditions is a typical large system.

The mathematical description of the composition of the railway transportation system is [9]

$$S_{\text{trans}} = (S_{\text{station}}, R_{\text{rail}}, T_{\text{train}}, F_{\text{faculty}}, R)$$

among them:

S_{trans} railway transportation system;

$S_{station}$ a collection of stations in the system;
R_{rail} a collection of railway lines in the system;
T_{train} a collection of trains in the system;
$F_{faculty}$ a collection of production personnel in the system, which determines the organizational structure of the dispatch management of the railway transportation system;
R The relationship between the various elements in the system. This relationship determines the transportation organization mode in the railway transportation system. The relationship between lines, stations, and trains is different, and the transportation organization mode is also different.

2. System boundaries

Considering the railway transportation system as the object of analysis, the passenger demand and natural environmental conditions constitute the environment of the system. The boundary of the system is the state node of the transport organization in the system that acts and ends the functional object. For the railway transportation system, the boundary of the system is the OD point of the railway transportation system relative to the passenger and cargo flow, that is, the station where the passenger and cargo flow are generated and disappeared, consisting of all the stations with passenger or cargo transportation operations in the railway network.

Incidents can cause station failures in railway transportation systems, so the boundaries of rail transport systems under emergent conditions may differ from those of rail transport systems under normal conditions. That is, under the conditions of an emergency, the boundary of the railway transportation system may be smaller than the boundary of the railway transportation system under normal conditions. Its mathematical description is: The boundary of the railway transportation system under normal conditions is $S_e = S_{station}$. A indicates a railway emergency, and the boundary of the railway transportation system is in the event of an emergency is $S_e^A = S_{station}^A$, among them, $S_e^A \subseteq S_e$. $S_{station}^A$ is a collection of stations with passenger or cargo transport operations in a rail transport system for emergency conditions.

3. System function and behavior

The function of the railway transportation system under normal conditions and emergency conditions is to realize the displacement of passengers and goods. The function of the railway transport system is determined by the interrelationship between the railway station lines, trains, and transport production personnel that make up the system. Under normal conditions, the railway transportation system achieves its functions under the guidance of various transportation plans. It can be seen that the transportation production plan determines the relationship between the components of the railway transportation system. Under the emergency conditions, the external environment of the railway transportation system has changed, and this change has caused the relationship between the various elements of the transportation system to change, so that its function is affected, that is, the original transportation plan cannot be completed. At this point, you need to re-adjust the transportation plan to restore its function or approach the original function of the system. Its mathematical description is as follows:

Railway transportation system $S_{trans} = (S_{station}, R_{rail}, T_{train}, F_{faculty}, R)$,$\forall R^0$, Make S_{trans} in the environment E^0 with function F^0.

Sudden events force the internal relationship of the railway transportation system R Change, that is $R^0 \rightarrow R'$, to make the function $F^0 \rightarrow F'$.

Looking for measures A to make $R' \rightarrow R''$,$S_{trans} = (S_{station}, R_{rail}, T_{train}, F_{faculty}, R'')$ make F' Approach F^0.

That is, the emergency affects the normal passenger and cargo transportation functions of the railway transportation system by affecting the changes in the internal elements of the railway transportation system. Measures are need to be taken to make the relationship to be changed again between the internal components of the system. This is embodied in the train operation organization scheme, so that the function of the railway transportation system under the emergency conditions approaches the railway transportation system functions before the emergency.

1.2.2 Transport Organization of Railway Transportation System

1. Railway transport organization definition

Many scholars have proposed the definition of railway transportation organization from different angles, such as macro and micro angles. Wu Feng believes that the comprehensive use of a variety of equipment to organize and coordinate the inside and outside technology and business of all aspects of the railway is the railway transport organization [10]. Yang Hao proposed that the transportation organization is a theory and technology [11] for the scientific, economic, rational allocation and utilization of transportation resources, developed in the production and operation practices of transportation enterprises. The management and optimization issues of the transportation organization were elaborated from various angles such as the use of vehicles, transportation networks and transportation production.

We believe that the transportation organization of the railway transportation system refers to the railway transportation industry uses scientific and technological means to scientifically, economically, and rationally allocate and utilize various transportation resources such as fixed and mobile facilities and equipment, time, space, and human resources, and to achive the displacement of people and objects quickly, reliably and economically.

This definition can truly reflect the essence of railway transportation organization, is not restricted by various transportation devices, and adapts to the changing process of transportation equipment in future railway transportation system.

The railway transportation system aims to achieve the displacement of people and things. How to achieve high efficiency and low cost transportation organization is a

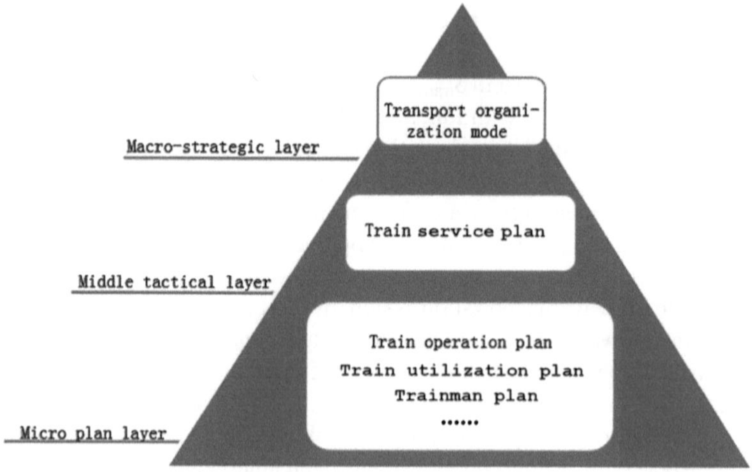

Fig. 1.1 Railway transport organization system

complex system engineering problem. Due to China's vast territory, uneven distribution of resources and productivity, railway transportation demand has been strong, and the contradiction between railway transportation capacity and transportation demand has long existed. Under the existing conditions, how to form an effective driving plan and maximize the use of transportation capacity is the key point and key consideration of the railway transportation organization at this stage.

2. Railway transportation organization system

From the perspective of resource allocation optimization, the railway transportation organization is a layer-by-layer process from macro to micro with the transportation schemes. The railway transportation organization system can be divided into the following levels [12] from macro to micro, as shown in Fig. 1.1.

(1) Transportation organization mode. The transport organization mode is a criterion that stipulates the most basic relationship between different types of trains, and road networks, that is, deciding whether a certain train can run on a certain line or not. The transport organization mode determines the spatial relationship between different types of trains and different types of railway lines. For high-speed railways, there are three modes: "full speed–transfer," "full speed–offline operation," and "mixed transportation" [13]. For the general railway, due to different freight trains and line grades, the transportation organization mode is more complicated.

(2) Train service plan. The train service plan is a technical document [14] in which the railway operation department assigns the train to the appropriate line or section based on the OD traffic passenger flow over a period of time. It mainly solves the problem of spatial resource allocation of mobile and fixed facilities such as trains and lines, including the train running path, running interval, logarithm of the train, stopping stations, maintenance skylight position and size, etc. The passenger flow and the character of passenger flow, the revenue of the railway department, the capacity

of the station and the passing capacity of the section, the class of the train, the number of the train, and the setting of the maintenance skylight constitute the main factors affecting the development of the train operation plan. The train service plan is the basis for the subsequent organizational plans such as the development of the train operation plan, the train utilization plan, and the crew plan.

(3) Train operation plan. The train operation plan is a technical document that assigns the train OD space allocation to the time dimension. It mainly solves the problem of time resource allocation of mobile and fixed facilities such as trains and lines [15]. Each train operation plan can be regarded as a kind of mapping in the time dimension of the train service plan, which mainly determines the time relationship between the trains and the lines. The specific content includes the order of the sections occupied by each train, the running time of the train in the section, the arrival and departure (or passing) time of the train at each station, and the stopping time of the train at the station, etc. [16]. Train running speed, station headway time constitute the main influencing factors for the preparation of train operation plan. The train operation plan is the basis for realizing train safety and punctual operation, and it is also an important content to verify whether the train service plan is reasonable.

(4) The plan of the train utilization (locomotive, highspeed train). The operation plan of train utilization is constrained by the train operation plan. It is a comprehensive plan to realize the train transfer and maintenance. It mainly decides the train in which station, what time, to take on which train number, and in what place, what time, carring out what type of maintenance, and so on. The train running diagram, the type and quantity of available highspeed trains, the relevant rules and regulations on the overhaul and repair process of the train maintenance and repair procedures, the selection of the train operation mode and the conditions of the maintenance base constitute the influencing factors for the train utilization plan. The train utilization plan is an important part of ensuring the implementation of the train operation plan.

(5) Trainman plan. The crew plan is a comprehensive plan that assigns trainman (groups) to trains to provide services for passengers, which is determined by the train operation plan. It mainly determines when and where the trainman (groups) leave for the train, etc. The crew plan is often combined with the train utilization plan.

Adopting a plan-oriented transportation organization is conducive to maximizing the rational use of transportation capacity. The planning system can be divided into the following four layers [17], as shown in Fig. 1.2.

The first level of the plan is oriented to the capacity planning level, which mainly includes the train basic operation diagram, the basic turnover chart of the locomotive and the highspeed train, and the train formation plan, which is generally formulated and managed by the transportation planning and transportation technology management department. The second level plans to face two aspects of market and equipment maintenance: The first is to realize the connection with the market, mainly including the passenger transportation marketing plan, freight marketing plan and technical plan, etc., which are formulated and managed by the marketing department. Second, the maintenance of equipment, mainly including maintenance plans for locomotives, highspeed trains, lines, signals, etc., which are formulated and managed by the equipment management department. The third-level plan is oriented to the dispatching

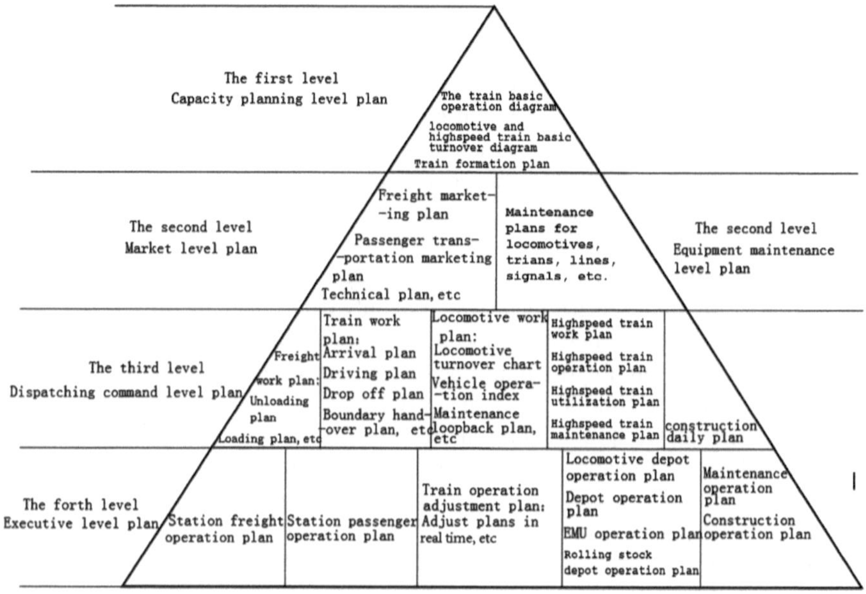

Fig. 1.2 Railway transportation planning system

command level, that is, the dispatching plan. The dispatching department at the two levels of the department and bureau will roll out daily according to the dynamic changes of daily demand, construction and maintenance, and frame the next day's transportation organization and production. The plan mainly includes four categories: freight work plan, train work plan, locomotive work plan, and construction daily plan. The fourth-level plan is the extension and refinement of the third-level scheduling plan. It is oriented to the execution level, and managed by the corresponding dispatch positions of the stations, which is the core content of the production operation plan of the station segment, and guarantees the realization of the production target of the dispatching plan.

In the four-tier railway transportation planning system, the third-level dispatching plan is oriented to the daily transportation organization and production command of the railway. It is the key to realizing the configuration of railway transportation resources and effective coordination of various businesses. Correct and reasonable preparation of the dispatching plan is to ensure good transportation order. Therefore, dispatching plans are the core and key in the four-tier railway transportation planning system.

3. Dispatching Command Organization Structure

The organizational structure of railway transportation is generally divided into several levels. At the same time, there is also a mode of centralized management at the first level. For example, the Japanese Shinkansen and the existing dispatching command system are separated from each other. The dispatching of Japan Railway takes the railway operation company as the unit and implements centralized management and first-level command. It adopts COSMOS comprehensive management system and is connected to the integrated disaster prevention system at the same time. It can obtain monitoring data or alarm information at the first time and take timely measures to deal with it [18, 19]. The German railway implements the three-level management of the dispatching command center—the dispatching command sub-center—the station attendant, and sets up one central dispatching command center and seven dispatching command sub-centers. The Frankfurt Central Dispatch Command Center is responsible for the operation of inter-regional and international passenger and freight trains, while coordinating the relationship between the seven dispatching command sub-centers [20]. The regional dispatching and commanding sub-centers are responsible for the daily transportation command work within the jurisdiction. The dispatching and commanding organization structure of the French railway is also divided into three levels, including the national dispatching center, the railway bureau dispatching center, and the basic units. The National Dispatch Center is mainly responsible for supervising the operation safety of the national railways, implementing macro-controls for railway bureaus and grass-roots units, and is responsible for the operation of major trunk lines and key trains, especially under the conditions of emergencies. The dispatching center of the railway bureau sets up various dispatching types such as train operation command, infrastructure maintenance, passenger transport, freight transport, electric power, and driver dispatch. The basic units mainly include stations, locomotive depots, passenger depots, and highspeed train depots.

The dispatching and commanding organization of China Railway is also divided into three levels, which are composed of China Railway Corporation, Railway Bureau, and station [17], as shown in Fig. 1.3.

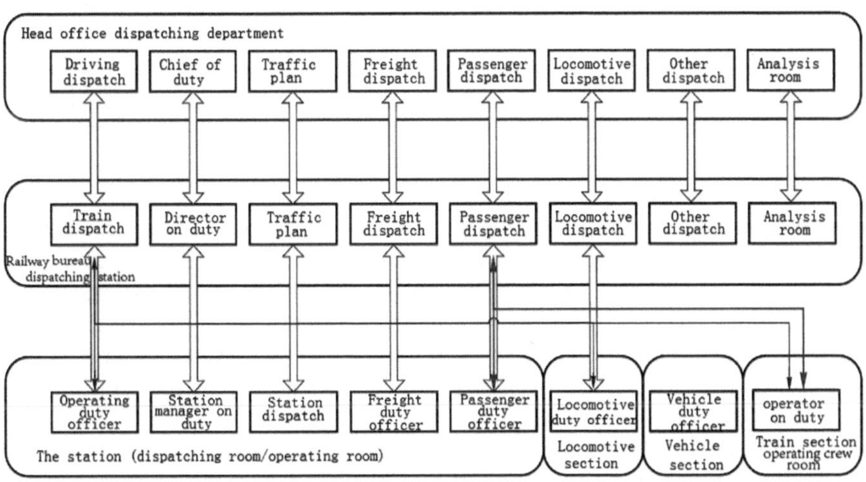

Fig. 1.3 China railway three-level dispatching command system

China's railway transportation dispatching and commanding work implements hierarchical management and centralized and unified command principles. The dispatching at all levels represents the leading organizations at all levels to direct daily transportation work. The main responsibilities of scheduling at all levels are summarized below.

(1) The head office dispatching department is mainly responsible for the daily passenger, freight and traffic organization work in the transportation organization and dispatching command. It also compilates and issues the whole road dispatching contour plan and the day (Shift) plan, supervising and inspecting the railway bureaus to complete the transportation production and management tasks according to the day plans in a balanced manner; and is responsible for approving the construction project plan for the daily I-level construction and the busy trunk line construction day plan. The responsibilities also include collecting the railway station dispatching work reports, checking the completion of daily transportation work, supervising and inspecting the railway bureaus according to the train formation plan, organize transportation according to the transportation production and operation plan, handling the problems arising from the interdepartmental demarcation station in time, and realizing the smooth flow of the boundary between the railway bureaus, mastering the loading and unloading vehicles of key users, ports and stations throughout the country, and mastering the departure and operation of passenger transport, military transport and key trains, and so on.

(2) In the transportation organization and dispatching command, the railway bureau is mainly responsible for the daily passenger, freight, and traffic organization work. It is responsible for compiling and releasing the railway station's freight, train, locomotive work plan and construction daily plan, timely collecting and reporting work report, organizing transportation stations and other transportation production units to work closely together to complete the transportation production and operation tasks, improving the plan delivery rate, organizing and adjusting the cargo flow and traffic flow within the railway bureau to complete the traffic flow adjustment plan and departure loading plan issued by China railway corporation in a balanced manner, organizing the balanced handover of trains at boundary stations to ensure the smooth operation of interoffice boundary stations, organizing and monitoring the train operation. It also supervises and checks the train organization plan, train operation plan, transportation production and operation plan and key requirements, to ensure the close connection between locomotive and train and realize the train operation according to the plan. Its responsibilities also include mastering the loading and unloading vehicles at various stations and major users and ports within the administration of the railway bureau, and the key trains of passengers, special and special transport, exceeding the limit and overweight, and vehicles carrying dangerous goods, etc.

(3) The station (dispatching room/operating room) is distributed in each major production station section, including station dispatch, freight duty officer, manager on duty (station), locomotive duty officer (locomotive section), vehicle duty officer (vehicle section), and so on. In the transportation organization and dispatching command, it undertakes the tasks of reporting the source of goods, freight flow,

traffic flow, and the actual status of trains, locomotives, and vehicles accurately and in time. It also participates in the preparation of the railway bureau scheduling plan, according to the daily (shift) plan issued by the railway bureau, correctly prepare the detailed work plan and phase adjustment plan for the station, and specifically organize the production activities for the station.

1.3 Technical Characteristics of China's High-Speed Railway Operation

1. Fast continuous operation speed

High speed is the core feature of high-speed railway. All countries in the world have improved the operation speed of trains through continuous innovation of technology. In April 2007, the French railway test speed reached 574.8 km / h, creating the world record at that time. After six times of speed increase, China entered the high-speed era, and the highest speed of high-speed trains continued to record. The Beijing-Tianjin intercity railway was designed at a speed of 350 km/h, and the CRH_3 train operated at a speed of up to 394.3 km. It was the world's operating speed in the early days of operation. One of the highest trains, the Wuhan-Guangzhou high-speed railway test speed reached 394.2 km, which set a record for the highest speed of the high-speed railway at the time when the two cars were reconnected. The Beijing-Shanghai high-speed railway was designed to operate at a speed of 350 km per hour, the highest operation speed is 380 km per hour, and the average commercial operation speed is 300 km. The CRH380A train has set the world's highest railway operation test record on the Beijing-Shanghai high-speed railway at an operating speed of 486.1 km per hour.

2. Long continuous operating hours

High-speed railways generally span multiple provinces and cities, and the lines are long. It is necessary to set multiple direct trains for a long period of time. The Beijing-Shanghai high-speed railway has a mileage of about 1318 km. The one-stop direct train runs between Beijing and Shanghai for 5–8 h. The Beijing-Guangzhou high-speed railway is the longest high-speed railway in the world, with a total length of 2118 km and 36 stations, consisting of Beijing-Shijiazhuang section, Shijiazhuang-Wuhan section, and Wuhan-Guangzhou section. The train runs continuously from Guangzhou North Railway Station to Wuhan Station for 3 h, to Changsha for 2 h. Shanghai-Nanjing Intercity Railway runs directly from Shanghai to Shanghai, with a total time of 73–75 min.

3. High train density

The high-speed railway has the characteristics of large density and public transportation. The Beijing-Tianjin intercity railway and the Shanghai-Nanjing intercity railway have been implemented the bus-likely operation schema at the beginning of the opening, and the train departure interval is 5 minutes in the morning and evening peak hours. During the Guangzhou Asian Games, the maximum number of trains on the Wuhan-Guangzhou high-speed railway was 160 per day, and the Guangzhou-Changsha section operated an train every 11 minutes on average, with a minimum interval of 5 min. The high-speed rail "bus-likely" operation mode allows passengers to purchase tickets at any time and get on the train at any time, which is really convenient for passengers to travel.

4. Large regional span and large differences in climate and geological conditions

China has a vast territory and complex climate and geological conditions. China's high-speed railway lines generally have long mileage, and many lines often span multiple time zones or latitudes. The climate and geological environment along the railway will vary greatly. For example, the Zhengzhou-Xi'an high-speed railway is 458 km long, and the line runs through the mountain of westen Henan province and the WeiRiver alluvial plain. 80% of the sections are loess subsidence areas, and the terrain and climatic conditions of each section are different. The Beijing-Shanghai high-speed railway has a total length of 1318 km, running through the three municipalities directly under the Central Government of Beijing, Tianjin, and Shanghai, and the four provinces of Hebei, Shandong, Anhui, and Jiangsu. The Beijing-Guangzhou high-speed railway has a total length of 2118 km, connecting the Bohai Economic Circle, the Central Plains Economic Zone, the Wuhan City Circle, the Changsha, Zhuzhou and Xiangtan City Clusters, and the Pearl River Delta Economic Zone. The two high-speed railways pass through the flat and open North China Plain, the undulating low mountain hills and the hilly plains. In addition to geological differences, the climate also has a large change. Although the whole line belongs to the monsoon climate, Beijing is a temperate monsoon climate, and Shanghai and Guangzhou have subtropical monsoon climates.

5. Railway network conditions changeable, transportation organization is difficult

The high-speed railway and passenger dedicated lines mainly undertake passenger transportation tasks. Important passenger hub stations are often connected with other railway lines, and the interweaving relationship with existing railway lines in the surrounding is very complicated. For example, the Beijing-Shanghai high-speed railway has a variety of trains, including one-stop direct trains, trains that only stop at large stations, and short-distance trains, and other railway lines in Beijing, Shanghai, Jinan, and other large passenger stations. Due to the complicated passenger flow structure and the varied railway network conditions, the development of the train operation plan is very complicated, and the transportation organization is difficult.

6. Various modes of transport organization

The determination of the organization mode of China's high-speed railway transportation is based on the principle of safe transportation. Taking into account the characteristics of passenger flow, railway technical conditions, train running speed, and other factors to ensure efficient and fast operation of the train, it has the following characteristics.

(1) The main trunk line transports passengers and cargo separately. The purpose of the construction and operation of the intercity railway is to alleviate the busy lines and meet the transportation demand. The main trunk lines with large transportation volume are diverted from passengers and goods, and the transportation capacity of the lines is maximized to improve the train operation quality.

(2) A direct train on a high-speed line is combined with a short-distance train. Implement the combination of short-distance and long-distance trains. A large number of passenger trains are operated between the central cities with a distance of 1000–2000 km. Reasonably designed parking stations are convenient for passengers to stop and transfer. The combined operation of long-distance and short-distance trains also appropriately eases the passenger demand during peak passenger flow.

(3) Intercity high-speed trains are publicized. 60% of passengers in large cities have stable official and commute travels. The passengers entering and leaving the station along the line have certain regularity. The vehicle has high requirements on the punctuality and the transfer mode is fixed. Therefore, intercity high-speed trains must be transited.

7. The system technology is complex and the security is difficult

High-speed railway is a complex system engineering with the comprehensiveness of a general system. Its successful implementation is the result of integrated innovation based on the rapid development of computer and application technology, microelectronics technology, new material technology, and so on. The operation of high-speed railway involves the improvement of railway infrastructure quality, the reliability of train operation control system, the performance innovation of high-speed trains, the improvement of track technology, etc. The technical composition of the system is very complicated, and it is necessary to establish a complete safety guarantee system and improve the safety performance of high-speed railways.

8. Sensitive to emergencies

Because the high-speed train runs fast and has long operating mileage, it often spans multiple provinces and cities. In addition, the intercity high-speed traffic density is high, and the departure interval is short. If the train is disturbed by unexpected events, the train can adjust the space is small, which is often affecting all subsequent trains and is highly sensitive to emergencies. High-speed railways run fast. In the event of an accident, it is easy to cause huge casualties and property losses, and the social impact is very bad. Therefore, it is necessary to establish a more complete high-speed railway emergency management plan, respond to emergencies in a timely manner, and form effective emergency measures, restore the normal operation order of trains as soon as possible, and reduce the losses caused by unexpected events.

1.4 High-Speed Railway Dynamic Emergency Management

1.4.1 The Impact of Emergencies on Railway Transportation

The railway is an open and large system. No matter it is an internal fault of the system or a natural disaster, the external events of the system will have an impact on the safe operation of the system. The impact of the emergency on the railway transportation will be explained through some specific events below.

1. German Eschede accident

On June 3, 1998, the German high-speed train ICE 884 departed from Munich station to Hamburg. After more than five hours of driving, the safety high-speed passenger train with 12 carriages carrying more than 400 passengers was travelling as usual. When driving to a road bridge in Escher, it rushed out of the track at a speed of 200 km/h. The 410-m-long body was completely derailed and hit the rail and bridge. The 300-ton double-line bridge collapsed completely, and the first carriage of the high-speed train flew into the air, and fell heavily on the ground, while the other carriages were twisted into a mass under the weight. In just over a minute, 101 people were killed and 88 people were injured. It caused the railway accident with the largest number of casualties in German history and even in the history of high-speed railways. After the accident, Germany stopped all trains.

2. Eurostar trains stopped due to blizzard

In November 2010, due to the blizzard attack, four trains of the Eurostar stopped at the same time in the Channel Tunnel due to power system failure. More than 2000 passengers were trapped in the Cross-Harbour Tunnel for more than 10 h, and tens of thousands of passengers were delayed, which causing serious consequences.

3. Japan Shinkansen stopped due to earthquake

In March 2011, an earthquake measuring 7.2 on the Richter scale occurred in the sea east of Honshu Island, Japan. Affected by the earthquake, the Northeast Shinkansen was once stopped for transportation to ensure safety, which not only caused traffic losses, but also had a negative impact on the economy.

4. Los Angeles train collision

On July 12, 2008, a serious train collision occurred in Los Angeles, USA. The reason for the incident was that the passenger train driver slammed a red light at a crossing that should have stopped, causing the train to collide with an oncoming truck. In this accident, A total of 26 people were killed and more than 130 were injured in this accident. This is the most significant accident in the US passenger trains in the past 20 years.

5. Ningbo-Wenzhou railway “7·23” Accident

On the evening of July 23, 2011, between Yongjia Station and Wenzhou South Railway Station, the D301 train from Beijing to Fuzhou and the D3115 train from Hangzhou to Fuzhou were damaged by lightning, causing the train control system to malfunction and lose the control of the center, which led to a rear-end collision, killing 40 people and injuring more than 200 people. The sudden incident left a shadow on people. People have questioned the safety of China’s high-speed railway which are extremely unfavorable to the development of China’s high-speed railway.

More railway safety incidents can be found in the literature [21]. However, reviewing the impact of emergencies on railway transportation, almost every emergency will cause major property losses and casualties. This series of accidents led us to think about the research on the theory and method of high-speed railway transportation organization under emergency conditions. The study of the theoretical methods of transport organization of high-speed railways under emergencies is bound to become the focus of the future. With the development of the economy, high-speed rail transportation has become the development direction of railways. Once an accident occurs in high-speed railways, it not only has a major impact on people’s lives and property, but also poses a serious threat to people’s physical and mental health.

1.4.2 Railway Emergency Management

1. Railway emergency management concept

Emergency management involves many fields such as operations research, management, sociology, psychology, information technology, and mathematical modeling. It is closely related to emergencies. Many scholars have conducted in-depth research in this field. Ji Lei pointed out that emergency management is in the process of responding to emergencies, in order to reduce the harm of emergencies, achieve the purpose of optimizing decision making, analyze the causes, processes and consequences of emergencies, and effectively integrate all aspects of society related resources, the process of effective early warning, control and handling of emergencies [22]. Bammidi and Moore believe that emergency management includes contingency planning [23] in addition to identifying, predicting, and responding to emergencies to reduce the social damage caused by emergencies or to reduce losses to acceptable levels. Zhou Huijuan believes that emergency management is a process of preventing, preparing, responding to and recovering from emergencies that in order to prevent them from happening or reducing their possible consequences and impacts. It covers the various processes of before, during and after an emergency, including precautionary measures taken in response to emergencies, response actions taken at the time of the incident, various after-care measures taken after the incident and reduction of damage [24]. Wang established a network mapping model [25] for emergency response and resource allocation. The main objectives of emergency management are to save lives, reduce the number of injured, and protect property and environmental safety. In short, the purpose of emergency management is to ensure public health and safety regardless of the cause of the emergency.

From the above definitions, emergency management is a complex system engineering. The main contents include: accident analysis, prediction and early warning, resource planning, organization, deployment, event processing (direct and background command and dispatch, specific implementation), the start-up, operation and dynamic adjustment of the plan, the management and release of information, and the construction of the emergency system.

Based on the nature of emergency management and the characteristics of railway transportation system, we propose the definition of railway emergency management as follows. Railway emergency management is based on the study of the occurrence mechanism and evolution law of railway emergencies, using advanced system modeling and analysis, management and decision support, modern information processing and other methods and technologies, according to the dynamic characteristics of the incident, all mobile equipment, fixed equipment, space, time, and manpower and other transportation resources and emergency resources related to railway emergency response are dynamically scheduled and configured, in order to restore the normal and safe operation of the railway as soon as possible, and reduce the personal injury and death and property loss caused by the railway emergencies.

Railway emergency management requires certain units and mechanisms of the railway to establish a certain management system and mechanism, and adopt a series of countermeasures to plan, organize, and direct the whole process of emergency (pre-event prevention, incident response, handling, and after-treatment). The basic purpose of coordination and control is to protect the safety of people's lives and property, and promote the harmonious and rapid development of railways.

It can be seen from the definition of railway emergency management that the main body of railway emergency management is the personnel, organizations, and institutions dealing with railway emergencies, and the objects are various railway-related emergencies.

2. Railway emergency management system

There are many kinds of railway emergencies, complicated situations, sudden emergencies, and large coverage. Railway emergency management involves all levels from the high rise to the grassroots. It involves railway business units such as electric trains and electric vehicles. It also involves different fields such as medical care, fire control, environmental protection, and military forces. This has brought many difficulties to emergency daily management and emergency rescue command. The only way to solve these problems is to establish a railway emergency management system with unified coordination and command, complete structure, complete functions, responsiveness, efficient operation, resource sharing, strong support, and implement standardized and standardized operation procedures [26]. Therefore, we try to construct a complete railway emergency management system framework and describe and explain its important content.

According to China's current emergency management system "one case with three systems," the railway emergency management system can be divided into five levels

[24]: railway emergency management legal system, railway emergency management system, railway emergency management mechanism, railway emergency plan and railway emergency management guarantee systems. They play their respective roles to guide and standardize the four phases of emergency management from different levels, namely prevention, preparation, response, and recovery. The content included in these five levels, and the relationship between the layers constitutes the complete framework of the railway emergency management system (REMS), as shown in Fig. 1.4.

The railway emergency management legal system includes laws and regulations related to emergency activities and railway emergency regulations. Railway emergency management laws, regulations, and related rules can clarify various types of norms under emergency conditions, supervise emergency management, ensure that various emergency measures are implemented, and make management more standardized and institutionalized.

The railway emergency management system is the macro-layer of the whole system. It mainly determines the organizational structure of emergency management, the division of responsibilities and authority of the organization, and the affiliation between various organizations. At the same time, it formulates railway emergency planning and points out the direction for railway emergency management activities. China's current railway emergency management system adopts a combination of horizontal and vertical modes. It is vertically divided into China Railway Corporation, Railway Bureau, Station Emergency Command Center and Emergency Office, and links various business units horizontally. The basic idea of China's railway emergency management system is "gradient responsibility and territoriality."

The railway emergency management mechanism is the meso-level of the entire system, and its main function is to make tactical decisions. Railway emergency management mechanisms include prevention and emergency preparedness mechanisms, detection/monitoring and early warning mechanisms, information reporting and notification mechanisms, emergency command coordination mechanisms, emergency response mechanisms, information release mechanisms, railway professional rescue mechanisms, rehabilitation and reconstruction mechanisms, and investigation and evaluation mechanism, emergency guarantee mechanism, training rehearsal mechanism, and plan management update maintenance mechanism. The railway emergency management mechanism is mainly to establish various emergency management methods and measures in the four stages of emergency management "prevention, preparation, response, and recovery," and to achieve institutionalization and procedural management.

The railway emergency plan is the microscopic layer of the entire system. The management of the emergency plan involves three main links: preparation and approval of the plan, revision and maintenance of the plan, and exercise of the plan. The main function of the emergency plan is to guide the operation and execution of the actual emergency response, which is conducive to more procedural and institutionalized emergency response. Each perfect emergency plan will clearly define the responsibilities of various organizations and personnel involved in the four phases

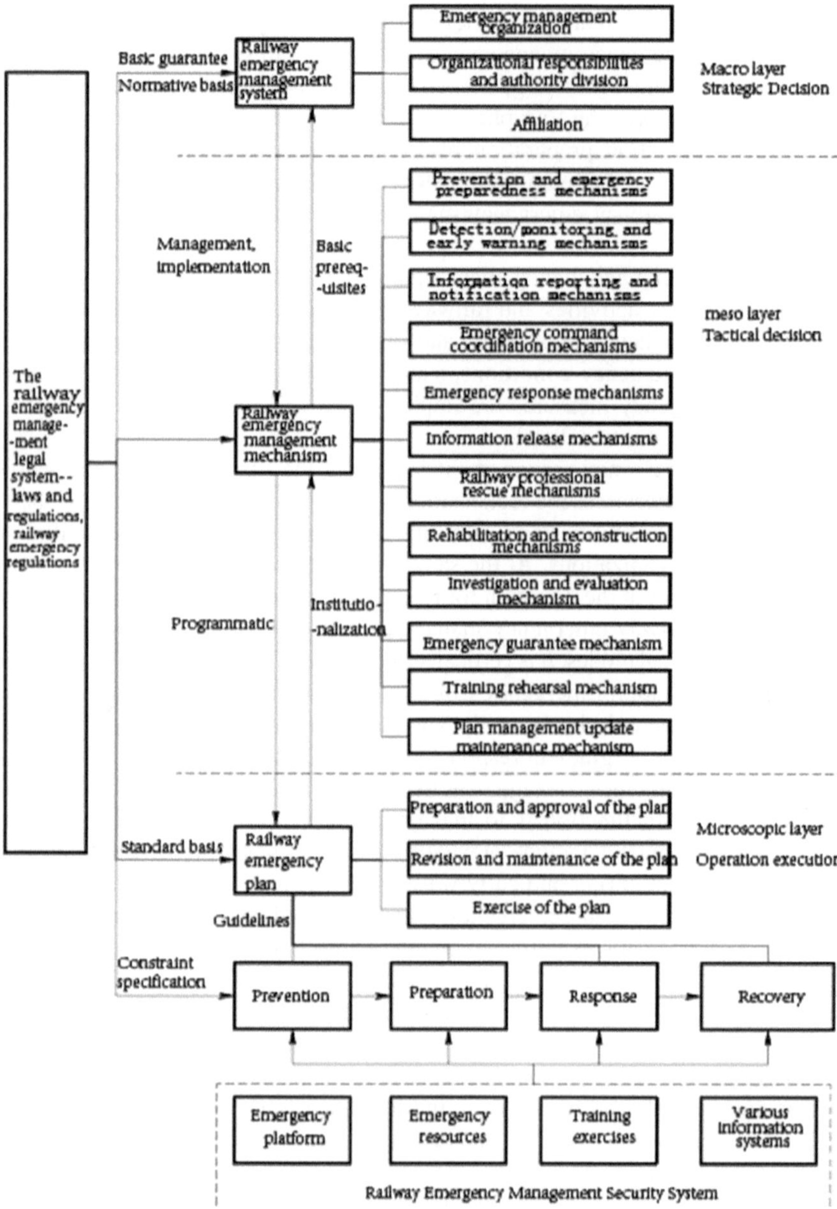

Fig. 1.4 Rems framework

of emergency management, the use of emergency resources, rescue methods, and institutional coordination.

The railway emergency management and protection system is an important supporting part of the entire emergency management system. Its main function is to ensure emergency response activities quickly, efficiently, and scientifically, with the help of certain scientific and technical means, through emergency platforms, emergency resources, training drills and various information systems, to develop and improve emergency management capabilities and levels.

The four stages of prevention, preparation, response, and recovery of railway emergency management are carried out under the norms, constraints, and support of the railway emergency management law, system, mechanism, plan, and guarantee system. The specific contents are shown in Fig. 1.5.

It can be seen from Fig. 1.5 that the main work contents of the prevention stage of railway emergency management include: risk assessment of railway hazard sources and accident occurrence points, tracking of environment and equipment through inspection and monitoring systems, and timing of equipment maintenance and the development of relevant safety procedures. These means and methods can help to eliminate railway accidents in the bud. The main work contents of the railway emergency management preparation stage include Clarifying the responsibilities of each emergency organization, preparing emergency plans, site selection of emergency service facilities, emergency resource allocation, comprehensive assessment

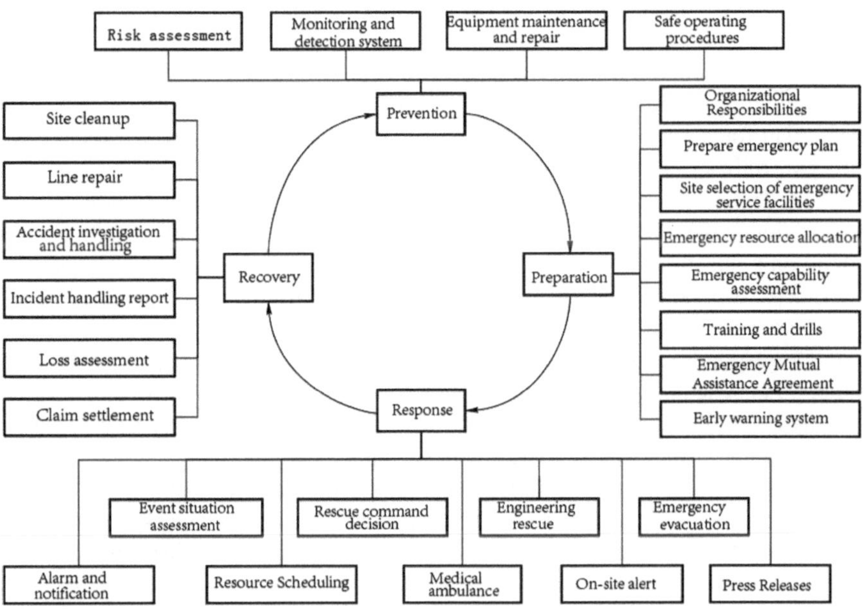

Fig. 1.5 Railway emergency management process content

of railway emergency response capabilities, emergency training and drills, and developing complete early warning system. When an emergency needs to respond in a timely manner, each emergency relate

d organization and personnel needs to conduct alarms and notifications, evaluate the situation, resource dispatch, rescue command decision, medical rescue, engineering rescue, on-site alert, emergency evacuation, and press release. After the emergency is disposed, it is necessary to carry out the recovery phase of the accident site cleaning, line repair, accident investigation and handling, accident handling report, loss assessment, and after-care claims.

1.4.3 The Relationship Between Railway Transportation Organization and Railway Emergency Management

The railway transportation organization process is an important part of railway emergency management under emergency conditions. The technical documents and plans from the macro to the micro-level in the process of railway transportation organization constitute the important management methods, scheduling strategies and adjustment measures in the four stages of prevention, preparation, response, and recovery of railway emergency management under the emergency conditions. The relationship between the two is shown in Fig. 1.6.

It can be seen from Fig. 1.6 that the railway emergency plan mainly involves the middle-level plans, such as passenger flow forecasting and distribution, line capacity estimation, driving route search, train operation plan, and train basic operation chart. Corresponding to the prevention phase of railway emergency management, it is the basic plan for maintaining normal railway transportation. Through reasonable and effective railway emergency plans, the railway accidents should be eliminated in the bud. The preparation of various emergency operation plans corresponds to the preparation stage of railway emergency management, which mainly includes train operation plan, stage adjustment plan, station operation plan, locomotive operation plan, depot operation plan, and Highspeed train operation plan and other micro-executive layer plans. The effective implementation of various emergency operation plans corresponds to the response stage of railway emergency management. Through the optimal scheduling of various transportation resources of the railway, the effective matching of train flow, passenger flow and logistics, the adverse effects caused by emergencies are reduced. The recovery phase of railway emergency management mainly includes various types of emergency maintenance plans involved in the transportation organization process, including various maintenance plans for various mobile and fixed equipment such as locomotives, vehicles, highspeed trains and lines. Real-time monitoring and evaluation, timely maintenance of various fixed and mobile equipment facilities of the railway, will help to restore normal transportation order as soon as possible.

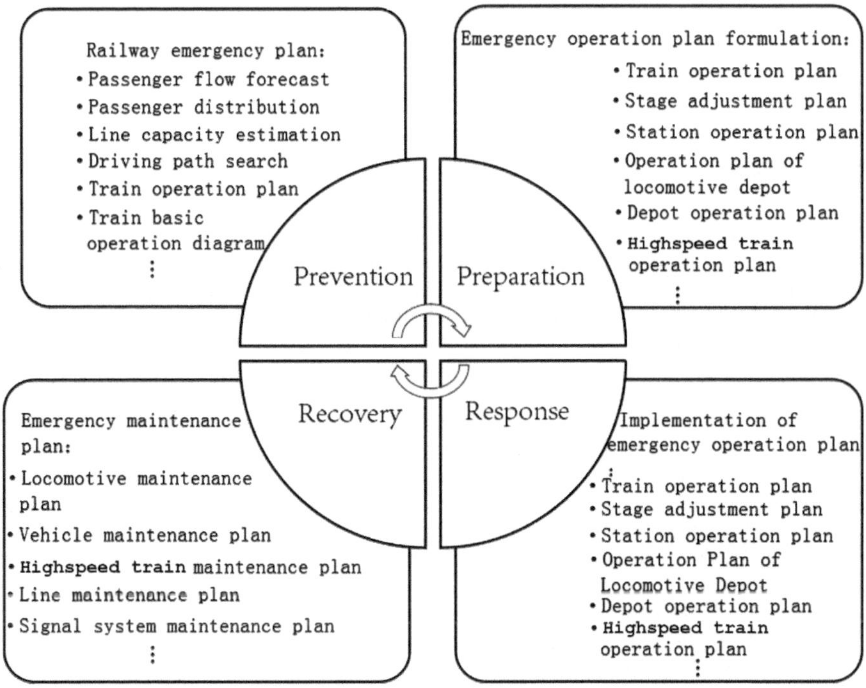

Fig. 1.6 Relationship between railway transportation organization and emergency management

In short, under the emergency conditions, a reasonable and efficient railway transportation organization plan is the basis for ensuring the effective implementation of emergency management, an objective basis for formulating emergency disposal measures, and a key content for improving the efficiency of emergency response.

At present, China's high-speed railways have been in operation in a network, and the lines affect each other. The relationship of lines is complex, and the dynamics and time–space correlation are stronger. Emergencies affect a wide range of space and are difficult to deal with. The study of the theoretical methods of high-speed railway transportation under emergency conditions can prevent possible emergencies in advance, improve the handling efficiency of emergencies, reduce the occurrence of emergencies, control the impact of emergencies, and minimize the loss caused by the incident.

1.5 Chapter Summary

This chapter mainly introduces the development status of the world high-speed railway and its transportation organization, summarizes the operational technical characteristics of China's high-speed railway, and finally introduces the basic concept of railway emergency management, focusing on the analysis of the relationship between railway transportation organization and emergency management.

References

1. International Union of Railways—UIC. General definitions of highspeed [EB/OL]. https://www.uic.org/highspeed#General-definitions-of-highspeed
2. Future of Rail 2050 [EB/OL]. https://www.arup.com/Homepage_Future_of_Rail.aspx
3. International Union of Railways—UIC. High speed lines in the world [EB/OL]. https://www.uic.org/spip.php?article573
4. Present and future of the European rail research: challenge 2050 [EB/OL]. https://www.errac.org/
5. A history of Japanese railways [EB/OL]. https://www.jrtr.net/history/index_history.html
6. Rail technical strategy [EB/OL]. https://www.rssb.co.uk/research-development-and-innovation/futurerailway
7. Haifeng Y (2004) Study on containers transport organization between railway network container freight stations. Southwest Jiaotong University, Chengdu (in Chinese)
8. Zhenjun W (2008) Transportation system engineering. Southeast University Press, Nanjing, p 28 (in Chinese)
9. Xuelei M (2011) Theories and methods on train operating in emergency. Beijing Jiaotong University, Beijing (in Chinese)
10. Editorial Board of China Railway Encyclopedia Editor-in-Chief, Transportation and Economy (2001) China railway encyclopedia: transportation and economic volume. China Railway Publishing House, Beijing, p 506 (in Chinese)
11. Hao Y (2007). Transport histology. China Railway Publishing House, Beijing, p 5 (in Chinese)
12. Li W (2012) Fuzzy random optimization for train operation in emergency. Beijing Jiaotong University, Beijing (in Chinese)
13. Qiyuan P (2007) Passenger transport line transport organization. Science Press, Beijing, pp 46–47 (in Chinese)
14. Li W, Limin J, Yong Q et al (2011) A two layer optimization model for high speed railway line planning. J Zhejiang Univ Sci A: Appl Phys Eng 12(12):902–912
15. Li W, Yong, Q, Jie X et al (2012) A fuzzy optimization model for high-speed railway timetable rescheduling. Discr Dyn Nat Soc
16. Lei N, Peng Z, Limin J et al (2008) High-speed railway transportation organization technology. Beijing Jiaotong University Press, Beijing, pp 62–66 (in Chinese)
17. Tao Z (2013) Research on the planning process modeling of railway bureau daily-dispatching-planning based on cooperative mechanism. Beijing Jiaotong University, Beijing (in Chinese)
18. Hong L, Jun L, Wendong L et al (2012) Status and enlightenment of Japan railway transport organization. Railway Transp Econ 34(3):64–72 (in Chinese)
19. Japanese railway technology today [EB/OL]. https://www.jrtr.net/technology/index_technology.html
20. Dechun X (2004) German railway dispatching command system. China Railway Sci 25(4):141–144 (in Chinese)
21. Safety Committee of China Railway Society, Editorial Committee of "Hundred Years of Railway Safety Memorabilia" (2009) Centennial railway safety memorabilia: 1876–2008. Shanghai Jiaotong University Press, Shanghai, pp 218–510 (in Chinese)
22. Lei Ji, Hong C, An C et al (2006) Emergency management of emergencies. Higher Education Press, Beijing (in Chinese)
23. Bammidi P, Moore LK (1994) Emergency management systems: a systems approach. In: IEEE international conference on systems, man and cybernetics, pp 1565–1570

24. Huijuan Z (2011) Preplan Management and resource allocation optimization in railway emergency management. Beijing Jiaotong University, Beijing (in Chinese)
25. Dan W, Chao Q, Hongwei W (2014) Improving emergency response collaboration and resource allocation by task network mapping and analysis. Safety Sci 12(70):9–18
26. Yong Q, Zhuo W, Limin J (2007) Research on the framework and application of rail transit emergency management system. Chin J Saf Sci 17(1):57–66 (in Chinese)

Chapter 2
The Overview of the High-Speed Railway Transport Organization in Emergency

In this chapter, the basic concepts and theories about the high-speed railway transport organization under emergency are introduced. These contents lay the foundation for comprehending the essence of the high-speed railway transport organization under emergency and provide theoretical basis and support for the methodologies described in the subsequent Chapters. Firstly, in this chapter, the essence of emergency and railway emergency is introduced. Secondly, the particularity of the high-speed railway transport organization under emergency is analyzed. Thirdly, we focus on the definition, organization principles and strategies, optimization objectives and constraints, and the process of transport organization under variety type of emergency. Finally, the complexity of transport organization under serious emergency is analyzed.

2.1 Emergency

Because of the suddenness and harmfulness of emergency, many researchers have studied the essence and extension of emergency from different aspects and accumulated valuable findings. The emergency is defined that the stability and controllability of the system are destroyed and the behavior of the system is abnormal, because of enormous changes in the internal and external environment of the system [1]. The emergency is elicited by certain chances accidents and is the process from the quantitative change to the qualitative change of the internal contradictions. It is difficult to predict the certain time, actual scale, actual situation, and influence of the emergency completely [2]. An emergency also is defined as an event that occurs abruptly and has abnormal characteristics, causes certain social influences which endanger social stability and interfere with normal social order [3]. The emergency is the general term for unexpected events, usually including various natural disasters, serious accidents, terrorists, and major mass riots, political and economic events [4].

L. Jia et al., *High-Speed Railway Operation Under Emergent Conditions*,
Advances in High-speed Rail Technology,
https://doi.org/10.1007/978-3-662-63033-4_2

There are enormous studies on the emergency. From the perspective of sociology, the literatures believe that emergencies occur in human society only. It is the various activities of human beings that lead to abrupt dangers which expose human survival and welfare to threats [5]. Some literatures believe that emergencies are an unpredictable situation, which often have an impact on the social moral ethics frameworks of human beings [6]. Through the analysis of SARS, it is pointed out that emergencies are a special situation that human beings need to apply unconventional means to tackle. It requires human beings to break some stereotypes and innovate in social mechanisms [7]. The international authoritative definition of the emergency is the interpretation of the "public emergency" by the European Court of Human Right: The emergency is a special, imminent crisis or dangerous situation that affects all citizens and poses a threat to the normal life of society as a whole [8]. Some studies of UK define an emergency as any situation that threatens people's health, life, property, or living environment [9]. The unpredicted events are also known as emergency in the USA. The definition can be broadly summarized as: supplemented by the federal government, announced by the President of the USA, in any occasion, in any situation, anywhere in the USA, serious events are addressed with the help of federal government to save lives, ensure public health, safety, and property or mitigate or divert major threats from disasters [10].

The above different definitions of emergency reflect the three general characteristics of emergencies: abruptness, danger, and urgency. According to the definition in the "Emergency Response Law of the People's Republic of China" (Emergency Response Law) which was implemented on November 1, 2007, the emergency occurs suddenly, which causing serious social hazards, natural disasters, accident disasters, public health events, and social security incidents that require emergency response measures. The emergencies are divided into four categories as follows [11]:

(1) Natural disasters. Mainly including flood and drought disasters, meteorological disasters, earthquake disasters, geological disasters, marine disasters, biological disasters, and forest grassland fires.

(2) Accident Disaster. Mainly including various types of safety accidents, transportation accidents, public facilities and equipment accidents, environmental pollution, and ecological damage events of industrial and mining enterprises.

(3) Public health events. Mainly including infectious disease epidemics, group unexplained diseases, food safety and occupational hazards, animal epidemics, and other events that seriously affect public health and life safety.

(4) Social security events. Mainly including terrorist attacks, economic security incidents, and foreign-related emergencies.

According to the property of the emergency, the degree of urgency, the developing situation, and influence scope, the emergency is divided into four levels by the Emergency Response Law, namely level I (extremely serious), level II (serious), level III (large), and level IV (general) which are represented by red, orange, yellow, and blue, respectively.

2.1.1 Railway Emergency

Railway emergency, as a subset of generalized emergencies, has specific properties and essence due to the different objects and disposal measures.

Definition 2.1 Railway emergency is the any events which can change the railway transport organization or transport planning. According to different criteria such as the causes, objective, and the influence, railway emergency can be divided into different categories.

1. Classification according to event's causes

According to the event's causes, railway emergencies can be divided into the following three categories.

(1) Natural disasters

Natural disasters in railway emergencies are caused by factors such as astronomy and geography, which damages the railway infrastructures, or causing abnormal fluctuations in railway passenger flow. Thus, the railway planning cannot be implemented.

According to the causes of natural disasters, natural disasters directly affecting railway transport mainly include inclement weather (such as heavy rain, heavy snow, storms), geological disasters (debris flows, landslides, collapses, etc.), and earthquake.

(2) Railway traffic accidents

The Regulations on Emergency Rescue and Investigation of Railway Traffic Accidents implemented on September 1, 2007 in China, defines railway traffic accidents as follows: the collision of railway rolling stocks with pedestrians, motor vehicles, non-motor vehicles, livestock and other obstacles, or conflicts, digression, fire, and explosions of train that effect railway daily ordinary operation.

According to factors such as the number of derailed trains and the interruption of railway traffic accidents, railway traffic accidents can be divided into four major categories: extremely serious, serious, larger, and general accidents. The number of the derailed passenger trains in busy main lines is more than 18, and interruption length is more than 48 h, which formulates extremely serious accidents. The number of derailed passenger vehicles is more than 2 and less than 18, and the interruption length of busy main railway lines is more than 24 h, or the interruption of railway traffic in branches is more than 48 h, which formulates serious railway traffic accidents. The interruption length of busy railway main lines is more than 6 h, or the interruption of railway traffic in branches is more than 10 h, which formulates larger railway traffic accidents. Moreover, the other is general accidents.

(3) Other public safety events

The emergencies, excepted for the above mentioned, involving public security and damaging railway infrastructures or causing the abnormal fluctuation of passenger flow, such as wars and terrorist attacks, are defined as other public safety events.

2. Classification according to affected objectives

Although the causes of railway emergency are various and railway infrastructure, mobile equipment, and railway operators and passengers are affected, the railway emergency finally leads to an imbalance between traffic capacity and traffic volume. According to change of traffic capacity and traffic volume, emergencies can be classified as follows:

(1) Transportation volume remains the same, transportation capacity decreases

The unpredicted events such as natural disasters and railway traffic accidents causing reduced capacity or completed blockage mainly conclude situations as follows:

(a) The turnouts of station bottlenecks and signal equipment malfunction which cause inefficient arrangement of siding tracks and increase of dwell time; finally, the capacity is reduced.

(b) The speed limitation in railway sections increases the running time, which causes the reduction of capacity.

(c) Damage of railway infrastructure, such as sub-grade settlement, track breakage, and bridge collapse, causes partial or complete section blockage, which restricts trains passing through as planning. Thus, trains need to wait for line recovery or run on circuitous routes.

(d) Train crash, fire, or collision with other obstacles which results in partial or complete blockage.

According to statistic of historic railway emergencies [12], imbalance between traffic volume and traffic capacity is the most common conditions, caused by unpredicted events.

(2) Transportation volume remains the same, and transportation capacity increases

When natural disasters occur, although it has no significant impact on railway transportation capacity, it affects other modes of transportation such as aviation and highways. This makes people who choose to travel by these modes of transportation turn to rail transportation, resulting in abnormal fluctuations in railway passenger flow.

(3) Simultaneous change of traffic capacity and volume

When the enormous passenger volume generated by holidays and decline in capacity caused by emergencies, it is necessary to not only insert trains to meet passenger demand but also reroute rescheduled trains to reduce the occupation of affected sections. This is the most complicated situation where multiple resources need to be integrated and reschedule the planning.

3. Classification according to influence degree

The influence of emergency can be represented by the length of railway emergency, the degree of fluctuation of passenger volume, and the degree of the traffic capacity. According to degree of influence, the railway emergency can be classified as follows:

(1) General railway emergency

The general railway emergency is the unpredicted events whose influence can be eliminated by adjusting the railway timetable. It concludes minor capacity loss of local tracks, small fluctuation of passenger volume, short duration of natural disasters, railway traffic accidents, and/or public safety events. The characteristics of the general railway emergencies are described as follows:

(a) Shorter duration of the emergency. The emergencies that affects the ordinary operation of the railway lasts for a short time, and the decline of line capacity can be recovered as soon as possible.

(2) The fluctuation of passenger flow is minor. The original planned trains can take the increased passenger flow caused by the emergency. According to data statistic in the literature [13], to guarantee passenger comfort, the utilization rate of train seats is generally 0.7–0.8 under ordinary daily operation. Under the case of the large fluctuations of passenger volume in holiday, the utilization rate of train seats comes up to 1.19. When general railway emergencies occur, how to transfer passengers to their destination is the real problem. Therefore, the utilization rate of train seats can be appropriately increased.

(3) The reduction of network capacity is less. The speed limitation is applied in some sections because of emergency, which results in scheduled trains can be operated as planning. After rescheduled timetable, the scheduled trains can be arrived at their terminal stations within a reasonable time, and the serious delays cannot occur.

Under this circumstance, the scheduled trains can carry the increase of passenger flow caused by emergencies, and there is no need to insert new trains. Moreover, because that the duration of the emergency is short and the loss of the line capacity is minor, it is unnecessary to adjust the number of trains, but the scheduled timetable needs to be retiming. Thus, the passenger transportation task can be completed within a reasonable time.

(2) Serious railway emergency

The impact of serious railway emergencies can be eliminated by rescheduling timetable, and serious railway emergencies include natural disasters, railway traffic accidents, and public safety events with less loss of line capacity, minor fluctuation of passenger flow, and short period. The characteristics of serious emergencies are listed as follows:

(a) Long duration of the emergency. The influence on ordinary operation lasts long, and the loss of line capacity can be restored as soon as possible.

(b) The minor fluctuation of passenger flow. Most sections have speed limitation, or more than one section happens blockages. Moreover, the topology of railway network is changed, which causes scheduled trains cannot be operated as original plan and scheduled transport tasks cannot be completed.

Under this circumstance, the scheduled trains can carry the increase of passenger flow caused by emergencies, and there is no need to insert new trains. Moreover, because that the length of the emergency is long and the loss of the line capacity is large, the rescheduled strategies such as train detours and train connection must be adopted to reduce the number of trains on original routes.

(3) Catastrophic railway emergency

The impact of the catastrophic emergency can only be dealt by major change of rescheduling strategies and transport planning, and the cross-industry emergency respond mechanism. Catastrophic railway emergencies include the natural disasters, railway traffic accidents, and public safety events with large-scale network capacity serious loss, great fluctuation of passenger flow, and long duration of impact. The characteristics are list as follows:

① Long duration of the emergency. The impact on ordinary operation lasts longer, and the loss of line capacity can be recovered at cost of longer time.
② Great fluctuation of passenger flow. The scheduled planning cannot carry the increase of passenger flow caused by the emergencies; moreover, a large number of passengers are stuck in stations.
③ Serious loss of line capacity. Multiple lines in the region take speed limitation, or many sections happen blockages. Thus, the topology of network is changed great.

Under this circumstance, the contradiction between traffic volume and capacity is significant. It is necessary to rely on cross-industry emergency respond mechanism, change the transportation organization strategy and reschedule the transportation organization plan, to alleviate the contradiction between transport volume and transport capacity.

2.1.2 The Characteristics of the Railway Emergency

According to above analysis of the railway emergency, the characteristics of the railway emergency are listed as follows:

1. Diversity of cause

According to the above analysis, due to different type of the railway emergency, the causes of the railway emergencies are different. The first category is all kind of natural phenomena that causes natural disasters, such as heavy rain, blizzard, strong winds, earthquakes, and mudslides which cause speed limitation or railway line blockages. Second the operational states of railway infrastructures directly account for the efficiency and safety of train services. Moreover, other personal factors, such as operations of dispatchers and drivers, other persons incursion, or vehicles routes, have an influence on operations.

2. The abruptness of the railway emergencies

In recent years, with the development of computer technology, aviation technology, geology, and other fields, the prediction of climate, humidity, and precipitation in a certain period of time and region has been achieved. It has found some regular patterns about natural disasters and climate change. However, for the most destructive natural disasters (such as mudslides, landslides, and earthquakes), the detailed happed time, port, and influences cannot be predicted. The natural disasters are unpredicted. On the other hand, due to the diversity of equipment states, different utilized periods, and different railway staffs, technical level, work efficiency and mental state of railway workers, the railway emergencies caused by equipment failure and human factors are more irregular.

3. The diversity of effected objectives

According to causes and degree of railway emergencies, the impacts are involved multiple industries and fields. On the one hand, with the development of railway emergencies, travel of passengers on multiple lines is influenced. On the other hand, when the serious harm occurs, the rescue is involved multiple departments such as military, polices, health care, and other government departments.

4. The harmful consequence of railway emergencies

Railway emergencies generally have serious social impact. From the perspective of passengers, it may cause large-scale delays and has an influence on passengers' travel. From the safe perspective, it may endanger the safety of life and property and cause serious damage of railway infrastructures, which reduces the people's confidence on railway transportation safety.

5. Urgency of event processing

When train delays occur, railway timetable should be rescheduled to reduce delay propagation and restore the train services as soon as possible. For the blockages caused by serious railway emergencies, the appropriate strategies such as rerouting or organizing emergency rescue should be applied to reduce the impact of emergencies on operations. In short, whatever type and level of emergencies occur, appropriate measures should be taken to eliminate the development of emergencies and reduce the loss and harm caused by emergencies.

6. Correlation between railway events

In the real operation environment, different emergencies usually interact with each other. For example, heavy rains can cause speed limitation and also cause landslides or mudslides which damages railway tracks failure. Moreover, the lightning may cause failure of the railway signal system, which brings much inconvenience to dispatchers. The correlation between emergencies has greatly increased the complexity of transport organization.

7. Diffusivity of influence of railway emergencies

The essence of railway transportation is that trains occupy resources conflict-free and complete the travel services. When the block sections are occupied unscheduled, the delays inevitably occur, especially in complete segment blockage. The delayed trains need to be rescheduled by running other railway lines to make full use of the other network capacities, in order to reduce losses caused by unexpected events. Therefore, a single emergency occurs in a certain section may cause a change of the train plans in a region.

8. Uncertainty of related attributions

The attributions of emergencies such as the time of occurrence, scope of influence, and duration are difficult to be predicted and described by exact numbers. The train running states are also fuzzy and random. Those characteristics increase the difficulty of transport organization under emergencies.

9. Differences of the content of transport organization

According to the degree of influence of railway emergencies on railway transportation system, the different response measures are taken. (1) When the minor emergencies occur, railway timetable needs to be only rescheduled. Thus, the influences cased by emergencies can be eliminated, and the transport tasks can be completed within planned time. (2) When the major emergencies occur, railway timetable and train routes need to be adjusted simultaneously. That is, the number of operated trains, stop pattern, train routes, and arrival/departure time should be rescheduled, which means that the railway line planning and timetable rescheduling are integrated. (3) When serious emergencies occur, it is often necessary to establish emergencies response department to carry out rescue measures as quickly as possible. The adjusted transport organization strategies should be applied to evacuate the passengers. Compared with above two cases, the contents of transport organization of the third case are more complicated.

2.2 Analysis of Problem Relevant to Railway Transportation Organization Under Emergencies

The essence of railway transport organization is to coordinate each department of railway system and optimize the structure of railway system. Therefore, the railway system is more robust when emergencies occur.

The railway transport plans under emergencies are comprehensive plans integrated the layouts of existing rail lines and high-speed rail lines, ordinary speed trains, and high-speed trains. The differences between the railway transportation organization under emergencies and under normal conditions are listed as follows.

2.2.1 The Different Objectives of Railway Transportation Organization

The railway transportation organization includes the transport mode selection, train service planning design, and railway timetable formulation. The differences of objective about railway transportation organization under emergencies and under normal conditions on each aspect are analyzed as follows.

1. Railway transport organization mode selection layer

The criteria of transport organization mode selection under normal conditions are the attraction range of passengers, the safety and punctuality of train services, and operation costs. The objective of transport organization mode selection has significant diversity. Internationally, in France, high-speed trains operated in ordinary lines to extend the high-speed train service range, which reduces the number of passenger transfers and improves the quality of high-speed railway services. The high efficiency of passenger-oriented transport organization about is focused in Germany and Japan to reduce the number of passenger transfers [14].

In China, Professor Hu Siji pointed out that two transport organization modes mainly applied to the Beijing-Shanghai high-speed railway corridor. Those modes include whole operated trains are homologous high-speed trains named "full high-speed trains," or operated mixed with middle-speed trains [15]. The transport organization mode of "high-speed transfer" proposed by Professor Hu is based on the safety, punctuality and comfort, that is, the objectives of high-speed railway transport organization is safety, punctuality, and comfort. Specifically, the average travel time of passengers, the average travel price of passengers, the comfort and convenience of passenger travel, the revenue of railway administers, the costs of operation (i.e., extension of infrastructures and purchase of rolling stocks), the residue capacity of railway network, and the healthy states of infrastructures are as the objectives of transport organization mode selection. For instance, the capacity utilization of the high-speed railway network, that is, the number of operated trains, is focused on transport mode selection [16, 17].

When railway emergencies occur, the main objective is whether transport tasks can be completed. Under the emergency conditions, the trains are transported to their destination as much as possible to reduce the influence of emergencies and the loss of social benefits. For example, when the heavy snow occurs in southern China at the Spring Festival in 2008, the Guangzhou Railway administers issued an emergency dispatch order in accordance with the requirements of the China Railway Corporation, and changed some passenger trains routes, that is, Beijing-Kowloon Railway, Shanghai-Kunming High-speed Railway, and other normal railway lines. With the construction of railway network, it is inevitable to change the original transport organization modes when emergencies occur to achieve the trains transit tasks as quickly as possible.

2. The train service plan layer

Under normal conditions, there are two objectives considering the design of the train service plan: (1) the revenue of the railway administers. When the price of the ticket is certain, the optimization objective is the minimization of operation costs. (2) The passenger-oriented objectives, such as passenger's satisfaction, passenger waiting time, passenger transit time, passenger travel time, and the passenger profit, are considered. In fact, the multi-objectives or bi-level programming models are formulated in many studies on the design of the train service plan.

For example in literature [17], the costs of high-speed railway train operation are divided into fixed cost and variable cost, and the design of train service plan model was constructed considering the cost of operation, the seat utilization of trains, and the total waiting time of passengers. The objective of design of the train service plan is the minimization of operation costs. Moreover, the original-destination passenger flow is the basis of design of the train service plan. Feng et al. [18] formulated a bi-level programming model where the profit of railway administers and passengers are considered in different layers. Lufeng et al. [19] proposed a bi-level programming model balanced the interest of authorities and demand of passengers. Chang et al. [20] multi-objective programming model where considering the costs of railway operation and travel time of passengers are proposed.

The essence of rescheduled the train service plan refers to reroute trains, and the objective is to transit tasks which can be completed as soon as possible when railway emergencies occur. Under the emergency conditions, the railway network capacities are reduced cased by emergencies, which results in large-scale delay propagation. The railway services are questioned, which causes serious direct or indirect economy loss. Therefore, how to complete the transit tasks as quickly as possible under the declined network capacities conditions is the main objective. The goals of reducing the traveling time and evacuating passenger quickly are consistent. Thus, the objective of design of the train line plan is to reroute trains appropriately and complete transit tasks as soon as possible when emergencies occur.

3. The railway timetable layer

Under the normal conditions, the total waiting time, the rate of late, total delays, and passenger satisfaction are mainly considered, when the railway timetable is designed. Bo et al. [21] proposed a periodic timetable graph model where the train schedule problem is viewed as a periodic job shop scheduling problem and the minimum waiting time of trains is as the objective. Furthermore, dwell time, arrival and departure headway constraints are considered in this model.

When the emergencies occur, the objectives of railway timetable rescheduling are different from railway timetable scheduling under normal conditions. Specifically, the objectives of rescheduling problem are maximization of punctuality rate and minimization of total delays. Moreover, there is strong capability to tackle the influence caused by emergencies.

2.2.2 *Different Constraints Caused by Emergencies*

Generally, railway emergencies have great influence on transport resources (mobile equipment and infrastructures) and travel characteristic of passengers.

1. Natural disasters

The natural disasters are the fatal and destroyed events generated by the variation of natural conditions caused by factors such as astronomy, geography, and human activities. It can be categorized as meteorological hydrological disasters, marine disasters, geological disasters, earthquake disasters, and biological (human) disasters. The natural disasters that directly endanger the railway include floods, collapses, landslides, mudslides, earthquakes, wind, sand, collapse, and freezing which belong to meteorological factors, geographical factors, geomorphological factors, geological factors, and human factors.

The frequency, diversity, and broad influence are characteristics of natural disasters in China. The outbreak areas of natural disasters have accounted for more than 20% of the total operation mileage [22]. According to statistics, railway natural disasters are mainly concentrated on sub-grade. Damaged sub-grades are nearly 90,000 and the cumulative length has exceeded 10,000 km. Moreover, bridges, culverts, tunnels, stations, and all other infrastructures are also severely affected by natural disasters. In China, natural disasters are divided into more than ten categories and multiple type disasters occur together. The railway blockages are caused by natural disasters is more than 100 times a year, and the cumulative time is 1000–2000 h, and the peak frequency reached 211 times.

Since railway lines are mainly concentrated in the east of Baotou-Lanzhou-Kunming line; the central and eastern regions are the most frequent areas of natural disasters in China's railways. High-speed railway lines such as the Beijing-Shanghai high-speed railway are concentrated in these areas. According to statistics, from 1973 to 1991, there were 889 natural disasters in Jiangsu and Anhui provinces, with an average annual rate of 46.78. Among them, Shanghai-Nanjing and Tientsin-Pukow railway are more than 50 times, and Huainan Railway Line is more than 100 times [22].

The proportion of west of the Beijing-Guangzhou railway lines has already accounted for 45%, and the railway disaster in the western region is also very serious. According to statistics, the occurring density of large mudslides is 0.4–1.4 individual/km such as Chengkun Line, Baocheng Line, and Bao Tian Line. 1,366 natural disasters occurred in the summer of 1981, and the cumulative length of the affected distances accounted for 88.13% of the total length [23]. Except for earthquakes that damaged railway infrastructures and caused train derailment, since 1949, only geological disasters have caused hundreds of major railway accidents and more than tens of thousands of cumulative blockage hours.

The railway system is affected by snow weather, which is a common phenomenon especially the countries located in the severe cold regions of northern hemisphere. The railway lines are often damaged by snow and ice disasters during the severe winter season. The turnout failure, declining performance of the rolling stocks, power supply system, and difficulty of operations caused by large-scale snow and ice disasters make railway lines blockage. The minor damage of the snow and ice disaster is the train services cancelation and delays. Furthermore, the major damages show large-scale railway lines blockages, serious facilities failures, and people injured. The power supply system, communication signal system, railway tracks, station facilities, and rolling stocks are damaged by snow and ice disasters in different degrees. At the beginning of 2008, southern segments of the Beijing-Guangzhou line, Shanghai-Kunming, Jiaoliu, Yuhuai, and other railway main lines were interrupted many times because of the snowstorm in southern China. The power supply system of Hunan province was damaged, and the Beijing-Guangzhou Corridor suffered serious damages. The number of trains effected by the disasters exceed 400. The maximum delays are more than 30 h in the Beijing-Guangzhou railway line, and more than 637 thousand people was stranded in Guangdong region, in which 34 thousand people was stranded in Guangzhou station [22].

The Wenchuan earthquake in Sichuan province caused railway tracks of main lines such as the Baocheng line, the Chengkun line, the Tianbao section of the Longhai line, the Chengyu line, the Xiangyu line, the Yang'an line, the Dacheng line, and branch lines such as the Chengwhen Line, Detian Line, and Guangyue lines damaged. Bridges, tunnels, culverts, and communication signals, traction power supply system, and infrastructures were also damaged. A freight train was derailed due to the collapse of the 109th Tunnel of Baocheng Line. The spatial geometry of the line changes, which damaged the lines smoothness. The relative displacement of the track is caused, the geometrical size is destroyed, and the structural stress state changes greatly. Those made blocks blockages. The sub-grades are damaged seriously caused by landslides and collapsed rocks. Ring-shaped or cross-shaped cracks appear in the pier, the concrete of the beam is peeled off, and the structure is cracked. The tunnel entrance is rocked and the structure of the gate is destroyed. The main transformer of the traction substation has different degrees of pressure bushing misalignment and oil leakage, bottom fuel injection and displacement problems, and catenary broken. Local cracking and collapse of the wall occurred in the signal building and the station building.

At 20:30:05 on July 23, 2011, Wenzhou City, Zhejiang Province, D301 train from Beijing South Railway Station to Fuzhou Station and D3115 train from Hangzhou Station to Fuzhou South Railway Station crashed due to bad weather conditions. Forty people were killed, and more than 200 were injured. Moreover, the traffic was interrupted for 32 h and 35 min, and the direct economic loss was 19,371,650 yuan.

The conflict between the multiple occurrences, randomness, regional distribution imbalance, and stability of railway transportation is serious. When natural disasters occur, the facilities of railway lines and stations damaged seriously. The direct results are that the interruption of the line or the operation of the station equipment is abnormal. The central role of the station in the railway network cannot be completed. For the train organization under the emergencies occur, the constraints include the declined through capacity of lines and arrival and departure capacities of stations.

2. Railway emergencies

According to the serious railway emergencies data from 1970 to 2008 [12], compared with normal conditions, the constraints of railway transport organization under abnormal conditions have two characteristics as follows, shown in Table 2.1.

(1) The rolling stocks were damaged due to railway emergencies, which declines the railway traffic capacity.

(2) The majority of railway accidents cause disconnection of the railway lines or disruption of traffic.

According to Table 2.1, the duration of line blockage is at least one hour and 14 min, and the maximum is 15 h. For the train transport organization under the emergency conditions, the topology of the railway network changes, and the train cannot be operated according to the scheduled path, which actually forms the train routes allocation constraint.

2.2.3 Difference Design Sequence of Transport Organization Plans

Under the normal conditions, the transport organization plans design sequence is the top-down, following the sequence: the transport organization mode design, railway line plan, railway timetable schedule, rolling stock schedule, and crew schedule. Under the abnormal conditions, the construction of transport organization is same as the normal conditions; however, the sequence of solving problem is changed. When railway emergencies occur, firstly, the railway timetable only adjusted is considered to achieve the scheduled plans. If not, railway line plan rescheduling, or transport organization mode rescheduling are also applied, and then the train timetable is rescheduled according to rescheduled up-layer plans. Similar to transport organization under normal conditions, the rolling stock schedule and crew schedule is based on train timetable. Therefore, the process of transport organization between normal and abnormal conditions is different.

2.3 Analysis of Different Transport Organization Under Emergency Conditions

According to the above analysis, emergencies will have different degrees of impact on different participants in the railway transportation system. As the operating environment of the system changes, the goals, constraints, and organizational content optimized by the transportation organization will change.

Table 2.1 1970–2008 China railway serious emergencies list[a]

Index	Time	Location	The number of persons injured or died	The damaged vehicles	The during time of blockage
1	7/12/1971	LiuliHe station of BJ-GZ HSR	14 persons died and 22 injured		1 h 40 min
2	16/10/1976	The 41 km point of GZ-SZ HSR	18 persons injured		24 h 59 min
3	16/12/1978	Yangzhuang Station of LZ-LYG Railway	106 persons died, 47 seriously injured, and 171 slight injured	3 persons vehicles destroyed	9 h 30 min
4	19/2/1980	Zhengzhou Station	1 person died and 8 injured	1 passenger vehicle destroyed	
5	22/1/1980	Zhuzhou station of BJ-GZ HSR	22 persons died and 4 injured	1 passenger vehicle destroyed and 1 damaged	
6	9/7/1981	Lizi Yida railway bridge of CD-KM Railway	130 persons died and 146 injured	2 rolling stocks, 1 luggage vehicle and 1 passenger vehicle destroyed	15 days
7	20/10/1981	Section between Eergeqi station and Chaoyang Village Station	3 persons died and 65 burned	1 passenger vehicle destroyed	2 h 50 min
8	14/5/1984	Section between Shensan station and Dahongqi Station	6 persons died and 22 injured	2 passenger vehicles destroyed and 1 damaged	1 h 14 min
9	18/12/1984	Section between Rongjiawan station and Huangxiuqiao Station	1 person died and 3 injured		
10	15/1/1986	Section between Baishidu station and Pingshi Station of BJ-GZ HSR	7 persons died, 11 seriously injured, and 27 slight injured		
11	18/2/1987	Zhaodong Station	6 seriously injured and 7 slight injured	1 vehicle destroyed	More than 1 h

(continued)

Table 2.1 (continued)

Index	Time	Location	The number of persons injured or died	The damaged vehicles	The during time of blockage
12	22/4/1987	Songhua river bridge HB-BA Railway	12 persons died and 44 injured	1 passenger vehicle destroyed	
13	18/7/1987	Mengmiao Station of BJ-GZ HSR	8 persons died, 30 seriously injured, and 39 slight injured	2 passenger vehicle destroyed	
14	7/1/1988	Matianxu Station of BJ-GZ HSR	34 persons died, 30 injured	2 passenger vehicle destroyed	
15	17/1/1988	Beiyin river station of LL-HB Railway	19 persons died, 25 seriously injured, and 51 slight injured		
16	24/1/1988	Section between Qiewu station and Dengjia village Station of CD-KM Railway	88 persons died, 62 seriously injured, and 140 slight injured		
17	24/3/1988	Kuangxiang Station of SH-KM Railway	28 persons died, 20 seriously injured, and 79 slight injured	2 rolling stocks destroyed and 1 rolling stock damaged	23 h
18	1/7/1988	Section between Anyang station and Baolianshi Station	6 persons died, 6 seriously injured, and 13 slight injured	1 passenger vehicle destroyed	
19	5/1/1989	Shimenkan Station of CQ-GY Railway	More than 20 persons injured		
20	30/4/1989	Xiaocongtou Station of SH-KM Railway	1 seriously injured and 20 slight injured		
21	26/6/1989	Section between Songjiang station and Xiexing Station	24 persons died, 11 seriously injured and 28 slight injured		4 h 7 min
22	13/6/1991	Section between Xinmaqiao station and Caolaoji Station of TS-PK Railway	28 persons injured		18 h 37 min

(continued)

Table 2.1 (continued)

Index	Time	Location	The number of persons injured or died	The damaged vehicles	The during time of blockage
23	18/8/1991	Dayaoshan tunnel of BJ-GZ HSR	More than 10 persons		
24	21/3/1992	Wulidun station of SH-KM Railway	15 persons died and 25 injured	2 rolling stocks destroyed and 9 vehicles destroyed	35 h
25	10/7/1993	Section between Xinxiangnan station and Qiliying Station of BJ-GZ HSR	32 crews died, 7 persons seriously injured, 4 persons slight injured; 8 passengers died, 2 passengers seriously injured, 35 passengers slightly injured	1 rolling stock destroyed, 3 passenger vehicles destroyed, 15 damaged, 1 freight vehicle destroyed, and 2 damaged	11 h 15 min
26	15/1/1994	Yuguanying station of Luobao line	7 persons died and 12 injured	1 rolling stock destroyed and 1 damaged	3 h 9 min
27	29/4/1997	Rongjiawan station of BJ-GZ HSR	126 persons died, 45 seriously injured, and 185 slight injured		
28	9/7/1999	Section between Hengyangbei station and Hengyang station	9 persons died, 15 seriously injured, and 25 slight injured	5 passenger vehicles destroyed, 4 damaged, 2 serious damaged, and 1 minor damaged	
29	11/4/2006	Section between linzhai station and Dongshui station of BJ-KL Railway	2 persons died, 18 injured		
30	28/2/2007	Section between Pearl Spring hong station and shanqu station of South Xinjiang Railway	3 persons died, 34 injured		
31	23/1/2008	Section between Anqiu station and Changyi station of QD-JN Railway	18 persons died, 9 injured		

(continued)

Table 2.1 (continued)

Index	Time	Location	The number of persons injured or died	The damaged vehicles	The during time of blockage
32	28/4/2008	Section between Zhou village station and Wang village station of QD-JN Railway	72 persons died, 416 injured		

[a]Notice: Beijing-Guangzhou high-speed railway (BJ-GZ HSR), Guangzhou-Shenzhen high-speed railway (GZ-SZ HSR), Lanzhou-Lianyungang Railway (LZ-LYG Railway), Chengdu-Kunming Railway (CD-KM Railway), Harbin-Bei'an Railway (HB-BA Railway), Lalin-Harbin Eas Railway (LL-HB Railway), Shanghai-Kunming Railway (SH-KM Railway), Congqing-Guiyang Railway (CQ-GY Railway), Tientsin-Pukow Railway (TS-PK Railway), Beijing-Kowloon Railway (BJ-KL Railway), Qingdao-Jinan Railway (QD-JN Railway)

According to the degree of impact of the emergency, the transportation organizations under emergency conditions are mainly divided into three types: transportation organizations under general emergency conditions, transportation organizations under serious emergency conditions, and transportation organizations under catastrophic emergency conditions. As shown in Fig. 2.1.

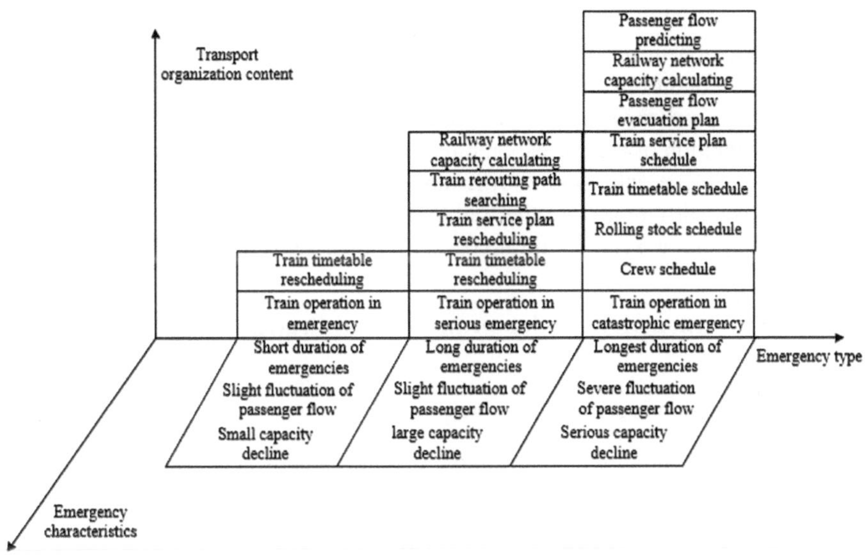

Fig. 2.1 Types of transport organizations under emergency conditions

2.3.1 Transport Organization in General Emergency Conditions

Definition 2.2 Transportation organization under general emergency conditions.

Train operation in emergency (TOE) under general emergencies refers to the adjustment of train operation plan under the condition of small loss of local line capacity, small fluctuation of passenger flow, and/or short duration and implement of the process.

According to the definition, the main content of the transportation organization under the general emergency situation is the traditional train operation adjustment problem. Generally, the goal is to reduce the delay time of the train, increase the punctuality rate of the train, and resume the operation of the train according to the figure as much as possible.

At present, many scholars have conducted a lot of research on this type of problem, so they will not repeat them.

2.3.2 Transport Organization in Serious Emergencies

Definition 2.3 Transportation organization in serious emergencies.

Train operation in serious emergency (TOSE) refers to the plan for train operation and the conditions of greater loss of local line capacity, less fluctuation of passenger flow, and/or longer duration of the emergency. The process of adjustment and implementation of the corresponding train operation plan at two levels.

According to the definition, TOSE has the characteristics of serious emergencies and is very different from the traditional train operation adjustment problem. It is located at the two levels of train operation plan adjustment and train operation adjustment.

1. **Transportation organization principles and strategies in serious emergencies**

(1) Transportation organization principles

The content of transportation organization in serious emergencies includes the adjustment of train service plan and train timetable adjustment. Therefore, in addition to the relevant principles of train operation adjustment, it also contains the principles of train operation plan adjustment, as follows.

(a) Try to make the trains with less delay, long-distance and high-level to maintain the original routes.

(b) The trains on original line have a high grade, and the trains detouring to other routes are low.

(c) Trains detouring to other paths can arrive at the station ahead of time, but they cannot depart in advance.

(d) Reconnected trains can arrive at the station ahead of time, but they cannot depart in advance.

(e) When the train detours, try to choose a route with a small number of crossing railway lines.

(f) In the scheduling strategy, according to the impact of various strategies on passenger travel and the risk of equipment conditions and technology when the strategy is implemented, the detour priority of the train is the highest, the reconnection of the train is the second, and the priority of the train suspension is the lowest.

(g) High-level trains can cross low-level trains.

(h) Low-level trains cannot cross high-level trains.

(i) The trains on this line have high grades, and the cross-line trains have low grades.

(j) On-time trains have a high grade, and serious late trains have a low grade.

(k) Trains can arrive at the station in advance, but they cannot depart early.

(l) Trains with special requirements may be given priority.

(m) Punctuality rate of important stations in large hubs is considered.

(n) High-grade passenger trains and low-grade freight trains.

(o) Freight trains can be stopped when necessary to reduce capacity occupancy.

(p) Low-level trains can be overtaken by up to three high-level trains at the same station.

(2) **Transportation organization strategy**

The transportation organization strategy under severe emergencies also includes two levels of train service plan adjustment and train timetable adjustment, mainly including the following content.

(a) According to the capacity loss of the affected line and the remaining capacity of other lines, the train can detour to other lines to reduce the occupation of the capacity of this line.

(b) Based on the loss of capacity of the affected lines and the remaining capacity of other lines, two short trains, similar stops, and the same type of trains can be reconnected to reduce the occupation of the capacity of this line.

(c) In the case of a large loss of capacity, after the detour of the train and the reconnection of the train, the line capacity is still insufficient to generate an effective transportation organization plan, and certain low-level trains may be canceled.

(d) When the train operation plan is adjusted, the following strategies can be adopted when adjusting the corresponding train arrival and departure times.

(a) According to the technical status of the locomotive, the technical level of the driver, the condition of the train formation, the allowable speed of the line and the natural environment, organize the accelerated operation of the train, reduce the operating hours of the interval, and restore the late train to the punctual operation [22].

(b) According to the layout of equipment in the station and the operation of the train, organize rapid operation of the station to reduce the time when the train stops.

(c) High-level trains can use crossing lines to cross low-level trains at stations or sections.

(d) On-time trains can use crossing lines to cross severely late trains at stations or sections to reduce late spreads.

(e) Change to the departure line to fully utilize the station's ability to receive and depart from the train.

(f) Changing the train will make Hehe crossing station to reduce the impact of late trains on other trains.

2. **Transportation organization optimization goals and constraints in serious emergencies**

In serious emergency conditions, the transportation organization includes train service plan adjustment and train timetable rescheduling. Therefore, its optimization goal must first take into account the macro-level social benefits, passenger satisfaction and the ultimate realization of transportation goals; we must also pay attention to the train delay time and train punctuality rate at the micro-level. In addition, in serious emergency conditions, the loss of line capacity is large and the duration is long. Most trains will be more or less late. At this time, it is meaningless to pursue the punctuality rate of all trains. Therefore, at the level of train timetable rescheduling the number of trains that are seriously delayed and the delay time of large hub stations can be reduced as much as possible, which is very different from the goal pursued by traditional train operation adjustment.

Considering serious emergencies, train detours, train reconnections, and train suspensions can be used to adjust train service plans. Therefore, in addition to the constraints of the train timetable adjustment level such as the train interval operation time, station operation time and tracking train interval time, the constraint conditions also include the capacity constraint of the roundabout route in the train service plan adjustment level, and the technology and equipment of train reconnection constraints, stop constraints for detour trains and reconnected trains, and train stop constraints.

3. **Transportation organization process in serious emergencies**

Compared with the traditional train operation adjustment, the TOSE process is more complicated, as shown in Fig. 2.2.

After an emergency occurs, dispathers should judge whether the railway line is interrupted firstly. If an interruption occurs, determine whether it can be repaired in a short time, and then choose to directly adjust the train timetable according to the length of the repair time, or adjust the train service plan after adopting the detour strategy. If the lines are conncted, according to the type, level, specific attributes and driving rules of the emergency, then calculate the line passing capacity, generate a detour route for the section with a great loss of capacity, and adopt the train detour strategy. If the newly created detour route still cannot compensate for the lost capacity, the train reconnection and train suspension strategies need to be adopted until an effective transportation organization plan can be formed.

In summary, the multi-level of optimization problems, the complexity of transportation organization principles and strategies, the diversity of optimization goals

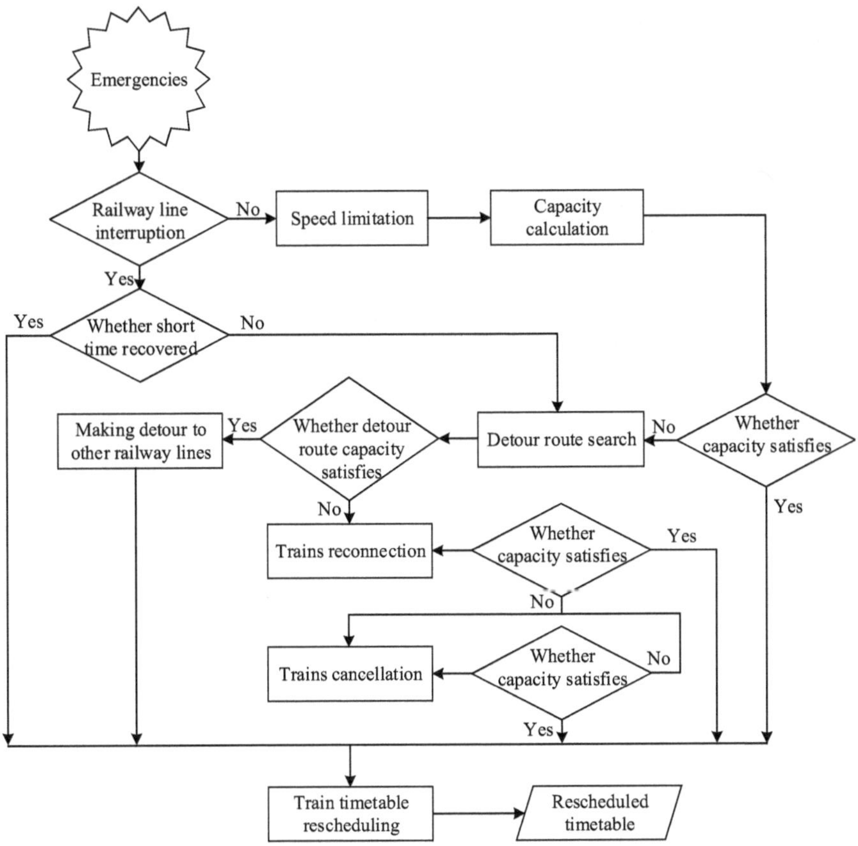

Fig. 2.2 Transportation organization process under severe emergencies

and constraints, and the complexity of transportation organization processes all determine that there is a significant difference between the transportation organization under severe emergency conditions and the traditional train operation adjustment.

2.3.3 *Transport Organization in Catastrophic Emergency*

Definition 2.4 Transport organization in catastrophic emergency.

Train operation in catastrophic emergency (TOCE) refers to the adjustment of transport organization strategy and the adjustment of transport organization plan under the conditions of severe loss of line capacity, large fluctuation of passenger flow, and/or long duration of the emergency.

According to the definition, TOCE often rises to the national level. It is necessary to establish an emergency linkage agency to mobilize emergency rescue resources outside the railway (such as local troops, police), to reduce passenger stranding and resume normal driving as soon as possible.

1. **Transportation organization principles and strategies in catastrophic emergencies**

Under the condition of catastrophic emergencies, transportation organizations aim at emergency rescue and passenger evacuation, and their transportation organization principles and strategies have changed significantly compared with the above two. The transportation organization principles include the following.

(a) The formulation of the transportation organization plan needs to be conducted under the unified command of the emergency linkage agency.

(b) The rescue train has the highest rank, followed by the rescue supplies train, and the other trains have a relatively low rank.

(c) When calculating the capacity, it is necessary to consider the line capacity of the rescue train and the rescue material train.

(d) The goal is to quickly evacuate passengers, regardless of economic costs.

The transportation organization strategy under the condition of catastrophic emergency includes the following contents.

(a) Provide guidance to passengers stranded at large hub stations and provide refund services.

(b) Predict the trend of passenger flow fluctuations, and re-establish the train operation plan.

(c) Trains can detour to other routes to make full use of the capabilities on the road network.

(d) As long as possible to run long marshaling trains to improve passenger transportation capacity.

(e) Train interchanges can be re-formulated, and train operation sections can be extended or shortened according to the type and level of emergencies.

2. **Transportation organization optimization goals and constraints in catastrophic emergencies**

The impact of catastrophic emergencies is very bad, and it is necessary to carry out rescue quickly to reduce the loss to a minimum. From the perspective of transportation organizations, the goal is to increase passenger turnover, evacuate stranded passengers as soon as possible, and resume normal driving.

In addition, under the condition of catastrophic emergencies, transportation organization involves multiple levels of optimization including emergency rescue, passenger flow prediction, train service plan preparation, and train timetable preparation. the upper-level solution provide constraints for the lower-level plan , and the lower-level plan also provides feasibility verification for the upper-level plan.

3. **Transportation organization process in catastrophic emergency**

TOCE involves multiple levels of content including emergency rescue, passenger flow prediction, train service plan preparation, and train timetable preparation.

After an emergency occurs, dispatchers should judge whether the railway line is interrupted firstly, then set up maintenance skylights according to the speed limit conditions and emergency rescue implementation requirements. And judge whether the rescue train needs to be operated, and then calculate the network capacity. It should be noted here that if the rescue train is operated, the capacity calculation needs to deduct the capacity occupied by the rescue train and generate a rescue train path. In addition, due to the impact of emergencies, some large hub stations may be stranded with a large number of passengers. At this time, passenger flow prediction and evacuation under emergencies are required, integrating OD passenger flows with urgent travel needs, generating train routes, calculating trains' service frequency, stop plan, passenger flow allocation, and then re-compile train operation plan, rolling stock plan and crew plan according to the train service plan. Figure 2.3 shows the key links in the transportation organization under the condition of catastrophic emergencies. The formulation of plans at various levels involves a large number of static and dynamic attributes of infrastructure, technical equipment, and emergencies. The transportation organization under normal conditions is a complex problem of nonlinear, multi-level, and cyclic optimization. Furthermore, under the condition of catastrophic emergencies, transportation organizations are involved in emergency rescue, emergency evacuation, etc., and the formulation of their plans is more complicated.

2.3.4 Relationship Between Transportation Organizations Under Different Types of Emergencies

According to the above research, the main content of the transportation organization under the general emergency situation is the traditional train timetable adjustment problem. At present, many scholars have conducted a lot of research on this type of problem, forming a relatively mature theoretical basis, solution and optimization system, which can provide a useful reference for the optimization of transportation organizations under severe emergencies.

According to the statistics of railway emergencies over the years, the occurrence of serious emergencies is relatively common and should be highly valued by relevant staff. According to the characteristics of serious emergencies and transportation organization principles and strategies, TOSE only involves the optimization of railway internal resources, and the transportation organization plan is relatively easy to implement; while TOCE needs to establish a cross-industry emergency linkage mechanism, involving a large number of off-railway resources and emergency rescue, the implementation of transportation organization plans is more difficult.

In summary, with the escalation of the impact of emergencies, the contents of their corresponding transportation organizations are becoming more and more complicated. There are inclusive links between transportation organizations under the conditions of three different types of emergencies, as shown in Fig. 2.4. TOE is an important

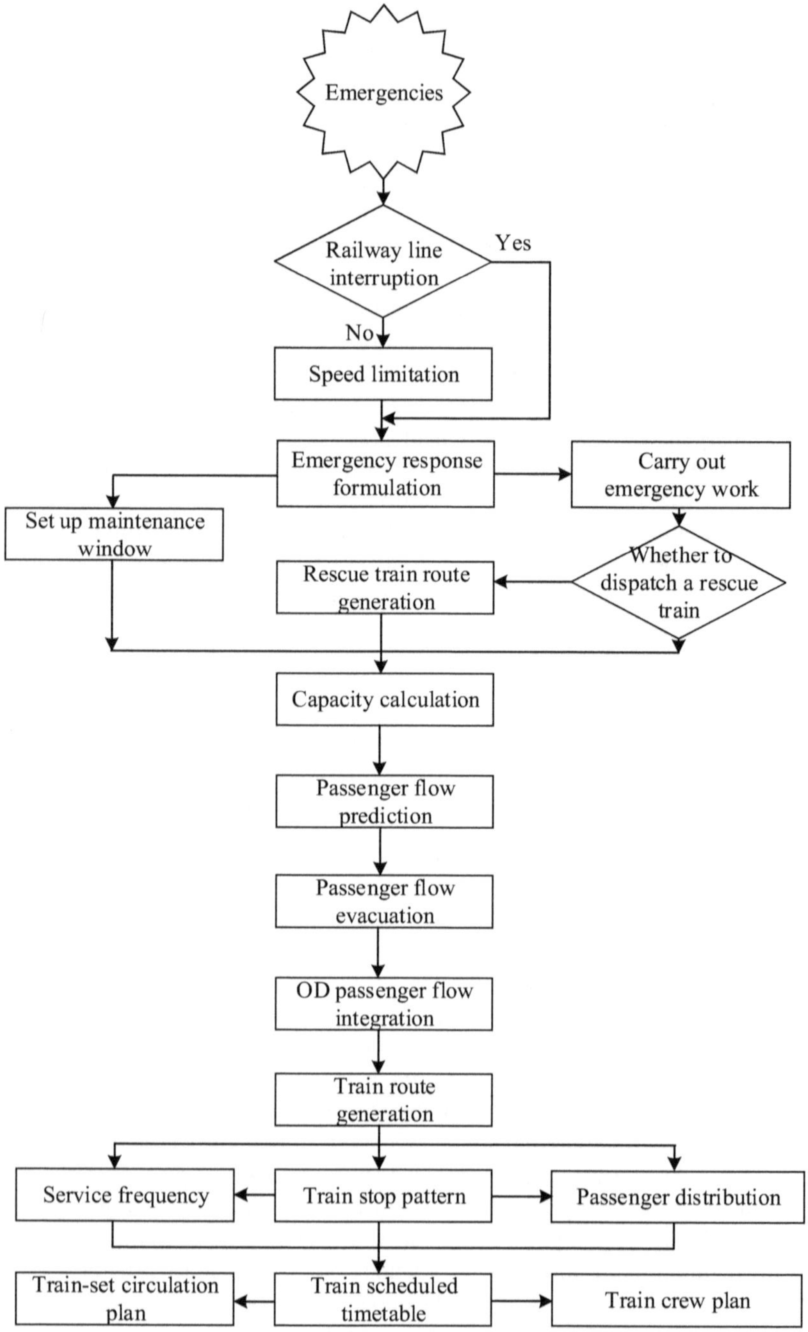

Fig. 2.3 Transportation organization in catastrophic emergencies

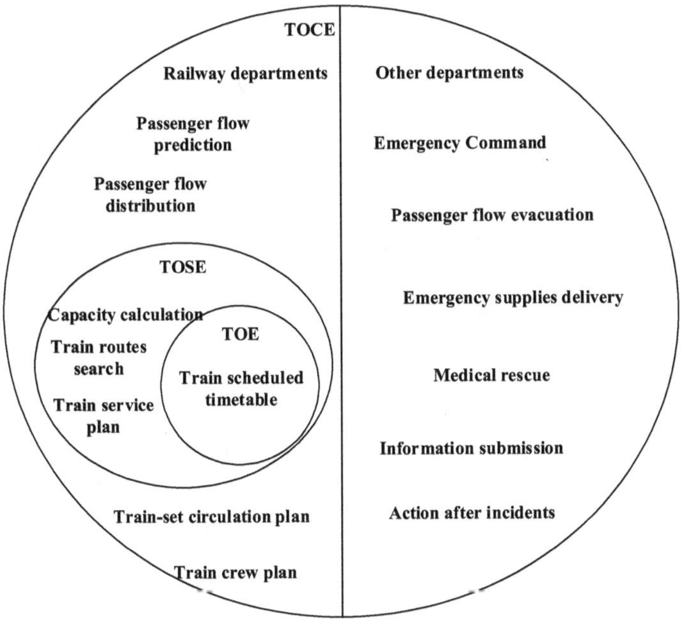

Fig. 2.4 Relationship between transportation organizations under different types of emergencies

part of TOSE; also TOSE is an important part of TOCE, which provides a basis for TOCE optimization. However, no matter what type of transportation organization is ultimately to generate a timetable that can be effectively implemented to guide the train to restore normal transportation order as soon as possible and reduce losses caused by emergencies.

2.3.5 *Analysis of Transport Organization Complexity Under the Serious Emergency Conditions*

According to the above analysis, when the serious emergencies occur, firstly, the train rerouting, trans-reconnection patterns changing, and train cancelation should be applied in train service plan layer. Secondly, train timetable is adjusted, which makes it possible to complete the transportation tasks while the shortages of railway network capacity. Compared with the traditional railway timetable rescheduling problem, the transportation organization under the serious emergency conditions involves more steps. More new characteristics of optimization problem are displayed, and it is necessary to explore new mathematical models and methods.

2.3.6 *Structure of Problem is Complex*

1. **Multi-layers**

According to the analysis in Sect. 2.2, transportation organization problems under serious emergencies involve multiple levels of optimization, including analysis of the impact of emergencies, capacity loss estimation, detour route search, train operation plan adjustment, train operation adjustment, etc. Macro- to micro-problems are at multiple levels.

2. **Mutual constraints between layers**

In the above problems, the upper optimization results are an important part of the constraints of the lower optimization, and the lower optimization results can be used to verify the rationality of the upper optimization and also an important constraint of the upper optimization. For example, the result of the adjustment of the train operation plan determines all the trains running on the route, as well as the attributes of train formation, train type, train class, etc. These are the input of the train operation adjustment problem, which determines the constraints of the train operation adjustment; at the same time, the train operation. The result of the adjustment can be used to evaluate the rationality of the upper-level optimization to change the detour route or reconnection strategy of the train to make the problem better overall.

3. **Mutual constraints within the same layer**

In the above problems, there are mutual constraints within certain levels, such as the detour of trains, the reconnection of trains, and the suspension of trains and other scheduling strategies. How to form a better combination of scheduling strategies to provide a basis for lower-level optimization is our focus.

2.3.7 *The Problem Object is Complex*

1. Multiple lines involved

Unlike the traditional train operation adjustment and optimization problem, which involves only one section or several sections of a line, the optimization problem involves the adjustment of the train operation plan, so that the train may detour to other routes, and this will inevitably involve the cost. For lines other than the line, the objects of capacity calculation and path search will also be expanded to other lines adjacent to the affected section.

2. Various fixed equipment

Train reconnection involves complex train dispatching in and out of the station, which needs to be coordinated within the station to disconnect lines, communication signal systems, traction power supply systems, etc., which increases the complexity of upper-level optimization.

3. Various train types

After the adjustment of the train operation plan, the train may detour to other routes. Due to the different route grades of the detour route, it may involve high-speed, medium-speed, low-speed, and passenger, freight and other types of trains, making the classification of train grades more For the complex.

2.3.8 Complex Environment Problem

1. Diverse environment

According to the different objects affected by the emergency, the state of the line and the attributes of the mobile device will change accordingly. Natural disasters such as mudslides and floods may cause line interruption and change the topology of the road network; natural disasters or signal equipment such as heavy rain and heavy snow. Failures may limit the speed of certain line sections; train failures may limit the speed of some trains themselves. The impact of emergencies makes the operating environment of the transportation organization system more complicated.

2. Dynamic

With the development of emergency status and the progress of rescue work, the operating environment of the railway transportation system is dynamically changing, which allows train speed limit, capacity calculation, and train path search to get different calculation results as the emergency develops. However, when adjusting the train operation plan, different strategies should be adopted according to the change of capacity and route, and rolling optimization should be combined with the adjustment of train operation. For different stages of emergency development, the remaining capacity of the line should be fully utilized to complete the transportation task.

3. Uncertainty

Due to the impact of emergencies, the relevant parameters, optimization goals, and constraints of the transportation organization optimization problem are all uncertain. For example, with the development of emergencies and the progress of rescue work, the change of train speed limit often has a certain randomness. At the same time, due to the difference in line characteristics, as well as the driver's driving behavior, equipment status, and natural climate, the train interval operating hours often change within a range, with obvious ambiguity. How to express these uncertain parameters more accurately and make them more in line with the actual operating environment is also our research focus.

2.4 Chapter Conclusion

This chapter elaborates on the definition and classification of emergencies and analyzes the differences between railway transportation organizations under emergencies and transportation organizations under normal conditions. According to the degree of impact of emergencies, transportation organizations under emergencies are divided into three types and correspond to different contents. There is an inclusion relationship between the three types. At the same time, this chapter also analyzes the complexity of transportation organization problems under severe emergencies, which lays the foundation for the next step of exploring research methods and problem-solving ideas.

References

1. Qin Q, Li T, Lu L E(2004) Emergency management and response. Xinhua Publishing House, Beijing (in Chinese)
2. De Ren S, Jie B, Wang Z (2003) Crisis Handbook. New World Press, Beijing (in Chinese)
3. Wang S (2007) Research on open government information in response to emergencies. Tianjin Normal University (in Chinese)
4. Gao S. The role of the media in government crisis management [EB/OL]. https://ruanzixiao.myrice.com/mtzzfwjzdzy.htm (in Chinese)
5. Keith MH, Jeffrey LC (2003) A social constructionist approach to crisis management: Allegations of sudden acceleration in the Audi 5000. Commun Stud
6. Coady C (2004) Terrorism, morality, and supreme emergency. Ethics
7. Brennan D, Ruth BM "It will happen again", what SARS taught businesses about crisis management. Manage Decis
8. Ming liang Q, Hong C, Hong Z, Ying S (2006) Current situation and prospect of research on emergency management of public emergency. Manage Rev 18(4):35–45 (in Chinese)
9. UK Government advice on definition of an emergency (2007)
10. Firenze B (2001) Labor safety system research. Saf Sci J 45(2):31–37
11. Standing Committee of the Tenth National People's Congress (2007) Law of the people's republic of China on emergency response. Law Press, Beijing (in Chinese)
12. Safety Committee of China Railway Society, Editorial Committee of "Centennial Railway Safety Events" (2009) Centennial railway safety events: 1876–2008. Shanghai Jiaotong University Press, Shanghai, pp 218–510 (in Chinese)
13. Haizhi W (2006) Research on passenger train operation plan of passenger dedicated line. China Academy of Railway Sciences, Beijing (in Chinese)
14. Bangmo He (1995) Transportation organization mode of Beijing-Shanghai high-speed railway. China Railway Sci 16(3):13–23 (in Chinese)
15. Siji Hu (1996) Full high-speed rapid transfer" is the best choice for the Beijing-Shanghai high-speed high-speed railway transportation organization mode. J China Railway Soc 18(A00):90–96 (in Chinese)
16. Qiyuan P, Haifeng Y, Deyong W (2004) Research on the Wuhan-Guangzhou high-speed railway transportation organization model. J Southwest Jiaotong Univ 39(6):703–707, 711 (in Chinese)
17. Bo W, Hao Y (2007) Research on the train operation plan of Beijing-Tianjin intercity railway based on periodic operation diagram. Journal of the China Railway Society 29(2):8–13 (in Chinese)
18. Feng S, Lianbo D, Liang H (2007) Bi-level programming model and algorithm for passenger train operation plan. China Railway Sci 28(3):110–115 (in Chinese)

19. Lufeng C, Rui S, Shiwei He et al (2013) Research on optimization of high-speed railway train operation scheme based on double-level programming model. Railway Transport Econ 35(6):18–23 (in Chinese)
20. Chang YH, Yeh CH, Shen CC (2000) A multi-objective model for passenger train services planning: application to Taiwan's high-speed rail line. Transp Res Part B 34(2):91–106
21. Bo W, Hao Y, Feng N et al (2007) Research on the model and algorithm of periodic operation chart preparation. J China Railway Soc 29(5):1–6 (in Chinese)
22. Qiyuan P (2007) Passenger dedicated line transportation organization. Science Press, Beijing, pp 46–47 (in Chinese)
23. Shuying X (1997) The causes and types of natural disasters in Chinese railways. J Railway Normal Univ 14(4):33–39, 47 (in Chinese)
24. Huawu He (2008) Impact of disasters on railways and their defense countermeasures. China Railways 10:1–8 (in Chinese)

Chapter 3
Estimation of Carrying Capacity of High-Speed Railway Section in Case of Emergency

The carrying capacity of the railway section is significantly different between emergency conditions and normal conditions. The railway transportation system will be affected to varying degrees, the types and levels of emergencies. For example, the original block mode may be degraded, due to the interruption of traffic in certain sections or the adoption of train speed limit measures, and the carrying capacity will also change dynamically with the development of the emergency. In addition, there is fuzziness in the carrying capacity under emergency conditions, due to factors such as line characteristics, train speed, signal equipment reaction time, and driver's behavior.

First, four methods of calculating the section carrying capacity are analyzed, which are graphic method, analytical method, optimization method, and simulation method. Then, according to the characteristics of "mixed mode of high- and medium-speed trains" of China's high-speed railway, the widely used train deduction coefficient method is introduced. In addition, the uncertainty and fuzziness of section carrying ability under emergencies are mainly studied, and the calculation method of section carrying ability based on fuzzy Markov chain is proposed, and a specific case is given.

3.1 Analysis on Calculation Method of Carrying Capacity in Railway Section

Railway carrying capacity refers to the maximum number of trains or train pairs of standard weight that can be released by railway fixed equipment in a unit time (usually one day and night) under the premise of a certain locomotive, vehicle type and certain transportation organization method [1]. According to the different types of infrastructure passed by trains, it can be divided into section carrying capacity and station carrying capacity. Section carrying capacity refers to the maximum number of trains

L. Jia et al., *High-Speed Railway Operation Under Emergent Conditions*,
Advances in High-speed Rail Technology,
https://doi.org/10.1007/978-3-662-63033-4_3

that a railway section can pass in a unit time (one day or night or several hours) under certain train types and transportation organization conditions. The influencing factors of the section carrying capacity mainly include the number of section main lines, railway signal system, train operating speed, length of blocked section, train type and train sequence, etc. The current methods used to assess capacity of the railway section include graphic methods, analytical methods, simulation methods, and optimization methods.

1. **Graphic Method**

The graphic method is a method of manually simulating the actual transportation production, and using characteristics to graphically represent and determine the passing capacity [1]. Although this method is feasible, it is usually used to verify the feasibility of train operation plan because of its heavy workload and different quality of the layout.

2. **Analytical Method**

The analytical method is a method of establishing a mathematical model based on the analysis of related factors such as existing fixed equipment and mobile equipment. This method can take all kinds of factors into account, and the workload is smaller than that of graphic method, and the train deduction coefficient method is mainly used in China now.

By calculating the section carrying capacity of a parallel diagram with only one kind of train, and determining the deduction coefficient of various trains under different running speed, headway and quantity requirements, the train deduction coefficient method obtains the section carrying capacity of the nonparallel diagram. In Ref. [2], for high-speed railway, the technical load coefficient was proposed to avoid the error caused by train operation and equipment failure, the deduction coefficient of high-speed train and the average deduction coefficient of high- and medium-speed train mixed operation were obtained, and then the section carrying capacity of nonparallel operation diagram of high-speed and high- and medium-speed mixed operation was calculated, respectively. Reference [3] proposed a calculation method for headway in the moving block mode and then achieved the calculation of the section carrying capacity by drawing the maximum train operation diagram of the section. Through determining the interval time and additional start stop time in different tracking modes, the section carrying capacity was calculated, and finally determined by the way of drawing train operation diagram [4]. In Ref. [5], the capacity of high-speed railway section is calculated by theoretical analysis of the deduction coefficient under three combinations of different speed levels and stop strategies and the train diagram drawn by graphic computer. In addition, the effect of different numbers of medium-speed trains on the carrying capacity of high-speed and medium-speed train mixed running lines is also studied. For high-speed trains on passenger dedicated lines under the condition of quasi-moving block, literature [6] adopted a continuous primary speed curve control mode and gave four calculation methods and related parameters of train headway, and used traction calculation to verify the accuracy. The paper calculated the carrying capacity of the parallel operation diagram based on the non-stop high-speed trains and obtained the carrying capacity of the passenger dedicated line by calculating the deduction coefficient of medium-speed trains and stopped high-speed trains. Aiming at the characteristics of time and periodicity of intercity railway passenger flow, based on the structural analysis of the train operation diagram, the method of combining the average minimum train interval method, and the low-speed train deduction coefficient method was used to calculate the intercity railway section carrying capacity [7].

Furthermore, in 2004, the International Union of Railways adopted a methodology UIC406 of calculating railway capacity [8]. The main flow is that the train timetable is first generated according to the basic line, and then the train running line in the train timetable is compressed to get the minimum headway, and the carrying capacity is calculated [9–11]. To evaluate the railway capacity, literature [12] proposed that the number of trains, average speed, stability, and homogeneity of the train schedule are the main factors and analyzed the importance of the choice of interval length. Reference [13] took the Spanish railway as an example to analyze the influence of train speed, stop mode, train homogeneity, block partition length, and robustness of train schedule on railway capacity.

3. **Optimization Method**

The main idea of optimization method is to evaluate the carrying capacity of the section by establishing a mathematical optimization model, and its goal is to enable the established railway section to run more trains to obtain the maximum carrying capacity. The literature on the subject dates to the 1970s. By determining the train departure time, literature [14] and [15] used the branch-and-bound method to solve the single-line train operation plan generation problem and calculated the section carrying capacity. Mixed-integer programming ([16]) and some evolutionary algorithms ([17, 18]) were also used to solve such problems. Reference [19] established an optimization model for the carrying capacity of the double-track railway's automatic block, mainly considering constraints such as skylight time, the ratio of continuous trains and the number of high- and medium-speed trains, among which the headway was a key factor.

4. **Simulation Method**

With the development of computer technology in recent years, some scholars have tried to use relatively mature computer simulation software to analyze the railway section carrying capacity, such as MULTIRAIL ([20]), SIMONE ([21]), RAILSYS ([22]), and OPENTRACK ([23]). The basic principle of the software is to build the railway infrastructure, set the parameters of the operation diagram, and realize the drawing and capacity evaluation of the train operation diagram by simulation.

From the above analysis, we know that mathematical models or simulations can only obtain approximate solutions for section carrying capacity, because often only the key factors are considered, but in practice the problem involves a variety of complex factors. In addition, no matter which method is used, it will involve factors such as the railway signal system, the length of the block partition, the train running speed, the headway, and the train stop mode. In reality, some factors will appear different degree of fuzziness or randomness due to the driver's behavior, equipment state, or emergencies.

3.2 Analysis of Train Deduction Factor

The carrying capacity of railway section is affected by the factors such as train stop and overtaking mode, and the degree of influence can be expressed by deduction coefficient. This section introduces the calculation method of train deduction coefficient.

1. **Maximum section carrying capacity of high-speed parallel diagram**

Under normal conditions (with only one train, no stop, and no overtaking), the maximum carrying capacity (column/day) of the section of the one-way parallel diagram can be expressed by Eq. 3.1.

$$N_{\max} = \frac{1\,440 \times 60 - t_{\text{maintenance}} - t_{\text{inefficacy}}}{I} \tag{3.1}$$

where $N_{\max}$ is the maximum section carrying capacity, $t_{\text{maintenance}}$ is the total comprehensive maintenance time during the day, $t_{\text{inefficacy}}$ is the additional lost time except maintenance time, and I is the headway.

2. **Deduction coefficient for the stopped high-speed train**

When the train stops at stations, its running time will increase due to the increase of the stop time and the additional time of starting and stopping, and it will also occupy more section carrying capacity (as shown in Fig. 3.1). Therefore, compared with the case of non-stop, the deduction coefficient will be generated when the train stops, and the calculation method is shown in Eq. 3.2.

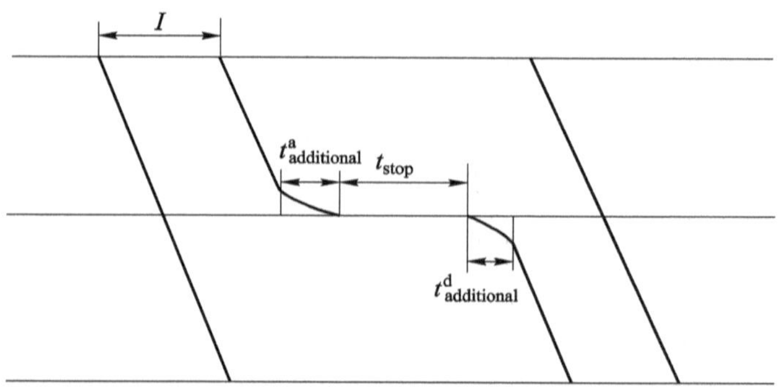

Fig. 3.1 Schematic diagram of the deduction coefficient for the stopped high-speed train

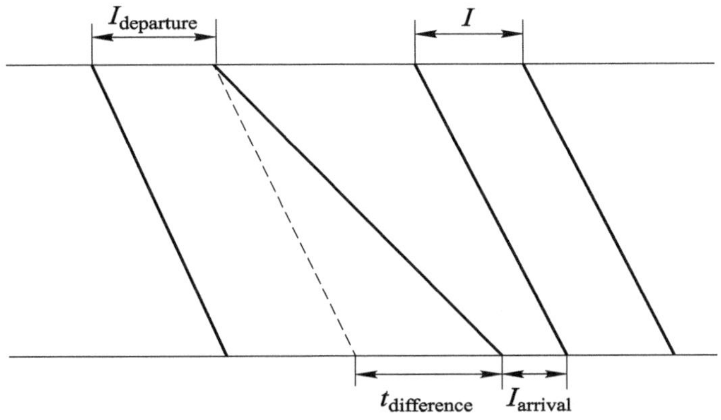

Fig. 3.2 Schematic diagram of the deduction coefficient for the non-stop medium-speed train

$$\varepsilon_{high}^{stop,high} = \frac{I + t_{additional}^{a} + t_{stop} + t_{additional}^{d}}{I} \tag{3.2}$$

where $\varepsilon_{high}^{stop,high}$ is the deduction coefficient for stopped high-speed train, $t_{additional}^{a}$ is the additional time for train arrival, $t_{additional}^{d}$ is the additional time for train departure, t_{stop} is the dwell time.

3. **Deduction coefficient for the non-stop medium-speed train**

In the case of no stop and no overtaking, medium-speed trains will occupy more section capacity, as shown in Fig. 3.2. The calculation method is shown in Eq. 3.3.

$$\varepsilon_{high}^{quasi} = \frac{I_{departure} + t_{diffenrence} + I_{arrival} - I}{I} \tag{3.3}$$

where $\varepsilon_{high}^{quasi}$ is the deduction coefficient for the non-stop medium-speed train, $I_{departure}$ is the departure headway between high- and medium-speed trains, $t_{diffenrence}$ is the running time difference between high- and medium-speed trains, $I_{arrival}$ is the arrival headway between high- and medium-speed trains.

4. **Deduction coefficient for the stopped medium-speed train**

In addition to the driving speed, if the medium-speed train stops at stations, it will occupy more carrying capacity, as shown in Fig. 3.3, and the calculation method is shown in Eq. 3.4.

$$\varepsilon_{high}^{stop,quasi} = \frac{I_{departure} + t_{diffenrence} + I_{arrival} + T_{stop} - I}{I} \tag{3.4}$$

$$T_{stop} = t_{additional}^{a} + t_{stop} + t_{additional}^{d} \tag{3.5}$$

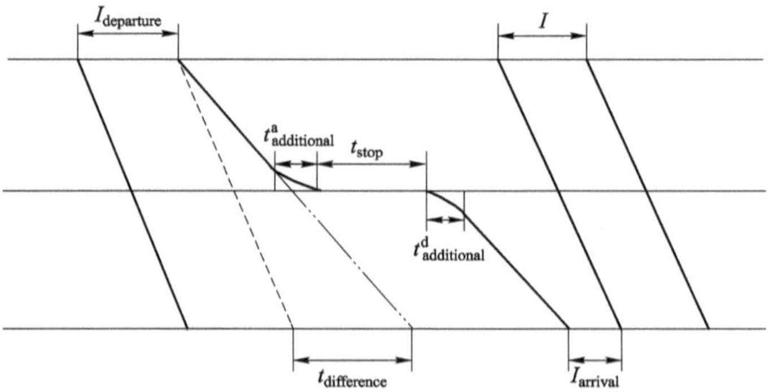

Fig. 3.3 Schematic diagram of the deduction coefficient for the stopped medium-speed train

where $\varepsilon_{\text{high}}^{\text{stop,quasi}}$ is the deduction coefficient for the stopped medium-speed train, T_{stop} is the additional time for medium-speed train dwelling.

4. **Deduction coefficient for medium-speed train waiting for high-speed train**

As shown in Fig. 3.4, the medium-speed train stops and waits for the high-speed train to overtake, and the calculation method of the deduction coefficient is shown in Eq. 3.6.

$$\varepsilon_{\text{high}}^{\text{stop, stopover}} = \frac{I_{\text{departure}} + t_{\text{diffenrence}} + I_{\text{arrival}} + t_{\text{stop}} - 2I}{I} \tag{3.6}$$

where $\varepsilon_{\text{high}}^{\text{stop, stopover}}$ is the deduction coefficient for medium-speed train waiting for high-speed train. And the stop time of the medium-speed train must meet the requirements of the headway between trains:

$$t_{\text{stop}} \geq I_{\text{departure}} + I_{\text{arrival}} \tag{3.7}$$

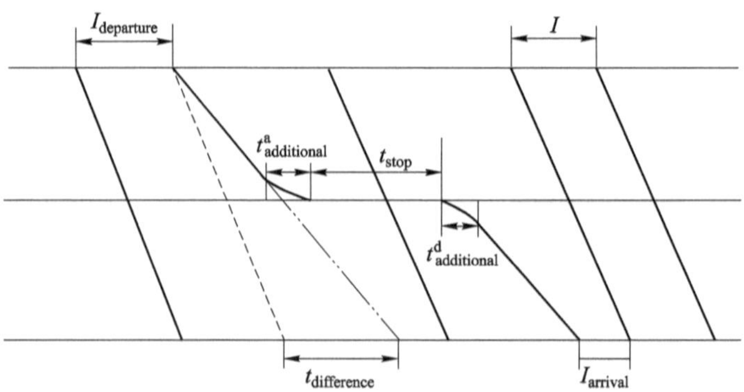

Fig. 3.4 Schematic diagram of the deduction coefficient for medium-speed train waiting for high-speed train

3.3 Factor Analysis of Railway Train Headway

The headway of railway train is very important in safe railway operation, which controls the capacity of railway sections. Usually, there are two headway block modes, i.e., the quasi-moving block mode and the fixed block mode, which are described in below sections.

1. **Train headway under quasi-moving block mode**

China's high-speed railway mainly adopts the quasi-moving block mode to control the train operation and adopts the target distance control mode. The target distance control mode determines the train braking curve according to the target distance, target speed, and the performance of the train itself. Instead of setting the speed level of each block area, the primary braking mode is adopted. The target point of the quasi-mobile block is the beginning of the blocking zone occupied by the preceding train. The operation in the blocking zone is not affected by the preceding train, and the braking point varies with the line parameters and the performance of the train (see Fig. 3.5). In this case, the time of train headway can be calculated as Eq. 3.8.

$$I = \frac{L_{\text{action}} + L_{\text{delay}} + L_{\text{brake}} + L_{\text{protect}} + L_{\text{comfortable}} + L_{\text{train}}}{v} \times 3.6 \tag{3.8}$$

where L_{action} is the distance related to response time, for example, when *Train A* (the train behind) receives a braking signal, the time is needed for a train driver to take an action, L_{delay} is the distance related to the time needed to launch the brake system, L_{brake} is the distance from break launch to *Train A* stopping, L_{protect} is the safety distance depending on rail conditions and train speeds, $L_{\text{comfortable}}$ is the distance of normal operation, L_{sum} is the total of the distance, and v is the average running speed of the train.

2. **Train headway under fixed block mode**

In the fixed block mode, train operation depends on the signal light control. Most of China's railways still adopt this control mode, and many literatures have analyzed the headway under this mode.

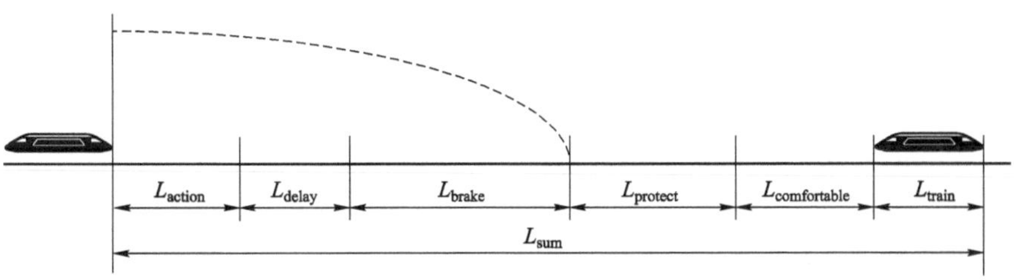

Fig. 3.5 Schematic diagram of the deduction coefficient for medium-speed train waiting for high-speed train

(1) In the three-pattern fixed block section, the headway interval of the train usually separates in three-pattern block areas [25] (see Fig. 3.6), and the calculation of the train headway expression shown as Eq. 3.9.

$$I = \frac{L_{\text{train}} + L'_{\text{subsection}} + L''_{\text{subsection}} + L'''_{\text{subsection}}}{v} \times 3.6 \tag{3.9}$$

where L_{train} is length of train; $L'_{\text{subsection}}$, $L''_{\text{subsection}}$, $L'''_{\text{subsection}}$ are three block section length, respectively.

(2) In the four-pattern fixed block section, the headway interval of the train usually separates in four-pattern block areas [2], as shown in Fig. 3.7. The train headway can be calculated by Eq. 3.10.

$$I = \frac{4L_{\text{subsection}} + L_{\text{train}}}{v} \times 3.6 \tag{3.10}$$

where $L_{\text{subsection}}$ is block section length.

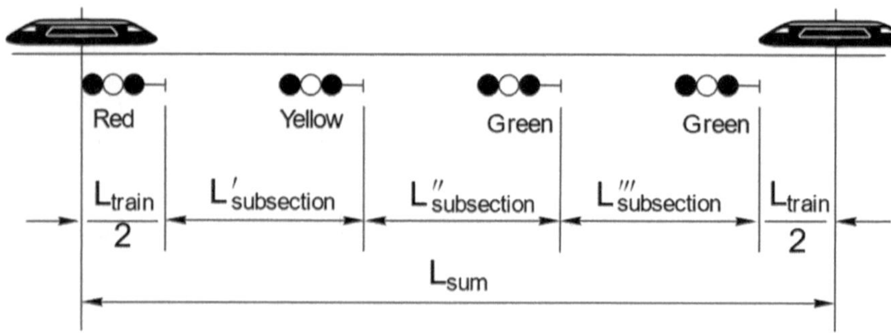

Fig. 3.6 Three-pattern fixed block mode

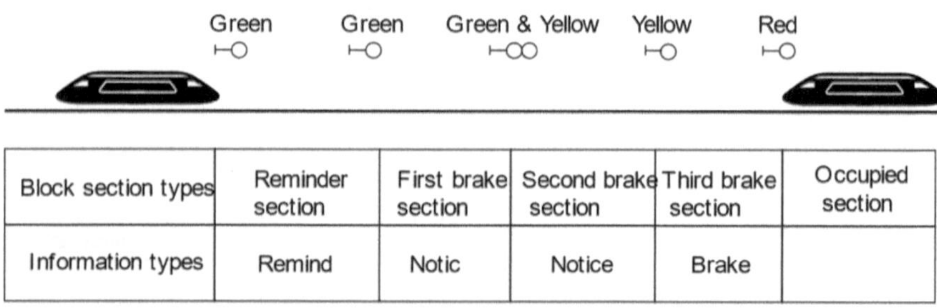

Fig. 3.7 Four-pattern fixed block mode

3.4 Calculation of Carrying Capacity of High-Speed Railway Section Under Emergency

3.4.1 Uncertainty Analysis of the Railway Section Capacity

1. **Uncertainty analysis of the train headway under quasi-moving block mode**

Equation 3.8 can be transferred to the speed–time expression as Eq. 3.11.

$$I = t_{\text{action}} + t_{\text{delay}} + \frac{v/3.6}{2a} + \frac{L_{\text{protect}} + L_{\text{comfortable}} + L_{\text{train}}}{v/3.6} \tag{3.11}$$

As can be seen from Eq. 3.11, to determine the parameters depends on, for example, rail conditions, train speeds, types of train control systems, types of the braking systems, and also the driver behaviors, etc. The t_{action} is affected by many factors such as the time of the train control system to response, the time of information transmission, and the reaction time of the driver. Landex [12] studied the t_{action} and believed that t_{action} which is around 9.5 s.

The t_{delay} is the response time of the train brake system. Again, t_{delay} is also a fuzzy value, which is generally considered as around 2.5 s [12].

The a is the braking deceleration speed and is determined based on the braking performance of the train and route conditions. Xianming [24] studied the corresponding deceleration speeds when a train is running on a downhill slope with 20%, plain section and approaching to the station, and a can choose 0.565 ms^{-2}, 0.75 ms^{-2}, and 0.5 ms^{-2}, respectively. Because of the differences of the train braking system and the route conditions, the value of a is fuzziness with the range from 0.5–0.75 ms^{-2}.

If the safety margin was considered, the value of L_{protect} can choose with the range of 80–150 m. The value of $L_{\text{comfortable}}$ changes with the differences of the control mode, the longer the value of $L_{\text{comfortable}}$, the drivers' behavior shows more calm down. However, it will increase the train headway at the meantime. Therefore, it will keep the comfortable driving and high usage of the ability by chosen 1–2 block length.

The average running speed v of the train is affected by many factors, such as line condition, driver's behavior, equipment state and external environment, so it is the most obvious parameter with fuzzy property. Especially under the condition of emergency, the speed limit by the railway department only specifies the maximum speed that the train can travel in a period of time and area, and it still has a certain gap with the average running speed of the train. At the same time, with the progress of equipment repair and transportation organization, the change of impact degree of emergency often has certain randomness, which also makes the accurate expression of average train speed more difficult.

In summary, the train headway in the quasi-moving block mode is uncertain, and its calculation formula is expressed as:

$$I = \tilde{t}_{\text{action}} + \tilde{t}_{\text{delay}} + \frac{\hat{v}/3.6}{2a} + \frac{\tilde{L}_{\text{protect}} + \tilde{L}_{\text{comfortable}} + L_{\text{train}}}{\hat{v}} \times 3.6 \tag{3.12}$$

In this study, superscript "~" expresses that a parameter is a fuzzy parameter, and the superscript "^" expresses a parameter which is a fuzzy random parameter.

2. **Uncertainty analysis of the train headway under quasi-moving block mode**

In fixed block mode, the train headway mainly depends on the length of the block section and the train operation speed. Normally, the three-pattern block section requires whose length of one block section to satisfy one train barking distance [26]. And the four-pattern block section requires two or three block sections. The length of the block sections is determined by the train braking distance, signaling indication system and the safety redundancy. However, the main factors that affect the braking distance are the traction machines, traction weight, route speed restriction, route slope, etc. These factors increase the uncertainty of the train headway under the fixed block mode. The length of the block section can choose from the range 1600–2600 m under the three-pattern mode and the 700–900 m under the four-pattern mode [24]. Therefore, under the three-pattern mode and four-pattern mode, the calculation function of the train headway with uncertainty can be transformed into Eq. 3.13 and Eq. 3.14, respectively.

$$I = \frac{L_{\text{train}} + \tilde{L}'_{\text{subsection}} + \tilde{L}''_{\text{subsection}} + \tilde{L}'''_{\text{subsection}}}{\hat{v}} \times 3.6 \tag{3.13}$$

$$I = \frac{L_{\text{train}} + 4\tilde{L}_{\text{subsection}}}{\hat{v}} \times 3.6 \tag{3.14}$$

3. **Uncertainty analysis of the section capacity**

The impact of disruptions and the maintenance ability of railway management departments are uncertain, which cause a random variation in train operation speed correspondingly. Therefore, the section capacity changes randomly in a period of time. Moreover, in a certain time of future, the railway network capacity is associated with the current equipment instead of the former condition. As the conclusion, the transformation of the section capacity has the non-aftereffect property for different rail lines and emergency types. The changing process of the section capacity status can be described as Markov chain (MC). Due to the fuzziness property of the section capacity, the capacity status changing process is considered as the fuzzy Markov chain (FMC) in this paper.

As shown in Fig. 3.8, the time axis is divided into M periods. Because of the different effect level of the special events in each period, the different emergency grade, the different degree of the maintenance, the train speed limitation is different. So, the capacity of each time period is also different. Then, the whole section capacity can be computed by Eq. 3.15, where $\tilde{N}^{(i)}_{\text{mix}}$ is fuzzy value of the section capacity in the ith period, by setting $\Delta t^{(i)} = t_i - t_{i-1}, i = 1, 2, \ldots, M$.

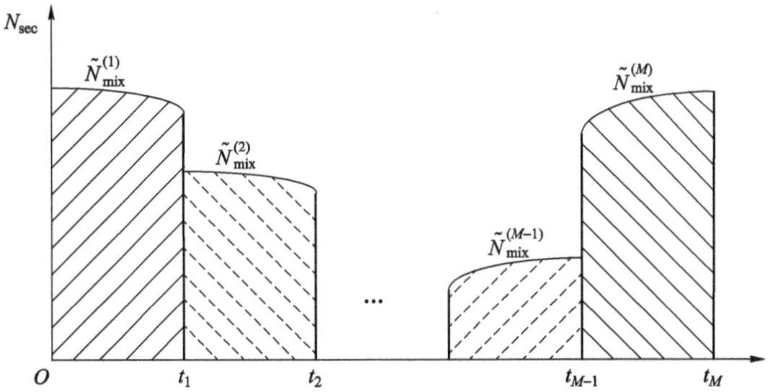

Fig. 3.8 Section capacity changing process in an emergency

$$N_{sec} = \sum_{i=1}^{M} \tilde{N}_{mix}^{(i)} \tag{3.15}$$

Note that, the division of time period is determined by the impact of the emergency and the recovery degree of the line capacity, so the division of time period is also fuzzy, which can be determined by the evolution process of the emergency or the speed limit strategy under the actual conditions.

3.4.2 *Fuzzy Markov Chain*

The fuzzy random variable (FRV) is a measurable function, which is the set from the probability space mapping to the fuzzy space. For example, for a given probability space $\{\Theta, \Lambda, P\}$, $\tilde{u}_1, \tilde{u}_2, \cdots, \tilde{u}_n$ are the fuzzy variable, if $\xi(\omega_i) = \tilde{u}_i$, $i = 1, 2, \cdots, n$, then, $\xi(\omega_i)$ is the fuzzy random variable [27].

Definition 3.1 For a given probability space (Θ, Λ, P), Θ is nonempty set, Λ is the σ algebra of Γ, P is the probability. The number of the fuzzy random variable $\{\tilde{X}(t), t = 0, 1, 2, \cdots, n\}$ is limited or countable. For the whole possible fuzzy conditions of all $\tilde{X}(t)$, there is one group of fuzzy set $\tilde{A} = \{\tilde{A}_0, \tilde{A}_1, \tilde{A}_2, \cdots, \tilde{A}_n\}$ corresponding to $\tilde{X}(t)$, make s to the possibility of one fuzzy event, if the possibility of $\tilde{X}(t+1) = \tilde{A}_{t+1}$ (variable $\tilde{X}(t+1)$ at the moment $t+1$ stay in the $\tilde{A}_{t+1}$ status) can only relate to $\tilde{X}(t)$, but not the status before the n, expressed as:
$s(\tilde{X}(t+1) = \tilde{A}_{t+1}|\tilde{X}(t), \tilde{X}(t-1), \cdots, \tilde{X}(0))= s(\tilde{X}(t+1) = \tilde{A}_{t+1}|\tilde{X}(t))$,

Thus, the sequence $\tilde{X}(t)$ is named as fuzzy Markov chain (FMC).

Signed $\boldsymbol{S}'$ as one-step transfer possibility matrix, which is denoted by Eq. 3.9.

$$S' = \begin{bmatrix} s_{0,0} & s_{0,1} & s_{0,2} & \cdots & s_{0,j} & \cdots \\ s_{1,0} & s_{1,1} & s_{1,2} & \cdots & s_{1,j} & \cdots \\ \vdots & \vdots & \vdots & & \vdots & \\ s_{i,0} & s_{i,1} & s_{i,2} & \cdots & s_{i,j} & \cdots \\ \vdots & \vdots & \vdots & & \vdots & \end{bmatrix} \tag{3.16}$$

The $i+1$-th row of the S' is the conditional probability of a given $\tilde{X}(i) = \tilde{A}_i$, i.e., $s_{i,j} = s(\tilde{X}(j) = \tilde{A}_j | \tilde{X}(i) = \tilde{A}_i)$. Noted that the FMC defined here still strictly abides by "no aftereffect."

The difference between FMC and Markov chain on probability space is the state space of variables. The object observed by the latter is a random variable, and all possible states are real numbers, while the object observed in the former is a FRV, and all possible states are fuzzy numbers.

3.4.3 Calculation of Section Capacity Fuzzy Value

Assuming in the continuous emergency time T, the speed type of the train is Q_{speed}, denoted as $v^{(1)}, v^{(2)}, \cdots, v^{(Q_{\text{speed}})}$. Divide the continuous time T into Q_c time periods, the correspond section capacity has Q_c kind of situations, denoted as $\tilde{c}_1, \tilde{c}_2, \cdots, \tilde{c}_i$, $i = 1, 2, \cdots, Q_c$, consisting all the possible fuzzy sets $\tilde{C}$ of the capacity changing Markov chain procession.

Firstly, the section capacity can be calculated in a time period by Eq. 3.17.

$$c = (60 \times 60 - t_{\text{mai}})/I \tag{3.17}$$

Here, c is the section capacity in each time period. t_{mai} is the comprehensive maintenance time. If different trains' speeds exist, the capacity can be calculated by the deduction coefficient method [28].

Because of the influence of other factors, the dispatcher cannot calculate the section capacity according to the train speed in real time, which means that the section capacity within a period of time often corresponds to a set of changing train speed, so $Q_{\text{speed}} \neq Q_c$.

The status transfer matrix of train speed can be denoted by Eq. 3.18.

$$P = \begin{bmatrix} P_{1,1} & P_{1,2} & \cdots & P_{1,Q_{\text{speed}}} \\ P_{2,1} & P_{2,2} & \cdots & P_{2,Q_{\text{speed}}} \\ \vdots & \vdots & & \vdots \\ P_{Q_{\text{speed}},1} & P_{Q_{\text{speed}},2} & \cdots & P_{Q_{\text{speed}},Q_{\text{speed}}} \end{bmatrix} \tag{3.18}$$

Here, $P_{i,j}$ is the probability of the train speed from $v^{(i)}$ to $v^{(j)}$.Then, set the status transfer matrix of the section capacity under the possible measurement can be calculated by Eq. 3.19.

$$S = \begin{bmatrix} s_{1,1} & s_{1,2} & \cdots & s_{1,Q_c} \\ s_{2,1} & s_{2,2} & \cdots & s_{2,Q_c} \\ \vdots & \vdots & & \vdots \\ s_{Q_c,1} & s_{Q_c,2} & \cdots & s_{Q_c,Q_c} \end{bmatrix} \tag{3.19}$$

Here, $s_{i,j}$ is the condition probability of the section capacity from $\tilde{c}_i$ to $\tilde{c}_j$ under the possible measurement [29], denoted as Eq. 3.20.

$$s_{i,j} = \left(\sum_{m=0}^{Q_{\text{speed}}} P_m \mu_{\tilde{c}_i}(m) \sum_{s=0}^{Q_{\text{speed}}} P_{m,s} \mu_{\tilde{c}_j}(s) \right) / P(\tilde{c}_i) \tag{3.20}$$

Here, $\mu_{\tilde{c}_i}(m)$ is the membership function of the section capacity of the time periods; $P(\tilde{c}_i)$ is the probability that the section capacity of the time periods takes the fuzzy value $\tilde{c}_i$; P_m is the probability that the train speed takes $v^{(m)}$; $P_{m,s}$ is the probability that the train speed changes from $v^{(m)}$ to $v^{(s)}$.

For the convenient of calculation, define matrix Q_1 and Q_2 by Eqs. 3.21 and 3.22.

$$Q_1 = \begin{bmatrix} \mu_{\tilde{c}_1}(0) & \mu_{\tilde{c}_2}(0) & \cdots & \mu_{\tilde{c}_N}(0) \\ \mu_{\tilde{c}_1}(1) & \mu_{\tilde{c}_2}(1) & \cdots & \mu_{\tilde{c}_N}(1) \\ \vdots & \vdots & & \vdots \\ \mu_{\tilde{c}_1}(N) & \mu_{\tilde{c}_2}(N) & \cdots & \mu_{\tilde{c}_N}(N) \end{bmatrix} \tag{3.21}$$

$$Q_2 = \begin{bmatrix} \frac{P_0\mu_{\tilde{c}_1}(0)}{P(\tilde{c}_1)} & \frac{P_1\mu_{\tilde{c}_1}(1)}{P(\tilde{c}_1)} & \cdots & \frac{P_N\mu_{\tilde{c}_1}(N)}{P(\tilde{c}_1)} \\ \frac{P_0\mu_{\tilde{c}_2}(0)}{P(\tilde{c}_2)} & \frac{P_1\mu_{\tilde{c}_2}(1)}{P(\tilde{c}_2)} & \cdots & \frac{P_N\mu_{\tilde{c}_2}(N)}{P(\tilde{c}_2)} \\ \vdots & \vdots & & \vdots \\ \frac{P_0\mu_{\tilde{c}_N}(0)}{P(\tilde{c}_N)} & \frac{P_1\mu_{\tilde{c}_N}(1)}{P(\tilde{c}_N)} & \cdots & \frac{P_N\mu_{\tilde{c}_N}(N)}{P(\tilde{c}_N)} \end{bmatrix} \tag{3.22}$$

where $N = Q_c$.

Therefore, we can get Eq. 3.23.

$$S = Q_2 P Q_1 \tag{3.23}$$

Under the possibility measurement, $\hat{N}_{\text{mix}}^{(1)}$ is the carrying capacity of the section corresponding to the first speed limit within the duration time T. If $\hat{N}_{\text{mix}}^{(1)} = \tilde{c}_1$, then:

$$\hat{N}_{\text{mix}}^{(2)} = \begin{cases} \tilde{c}_1, s_{1,1} \\ \tilde{c}_2, s_{1,2} \\ \vdots \\ \tilde{c}_{Q_c}, s_{1,Q_c} \end{cases} \tag{3.24}$$

The conditional possibility of $\hat{N}_{\text{mix}}^{(2)}$ taking the fuzzy value $\tilde{c}_i$ is $s_{1,i}$, $i = 1, 2, \cdots, Q_c$. For all periods, section capacity can be generalized as Eq. 3.25.

$$\hat{N}_{\text{mix}}^{(i)} = \begin{cases} \tilde{c}_1, s_{h,1} \\ \tilde{c}_2, s_{h,2} \\ \vdots \\ \tilde{c}_{Q_c}, s_{h,Q_c} \end{cases} \quad i = h + 1 \tag{3.25}$$

Therefore, the section capacity of the divide-period in limited speed condition $\hat{N}_{\text{mix}}^{(i)}$ is the fuzzy random variable.

The section capacity in the whole time T is the sum of all periods of capacity, denoted by Eq. 3.26.

$$N_{\text{sec}} = \sum_{i=1}^{Q_c} \tilde{N}_{\text{mix}}^{(i)} = \begin{cases} \sum_{i=1}^{Q_c} \tilde{c}_1, \prod_{i=1}^{Q_c} s_{i,1} \\ \sum_{i=1}^{Q_c-1} \tilde{c}_1 + \tilde{c}_2, \prod_{i=1}^{Q_c-1} s_{i,1}, s_{Q_c,2} \\ \sum_{i=1}^{Q_c-1} \tilde{c}_1 + \tilde{c}_3, \prod_{i=1}^{Q_c-1} s_{i,1}, s_{Q_c,3} \\ \vdots \\ \sum_{i=1}^{Q_c-1} \tilde{c}_{Q_c} + \tilde{c}_{Q_c-1}, \prod_{i=1}^{Q_c-1} s_{i,Q_c}, s_{Q_c,Q_c-1} \\ \sum_{i=1}^{Q_c} \tilde{c}_{Q_c}, \prod_{i=1}^{Q_c} s_{i,Q_c} \end{cases} \tag{3.26}$$

It is noticeable to see that in possible measurement, the section carrying capacity is still a fuzzy random value in the continuous duration T under the emergency condition.

In summary, the steps to solve the section capacity within the duration T under the emergency conditions are as follows.

(1) Give the transfer probability of train speed under emergency conditions;
(2) Calculate the fuzzy value of section capacity under different speed limit conditions;
(3) Calculate the conditional transfer probability of the section capacity, according to the train speed transfer probability;
(4) Set the first time period section capacity as $\tilde{N}_{\text{mix}}^{(1)} = \tilde{c}_1$, and according to the conditional transfer probability obtained in step (3), get the section capacity of all subsequent time periods in turn;
(5) Accumulate the section capacity obtained in step (4) to obtain the final section capacity within the duration T.

3.4.4 Sensitivity Analysis

The FMC-based section capacity calculation method mainly depends on the fuzzy value calculation of the headway and the condition transfer probability of the capacity. In moving block mode, the trains comfortable driving distance (can take one to two block lengths based on the train speed) are the key contributors compare with other two items. In addition, the parameters of the train length, the train deceleration, and the block length can be obtained by general measurement. Therefore, the main sensitive parameter of the train headway is the train speed in the moving block mode.

In the fixed block mode, from Eqs. (3.13) and (3.14), the sensitive parameter of the headway is the train speed as well. Moreover, the condition transfer probability of section capacity depends on the train speed status transfer probability. Therefore, considering limited speed condition, the estimation of the train operation speed and the speed changing probability is the sensitive parameter of section capacity calculation in this paper. The speed transfer matrix responds the dynamic changes of the operation environment in the emergency condition. By preceding the statistics of the train speed in reality, the frequency of the speed changes forms the transfer probability of the different speed. This value can be obtained by three ways: ① record and count the train operation status of the occurred unexpected events in history; ② statistics based on train drivers' experience and the operation dispatcher experience; ③ proceed the identification according to the safety operation specification in emergency, such as if the wind speed is 25–30 ms^{-1}, the train limited speed is 70 kmh^{-1}; if the depth of the snow on the surface of the railway is 19–22 cm, the train limited speed is 160 $\text{km}\cdot\text{h}^{-1}$.

3.5 Case Study

Beijing-Shanghai high-speed railway is taken as an example to prove the validity of the method for calculating the section capacity of high-speed railway under the emergency conditions proposed in this chapter.

1. **Emergency Scenario**

Beijing-Shanghai high-speed railway, as the main artery of China, plays an important role in promoting the development of the national high-speed railway network, and Shanghai-Nanjing section is the busiest section. The details of design and train schedules of Shanghai to Nanjing section can be found in [28].

In Zhenjiang West to Wuxi East section, due to the heavy wind situation, speed restriction is implemented. The wind speed is expected to keep at about 20 ms^{-1} for 2 h. Then, the wind speed would increase at about 24 ms^{-1} for 5 h. At last, the wind speed would decrease and back to normal operation environment in 2 h. Besides, other sections and stations are not affected by the emergency.

2. **Section Capacity Calculation**

According to emergency development, the section capacity changing process can be divided into three stages based on the wind speed variation. Based on the railway traffic safety specification, in the above scenario, there are three kinds of train operation speed, 160, 70 kmh^{-1} and normal speed. Firstly, calculate the section capacity in different limited speeds. In this period, it only involved of the medium- and high-speed trains; therefore, when the train recovered to normal operation, the average speed of the train can be expressed as the fuzziness value $\tilde{v}_0 = (330, 340, 350\,\text{kmh}^{-1})$. In addition, the average speed of the train changes due to the environment, the operation equipment conditions, and the drivers' driving skills. Normally, the average speed changes in one range and smaller than the limited speed. Therefore, when the limited speed is 160 km·h^{-1}, the average speed of a train can be set as $\tilde{v}_1 = (130,$ $140, 155\ \text{kmh}^{-1})$. When the limited speed is 70 kmh^{-1}, the average speed can be set as $\tilde{v}_2 = (50, 60, 65\ \text{kmh}^{-1})$.

(1) Calculate the section capacity fuzzy value in different speed restriction condition

We can set t_{action} as 9.5 s, t_{delay} as 2.5 s, a as 0.7 ms^{-2}, $L_{protect}$ as 150 m. The value of $L_{comfortable}$ relates to the train speed, when the train is with high speed, it can take the length of two block section, which is about 4000 m; when the train is running in lower speed, $L_{comfortable}$ can take the length of one block section, which is about 2000 m. L_{train} can be set to the conservative value 400 m. Then, put all of the parameters into the Eq. 3.11, we getting the following results.

When the train operated normally, the train headway $\tilde{I}_0$ is (124, 127, 131). Section capacity $\tilde{c}_0$ is (27, 28, 29). When the limited speed is 160 km·h^{-1}, the train headway $\tilde{I}_1$ is (143,156, 168), $\tilde{c}_1$ is (21, 23, 25). When the limited speed is 70 km·h-1, $\tilde{I}_2$ is (163,187, 208), $\tilde{c}_2$ is (17, 19, 22).

(2) Calculate the capacity transfer conditional probability in emergency condition.

Assume the first stage of the train's limited speed is 160 km·h^{-1}. We can get the first stage section capacity $\hat{N}_{\text{mix}}^{(1)} = 2\tilde{c}_1 = (42, 46, 50)$.

In order to make the calculation convenience, assume $v_0 = 330$, $v_1 = 340$, $v_2 =$ 350, $v_3 = 130$, $v_4 = 140$, $v_5 = 155$, $v_6 = 50$, $v_7 = 60$, $v_8 = 65$. The probabilities of different speed p_{v_i} are set to the same, $p_{v_i} = 1/9$, $i = 1, 2, \cdots, 8$. The capacity transfer condition probability of the second stage and the third stages can be obtained based on Eqs. 3.27 and 3.28.

$$\boldsymbol{P}_1 = \begin{bmatrix} 0.1 & 0.05 & 0.05 & 0.1 & 0.4 & 0.3 & 0 & 0 & 0 \\ 0.05 & 0.1 & 0.05 & 0.2 & 0.3 & 0.3 & 0 & 0 & 0 \\ 0.05 & 0.05 & 0.1 & 0.2 & 0.2 & 0.4 & 0 & 0 & 0 \\ 0.1 & 0.05 & 0 & 0.5 & 0.2 & 0.1 & 0 & 0 & 0.05 \\ 0 & 0 & 0.05 & 0.1 & 0.7 & 0.15 & 0 & 0 & 0 \\ 0.05 & 0.05 & 0.1 & 0.2 & 0.2 & 0.4 & 0 & 0 & 0 \\ 0 & 0 & 0 & 0.05 & 0 & 0 & 0.5 & 0.2 & 0.25 \\ 0 & 0 & 0 & 0.05 & 0.05 & 0 & 0.3 & 0.5 & 0.1 \\ 0.01 & 0 & 0 & 0 & 0.05 & 0.04 & 0.1 & 0.2 & 0.6 \end{bmatrix} \tag{3.27}$$

$$\boldsymbol{P}_2 = \begin{bmatrix} 0.4 & 0.2 & 0.1 & 0.05 & 0.05 & 0.2 & 0 & 0 & 0 \\ 0.1 & 0.7 & 0.1 & 0 & 0 & 0.1 & 0 & 0 & 0 \\ 0.1 & 0.1 & 0.8 & 0 & 0 & 0 & 0 & 0 & 0 \\ 0.2 & 0.2 & 0.2 & 0.1 & 0.2 & 0.1 & 0 & 0 & 0 \\ 0.3 & 0.2 & 0.15 & 0.1 & 0.25 & 0 & 0 & 0 & 0 \\ 0.2 & 0.2 & 0.3 & 0 & 0.1 & 0.2 & 0 & 0 & 0 \\ 0 & 0 & 0 & 0.7 & 0.1 & 0.2 & 0 & 0 & 0 \\ 0 & 0 & 0 & 0.5 & 0.3 & 0.2 & 0 & 0 & 0 \\ 0 & 0 & 0 & 0.5 & 0.2 & 0.2 & 0 & 0 & 0.1 \end{bmatrix} \tag{3.28}$$

According to the capacity fuzzy value calculated above and Eq. 3.21, the membership matrix $\boldsymbol{Q}_1$ of the different velocity relative to the section capacity is calculated as follows:

$$\boldsymbol{Q}_1 = \begin{bmatrix} 1 & 0 & 0 \\ 1 & 0 & 0 \\ 1 & 0 & 0 \\ 0 & 0.9 & 0.1 \\ 0 & 1 & 0 \\ 0 & 1 & 0 \\ 0 & 0 & 1 \\ 0 & 0 & 1 \\ 0 & 0.2 & 0.8 \end{bmatrix} \tag{3.29}$$

The probabilities of the three section capacities are $P(\tilde{c}_0) = 1/3$, $P(\tilde{c}_1) = 3.1/9$ and $P(\tilde{c}_2) = 2.9/9$. According to Eq. 3.21, the value of the matrix $\boldsymbol{Q}_2$ can be obtained as follows:

$$\boldsymbol{Q}_2 = \begin{bmatrix} \frac{1}{3} & \frac{1}{3} & \frac{1}{3} & 0 & 0 & 0 & 0 & 0 & 0 \\ 0 & 0 & 0 & \frac{0.9}{3.1} & \frac{1}{3.1} & \frac{1}{3.1} & 0 & 0 & \frac{0.2}{3.1} \\ 0 & 0 & 0 & \frac{0.1}{2.9} & 0 & 0 & \frac{1}{2.9} & \frac{1}{2.9} & \frac{0.8}{2.9} \end{bmatrix} \tag{3.30}$$

According to Eq. 3.23, the condition probability of the section capacity transition between the second stage and the third stage can be obtained, and its transfer matrixes are denoted as S^1 and S^2.

$$\boldsymbol{S}^1 = \begin{bmatrix} 0.200\ 00.783\ 30.016\ 7 \\ 0.124\ 80.789\ 00.086\ 1 \\ 0.007\ 90.156\ 60.835\ 5 \end{bmatrix} \tag{3.31}$$

$$S^2 = \begin{bmatrix} 0.866\ 70.131\ 70.001\ 7 \\ 0.609\ 70.375\ 80.014\ 5 \\ 0.020\ 70.901\ 70.077\ 6 \end{bmatrix} \tag{3.32}$$

(3) Sum divided-period capacities

According to section capacity calculation method based on FMC in Sect. 3.4.3, the carrying capacity of the second stage and third stages can be expression as the fuzziness random variable:

$$\hat{N}^{(2)}_{\text{mix}} = \begin{cases} 5\tilde{c}_0, s_{1,0} \\ 5\tilde{c}_1, s_{1,1} \\ 5\tilde{c}_2, s_{1,2} \end{cases} = \begin{cases} (135, 140, 145), 0.124\ 8 \\ (105, 115, 125), 0.789\ 0 \\ (85, 95, 110), 0.086\ 1 \end{cases} \tag{3.33}$$

$$\hat{N}^{(3)}_{\text{mix}} = \begin{cases} 2\tilde{c}_0, s_{1,0} \\ 2\tilde{c}_1, s_{1,1} \\ 2\tilde{c}_2, s_{1,2} \end{cases} = \begin{cases} (54, 56, 58), 0.609\ 7 \\ (42, 46, 50), 0.375\ 8 \\ (34, 38, 44), 0.014\ 5 \end{cases} \tag{3.34}$$

Therefore, the total capacity for nine hours is calculated as follows:

$$\hat{N}_{\text{sec}} = \sum_{i=1}^{3} \hat{N}^{(i)}_{\text{mix}} = \begin{cases} (231,242,253), 0.076\ 1 \\ (219,232,245), 0.046\ 9 \\ (211,224,239), 0.001\ 8 \\ (201,217,233), 0.481\ 1 \\ (189,207,225), 0.296\ 5 \\ (181,199,219), 0.011\ 4 \\ (181,197,218), 0.052\ 5 \\ (169,187,210), 0.032\ 4 \\ (161,179,204), 0.001\ 3 \end{cases} \tag{3.35}$$

The expectancy value of the comprehensive capacity is 215.197. The results show that in current scenario the section capacity is about 215. The possible maximum and minimum values are 242 and 180, respectively. This provides alternative offers for the managers to make the decision depended on the different preferences. If the optimistic strategy was chosen, the value can set to 242. However, if the operator preferred to the conservative strategy, the main target is to meet the basic demand of the transportation, the value can set to 180. Generally, the compromise strategy is adopted, which means the value sets to 215.

3.6 Chapter Summary

Section capacity calculation involves multiple factors with complicated relationships. In emergency conditions, the factors of the calculation capacity present the characteristics of dynamic, fuzziness, randomness, no aftereffect, etc. In the reality, the capacity calculation cannot include all of the factors. FMC-based capacity calculation method can be more fault-tolerant of all kinds of sensitive factors and the uncertainties. In addition, this method provides variable choices for the managers, so that can suit to optimistic or conservative strategies.

The research in the future is mainly concentrated in two aspects: ① the train speed transfer matrix is the key factor to realize the section capacity calculation, the value of the transfer matrix needs to be confirmed in the practical operation environment; ② there is a big difference of train operation objectives, strategies, and principles between in the emergency and in the normal conditions. Line planning, timetable rescheduling, and rolling-stock rebalancing are yet to consider together in emergency condition. Our ultimate goal is to design and develop a real-time decision support system in the future, which will decrease the influence of the emergency effectively and recover the train operation quickly.

References

1. Editorial board of China Railway Encyclopaedia, Editorial Board of Transportation and Economy (2001) China railway encyclopaedia: transportation and economy. China Railway Publishing House, Beijing, p 415. (in Chinese)
2. Hao S, Zhu J (1995) Calculation and analysis of high-speed railway section capacity. China Railways. 22(5):26–30 (in Chinese)
3. Ji J, Yang Z (1992) An approach to the organization of train operation and the method of computing section carrying capacity in the condition of movable block system. J China Railway Soc 14(1):38–46 (in Chinese)
4. Yang Z, Yang Y, Sun Q et al (1995) A study on parameters for calculation block section carrying capacity and train's coefficient of removal on Beijing-Shanghai high-speed railway [J]. J Northern Jiaotong Univ 19(A01):1–8 (in Chinese)
5. Zhao L (2001) Calculation and analysis of carrying capacity of high-speed railway section. China Railway Sci 22(6):54–58 (in Chinese)

6. Sun S, Tian C, Chen Z (2008) Analysis and calculation of the carrying capacity on passenger dedicated lines. China Railway Sci 29(5):54–58 (in Chinese)
7. Hu Y, Wang J (2011) Discussion on calculation method of section passing capacity on intercity railway. Railway Transp Econ 33(2):1–5 (in Chinese)
8. UIC (2004) UIC 406 capacity. International Union of Railways, Paris
9. Höllmüller J, Klahn V (2005) Implementation of the UIC 406 capacity calculation method at Austrian railways (ÖBB). In: Hansen IA, Dekking FM, Goverde RMP et al Proceedings of the 1st international seminar on railway operations modelling and analysis. The Netherlands, pp 236–247
10. Alex L, Andersh K, Schittenhelm B et al (2006) Practical use of the UIC 406 capacity leaflet by including timetable tools in the investigations. In: Allan CAJ, Brebbia CA, Rumsey AF, et al. Proceedings of the 10th International conference on Computers in Railways. Prague
11. Wahlborg M (2004) Banverket experience of capacity calculations according to the UIC capacity leaflet. In: Allan CAJ, Brebbia CA, Hill RJ et al Proceedings of the 9th international conference on computers in railways. Dresden, pp 665–673
12. Alex l, Anders HK, Bernd S et al (2006) Evaluation of railway capacity. In: Annual transport conference at Aalborg University. Aalborg, pp 1–22
13. Abril M, Barber F, Ingolotti L et al (2008) An assessment of railway capacity. Transp Res Part E 44:774–806
14. Assad A (1980) Models for rail transportation. Transp Res Part A 14:205–220
15. Szpigel B (1972) Optimal train scheduling on a single track railway. Oper Res 72:343–352
16. Jovanovic D, Harker P (1991) Tactical scheduling of rail operations: the SCAN I system [J]. Transp Sci 25:46–64
17. Carey M, Lockwood D (1995) A model, algorithms and strategy for train pathing. J Oper Res Soc 46:988–1005
18. Higgins A, Kozan E, Ferreira L (1996) Optimal scheduling of trains on a single line track. Transp Res Part B 30:147–161
19. Jiang X, Wu Q, Tu P (2008) Study on carrying capacity calculation of automatic block section of double-tracked railways. Railway Transp Econ 30(11):84–87 (in Chinese)
20. Multimodal applied systems, multirail and fasttrack II [EB/OL]. https://www.multimodalinc.com
21. Incontrol Enterprise Dynamics. Simone: simulation model for networks [EB/OL]. https://incontrol.nl/files/simone.pdf
22. University of Hannover and RMCon Rail Management Consultants. RailSys. https://www.simpleware.com.cn/en/Product/Product_RailSys.html
23. Open track railway technology, railway simulation [EB/OL]. https://www.opentrack.ch
24. Shi X (2005) Research on the headway time of passenger train in China. China Railway Sci 7:32–35 (in Chinese)
25. Hao Y (2001) Railway transport organization. China Railway Publishing House, Beijing, pp 281–285. (in Chinese)
26. Zhang B, Wei Q (2009) Arrangement of through signal within automatic block section. Railway Signal Commun 45(11):1–4 (in Chinese)
27. Liu B, Peng J (2005) Course of uncertainty theory. Tsinghua University Press, Beijing, pp 178–179. (in Chinese)
28. Li W (2012) Fuzzy stochastic optimization method for railway traffic organization under emergency conditions. Beijing Jiaotong University, Beijing (in Chinese)
29. Pardo JM, Fuente D (2010) Fuzzy Markovian decision processes: application to queueing systems. Comput Math Appl 60(9):2526–2535

Chapter 4
Rerouting Path Generation in Emergency

Generally, there are three types of traffic paths: shortest path, specific path, and detour path [1]. The shortest path refers to the path with the shortest transportation distance among multiple paths existing between any arrival and departure station, which is adopted by passenger trains under normal operation conditions. A specific path is a designated path for certain freight traffic, such as ice filling of refrigerated vehicles and transportation of wide goods. Detour paths refer to some paths temporarily designated in daily transportation work due to changes in railway operating conditions. Under emergency conditions, the rail network capacity is seriously inadequate, and it is necessary to find paths that have alternative effects on the section affected by the emergency to constitute a detour path set (train path set). Therefore, before the adjustment of train operation, the detour path set is searched based on the determined emergency impact section.

The current research on train running path can be attributed to different types of shortest path problems. According to whether the passing time is considered, it can be divided into two categories: one is the traditional shortest path search, the most commonly used solution algorithm is Dijkstra algorithm or its improved algorithm; the other is the more complicated time-dependent shortest path, that is, the transit time of the arc segment is related to the time to reach the arc.

Reference [1] proposed Dijkstra algorithm in 1959, which makes each node be searched only once. Although the Bell_Ford algorithm was also a labeling method, its search speed was relatively slow [2, 3]. Reference [4] proposed a complexity computational framework that can be applied to the above multiple algorithms. Reference [5] and [6] fused the ideas of the above algorithm, divided the labeled nodes into two sets, and traversed all the nodes in turn according to the limit factor.

With the development of parallel computing technology, a series of parallel shortest path algorithms have been produced, which can be divided into two categories [7]: the algorithms proposed by Tseng [8], Paige [9], and Chandy [10], which are independent of computer type, and the algorithms proposed by Habbal [11] and Dey [12], which are aimed at computer characteristic type. The difference between

L. Jia et al., *High-Speed Railway Operation Under Emergent Conditions*,
Advances in High-speed Rail Technology,
https://doi.org/10.1007/978-3-662-63033-4_4

the two methods lies in the choice of network segmentation method and basic shortest path method.

Time-dependent shortest path (TDSP) is one of the derivative problems of shortest path (SP). Compared with traditional SP, TDSP has more practical application value in communication network and transportation network [13]. Reference [14] theoretically proved the shortcomings of the traditional shortest path algorithm in a time-dependent network, then constructed a time-dependent network model, and proposed the SPTDN algorithm to solve this problem. Reference [15] proposed a stochastic time-dependent network model, pointed out that K expectation is the formal description of the shortest path problem, and deduced the probability density function and path expectation cost calculation method for the node arrival time for the bus network. In [16], the optimization model of the shortest path problem of time-dependent network was established without limiting the nature of marginal cost function in the network, and a double-layer optimization intelligent algorithm based on network vertex priority coding was given.

In addition, some intelligent algorithms are also used in the shortest path search, such as genetic algorithm [16], ant colony algorithm [17–19], network shortest path algorithm with fuzzy arc length [20]. These studies provide a basis and useful reference for this chapter.

In the rerouting path set search model under emergency conditions, the railway network is first modeled, and then path origin–destination set under emergency conditions is determined. In this chapter, two path search models are established: one is based on sufficient capacity and the other is based on capacity time density. The path search model based on sufficient capacity fully considers the comprehensive capacity constraints of the new path under the condition of emergency, while the path search model based on capacity time density mainly considers the time reliability constraints of the path set. In the actual railway operation, the dispatcher can determine the path search scheme suitable for the site according to the development of the emergency.

4.1 Influence Factors Analysis

Rerouting path in emergency conditions is the powerful supplement to the original route of the railway network. The carrying capacity of other lines in the railway network can be effectively used to improve the transport service level in emergency.

The main influence factors of the rerouting route generation in emergency are as shown below.

(1) The generation of rerouting route is based on the railway network. Due to the influences of the emergency, the capacity of some stations or sections may decreases in a large extent. It may even break down in some extreme situations, which cannot satisfy the passenger transport demand. In order to reduce the number of stranded passengers, the capacity of related stations and sections of railway network would need to be used.

(2) Characteristics of spatial attributes. The railway network itself has spatial attributes. For example, each station has spatial coordinates and center mileage, and the section is generally demarcated by stations at both ends or kilometer posts, and the passenger flow demand is generally given in OD form.

(3) The links between the stations. In railway injection area (usually in metropolis), connection links are built between different stations to make the passengers transport convenience. Besides the normal train paths, fully used the connection links between the stations can provide more options for the train rerouting strategy in emergency.

(4) Train traction mode. Nowadays, most of the railways are power by electric. However, there are still a small part of railways that are towed by diesel locomotives. Ensuring the same traction mode is an important constraint of the train cross-line operation.

(5) Passengers rerouting length. The upper bound of the passengers' trip length exists in reality. Different kinds of incidents will cause different levels of damage. Passengers with different trip aims, occupations, and consumption abilities accept the rerouting distance in different degree.

(6) Detour line capacity. Connectivity is one of the most important constraints for choosing rerouting route. The capacity of detour line is also an important parameter to determine whether the rerouting strategy is feasible. For maintaining other routes away from the impact of emergency and operate normally, dispatchers generally rule out the section with small residual capacity.

(7) Detour line train distribution. It is necessary to pay attention to the running time of the train on the detour line, which means that the time distribution of the remaining capacity can affect the train's choice of path. This factor is also considered in order not to affect the normal operation of trains on other lines as much as possible.

Train rerouting path search process has the characteristics of time and space complexity, diversity of traction methods, limited capacity and detour length, etc. It is different from the search process of topology shortest path. It is also different from time-dependent path search problems with accurate operation time and obvious time division.

4.2 Rerouting Path Search Model Based on Sufficient Capacity

At present, the algorithm of shortest path generation on the network can provide strong basic theoretical support for the research of railway transportation dispatching command problem. Therefore, based on the traditional K-shortest path algorithm, an improved K-shortest path algorithm adapted to the characteristics of the rerouting path search under emergency conditions is proposed.

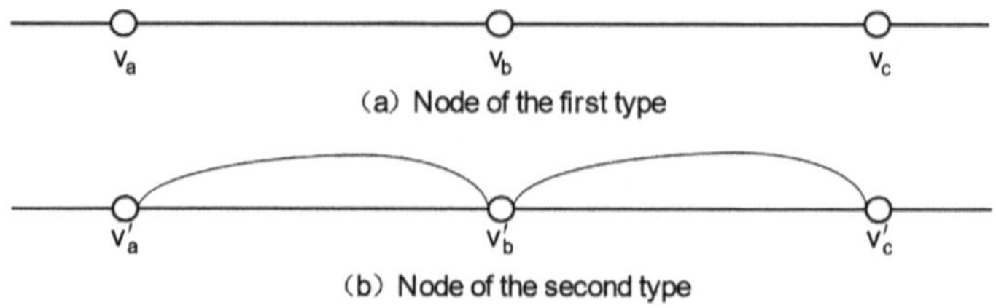

Fig. 4.1 Two types of nodes

4.2.1 Network Modeling

The railway network is established, in which the station is taken as the node and the section between stations is taken as the arc. Here, stations located on different lines in the same city are regarded as the same node. If there are connecting lines between different lines in the hub station, it shows that the station can provide cross-line operation for the train.

Therefore, there are two types of nodes: one is the non-cross-line node representing a single station, and the number of arcs connected to another node is 1 (see Fig. 4.1a); the other is the cross-line node representing a hub station with connecting lines (see Fig. 4.1b).

The railway network is described as $G(V, E)$, $V = V^0 \cup V'$, where V^0 is the set of first type nodes, V' is the set of second type nodes, and E is the set of arcs.

Emergencies may affect nodes and arcs in the topology of the railway network. When an emergency occurs at a node, the receiving and departure capacity of the station will be reduced. In serious cases, the emergency may even cause the node to fail completely, then all arcs connected to it will also fail. As shown in Fig. 4.2b, when node v_b fails due to an emergency, the connected arc (v_a, v_b), (v_b, v_c) and (v_b, v_d) also fail. When an emergency occurs at an arc, the section carrying capacity will be reduced. In serious cases, the emergency causes the speed limit value of the line section to fall to zero; that is, the arc fails. Figure 4.2c shows the failure of arc (v_b, v_c).

4.2.2 K-shortest Path Algorithm

For graph $G(V, E)$, let p be a path between nodes v_i and v_j, whose length is $d(p)$. The set $R(G, v_i, v_j)$ composed of all the different paths between v_i and v_j is called the path set between v_i and v_j on G.

$$R(G, v_i, v_j) = \{p | p \text{ is the paths between } v_i \text{ and } v_j\} \quad (4.1)$$

If $R' \subseteq R$, the length of each path in R' is less than ξ:

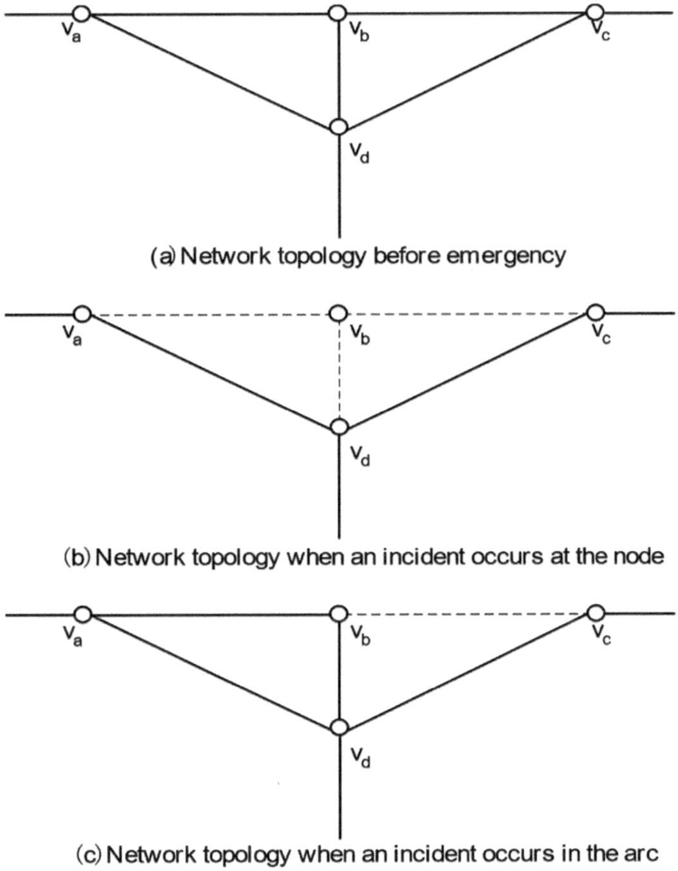

Fig. 4.2 Schematic diagram of the impact of emergencies on the topology of the railway network

$$R'(G, v_i, v_j) = \{p|d(p) \leq \xi\} \tag{4.2}$$

R' is called ξ-detour path set. Arrange the elements in $R(G, v_i, v_j)$ according to the path length from small to large:

$$R(G, v_i, v_j) = \{p_1, p_2, \ldots, p_N|d(p_1) \leq d(p_2) \leq \cdots \leq d(p_N)\} \tag{4.3}$$

where p_K is the K-shortest path between v_i and v_j, then the K-shortest path set is:

$$R(G, v_i, v_j) = \{p_1, p_2, \ldots, p_K|d(p_1) \leq d(p_2) \leq \cdots \leq d(p_K)\} \tag{4.4}$$

$$R(G, v_i, v_j) = \{p_l|l \leq K\} \tag{4.5}$$

In this way, the problem of generating a ξ-detour path set can be converted into a K-shortest path set problem to be solved.

Generally, any path between two vertices on a graph $G(V, E)$ must be the first shortest path of the same vertex on one of its subgraphs $G'(V, E')$, so the Kth shortest path problem between two vertices on a graph $G(V, E)$ can be converted into the first shortest path problem for solution. According to reference [21], the first shortest path p_1 can be separated from the path set $R(G, v_i, v_j)$ to obtain the first and non-first shortest path set $\{p_2, p_3, \ldots, p_N\}$. In this way, all paths can be converted into the first shortest path on a subgraph, and then sorted according to the length of each path, the solution of the K-shortest path problem can be obtained.

The steps of K-shortest path generation algorithm based on Dijkstra algorithm are as follows [22].

Step 1. The shortest path is obtained by Dijkstra's algorithm, and the shortest path, the shortest path length of each point, and the shortest path length are stored in the arrays Paths, Dists, and MinDis, respectively.

Step 2. Calculate the neighbors of the paths stored in the shortest path array and store them in array Neirs.

Step 3. Calculate the distance of the path $v_s - v_t - v_j - v_e$ passing through the nth shortest path through the neighboring point v_t and store it in the array TempDists, where v_j is the node on the nth shortest path adjacent to the neighboring point.

Step 4. Sort the distance value in array TempDists, calculate the path corresponding to the shortest distance and store it in array Paths, and add 1 to the number of shortest paths.

Step 5. If the number of shortest paths reaches K, then stop the calculation, otherwise go to Step 2.

4.2.3 Improved K*-shortest Path Algorithm Under Emergency*

There are two parts in this section, including path origin–destination set searching, and path searching based on sufficient capacity.

4.3 Path Origin–destination Set Searching

The problem mainly includes the following problems [23]:

(1) Nodes or edges affected by emergencies may not have the ability to cross the line, so it is necessary to pay attention to whether there are connecting lines between different types of lines at a hub station.

(2) There may be several detours in the train route concentration, so the length of detours needs to be limited. On the one hand, it makes the path generation more efficient; on the other hand, it avoids the impact on the unaffected sections.

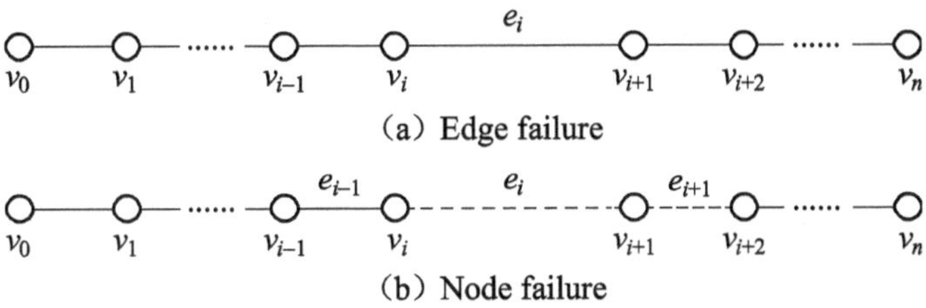

Fig. 4.3 Schematic diagram of origin and destination points of path generation under emergency conditions

(3) Cross-line trains should return to the established line for operation as soon as possible; that is, the origin and destination points of the route should meet the requirements for all kinds of trains to return to their established route.

In view of the analysis of the above problems, the following provisions are made in this section for the generation of train routing set under the condition of emergency.

(1) When an emergency occurs at a hub station with connecting lines, this node and the next node that can cross the line are taken as the origin point and destination point of the detour.

(2) When an emergency occurs on an arc or a station that does not have the ability to cross the line, the previous and the next node that can cross the line shall be taken as the origin point and destination point of the detour.

The determination of the origin–destination set of the detour must consider the nodes and edges affected by the emergencies [24]. The search process is as follows.

(1) Analyze the place where the emergency occurs and the affected part (edge or node) of the network. If it is an edge, go to step (2), if it is a node, go to step (3).

(2) As shown in Fig. 4.3a, arc e_i fails due to an emergency, and nodes v_i and v_{i+1} are nodes at both ends, performing the node judgment process. If both v_i and v_{i+1} are cross-line nodes, $v_i = P$, $v_{i+1} = Q$, where P is the starting station and Q is the terminal station. If at least one point in v_i and v_{i+1} does not have the ability of crossing lines, then search the nearest node v_{i-1} and v_{i+2} at both ends of v_i and v_{i+1}, respectively, then $v_{i-1} = P$, $v_{i+2} = Q$.

(3) As shown in Fig. 4.3b, node v_{i+1} is affected by an emergency. Search nodes v_i and v_{i+2} which are closest to node v_{i+1}, then $v_i = P$, $v_{i+2} = Q$.

4.4 Rerouting Path Search Based on Sufficient Capacity

Under emergency conditions, the total capacity of all routes between the origin station and the destination station shall not be less than the number of trains to pass in the adjustment period. Therefore, the generation of detour path should not only consider the path length, but also fully consider the comprehensive capacity of the new path set.

Definition 4–1 C-sufficient path set

The generalized cost w is assigned to each arc e in the graph $G(V, E)$, and the arc becomes e', forming the graph $G'(V, E')$. Let p be a path between notes v_i and v_j, with the length of $\mathrm{gd}(p)$ and the capacity of c_p. If there is $R' \subseteq R$, making $\sum_{p \in R'} c_p \geq N$, where N is the set target capacity, then R' is called C-sufficient path set.

Under the condition of emergency, the method and step of calculating C-sufficient path set are the same as K-shortest path, but the cycle condition are different. The condition for stopping the cycle of C-sufficient path set is that the total capacity of the generated path has reached or exceeded the number of trains to be allocated for the path.

The carrying capacity of the path takes the value of the arc with the smallest carrying capacity on the path. In the process of generating train path set, when the post generated path and the existing path share the same section, the capacity of the post generated path may be exactly the remaining capacity of the common arc. Then, the comprehensive capacity of the train path set is equal to or less than the simple sum of the capacity of each path [24].

The generating steps of C-sufficient path set are as follows:

Step 1. Parameter initialization, including target capability N.

Step 2. The shortest path is obtained by Dijkstra's algorithm, and the shortest path, the shortest path length of each point, and the shortest path length are stored in the arrays Paths, Dists, and MinDis, respectively.

Step 3. Calculate the neighbors of the paths stored in the shortest path array and store them in array Neirs.

Step 4. Calculate the distance of the path $v_s - v_t - v_j - v_e$ passing through the nth shortest path through the neighboring point v_t and store it in the array TempDists, where v_j is the node on the nth shortest path adjacent to the neighboring point.

Step 5. Sort the distance value in array TempDists, calculate the path corresponding to the shortest distance and store it in array Paths, and add 1 to the number of shortest paths.

Step 6. Calculate the passing capacity of path p. The remaining capacity of the smallest arc of each section of the path p is taken as the carrying capacity of the path p and is added to the comprehensive passing capacity C_{sum} of the path set.

Step 7. If the comprehensive capacity of the path set reaches N, then stop the calculation, otherwise go to Step 3.

4.5 Rerouting Path Search Model Based on Capacity Time Density

In order to obtain a train path set more in line with the actual operating environment, in addition to capacity constraints, this section also considers the time factor and proposes a time credibility based route search model (TCRSA).

4.5.1 Network Modeling

According to the factors above, we can establish a time-dependent network $G = (V, E, T)$, in which V, E, T are node set, segment set, and pathway search period time separately.

In network G, the stations are the nodes denoted as $v = (\mathrm{No}_v, \mathrm{Nm}_v, \mathrm{Sn}_v)$, here, No_v, Nm_v, and Sn_v mean node number, station name, and injection that stations belonged. The arc in the G denoted as $e = (\mathrm{No}_e, D_e, A_e, \hat{C}_e, \mathrm{Ty}_e, E_e, \mathrm{Des}_e)$, the elements in e mean arc number, the initial station, the terminal station, carrying capacity, arc type, traction ways and capacity time density in period T. If e is a connection link between two stations in one injection, then $Ty_e = 0$, otherwise, $Ty_e = 1$; $E_e = 0$ means the electric traction way, $E_e = 1$ means the diesel traction way. The definition of the capacity time density will be described below.

$T = [t_\mathrm{s}, t_\mathrm{e}]$ means the time range of the path searching, t_s and t_e stands for the start time and the finish time of the event influence. To make the capacity of the whole network fully used, the finish time of the path search can be postponed according to the event influence level.

In emergency condition, the train operation environment is very complicated. Besides section capacity, train operation speed and the train rerouting path lengths in emergency have both effects on the train safety operation. So, two integrated factors to consider these factors comprehensively, capacity time intensity and density, are proposed as below.

Definition 4–2 Capacity Time Intensity (CTI) of the route segment refers to the achieved capacity per unit time in a certain confidence level, which means the ratio of the carrying capacity and the average carrying time. Due to the different decisions of the dispatcher, this value can be chosen as the optimistic value, the pessimistic value, and the expectation values, correspondingly named as the optimistic time intensity $\mathrm{Tes}_e^\mathrm{o}$, pessimistic time intensity $\mathrm{Tes}_e^\mathrm{p}$, and expectation time intensity $\mathrm{Tes}_e^\mathrm{e}$, the expressions are separately as follows:

$$\mathrm{Tes}_e^\mathrm{o} = \frac{C_e^\mathrm{o}(\gamma, \alpha)}{l_e/E(v)} \tag{4.6}$$

$$\mathrm{Tes}_e^\mathrm{p} = \frac{C_e^\mathrm{p}(\gamma, \alpha)}{l_e/E(v)} \tag{4.7}$$

$$\mathrm{Tes}_e^\mathrm{e} = \frac{E(C_e)}{l_e/E(v)} \tag{4.8}$$

Here, $C_e^\mathrm{o}(\gamma, \alpha) = \sup\{r|\mathrm{Ch}\{C_e \ge r\}(\gamma) \ge \alpha\}$ is the optimistic value of the section capacity. $C_e^\mathrm{p}(\gamma, \alpha) = \inf\{r|\mathrm{Ch}\{C_e \le r\}(\gamma) \ge \alpha\}$ is the pessimistic value of the section capacity. $\mathrm{Ch}\{C_e \le r\}(\gamma) = \sup\{\beta|\mathrm{Pos}\{\omega \in \Omega|\mathrm{Cr}\{C_e(\omega) \le r\} \ge \beta\} \ge \alpha\}$ is the opportunity of the event $C_e \le r$ in the confidence level γ. $E(C_e)$ is the section capacity expectation. l_e is the length of the segment arc e. $E(v)$ is the average speed expectation in e. $\gamma, \alpha \in (0, 1]$.

Definition 4–3 Capacity Time Density (CTD) of the route segment refers to the time need that achieved per unit capacity in a certain confidence level, which means the ratio of the average carrying time and the carrying capacity, signed as Des_e, so $\mathrm{Des}_e = 1/\mathrm{Tes}_e$.

4.5.2 *Origin and Destination Point Search*

Some related parameters in the model are listed in the below.

seq: affected stations and section sequence set.

seq_e: terminal station set of the path search.

seq_s: initial station set of the path search.

$\lambda(v)$: CTD sum of all segment arc between the node v and initial node.

V_{sor}: permanent sign node set, which means all the node in the set has already found the reliable path to reach the initial node.

w_k: node sign for already find reliable path set.

Rs_s: distance constraint between origin node for the path searching and the influenced station.

Rs_l: the maximum length constraint of the rerouting path.

Rs_c: the minimum capacity constraint of the path search.

Len_v: the path length between the node v and the initial point.

Route: reliable rerouting path set.

4.6 Origin and Destination Point Search

Firstly, ensure the origin–destination point of the pathway search, specific stages shown below.

Step 1.1 Sort the affected stations and the sections according to the center distance, to form a sequence seq.

Step 1.2 Take two affected terminal objects from the sequence, named as Obj_s and Obj_e, if the route was downward, then, the center distance of Obj_s is less than the center distance of Obj_e; if not, then the opposite. Then, ensure the initial station of the path search, if the Obj_s is station, then process the Step 1.3; or else, process the Step 1.4.

Step 1.3 If Obj_s is the node v_{i-1}(shown in Fig. 4.4a), then, the segment (e_{i-2} e_{i-1}) that connected to v_{i-1} are invalid. In this situation, it is required to find the new injection node with crosslink capacity near the left side of v_{i-1}. This means if the route is downward, find the node that center distance less than v_{i-1}; otherwise, find the node that center distance more than v_{i-1}. In reality, the experienced dispatcher often chooses several station with the cross-line capacity near to v_{i-1}. Therefore, in this algorithm the train stations distance constraint set to Rs_s, which means that all

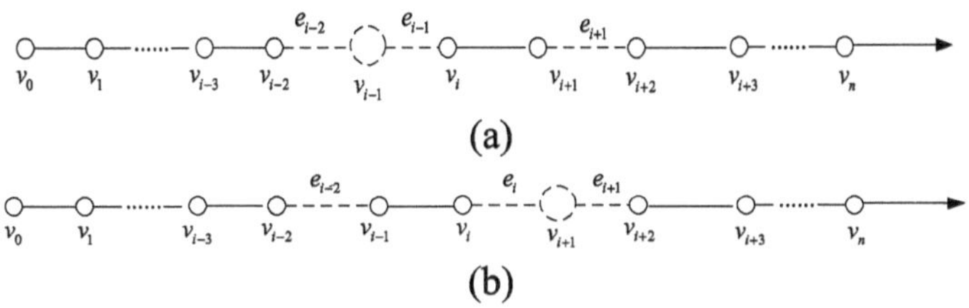

Fig. 4.4 Origins and destinations of the train path in emergency

the crosslink stations whose distance to v_{i-1} is less than Rs_s are selected to form the sequence seq_s.

Step 1.4 If Obj_s is the edge e_{i-2}(shown in the Fig. 4.4b), then it need to find the node set at the left side of e_{i-2} with the crosslink capacity. This process is similar to the steps in the Step 1.3. First set the station distance constraint Rs_s. Then, choose all the crosslink stations whose distance to v_{i-2} is less than Rs_s to form the sequence seq_s.

Step 1.5 Choose the terminal station set, the process is similar to Step 1.3 and Step 1.4. Find the crosslink node on the right side of the affected node or arc to generate the set seq_e.

4.7 Rerouting Path Search Algorithm

The process of the rerouting path search algorithm based on the capacity time density is below.

Step 2.1 $i = 0$, $j = 0$.

Step 2.2 Take the ith element ss_i of seq_s.

Step 2.3 Take the jth element se_j of seq_e.

Step 2.4 Put the initial point ss_i to set V_{sor}, i.e., $V_{\text{sor}} = \{ss_i\}$; $\lambda(ss_i) = 0$; $\lambda(v) = \infty, \forall v \in V, v \neq ss_i$; $\overline{V}_{sor} = V - \{ss_i\}$; $w_0 = ss_i$; $\text{Len}_{ss_i} = 0$.

Step 2.5 For each $v \in \overline{V}_{sor} \cap Na(w_k) \cap (E_{(w_k,v)} = E_{(w_{k-1},w_k)})$, if the carrying capacity of node v is less than the capacity constraint Rs_c, then, ignore the node v; if the carrying capacity of the segment (w_k, v) is less than the capacity constraint Rs_c, then, ignore the edge (w_k, v); otherwise, $\lambda(v) = \min\{\lambda(v), \lambda(w_k) + \text{Des}_{(w_k,v)}\}$, which means the less of the CTD (the less time to achieved per unit capacity), the easier of the segment is to be selected.

Step 2.6 Label the point v_{k+1} with minimum $\lambda(v)$ by w_{k+1}, and put v_{k+1} into the set V_{sor}, then, $\overline{V}_{\text{sor}} := \overline{V}_{\text{sor}} - \{v_{k+1}\}$; $\text{Len}_{w_{k+1}} = \text{Len}_{w_k} + \text{Len}(w_k, v_{k+1})$.

Step 2.7 Repeat step 2.5 and step 2.6 until ① if $\text{Len}_{w_{k+1}}$ is more than the maximum constraint Rs_l of the path length, turn to step 2.8; ② if the terminal point se_j has already in the set V_{sor}, turn to step 2.10.

Step 2.8 $j := j + 1$, repeat step 2.3 to step 2.7, until the last element of seq_e has been took out, then, turn to step 2.9.

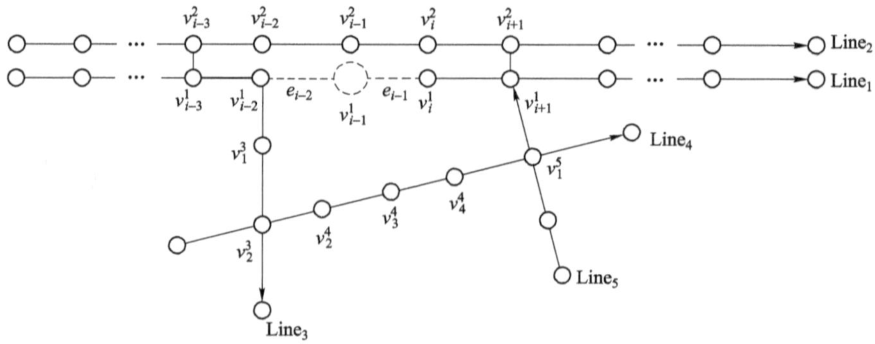

Fig. 4.5 Examples of path search in emergencies

Step 2.9 $i := i + 1$, repeat step 2.2 to step 2.7, until the last element of seq_s has been took out.

Step 2.10 One rerouting pathway with the minimum CTD was generated from ss_i to se_j, sign as rte, put rte into the set Route, turn to step 2.8.

As shown in Fig. 4.5, assuming that an emergency occurs at station v_{i-1}^1 of Line$_1$, according to the general path search algorithm, the nearest cross-line nodes v_{i-2}^1 and v_{i+1}^1 with v_{i-1}^1 are first searched, and then the detour $\text{rte}_1 = v_{i-2}^1\, v_1^3\, v_2^3\, v_2^4\, v_3^4\, v_4^4\, v_1^5\, v_{i+1}^1$ is generated. Because the general algorithm limits the origin and destination points to the nearest cross-line node, only path rte$_1$ or more complex path can be generated. If path rte$_1$ is adopted, the train needs to cross three railway lines, and the dispatcher needs to coordinate the trains on four lines, which increases the risk of transportation organization. Using the algorithm proposed in this section, the origin node set $\text{seq}_s = \{v_{i-2}^1, v_{i-3}^1\}$ and the destination point set $\text{seq}_e = \{v_{i+1}^1\}$ of the cross-over line are found first, then path $\text{rte}_2 = v_{i-2}^1 - v_{i-3}^1 - v_{i-3}^2 - v_{i-2}^2 - v_{i-1}^2 - v_i^2 - v_{i+1}^2 - v_{i+1}^1$ can be generated in addition to path rte$_1$. Route rte$_2$ makes full use of the connecting lines inside the station, i.e., arc (v_{i-3}^1, v_{i-3}^2) and (v_{i+1}^2, v_{i+1}^1), and only one route is bypassed by rte$_2$, which greatly reduces the risk of organization and management.

4.7.1 Analysis of Time Complexity and Characteristics of TCRSA

In order to further prove the effectiveness of TCRSA, the time complexity and characteristics of the algorithm are analyzed.

1. Time Complexity Analysis of TCRSA
The upper bound of the time complexity of path search for a pair of origin and destination nodes is $O(n^2)$, where n is the number of nodes in the network. The time complexity of TCRSA is $O(\text{Size}_{\text{seq}_e} \cdot \text{Size}_{\text{seq}_s} \cdot n^2)$,where $\text{Size}_{\text{seq}_s}$ and $\text{Size}_{\text{seq}_e}$ are the number of elements in the start and end set respectively. Due to

the limitation of Rs_S, the values of $Size_{seq_s}$ and $Size_{seq_e}$ are not very large, which are generally single digits. Therefore, the time complexity of TCRSA can be considered as $O(n^2)$.

2. Characteristics Analysis of TCRSA
According to the above analysis, the TCRSA has the following characteristics.

(1) Limitations on the length of detour paths under different emergencies are considered.
(2) The minimum capacity limit is added. When the remaining capacity of the section or station is very small, it will not participate in the generation of train detour path, so as to ensure that the train on the cross-line operates as the original diagram.
(3) Traction mode of the train is incorporated into the path search process to ensure the normal running of the train.
(4) The connecting line between stations also participates in route search as a kind of arc.
(5) In addition to the nodes adjacent to the affected stations and sections, the upstream and downstream cross-line nodes may also serve as the starting and ending points for the path search. This makes the detour more in line with the actual scheduling rules.
(6) The capability time density is used as the basis of the path search rather than the shortest mileage in the general path search algorithm. This index combines line length, train running speed, and line capacity during the period of emergency impact to obtain a train running path that is more in line with the actual operating environment. In addition, the capability time density can adopt optimistic value, pessimistic value and expected value, respectively, which fully reflects the dispatcher's different decision orientation in view of the development situation of emergency.

4.8 Chapter Summary

This chapter first introduces the existing path search methods, analyzes the basic research ideas and influencing factors of train path set search under the condition of emergency, and puts forward two solving models. First, we build a path search model based on sufficient capacity, which mainly improves the traditional Dijkstra algorithm to meet the generation demand of train running path under the condition of emergency. The model emphasizes the comprehensive carrying capacity of path search and avoids the problem of insufficient comprehensive carrying capacity of new path set obtained by search. In addition, a path search model based on the capability time density is proposed for the characteristics of the train path set, such as the sensitive running time, the fuzzy random passing capacity, and the spatial attribute. The model focuses on path search under the condition of the capacity realized in unit time under a certain confidence level, which reflects the time reliability of the new path set. At the same time, the content of this chapter is also the premise of compiling the opening plan, which lays the foundation for the next research.

References

1. Dijkstra EW (1959) A note on two problems in connexion with graphs. Numer Math 1:269–271
2. Bellman RE (1958) On a routing problem. Q Appl Math 16:87–90
3. Ford LR Jr, Fulkerson DR (1962) Flows in networks. Princeton University Press, Princeton
4. Allottino S (1984) Shortest-path methods: complexity, interrelations and new propositions. Networks 14:257–267
5. Glover F, Glover R, Klingman D (1984) Computational study of an improved shortest path algorithm. Network 14:25–37
6. Glover F, Klingman D, Phillips N (1985) A new polynomially bounded shortest paths algorithm. Oper Res 33:65–73
7. Anning NI, Zhicai J, Linjie G (2006) An overview of research on parallel shortest path algorithm in transportation network. J Highw Transp Res Dev 23(12):128–132 (in Chinese)
8. Tseng P, Bertsekas DP (1990) Parallel algorithms for network flow and other problems. SIAM J Control Optim 28(3):678–710
9. Paige RC, Kruskal CP (1985) Parallel algorithms for shortest path problems. In: Proceedings of the international conference on parallel processing, 14–20
10. Chandy KM, Misra J (1982) Distributed computation on graphs: shortest path algorithms. Commun ACM 25(11):833–837
11. Habbal M (1994) A decomposition algorithm for the all-pairs shortest path problem on massively parallel computer architectures. Transp Sci 28(3):273–290
12. Dey S, Srimani PK (1989) Fast parallel algorithm for all pairs shortest path problem and its VLSI implementation. IEE Proc Part E: Comput Dig Tech 136(2):85–89
13. He R, Li Y (2008) Model of time-dependent network paths and two-level optimal intelligent algorithm. J China Railw Soc 30: 32–37 (in Chinese)
14. Guozhen T, Wen G (2002) Shortest path algorithm in time-dependent networks. Chin J Comput 25:165–172 (in Chinese)
15. Chen Jingrong Yu, Jianning LY (2012) Network path optimization since multi-attribute random events. J Southwest Jiaotong Univ 47(2):291–298 (in Chinese)
16. Xiaojun K, Maocai W (2008) Shortest path algorithm implements based on genetic algorithm. Comput Eng Appl 44(23):22–23 (in Chinese)
17. Shijie S, Gaofeng L, Zhongyou Z et al (2010) An improved ant colony algorithm solving the shortest path and TSP problem. Comput Technol Dev 20(4):144–147 (in Chinese)
18. Changlin J (2012) Research and implementation of urban road network dynamic shortest path search based on distributed ant colony algorithm. Beijing Jiaotong University, Beijing (in Chinese)
19. Ghoseiri K, Nadjari B (2010) An ant colony optimization algorithm for the bi-objective shortest path problem. Appl Soft Comput 10(4):1237–1246
20. Ji XY, Iwamura K, Shao Z (2007) New models for shortest path problem with fuzzy arc lengths. Appl Math Model 31(2):259–269
21. Dengfeng C, Dengrong Z (2002) Algorithm and its application to N shortest paths problem. J Zhejiang Univ (Eng Sci) 36(5):531–534 (in Chinese)
22. Bai Yiduo Hu, Peng XL et al (2009) A kth-shortest path algorithm based on k-1 shortest paths. Geomat Inf Sci Wuhan Univ 34(4):492–494 (in Chinese)
23. Meng XL, Jia LM (2009) Paths generating in emergency on China new railway network. J Beijing Inst Technol 19(S2):84–88
24. Xuelei M (2011) Research on theory and method of vehicle operation organization under emergency conditions. Beijing Jiaotong University, Beijing (in Chinese)

Chapter 5
High-Speed Railway Line Planning Under Emergency Conditions

Different transportation organizations will be adopted under different levels of railway emergency events (refer to Chap. 2); for instance, serious emergencies involve the formation of line plan and train operation plan. This chapter aims to comprehensively analyze the train operation adjustment problem under emergency conditions, which can be ignored if the impact of the emergency is small, laying the foundation for the study of transportation organization problems under serious emergencies. The line plan embodies the allocation and application of high-speed railway resources. It is a bridge connecting transportation demand and transport resources, providing important support and basis for the train running planning and daily transportation production. Consequently, the key issue in the context of rapid development of high-speed railway is that the formation of the line plan is suitable for China's national conditions under emergencies.

Nowadays, in the study of the establishment and optimization of the line plan, the influence of factors such as train route, stop schedule plan, passenger transfer plan, OD passenger assignment, and passenger delay time on the model and algorithm is mainly considered. The line plan is a single- or multi-level target programming model that mostly takes the railway revenue, passenger travel cost, and passenger turnover as optimization targets. To solve model, there are two main methods, one is mathematic calculation method, and the other is heuristic method, such as simulated annealing algorithm (SA) and genetic algorithm (GA).

Based on the socioeconomic analysis, Deng [1] established a bi-level programming model for comprehensive optimization of passenger train line plan from the interests of both railway transport enterprise and passengers. According to the scale and characteristics of the problem, an optimization algorithm based on SA was designed to solve the large-scale line plan. By analyzing the influencing factors of railway passenger transportation demand related to passenger train line plan, Shi et al. [2] constructed the passenger transportation elasticity demand function relationship and attributed the selection of passenger transfer plan to the user equilibrium

L. Jia et al., *High-Speed Railway Operation Under Emergent Conditions*,
Advances in High-speed Rail Technology,
https://doi.org/10.1007/978-3-662-63033-4_5

assignment that under elastic demand. Combined the Stackelberg game relationship between railway enterprise and passengers, a bi-level programming model of passenger train line plan based on elastic demand was built, and an optimization algorithm based on SA was designed. Considering the stop mode of passenger dedicated line, the train stop schedule plan optimization model was established by Huang [3]. The model aimed to minimize the generalized cost of passenger travel and the remaining passenger turnover, taking into account the passenger's economic interests and social benefits. Based on the optimization model of train stop plan, the goal of maximizing the operating profit of passenger dedicated line and minimizing the generalized resistance of train running path was introduced, and a harmony search algorithm based on group diversity maintenance was proposed. Through analysis of the characteristic of the line plan of passenger dedicated line and its establishment process, Xu and Zou [4] proposed a method to determine and optimize the line plan by combining train flows, gave the concept of train flows combining, and gave the combining principle based on the study of the influence of the transportation organization by the train flows combination. The data structure of the combining train flows and related computer algorithm were designed and verified with examples. Zha and Fu [5] adopted a combination of quantitative and qualitative analysis. According to the OD flow of passengers on the railway network, in order to facilitate passenger travel, the determination of the direct passenger train line plan was reduced to the maximum weight matching problem of the bipartite graph, and an optimized initial scheme was obtained through calculation. On this basis, the economic favorable evaluation of the scheme was carried out, and the line was properly adjusted to make the line load distributed uniform and deduce the waste of carrying capacity, thus forming a satisfactory result. Du et al. [6] applied the concept of node importance to the planning of the line plan. Through the calculation of the evaluation index of the node importance, the three-level hierarchical structure of node importance of the cities along the passenger dedicated line was built. The first-level node was used as the departure and arrival points, the unfixed interval stop was adopted by the node on the second-level, and the service mode of the low-level trains that stopping at station by station was adopted by the node on the third level. The multi-objective mathematical programming model of the train plan was established, which was solved by using LINGO, and the optimization algorithm of stop schedule of the line plan was achieved. Zhang et al. [7] analyzed the high-speed train stop schedule and established the multi-objective 0–1 programming mathematical model of the high-speed line plan with the target of the minimum dwell time of passengers, the least waste of trains, and the minimum number of passengers who did not meet the demand. Moreover, the influence of the high-speed train stop was analyzed, the multilayer 0–1 programming solving method was designed according to the number of the stop times, and the problem was solved by combining the ordered combination tree method. According to the forecast of the departure and arrival passenger flow at each section of the planned Beijing-Shanghai high-speed railway, a satisfactory solution was obtained. Jia and Guan [8] established a multi-objective programming model for line plan in accordance with the actual and period operation mode of China's passenger dedicated line and transformed it into a single-objective programming model with priority structure. The GA embedded with game process and multivariable cascade coding was designed, and the validity of the model and algorithm was verified by an example. In the study of Dong et al. [9], the cost-capacity operation network was constructed for any line plan, the optimal passenger flow allocation scheme was described by the minimum cost flow of the network, and the minimum cost of the passenger flow allocation was used as the individual fitness function. Then, the individual elite was generated in consideration of the constraints of the maintenance of train cost and the fixed number of passengers, and the collaborative symmetric population crossover genetic algorithm for solving the optimization problem of the one-way passenger train line plan for the linear passenger dedicated line was designed and further extended to the mesh passenger dedicated lines.

Chang et al. [10] used the fuzzy mathematical programming method to study the high-speed railway line plan in Taiwan and established a multi-objective optimization model for the high-speed passenger train line plan on a railway corridor, aiming to minimize the total operation cost and total passenger travel time. Due to the small scale of the railway network and the high density of passenger flow, the train plan represented the characteristic of public transport operation mode. The 0–1 variable was used to describe whether a train of fixed service frequency and marshaling is selected, and the model is solved by the branch-and-cut method [11, 12]. Goossens et al. [13] built a three-category railway network by constructing a "type graph" to establish a line plan optimization model of multi-type railway network. The model described the passenger selection behavior of train operation route as a multi-commodity flow problem, with the goal of rational utilization of transportation capacity, constrained by passenger demand and flow conservation, and finally transformed into linear programming model. Chakroborty and Dwivedi [14] proposed an evolutionary optimization technique based on GA to solve the optimal transfer route problem on the railway network. An initial generation model of the route set was presented to build the path with fixed number, and the merits of the path were measured by passenger's satisfaction and travel time.

Game theory had also been used in the optimization of the train plan. A framework based on game theory was proposed to solve the problem of the line plan in the section failure mode [15]. Once a certain section failed, other routes and transportation modes could be used as supplementary to complete the passenger transportation, thus forming a non-cooperative game of two persons under complete information. Schŏebel and Schwarze [16] established the game model of the line plan optimization by using different lines as the game agent. The goal of each line was to minimize its own late time, and the minimum late time of the comprehensive transportation network was taken as the final equilibrium state.

All the literatures mentioned above aim at researching a certain factor, single characteristic or a certain solution method of the problem. The theory system of train operation organization optimization, which is systematic, aimed to network and emergence, fuzzy mixed with random, and more in line with the actual environment, has not been established yet. Compared with previous studies, a bi-level programming model, which is more comprehensive and more applicable, is established in this chapter. The upper level (stop schedule optimization) aims to minimize the total

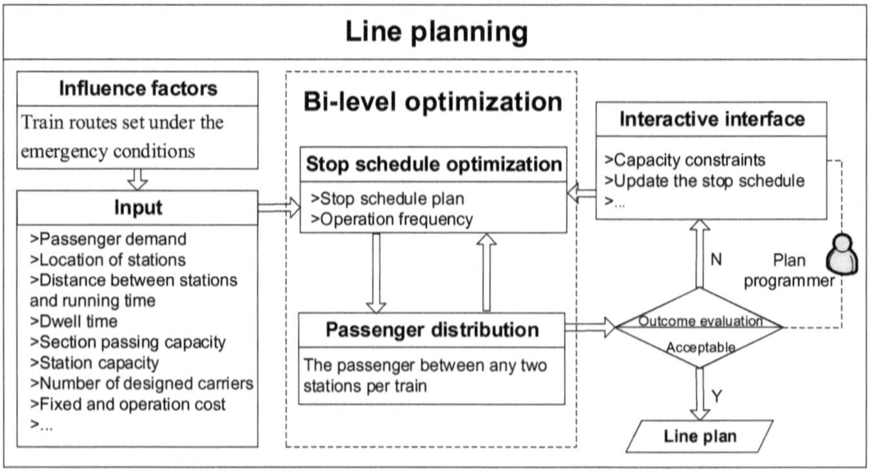

Fig. 5.1 Process of line planning in emergency

operation cost and the number of the unserved passengers, while the lower level (passenger flow allocation) takes the maximum passenger service and the minimum travel time of all passengers as the optimization target. The constraints under the emergency conditions are considered in each level of the programming model, and the stop schedule optimization and the passenger flow allocation under the emergency conditions are achieved.

5.1 Line Planning Model

Line plan is a technical document that the railway managers assign the train to the appropriate line or section based on the OD passenger flow in a period of time [17]. It mainly determines the spatial relationship between the train and the line, including the train route, the number of trains and the stop pattern et al. According to the characteristics of the line plan under emergencies, combined with the train routes set, a bi-level programming model of train stop pattern optimization and passenger flow allocation is established in this chapter, as shown in Fig. 5.1.

The notations are defined as follows.

T	Time period for the development of line plan, such as a day
S	Number of stations in the rerouting route set under emergency
E	Number of sections in the rerouting route set under emergency
L	Train types, including high-speed trains, medium-speed trains, etc., in this chapter
$L_{o,d}$	The length of the path between station o and station d
$P_{o,d,l}$	Running time of the type l train between station o and station d

(continued)

(continued)

T_i^s	Dwell time at station i
C_i^s	Passing capacity of station i during special operation period T
C_j^E	Passing capacity of section j during special operation period T
C_l	Average number of designed carries of type l train
K	Attendance rate, in the case of no over staffing, $K < 1$, the value of K can be appropriately increased under emergency
C_i^d	A sign that identifies the initial departure and arrival ability of the station i. If the station i has the ability, $C_i^d = 1$; otherwise, $C_i^d = 0$
D	Fixed cost of train operation
F	Cost for a train to operate one kilometer
M	Total service frequency of the stop schedule, which can be set as a large enough number to meet the passenger demand

5.1.1 Stop Pattern Schedule Optimization

According to the characteristics of the problem, the upper level, stop pattern schedule optimization, is researched in this chapter. The decision variables, objective functions, and constraints are as follows.

1. Decision variables

y_i: Service of train j. If train j is in the train plan, then $y_i = 1$; otherwise, $y_i = 0$.
$x_{j,i}$: Station service of train j at station i. If train j stops at station i, then $x_{j,i} = 1$; otherwise, $x_{j,i} = 0$.
$o_{j,i}$: Train originating station. If station i is the departure station of train j, then $o_{j,i} = 1$; otherwise, $o_{j,i} = 0$.
$d_{j,i}$: Train terminal station. If station i is the terminal station of train j, then $d_{j,i} = 1$; otherwise, $d_{j,i} = 0$.
$u_{j,o,d}$: Number of passengers on train j between station o and d.
$z_{j,l}$: Train type. If train j is type l, then $z_{j,l} = 1$; otherwise, $z_{j,l} = 0$.
$s_{j,i}$: Service of station. If station i is in the stop pattern schedule of train j, then $s_{j,i} = 1$; otherwise, $s_{j,i} = 0$.
$e_{j,k}$: Service of section. If section k is in the stop pattern schedule of train j, then $e_{j,k} = 1$; otherwise, $e_{j,k} = 0$.

2. Objective functions

(1) The total operating costs are minimized.

$$C_{\text{cost}} = \min\left(\sum_{j=1}^{M} D \cdot y_j + \sum_{j=1}^{M}\sum_{o=1}^{S-1}\sum_{d=o+1}^{S} \left(y_j \cdot F \cdot L_{o,d} \cdot o_{j,o} \cdot d_{j,d}\right)\right) \tag{5.1}$$

(2) The number of unserved passengers is minimal.

$$C_{\text{passenger}} = \min \sum_{l}\sum_{o=1}^{S-1}\sum_{d=o+1}^{S}\left(P_{o,d,l} - \sum_{j=1}^{M}\left(u_{j,o,d} \cdot z_{j,l} \cdot y_j\right)\right) \tag{5.2}$$

3. Constraints

(1) Constraints on the station ability for initial departure of trains.

$$C_i^d \geq o_{j,i}, \quad i = 1, 2, \ldots, S-1 \quad j = 1, 2, \ldots, M \tag{5.3}$$

(2) Constraints on the station ability for terminal arrival of trains.

$$C_i^d \geq d_{j,i}, \quad i = 2, 3, \ldots, S \quad j = 1, 2, \ldots, M \tag{5.4}$$

(3) The original station of the train constraints.

$$\sum_{j=1}^{M}\sum_{i=2}^{S-1}\sum_{i1=0}^{i-1}\left(y_j \cdot o_{j,i} \cdot x_{j,i1}\right) = 0 \tag{5.5}$$

$$\sum_{j=1}^{M}\sum_{i=2}^{S-1}\sum_{i1=0}^{i-1}\left(y_j \cdot o_{j,i} \cdot s_{j,i1}\right) = 0 \tag{5.6}$$

(4) The terminal station of the train constraints.

$$\sum_{j=1}^{M}\sum_{i=1}^{S-1}\sum_{i1=i+1}^{S}\left(y_j \cdot d_{j,i} \cdot x_{j,i1}\right) = 0 \tag{5.7}$$

$$\sum_{j=1}^{M}\sum_{i=1}^{S-1}\sum_{i1=i+1}^{S}\left(y_j \cdot d_{j,i} \cdot s_{j,i1}\right) = 0 \tag{5.8}$$

(5) One original station of a train constraints.

$$\sum_{j=1}^{M}\sum_{i=1}^{S}\left(y_i \cdot o_{j,i}\right) = 1 \tag{5.9}$$

(6) One terminal station of a train constraints.

$$\sum_{j=1}^{M}\sum_{i=1}^{S}\left(y_i \cdot d_{j,i}\right) = 1 \tag{5.10}$$

(7) The train has to stop at its original station.

$$o_{j,i} < x_{j,i} \quad j = 1, 2, \ldots, M \quad i = 1, 2, \ldots, S \tag{5.11}$$

(8) The train has to stop at its terminal station.

$$d_{j,i} < x_{j,i} \quad j = 1, 2, \ldots, M \quad i = 1, 2, \ldots, S \tag{5.12}$$

(9) The train has to pass through the scheduled station.

$$s_{j,i} > x_{j,i} \quad j = 1, 2, \ldots, M \quad i = 1, 2, \ldots, S \tag{5.13}$$

(10) Station capacity constraint.

$$C_i^s \geq \sum_{j=0}^{M}\left(y_j \cdot s_{j,i}\right) \quad i = 1, 2, \ldots, S \tag{5.14}$$

(11) Section capacity constraint.

$$C_i^E \geq \sum_{j=0}^{M} \left(y_j \cdot e_{j,k}\right) \quad k = 1, 2, \ldots, E \tag{5.15}$$

5.1.2 Passenger Allocation Optimization

Passenger allocation optimization under emergency can be performed based on the given stop pattern schedule and service frequency, and a mixed-integer programming model can be established according to the characteristics of the problem.

1. Objective functions

(1) Passenger service turnover is maximal.

$$C_{\text{RPK}} = \max \sum_{j=1}^{M} \sum_{o=1}^{S-1} \sum_{d=o+1}^{S} \left(y_j \cdot u_{j,o,d} \cdot L_{o,d}\right) \tag{5.16}$$

(2) Total travel time of all the passengers is minimal.

$$C_{\text{time}} = \min \sum_{j=1}^{M} \sum_{o=1}^{S-1} \sum_{d=o+1}^{S} \left(y_j \cdot u_{j,o,d}\right) \cdot \left(\sum_{l} \left(T_{o,d,l} \cdot z_{j,l}\right) + \sum_{o1=o+1}^{d-1} \left(x_{j,o1} \cdot T_{o1}^{s}\right)\right) \tag{5.17}$$

2. Constraints

(1) Passenger demand constraints.

$$\sum_{o=1}^{S-1} \sum_{d=o+1}^{S} \left(\sum_{j=1}^{M} \left(y_j \cdot u_{j,o,d} \cdot z_{j,l}\right) \leq P_{o,d,l}\right) \quad l = 1, 2 \tag{5.18}$$

(2) Constraints on the number of passengers on the train.

$$\sum_{o=1}^{k} \sum_{d=k+1}^{N} \left(y_j \cdot u_{j,o,d}\right) \leq \sum_{l} \left(C_l \cdot K \cdot z_{j,l} \cdot y_j\right) \quad j = 1, 2, \ldots, M \; k = 1, 2, \ldots, S-1 \tag{5.19}$$

5.2 Hybrid Intelligent Algorithm

At present, there are a large number of literatures investigate methods for solving the bi-level programming, which can be divided into two types according to the algorithm mechanism. One is to transform the bi-level programming into other forms on the basis of the characteristics of the problem, such as multi-objective planning [18, 19], single-level integer programming [20], Stackelberg game problem [21], and so on. The traditional deterministic algorithms are usually used to solve the transformed model, such as branch-and-bound method, pole search method, penalty function method, and gradient method. These methods search for local extremum points according to a certain deterministic strategy and try to reach a certain global optimum point based on the existing local extremum points. Although using these methods to find the optimum solution is theoretically feasible, it only applies to the problem with fewer objects. The other is a random search method that mimics the intrinsic mechanism of some biological evolution or physical phenomena in nature, such as simulated annealing algorithm, genetic algorithm, particle swarm optimization, and ant colony algorithm. Random factors are added to the search process of these algorithms to approach the global optimal solution with a certain probability, which have great advantages for solving the large-scale combination optimization problems [22].

The line planning problem of high-speed railway under emergency conditions is designed as a bi-level programming model in this chapter. The upper level is designed as a multi-objective mixed-integer programming model, and the lower level is designed as a linear integer programming model after a given stop pattern schedule and service frequency. Aiming at the characteristics of multi-objective, discrete, nonlinear, and large-scale combination optimization, the hybrid intelligent algorithm for plant multi-direction growth simulation algorithm combined with branch-and-bound method is proposed in this section.

5.2.1 Plant Growth Simulation Algorithm

Plant growth simulation algorithm (PGSA) is an intelligent algorithm on basis of the plant phototropism growth [23]. In the PGSA, the feasible solution region of the integer programming is taken as the plant growth environment, and its global optimal solution is regarded as the light source for plant growth. Furthermore, PGSA can realize the iteration process and search the optimal solution based on the plant growth phototropism in different light intensity environment.

1. Main procedures

Main procedures of the basic PGSA can be summarized as follows [23].

(1) Find an initial feasible solution and take it as the new growing node *G*.

(ii) Make straight lines paralleling with axes as the new growing node G, which can be divided by certain iterative length λ, and another new growing branches set S' will be gained.
(iii) For generating the next growing branch in the set S', calculate its objective function value, and take the best one satisfied the requirements of various constraints if necessary, as the optimal one.
(iv) Calculate the distance between the initial feasible solution and new growing branches of the set S', and redistribute the growth phytohormone among new growing branches in the set S'.
(v) Select an element from the set S', and take it as the next new growing node G. Then, set a new shorter iterative length λ, and repeat above processes till that there is no a new growing node to be found. Thus, the best solution generated from the last iteration will become the optimal one.

2. Advantages of algorithm

PGSA has been widely used in many fields, i.e., operational research, engineering planning, management, and optimization. Li and Wang [24] experimentally proved that PGSA has higher computational accuracy than SA and ant colony algorithm (AC) for solving Steiner minimum tree (SMT), an NP-complete problem. Guney et al. [25] used PGSA to control the electromagnetic interference, which got better results and higher efficiency than modified threshold ant colony optimization (MTACO), bat algorithm (BA) and bacteria foraging algorithm (BFA). Wang and Cheng [26] employed PGSA to solve the optimal planning problem of a power transmission network in a single stage, which achieved the best results compared with GA and particle swarm optimization (PSO). Yang et al. [27] solved the reactive power optimization in power system using PGSA, and the algorithm showed better stability than standard GA and PSO. In conclusion, the advantages can be summarized as follows:

(1) Faster convergence and global search ability compared with standard GA, PSO, and so on.
(2) The iterative steps are simple and easy to be programed.
(3) Suit solving large-scale, nonlinear integer programming problems.

3. Disadvantages of algorithm

(1) Growth phytohormone distribution mode cannot conform to the natural phototropism mechanism.

In the basic PGSA, there are several main concepts, i.e., plant root, trunk, branches, and growing nodes. The root is equivalent to the initial feasible solution of an optimization problem. The trunk and branches mean intermediate solutions in the iterative searching process starting from the root and its initial solution to optimal solutions. A growing node is a new growth branch with the ability of growing, and the growing node is also named a new growth branch. According to the plant phototropism growth pattern, the growth ability of each branch should be decided by the growth phytohormone; the higher is the phytohormone, the faster branch grows. The phytohormone will be redistributed among each growing node due to changed environment caused by the new growing nodes. PGSA determines the value of the phytohormone, that is, the grow probability, based on the relative position of the node to the initial root. However, due to the dependence on the initial solution per iteration, the searching process of the optimal solution may be restricted within certain local feasible solution regions.

(2) Single growth direction for new branches.

In the basic PGSA, the growing angle of new branches is defined as 90°, which means that the value of a dimension parallel with the coordinate axes in the solution region will be changed only. This is not only inconsistent with the natural properties of the plant growth in the real world, but also affects its computation speed, the quality of optimal solutions, and solving efficiency.

5.2.2 Plant Multi-direction Growth Simulation Algorithm

In order to overcome drawback mentioned above, an improved method (plan multi-direction growth simulation algorithm, PMGSA) will be put forward.

(1) Improve the phytohormone distribution method.

The value of phytohormone is decided by the light intensity condition in plant system. Although the position of the light source (optimal solution) is unknown in the iterative process, the plant phototropism growth pattern implies that the better solution is closer to the light source. This is more consistent with the natural mechanism of plant growth compared with PGSA.

(2) Increase new growing directions.

In fact, the growth direction for new branches is randomly changed based on its position. Therefore, a random selection mechanism is used for determining the growing direction of new branches to improve the global search ability.

1. Main procedures

Taking integer programming as an example, the main procedures of PMGSA are as follows.

$$\begin{cases} \min f(x) \\ \text{s.t.} \\ x \in X, x = (x_1, x_2, \ldots, x_n) \\ x \text{ is an integer vector} \end{cases}$$

Step 1: Initialization. Input related parameters, and determine the initial feasible solution $\boldsymbol{x}^0(\boldsymbol{x}^0 \in \boldsymbol{X})$ and the iterative length λ^k(a positive integer, *k* is the time of iteration). At the initial phase, the initial feasible solution should be taken as the new growing node $\boldsymbol{x}^G$. $k = 0$, $\boldsymbol{x}_{\min} = \boldsymbol{x}^0$, $f_{\min} = f(\boldsymbol{x}^0)$, $\boldsymbol{x}^G = \boldsymbol{x}^0$, $\mathbf{S} = \{\boldsymbol{x}^0\}$, $\Omega(\mathbf{S}) = 1$ and the iterative length is λ^k.

Step 2: Generate new branches at the new growing node $\boldsymbol{x}^G$ according to different rotation angles.

(i) For the fixed vertical growth direction, straight line (also namely as trunks) paralleling with axes through the new growing node $\boldsymbol{x}^G$ should be made firstly with the constraints of bounded closed set and other related-requirements (constraints should be met when the algorithm is used to solve the nonlinear integer programming with constraints). Then, take $\boldsymbol{x}^G$ as the center, and divide these line or trunks with the iterative length λ^k with both negative and positive directions into new growing branches or nodes $s'_{i_1 j_1}$ $(i_1 = 1, 2, \ldots, 2n; \quad j_1 = 1, 2, \ldots, m_{i_1})$. m_{i_1} is the number of new growing branches or nodes at the i_1-st trunk.

(ii) In addition to the above rotation angle, some plant, i.e., vines, may have multiple growth directions according to the different position of the node, so a new trunk with the symmetrical directions of these branches, which are randomly selected from n directions, can be found. Then, take x^G as the center, and divide these line or trunks by the iterative length λ^k with both directions into new growing branches or nodes $s'_{i_{2n+1} j_1}$ $(j_1 = 1, 2, \ldots, m_{i_{2n+1}})$. There are n random growing directions of new growing branches. From the above steps, the new growing branches set S′ is generated.

Step 3: Determination of the best solution at the current moment. Calculate the objective function value of each new growing branches or nodes, and take the best one $s'_{p,q}$ as the optimal one.

$$f^*(s'_{p,q}) = \min\{f(s'_{i_1 j_1}), \quad i_1 = 1, 2, \ldots, 3n; \quad j_1 = 1, 2, \ldots, m_{i_1}\}$$

If $f^*(s'_{p,q}) < f_{\min}$, there is $x_{\min} = s'_{p,q}$, $f_{\min} = f^*(s'_{p,q})$;

If $f^*(s'_{p,q}) = f_{\min}$, there is $x_{\min} := x_{\min} \cup \{s'_{p,q}\}$, $f_{\min} = f^*(s'_{p,q})$, which may find more than one optimal solutions and increase the diversity of the global optimal solutions.

Step 4: Calculation of the growth probability and redistribution the growth phytohormone among all new growing branches or nodes.

Set $S := S \cup S'$, $F = f_{\min} + \sum (f(s) - f_{\min})/\Omega(S)$.
For every $s \in S$,
If $F \leq f(s)$, there is $P_s = 0$, $S := S \backslash \{s\}$;
If $F > f(s)$, there is $P_s = (F - f(s))/\left(\sum_{s \in S} (F - f(s))\right)$.

Using F to replace the initial solution in PGSA not only conforms to the natural mechanism of plant phototropism, but also avoids the disadvantages of PGSA's dependence on the initial solution. In addition, it ensures that the growth probability after adding new branches still satisfies the formula: $\sum_{s \in S} P_s = 1$.

Step 5: Select the next new growing node $\boldsymbol{x}^G$. Generate a random number η, $\eta \in (0, 1)$, and select the node s_t that satisfies the condition $\sum_s^{t-1} p_s < \eta < \sum_s^t p_s$, $t \in [1, \Omega(S)]$ as the new growth node. Then, judge the iteration terminal conditions. If the iteration can be terminated go to Step 6; otherwise, go to Step 2.

Step 6: Output and terminate the algorithm. Set $\boldsymbol{x}_{\min}$ as the global optimal solution and $f_{\min}$ as its optimal objective functions value.

According to the scale and the characteristics of optimal problems, the iteration terminal conditions in the new proposed PMGSA can be set as follows:

(i) The certain iteration time or total calculation time is set in advance;
(ii) Reaching to the calculation accuracy;
(iii) Reaching to the number of iteration when the best solution is unchanged;
(iv) No more new branches are generated;
(v) Any synthesis of above conditions.

2. Convergence analysis of the PMGSA

A new growth branch is only determined by its current branch, which means that new growth branches are independence from the initial and their former generation branches. Also, the conditional transition probability of the searching process from one branch to another different growth branch will not be influenced by its former branches at any time. Therefore, the optimal solution searching process of the proposed PMGSA in this chapter can be regarded as a Markov process, and the convergence of the PMGSA can be further discussed by theory of Markov process.

For the proposed improved PGSA, both the multi-direction of new growth branches and the selection of these new growth branches are random and independent. Thus, a new plant branch is only related to its current branch, and there is no aftereffect. Let E as the possible situation of a new growth branch, and such situation can be further divided into two different sub-situations. One (E_0) means the optimal solutions. The other (E_n) means the state that there are not optimal solutions in these new growth branches. In addition, the relationships between these two sub-situations E_0 and E_n can be shown as follows:

(1) $E = E_0 \cup E_\mathrm{n}$;
(2) $E_0 \cap E_\mathrm{n} = \emptyset$.

Besides, each iteration or optimal solution searching process of the proposed PMGSA mainly consists of three procedures, i.e., (i) the generation process of new growth branches (also namely potential new solutions generation process), (ii) the selection process of these new growth branches (namely new solutions selection process), and (iii) the growth phytohormone redistribution process of these branches (namely the solution regions selection process). Above three procedures are carried out one by one in a cycle way. Thus, there are three transitions from one condition

($i \in$ E) to another ($j \in$ E) within the branch region for each iteration, whose conditional transition probability can be defined as g_{ij}, d_{ij}, and h_{ij}, respectively. Their transfer matrix can be further recorded as $\boldsymbol{G} = \{g_{ij}\}$, $\boldsymbol{D} = \{d_{ij}\}$, and $\boldsymbol{H} = \{h_{ij}\}$. The condition transformation matrix of the growth branches of the PMGSA in this paper can be calculated by $\boldsymbol{P} = \boldsymbol{D} \cdot \boldsymbol{G} \cdot \boldsymbol{H} = \{p_{ij}\}$.

According to the above process, we can know that the $g_{ij} > 0, d_{ij} > 0, h_{ij} \geq 0$. Therefore, the new branch region forming from the PMGSA can be recorded as homogenous Markov chain $\{S(k), k > 0\}$ with the characteristics shown as follows:

$$S(k) = S(0) \cdot (\boldsymbol{P})^k \tag{5.20}$$

$$S_j(k) = \sum_{i \in \mathbf{E}} \left(S_i(0) \cdot (p_{ij})^k\right), \quad k = 0, 1, 2, \ldots \tag{5.21}$$

Thus, the condition transition probability of the searching process can be calculated by

$$p_{i,j} > 0, \quad i \in \mathbf{E}, \quad j \in \mathbf{E_0} \tag{5.22}$$

$$p_{i,j} = 0, \quad i \in \mathbf{E_0}, \quad j \in \mathbf{E}_{\boldsymbol{n}} \tag{5.23}$$

According to the theory of Markov process, for $\forall i \in \mathbf{E}, \forall j \in \mathbf{E}_{\boldsymbol{n}}$, there is always a positive integer M (the value of the positive integer M depends on the scale of programming and its qualities), and when an iteration $k > M$, $(p_{ij})^k \to 0$ can be achieved. Thus,

$$\lim_{k \to \infty} S_j(k) = \lim_{k \to \infty} \sum\nolimits_{i \in \mathbf{E}} \left[S_i(0) \cdot (p_{ij})^k\right] = 0 \tag{5.24}$$

The probability of the new branch region excluding the optimal branch is 0 when $k > M$, which also means that a global optimal solution is alway gained by the proposed PMGSA. In other words, the proposed algorithm can find the optimal solution with the probability of 1. Thus, a theorem on the convergence of the PMGSA can be gained

Theorem 5.1: There must be a positive integer M for the PMGSA applied to solve nonlinear integer programming, which makes the following statement become truth: Their solutions can converge to a global optimal one with the probability of 1 when $k > M$.

3. **Experiments and discussions**

In order to verify the performance of the proposed PMGSA, several nonlinear integer programming experiments without constraints and with constraints are given to show its efficiency and global search capability compared to other algorithms for the same experiments.

(1) Case 1: Nonlinear integer programming without constraints

As for the model given by Zhu and Li (2005) [28], a nonlinear integer programming without constraints can be represented as follows:

$$\min \sum_{i=1}^{n} \left(x_i^4 - 4.9x_i^2\right) \tag{5.25}$$

$$s.t. |x_i| \leq 5, x_i \in \mathbf{Z} \tag{5.26}$$

Zhu [28] solved the case of $n = 2$ and $n = 4$ using the improved filled function method, and the number of iterations is 57 and 265 times, respectively. The number of optimal solutions is 1, which is determined by the algorithm process, and the objective function value is -11.7. Li [23] used the PGSA to solve the case of $n = 3$, and three global optimal solutions ($\boldsymbol{x}^* = \{(-1, -1, 1), (1, 1, 1), (-1, 1, 1)\}$) were obtained after 51 iterations. With the proposed PMGSA in this chapter, eight global optimal solutions can be further gained after ten iterations ($n = 3$), shown as follows:

$$x^* = \{(1, -1, -1), (-1, -1, -1), (-1, 1, 1), (-1, -1, 1), (1, 1, -1), (-1, 1, -1), (1, 1, 1), (1, -1, 1)\}$$

The plant growth process and its iterations process of the proposed improved PGSA are shown in Fig. 5.2 and Table 5.1, respectively.

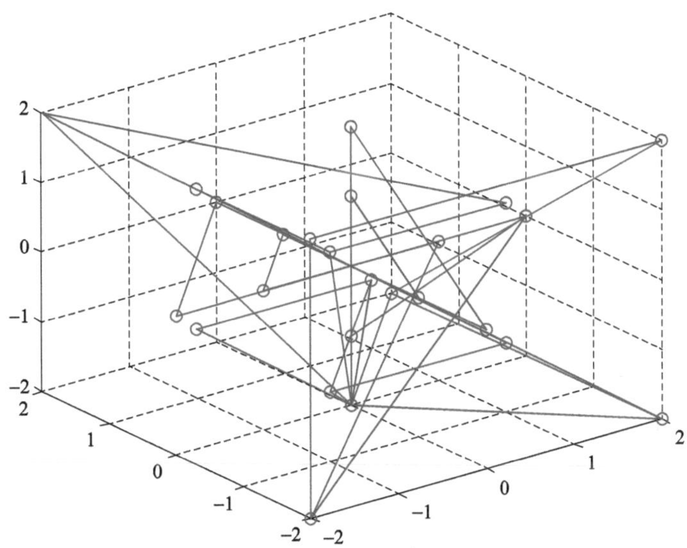

Fig. 5.2 Case 1: plant growth process of the PMGSA

Table 5.1 Case 1: iterations process of the PMGSA

IT	GL	OV	GP	FR	RN	IT	GL	OV	GP	FR	RN
0	0, 0, 0	0.00	1.0000	1.0000		5	2, 2, 2	−10.80	0.0811	0.5000	
1	2, 2, 2	−10.80	0.2000	0.2000	0.6156		−2, 2, −2	−10.80	0.0811	0.5811	
	−2, 2, −2	−10.80	0.2000	0.4000			−2,−2,−2	−10.80	0.0811	0.6622	
	−2,−2,−2	−10.80	0.2000	0.6000			1, 0, 0	−3.90	0.0293	0.6914	
	2, 0, 0	−3.60	0.0667	0.6667			−1, 0, 0	−3.90	0.0293	0.7207	
	−2, 0, 0	−3.60	0.0667	0.7333			0, 1, 0	−3.90	0.0293	0.7500	
	0, 2, 0	−3.60	0.0667	0.8000			0, −1, 0	−3.90	0.0293	0.7793	
	0, −2, 0	−3.60	0.0667	0.8667			0, 0, 1	−3.90	0.0293	0.8086	
	0, 0, 2	−3.60	0.0667	0.9333			0, 0, −1	−3.90	0.0293	0.8378	
	0, 0, −2	−3.60	0.0667	1.0000			2, 0, 0	−3.60	0.0270	0.8649	
2	−2, −2, 2	−10.80	0.1429	0.1429	0.2760		−2, 0, 0	−3.60	0.0270	0.8919	
	2, −2, −2	−10.80	0.1429	0.2857			0, 2, 0	−3.60	0.0270	0.9189	
	2, 2, 2	−10.80	0.1429	0.4286			0, −2, 0	−3.60	0.0270	0.9459	
	−2, 2, −2	−10.80	0.1429	0.5714			0, 0, 2	−3.60	0.0270	0.9730	
	−2,−2,−2	−10.80	0.1429	0.7143			0, 0, −2	−3.60	0.0270	1.0000	
	2, 0, 0	−3.60	0.0476	0.7619		6	−1, 1, 1	−11.70	0.0747	0.0747	0.4786
	−2, 0, 0	−3.60	0.0476	0.8095			1, −1, −1	−11.70	0.0747	0.1494	
	0, 2, 0	−3.60	0.0476	0.8571			1, 1, −1	−11.70	0.0747	0.2241	
	0, −2, 0	−3.60	0.0476	0.9048			−1,−1,−1	−11.70	0.0747	0.2989	
	0, 0, 2	−3.60	0.0476	0.9524			2, 2, −2	−10.80	0.0690	0.3678	
	0, 0, −2	−3.60	0.0476	1.0000			−2, −2, 2	−10.80	0.0690	0.4368	

(continued)

Table 5.1 (continued)

IT	GL	OV	GP	FR	RN	IT	GL	OV	GP	FR	RN
3	2, 2, −2	−10.80	0.1250	0.1250	0.3144		2, −2, −2	−10.80	0.0690	0.5057	
	−2, −2, 2	−10.80	0.1250	0.2500			2, 2, 2	−10.80	0.0690	0.5747	
	2, −2, −2	−10.80	0.1250	0.3750			−2, 2, −2	−10.80	0.0690	0.6437	
	2, 2, 2	−10.80	0.1250	0.5000			−2,−2,−2	−10.80	0.0690	0.7126	
	−2, 2, −2	−10.80	0.1250	0.6250			1, 0, 0	−3.90	0.0249	0.7375	
	−2,−2,−2	−10.80	0.1250	0.7500			−1, 0, 0	−3.90	0.0249	0.7625	
	2, 0, 0	−3.60	0.0417	0.7917			0, 1, 0	−3.90	0.0249	0.7874	
	−2, 0, 0	−3.60	0.0417	0.8333			0, −1, 0	−3.90	0.0249	0.8123	
	0, 2, 0	−3.60	0.0417	0.8750			0, 0, 1	−3.90	0.0249	0.8372	
	0, −2, 0	−3.60	0.0417	0.9167			0, 0, −1	−3.90	0.0249	0.8621	
	0, 0, 2	−3.60	0.0417	0.9583			2, 0, 0	−3.60	0.0230	0.8851	
	0, 0, −2	−3.60	0.0417	1.0000			−2, 0, 0	−3.60	0.0230	0.9080	
4	2, 2, −2	−10.80	0.1250	0.1250	0.2291		0, 2, 0	−3.60	0.0230	0.9310	
	−2, −2, 2	−10.80	0.1250	0.2500			0, −2, 0	−3.60	0.0230	0.9540	
	2, −2, −2	−10.80	0.1250	0.3750			0, 0, 2	−3.60	0.0230	0.9770	
	2, 2, 2	−10.80	0.1250	0.5000			0, 0, −2	−3.60	0.0230	1.0000	
	−2, 2, −2	−10.80	0.1250	0.6250		…	…	…	…	…	…
	−2,−2,−2	−10.80	0.1250	0.7500		10	1, −1, 1	−11.70	0.0575	0.0575	
	2, 0, 0	−3.60	0.0417	0.7917			1, 1, 1	−11.70	0.0575	0.1150	
	−2, 0, 0	−3.60	0.0417	0.8333			−1, −1, 1	−11.70	0.0575	0.1726	
	0, 2, 0	−3.60	0.0417	0.8750			−1, 1, −1	−11.70	0.0575	0.2301	

(continued)

Table 5.1 (continued)

IT	GL	OV	GP	FR	RN	IT	GL	OV	GP	FR	RN
	0, −2, 0	−3.60	0.0417	0.9167			−1, 1, 1	−11.70	0.0575	0.2876	
	0, 0, 2	−3.60	0.0417	0.9583			1, −1, −1	−11.70	0.0575	0.3451	
	0, 0, −2	−3.60	0.0417	1.0000			1, 1, −1	−11.70	0.0575	0.4027	
5	1, 1, −1	−11.70	0.0878	0.0878	0.4243		−1, −1, −1	−11.70	0.0575	0.4602	
	−1,−−1,−1	−11.70	0.0878	0.1757			2, 2, −2	−10.80	0.0531	0.5133	
	2, 2, −2	−10.80	0.0811	0.2568			−2, −2, 2	−10.80	0.0531	0.5664	
	−2, −2, 2	−10.80	0.0811	0.3378			…	…	…	…	
	2, −2, −2	−10.80	0.0811	0.4189			0, 0, −2	−3.60	0.0177	1.0000	

As shown in Table 5.1, IT, GL, OV, GP, FR, and RN denote iteration times, growth location, objective value, growth probability, feasible region, and random number, respectively. When IT is 6, four optimal solutions can be gained, which means that the proposed improved PGAS is efficient.

(2) **Case 2: Nonlinear integer programming experiment with constraints**

As for the model given by Meng et al. [29] and Li [23], a nonlinear integer programming with constraints can be represented as follows:

$$\min\ (x_1 - 1)^2 + (x_2 - 2)^2 + (x_3 - 4)^2 \tag{5.27}$$

$$s.t. - x_1 + x_2 + x_3 \leq 3.5 \tag{5.28}$$

$$x_1 + x_2 - x_3 \leq 6 \tag{5.29}$$

$$x_1, x_2, x_3 \geq 0, x_i \in \mathbf{Z}, i = 1, 2, 3 \tag{5.30}$$

As for such nonlinear integer programing with constraints, Meng et al. [29] calculated only two global optimal solutions ($\boldsymbol{x}^* = \{(1, 1, 3), (2, 2, 3)\}$) by the improved penalty function method. Also, three global optimal solutions ($\boldsymbol{x}^* = \{(2, 1, 4), (1, 1, 3), (2, 2, 3)\}$) were calculated by the basic PGSA after 37 iterations (Li et al. 2005). With the proposed PMGSA, the same three global optimal solutions can be gained only after eight iterations with the optimal objective function value of 2. The plant growth process and its iterations process of the PMGSA are shown in Fig. 5.3 and Table 5.2, respectively.

Although the same optimal objective function values can be gained both by the basic PSGA and the PMGSA, the iterations of the proposed improved PGSA are less than the former (8 < 37), which further means that the improved PGSA has higher efficiency and global search capability for such nonlinear integer programming.

(3) **Case 3: Nonlinear integer programming with constraints**

The nonlinear integer programming with constraints is represented as follows

$$\min\ x_1^2 + 2x_2^2 + x_3^2 + 2x_4^2 - 27x_1 - 31x_2 - 41x_3 - 34x_4 \tag{5.31}$$

$$s.t. x_1^2 + 2x_2^2 + 2x_3^2 + x_4^2 - x_1 + x_2 - x_3 \leq 61 \tag{5.32}$$

$$2x_1^2 + 3x_2^2 + 3x_3^2 + x_4^2 - x_2 - 2x_3 + x_4 \leq 90 \tag{5.33}$$

$$x_1^2 + 2x_2^2 + 2x_4^2 + 2x_1 - x_2 + 3x_4 \leq 43 \tag{5.34}$$

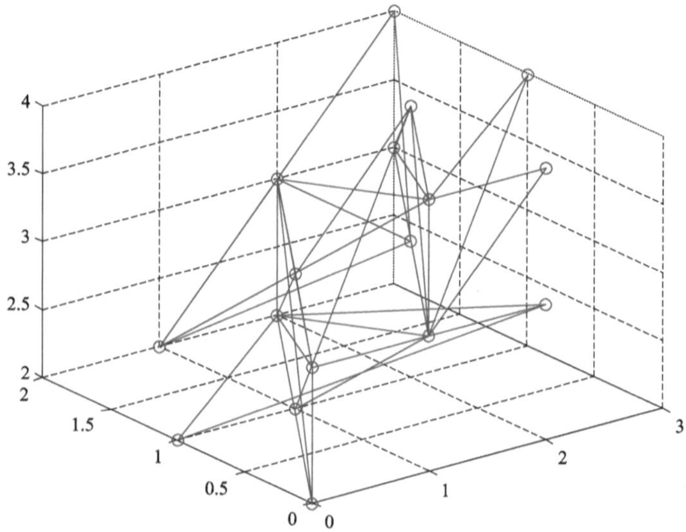

Fig. 5.3 Case 2: plant growth process of the PMGSA

$$0 \leq x_i \leq 20, i = 1, 2, 3, 4, x_i \in \mathbf{Z} \tag{5.35}$$

As for the nonlinear integer programing with constraints, with a hybrid algorithm based on GA [30], an optimal solution ($\boldsymbol{x}^* = \{(3, 4, 3, 2)\}$) was gained, whose objective function is −338. With the improved PGSA put forward by Luo et al. [31], the same optimal solution generated by the algorithm designed by Xiao and Zhang [30] was gained after 33 iterations. However, with the PMGSA, another two better optimal solutions can be gained. One is (2, 1, 5, 3) with the objective function value of −343, and the other is (2, 2, 5, 2) with the objective function value of −344. The iteration process is shown in Table 5.3.

Therefore, compared with the hybrid algorithm based on GA and the proposed improved PGSA, the proposed PMGSA in this chapter can enlarge the solution searching region and improve the convergence speed for solving nonlinear integer programming with constraints. Table 5.4 shows different calculation results generated by these algorithms.

As shown in Table 5.4, compared with other algorithms solving nonlinear integer programming with constraints or without constraints, the PMGSA can gain more global optimal solutions with less iterations and higher efficiency. The improved phytohormone reallocation mode and the random multi-direction of new branches growth are introduced to overcome the drawback of the basic PGSA, i.e., the dependence on the initial solution and the single new branch growth direction in the iteration process. Therefore, the PMGSA is more suitable for solving the bi-level programming for the train plan formation under emergency conditions, because of the better global searching capability and the convergence speed.

Table 5.2 Case 2: iterations process of the PMGSA

IT	GL	OV	GP	FR	RN	IT	GL	OV	GP	FR	RN
0	0, 0, 0	0	1.0000	1.0000		5	0, 1, 2	6	0.0096	0.9807	
1	0, 0, 3	6	1.0000	1.0000	0.8098		2, 0, 3	6	0.0096	0.9903	
2	2, 2, 2	5	0.4722	0.4722	0.5834		0, 0, 3	6	0.0096	1.0000	
	0, 0, 3	6	0.3888	0.8611		6	1, 1, 3	2	0.2857	0.2857	0.8659
	0, 0, 2	9	0.1388	1.0000			2, 2, 3	2	0.2857	0.5714	
3	2, 2, 3	2	0.4615	0.4615	0.7173		1, 2, 2	4	0.1428	0.7142	
	2, 2, 2	5	0.2307	0.6923			1, 0, 4	4	0.1428	0.8571	
	2, 0, 3	6	0.1538	0.8461			1, 0, 3	5	0.0714	0.9285	
	0, 0, 3	6	0.1538	1.0000			2, 2, 2	5	0.0714	1.0000	
4	2, 2, 3	2	0.3913	0.3913		7	1, 1, 3	2	0.3499	0.3499	0.0484
	1, 0, 4	4	0.2173	0.6086	0.9570		2, 2, 3	2	0.3499	0.6999	
	1, 0, 3	5	0.1304	0.7391			2, 1, 3	3	0.2000	0.8999	
	2, 2, 2	5	0.1304	0.8695			1, 2, 2	4	0.0500	0.9500	
	0, 1, 2	6	0.0434	0.9130			1, 0, 4	4	0.0500	1.0000	
	2, 0, 3	6	0.0434	0.9565		8	2, 1, 4	2	0.2232	0.2232	
	0, 0, 3	6	0.0434	1.0000			1, 1, 3	2	0.2232	0.4464	
5	1, 1, 3	2	0.3173	0.3173			2, 2, 3	2	0.2232	0.6696	
	2, 2, 3	2	0.3173	0.6346	0.5182		2, 1, 3	3	0.1428	0.8124	
	1, 0, 4	4	0.1634	0.7980			3, 2, 4	4	0.0624	0.8749	
	1, 0, 3	5	0.0865	0.8846			1, 2, 2	4	0.0624	0.9374	
	2, 2, 2	5	0.0865	0.9711			1, 0, 4	4	0.0624	1.0000	

5.2.3 Train Plan Formation Under Emergency Conditions Based on the PMGSA

According to the characteristics of the bi-level programming for the line planning under emergency, the hybrid intelligent algorithm for plant multi-direction growth simulation algorithm combined with branch-and-bound method is proposed in this section.

1. Generation the stop pattern schedule

Taking only a few major stations with the ability to be the original or terminal station into account, the stop pattern schedule must comply to the rules of the Chinese railway transport organization, and the service frequencies of different types of stop pattern schedule are designed as a decision variable, which greatly reduces the dimension of the variables and reduces a large number of origin departure and terminal arrival constraints.

According to the characteristics of Chinese railway transport organization, four types of stop pattern schedule are generally are as follows [32]:

Table 5.3 Case 3: iterations process of the PMGSA

IT	GL	OV	GP	FR	RN	IT	GL	OV	GP	FR	RN
1	0,2,2,2	−192	0.4424	0.4424	0.3796	5	0,0,0,3	−84	0.0365	0.8774	
	0,0,2,0	−78	0.1797	0.6221			0,0,2,0	−78	0.0339	0.9113	
	0,0,0,2	−60	0.1382	0.7604			0,0,0,2	−60	0.0261	0.9374	
	0,2,0,0	−54	0.1244	0.8848			0,2,0,0	−54	0.0235	0.9609	
	2,0,0,0	−50	0.1152	1.0000			2,0,0,0	−50	0.0217	0.9826	
2	0,2,2,2	−192	0.3569	0.3569	0.8759		0,0,1,0	−40	0.0174	1.0000	
	2,2,0,0	−104	0.1933	0.5502		6	2,2,3,0	−218	0.0532	0.0532	0.9699
	0,0,2,0	−78	0.1450	0.6952			2,2,2,1	−214	0.0522	0.1054	
	0,0,0,2	−60	0.1115	0.8067			3,2,2,0	−204	0.0498	0.1552	
	0,2,0,0	−54	0.1004	0.9071			2,3,2,0	−203	0.0495	0.2047	
	2,0,0,0	−50	0.0929	1.0000			2,0,4,0	−198	0.0483	0.2531	
3	0,2,2,2	−192	0.1912	0.1912	0.3968		2,3,1,1	−197	0.0481	0.3011	
	0,2,2,0	−132	0.1315	0.3227			0,2,2,2	−192	0.0469	0.3480	
	2,0,2,0	−128	0.1275	0.4502			2,0,2,2	−188	0.0459	0.3939	
	0,2,0,2	−114	0.1135	0.5637			2,2,2,0	−182	0.0444	0.4383	
	2,2,0,0	−104	0.1036	0.6673			4,0,2,0	−170	0.0415	0.4797	
	0,4,0,0	−92	0.0916	0.7590			1,1,3,0	−169	0.0412	0.5210	
	0,0,2,0	−78	0.0777	0.8367			1,2,2,0	−158	0.0386	0.5595	
	0,0,0,2	−60	0.0598	0.8964			2,1,2,0	−157	0.0383	0.5979	
	0,2,0,0	−54	0.0538	0.9502			2,2,1,0	−144	0.0351	0.6330	
	2,0,0,0	−50	0.0498	1.0000			1,1,2,0	−133	0.0325	0.6654	
4	2,0,4,0	−198	0.1084	0.1084	0.9025		0,2,2,0	−132	0.0322	0.6977	
	0,2,2,2	−192	0.1051	0.2136			2,0,2,0	−128	0.0312	0.7289	
	2,0,2,2	−188	0.1030	0.3165			0,0,3,0	−114	0.0278	0.7567	
	2,2,2,0	−182	0.0997	0.4162			0,2,0,2	−114	0.0278	0.7845	
	4,0,2,0	−170	0.0931	0.5093			0,0,2,1	−110	0.0268	0.8114	
	0,2,2,0	−132	0.0723	0.5816			0,1,2,0	−107	0.0261	0.8375	
	2,0,2,0	−128	0.0701	0.6517			1,0,2,0	−104	0.0254	0.8629	
	0,2,0,2	−114	0.0624	0.7141			2,2,0,0	−104	0.0254	0.8882	
	2,2,0,0	−104	0.0570	0.7711			0,4,0,0	−92	0.0224	0.9107	
	0,4,0,0	−92	0.0504	0.8215			0,0,0,3	−84	0.0205	0.9312	
	0,0,0,3	−84	0.0460	0.8675			0,0,2,0	−78	0.0190	0.9502	
	0,0,2,0	−78	0.0427	0.9102			0,0,0,2	−60	0.0146	0.9649	
	0,0,0,2	−60	0.0329	0.9430			0,2,0,0	−54	0.0132	0.9780	
	0,2,0,0	−54	0.0296	0.9726			2,0,0,0	−50	0.0122	0.9902	
	2,0,0,0	−50	0.0274	1.0000			0,0,1,0	−40	0.0098	1.0000	

(continued)

Table 5.3 (continued)

IT	GL	OV	GP	FR	RN	IT	GL	OV	GP	FR	RN
5	2,0,4,0	−198	0.0860	0.0860	0.2686	…	…	…	…	…	…
	0,2,2,2	−192	0.0834	0.1695		21	1,2,5,2	−320	0.0181	0.0181	0.1842
	2,0,2,2	−188	0.0817	0.2512			2,1,5,2	−319	0.0181	0.0362	
	2,2,2,0	−182	0.0791	0.3303			2,1,5,1	−291	0.0165	0.0527	
	4,0,2,0	−170	0.0739	0.4042			2,1,4,2	−287	0.0163	0.0689	
	0,2,2,0	−132	0.0574	0.4615			2,2,5,0	−284	0.0161	0.0850	
	2,0,2,0	−128	0.0556	0.5172			…	…	…	…	
	0,0,3,0	−114	0.0495	0.5667			0,1,0,0	−29	0.0016	1.0000	
	0,2,0,2	−114	0.0495	0.6163		22	2,2,5,2	−344	0.0179	0.0179	
	0,0,2,1	−110	0.0478	0.6641			2,1,5,3	−343	0.0178	0.0357	
	0,1,2,0	−107	0.0465	0.7106			3,1,4,3	−333	0.0173	0.0530	
	1,0,2,0	−104	0.0452	0.7558			1,2,5,2	−320	0.0166	0.0696	
	2,2,0,2	−104	0.0452	0.8010			…	…	…	…	
	0,4,0,0	−92	0.0400	0.8409			0,1,0,0	−29	0.0015	1.0000	

Table 5.4 Results comparisons among different algorithms

Experiment	Solving algorithms	OV	IT	Number of optimal solutions
Case 1	Filled function method [28]	−11.70	≥ 57	1
	Basic PGSA (Li et al. 2005)	−11.70	51	3
	PMGSA	−11.70	10	8
Case 2	Penalty function method [29]	2	–	2
	Basic PGSA (Li et al. 2005)	2	37	3
	PMGSA	2	8	3
Case 3	Hybrid algorithm [30]	−338	–	1
	Another improved PGSA [31]	−338	33	1
	PMGSA	−344	22	1

(i) Non-stopping-schedule (NSS). When the train runs between two major stations, high-speed short marshaling mode is generally adopted; if the train runs across more than one major station, high-speed long marshaling mode is generally used. This type of schedule has the shortest travel time and the highest priority. Assume there are a total of A_1 options.

(ii) Stop at major stations (SMS). The train runs across more than one major station and only stops at major stations. It is generally operated by high-speed long marshaling, which has shorter travel time and high priority. Assume there are A_2 options.

(iii) Stop at staggered stations (SSS). When the train runs across more than one major station, it not only stops at major stations, but also stops at some intermediate stations, using long marshaling mode. Most medium-distance passengers' demand can be satisfied, assuming the number of this mode is A_3.

(iv) Stop at all stations (SAS). The train runs between two major stations and stops at each intermediate station. This kind of stop schedule can meet the needs of short-distance passengers and has the lowest priority. Assume that there are A_4 options.

If N ($N = A_1 + A_2 + A_3 + A_4$) is the number of variables of the upper level, the dimensions of each growth node in the algorithm are also the same N. Then, the stop schedules for K stations can be described as a matrix $\mathbf{X}$ of $K \cdot N$, where each row is a stop schedule, and each column is the corresponding station (see Fig. 5.4). If station i is the planned stop station in schedule j, $x_{j,i} = 1$; otherwise, $x_{j,i} = 0$. The service frequency is represented by the growth nodes of U N-dimensional vectors, where each element is an integer and corresponds to a stop schedule. If the value of the element is 0, it means that the stop scheme is not adopted. Note that the value of the N is variable because the number of the growth nodes in each iteration of the algorithm is uncertain.

2. Main procedures based on the PMGSA

The main procedures of solving the bi-level programming model for line plan under emergency based on the PMGSA are as follows.

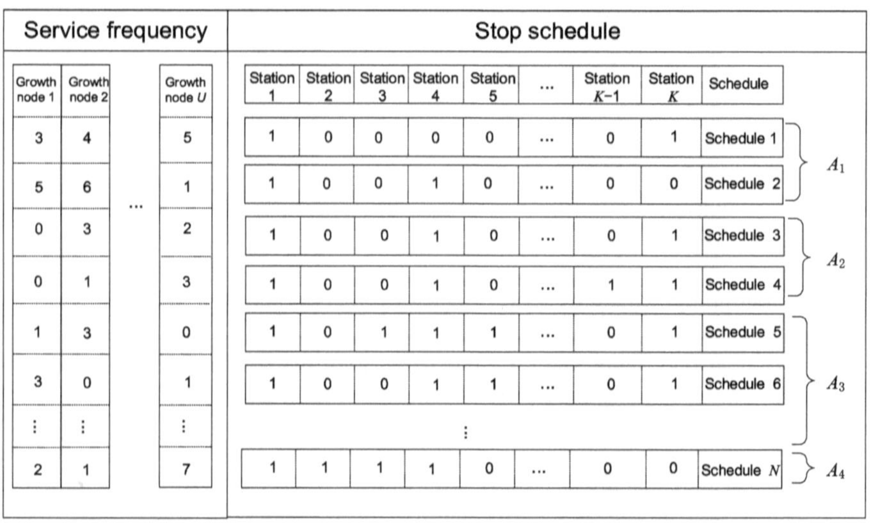

Fig. 5.4 Stop schedules and service frequency

Step1: Initialization. Input data of passenger demand, station, section length, operation time, etc., determine the decision variable x and the corresponding relationship between variable value and stop schedule.

Step2: At the initial phase, the initial feasible solution $\boldsymbol{x}^0\left(\boldsymbol{x}^0 \in \boldsymbol{X}_{\mathrm{I}}\right)$ should be set as the new growing node $\boldsymbol{x}^G$. $k = 0$, $\boldsymbol{x}_{\min} = \boldsymbol{x}^0$, $f_{\min} = f\left(\boldsymbol{x}^0\right)$, $\boldsymbol{x}^G = \boldsymbol{x}^0$, $\mathbf{S} = \left\{\boldsymbol{x}^0\right\}$, $\Omega(\mathbf{S}) = 1$, and the iterative length is λ^k.

Note that, $f\left(\boldsymbol{x}^0\right)$ not only includes the cost of the stop pattern schedule, but also the feedback of lower level (passenger flow distribution) under the schedule $\boldsymbol{x}^0$, so this step should carry out step 7.

Step 3: Generation of new branches.

Step 3.1: For the fixed vertical growth direction. Straight line paralleling with axes through the new growing node $\boldsymbol{x}^G$ should be made firstly. Then, set $\boldsymbol{x}^G$ as the center, and divide these line or trunks with the iterative length λ^k with both negative and positive directions into new growing branches or nodes $s'_{i_1 j_1}$ $\left(i_1 = 1, 2, \ldots, 2n;\ j_1 = 1, 2, \ldots, m_{i_1}\right)$.

Step3.2: For the random growing directions. Three directions should be selected firstly in a random way from the growing dimension directions (n), and a trunk can be further generated with the symmetric rotation growing principle. Then, set $\boldsymbol{x}^G$ as the center, and divide with λ^k into new growing branches or nodes $s'_{i_{2n+1} j_1}$, $j_1 = 1, 2, \ldots, m_{i_{2n+1}}$. There are n random growing directions of new growing branches.

Note that, considering that the decision variables represent the service frequency of different stop schedule, the iterative length λ can be randomly selected in the feasible region $(0,\ H]$ to ensure the diversity of solution spaces. In addition, if the solution does not satisfy the capacity constraints, reselect the direction or iterative length, and grow a new branch.

Step 4: Determination the best-so-far solution. Calculate the objective function value of each new growing branches or nodes, and take the best one $s'_{p,q}$ as the optimal one.

$$f^*\left(s'_{p,q}\right) = \min\left\{f\left(s'_{i_1 j_1}\right), i_1 = 1, 2, \ldots, 3n;\ j_1 = 1, 2, \ldots, m_{i_1}\right\}$$

If $f^*\left(s'_{p,q}\right) < f_{\min}$, there is $x_{\min} = s'_{p,q}$, $f_{\min} = f^*\left(s'_{p,q}\right)$;

If $f^*\left(s'_{p,q}\right) = f_{\min}$, there is $x_{\min} := x_{\min} \cup \left\{s'_{p,q}\right\}$, $f_{\min} = f^*\left(s'_{p,q}\right)$.

Step 5: Calculation of the growth probability.

Set $S := S \cup S'$, $F = f_{\min} + \sum (f(s) - f_{\min})/\Omega(S)$.

For every $s \in S$,

If $F \le f(s)$, there is $P_s = 0$, $S := S \backslash \{s\}$;

If $F > f(s)$, there is $P_s = (F - f(s))/\left(\sum\limits_{s \in S} (F - f(s))\right)$.

Step 6: Select the next new growing node $\boldsymbol{x}^G$. Generate a random number η, $\eta \in (0, 1)$, and select the node s_t that satisfies the condition $\sum_s^{t-1} p_s < \eta < \sum_s^t p_s, t \in [1, \Omega(S)]$ as the new growth node. Let $\boldsymbol{x}^G = s_t,\ k = \mathrm{k} + 2$, go to Step 3.
Step 7: Allocation of the passenger flow by the branch-and-bound method. Note that, the branches are selected according to passenger flow characteristics in order to improve the efficiency of the problem.

To simplify the expression, UB, NF, P_0 denote the upper bound of the optimization results, the lower bound of the problem to be solved, and original passenger flow distribution, respectively. $S(P_i)$ is the feasible set of problem P_i in mixed-integer programming, and $\overline{x}$ represents the feasible solution of $S(P_i)$. $S(\overline{P}_i)$ is the feasible set of the relaxation problem $\overline{P}_i$.

Step 7.1: Let $NF = \{0\},\ \overline{x} = \phi,\ UB = +\infty$;
Step 7.2: Solve the relaxation problem $\overline{P}_v(v \in NF)$. Let the optimal solution be denoted by x^v, and the optimal value be denoted by f_v. If $f_v \geq UB$, then $NF := NF\backslash\{v\}$ and go to Step 7.5; otherwise, go to Step 7.3.
Step 7.3: If $f_v < UB$ and $x^v \in S(P_0)$, then let $UB = f_v, \overline{x} = x^v, NF := NF\backslash\{v\}$, and go to Step 7.5; otherwise, go to Step 7.4.
Step 7.4: If $f_v < UB$ and $x^v \notin S(P_0)$, the problem needs to be branched. The higher the priority, the longer the distance, the more preferential the schedule is to be branched. $S(P_v)$ is divided into two subsets $S(P_{v_1})$ and $S(P_{v_2})$. Let $NF := (NF\backslash\{v\}) \cup \{v_1, v_2\}$, then go to Step 7.2.

Note that, compared with the traditional method of branching randomly, the above method not only speeds the procedure of problem solving, but also preferentially meets the transportation demands of high-priority and long-distance passengers.

Step 7.5: If set NF is all visited, the calculation ends, and $\overline{x}$ is the optimal solution of the lower model. Otherwise, go to Step 7.2.
Step 8: End the algorithm. According to the scale of the problem, the calculation is terminated when the optimal solution remains unchanged in the iteration or the number of iterations reaches a certain number, and then the optimal line plan output. The algorithm flowchart is shown in Fig. 5.5.

5.2.4 Characteristics of the Algorithm

(1) The service frequency of different stop pattern schedule is used as the decision variable, which greatly reduces the number of the variables, and satisfies a large number of unique constraints for original departure and arrival, simplifying the problem scale and solving difficulty.
(2) The convergence speed of the algorithm is greatly improved, and the increase of the diversity of the solution space can avoid the result falling into the local optimal solution because of the introduction of the new search direction.

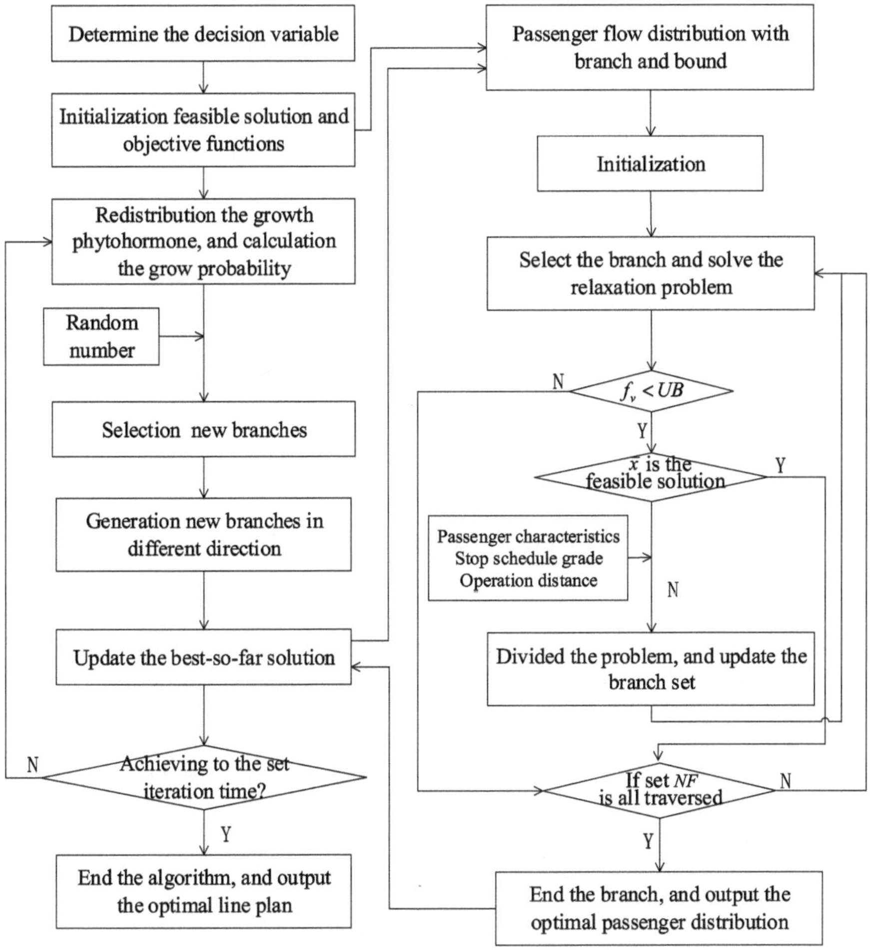

Fig. 5.5 Plant growth chart of bi-level programming

(3) The nodes with growth ability and their growth phytohormone are determined according to the optimal solution for each iteration in the PMGSA procedure, which is closer to the natural characteristic of the plant and avoids the disadvantages of relying on the initial solution compared with the basic PGSA.

(4) The iteration is set to a random number according to the characteristics of the line planning under emergency. Therefore, more strategies can be generated at a faster speed, and the bi-level programming can be converged more stably to the global optimal solution.

(5) Compared with the traditional method of traversing branches randomly, the lower level in the proposed PGMSA not only preferentially meets the transportation demands of high-grade and long-distance passengers, but also speeds up the problem solving.

(6) The hybrid intelligent algorithm for plant multi-direction growth simulation algorithm based on embedded branch-and-bound method is used to realize the overall optimization of the stop pattern schedule optimization and passenger distribution under emergency conditions. Therefore, the algorithm is suitable for solving the problem of line planning under emergency, which has the characteristics of complex system structure, complex object relationship, and complex environment and diverse behavior.

5.3 Case Study

The Taiwan high-speed railway and the Beijing-Shanghai high-speed railway are taken into consideration to illustrate the effectiveness of the model and algorithm.

5.3.1 Taiwan High-Speed Railway

As for the Taiwan high-speed railway, a comparative study compared with [10] is carried out in this section. The total distance of the Taiwan high-speed railway is 340 km, connecting two major cities, Taipei and Kaohsiung. There are seven stations on the line, of which three stations are able to be original and terminal stations.

According to the model proposed in this chapter, the stop pattern schedule is set as follows: $NSS = 3, \quad SMS = 1, \quad SSS = 14, \quad SAS = 3$, so the dimension of the plant growth node is 21. The iteration procedure of the solution is shown in Fig. 5.6. After 40 branch calculations, the optimal solution converges to 18 277, and the line plan is shown in Fig. 5.7. The total service frequency is 11, i.e., a total of 11 trains are operated, of which the last train departs from the fourth station. The generated line plan not only satisfies the passenger demand, but also avoids the disadvantages of reference [10] that the train can only depart from the first station, and reducing the wasted capacity. The total calculation time is 67 s. Therefore, the feasibility of the model and algorithm of the proposed bi-level programming for line plan is illustrated.

5.3.2 Beijing-Shanghai High-Speed Railway

The Beijing-Shanghai high-speed railway with complex transportation organization mode is taken into account to further verify the effectiveness of the algorithm.

1. Basic data

The Beijing-Shanghai high-speed railway has a total length of 1318 km, running through three municipalities (Beijing, Tianjin, and Shanghai) and four provinces (Hebei, Shandong, Anhui, and Jiangsu), connecting two major economic zones of

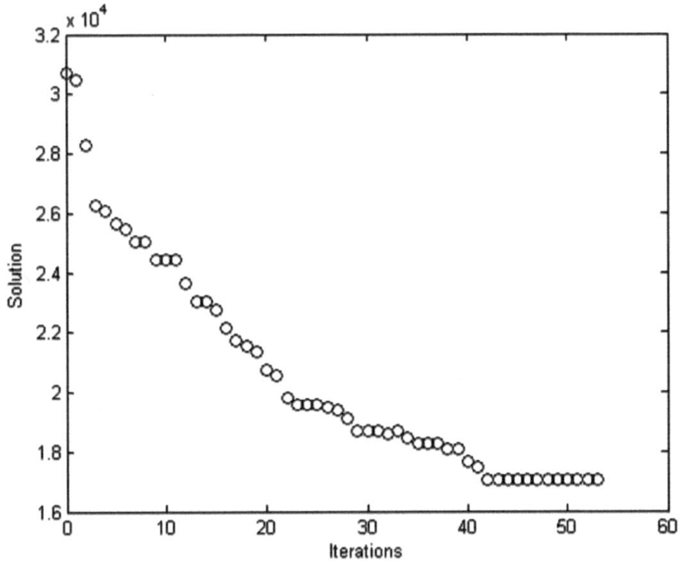

Fig. 5.6 Iteration procedure of the solution of the Taiwan high-speed railway (HSR)

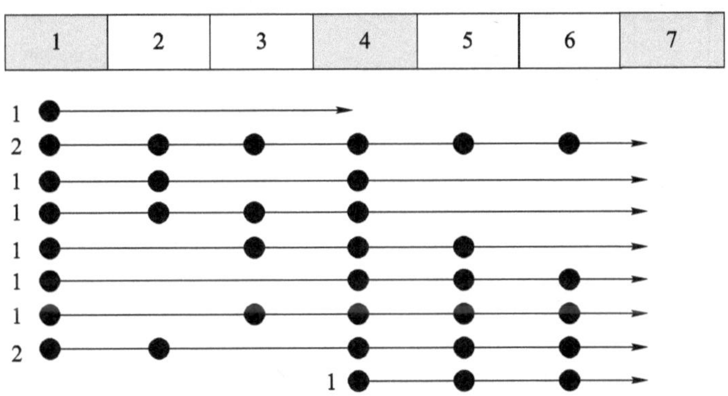

Fig. 5.7 Optimal line plan of the Taiwan high-speed railway (HSR)

Bohai Sea and Yangtze River Delta. It is the main artery of our country, which plays an extremely important role in promoting the development of the national high-speed railway network. The Beijing-Shanghai high-speed railway and its surrounding railways are shown in Fig. 5.8, and the two nearest railway lines are Shanghai-Nanjing intercity railway and Beijing-Shanghai existing ordinary speed railway. The three types of railway line, the existing ordinary speed railway, the intercity railway, and the high-speed railway, are represented by solid lines, dotted lines, and dashed lines, respectively. In this section, only the transportation organization in the direction from

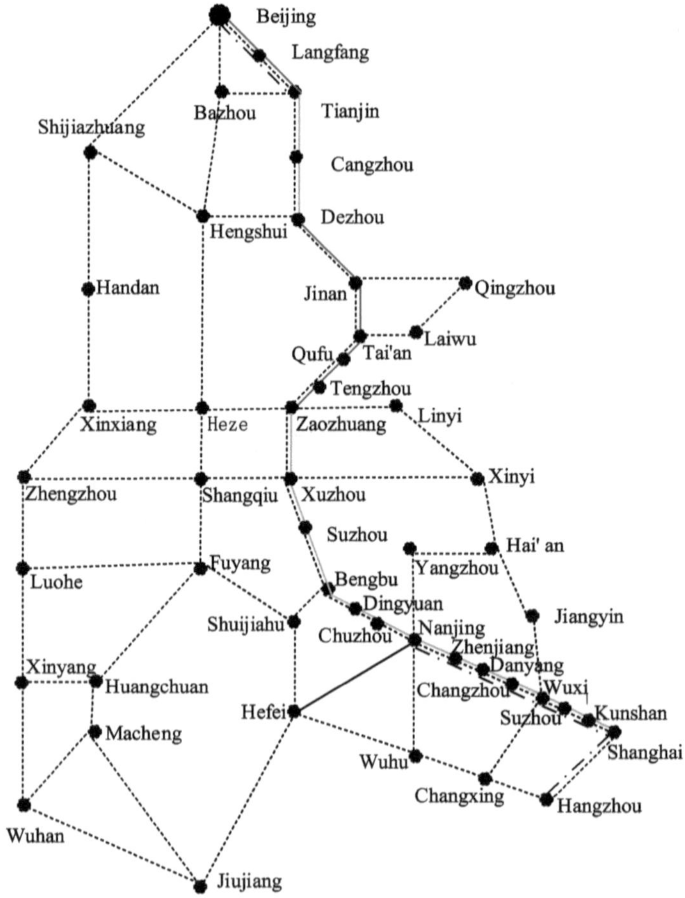

Fig. 5.8 Part of China railway network

Beijing to Shanghai is considered, because the Beijing-Shanghai high-speed railway is operated on double lines.

The Beijing-Shanghai high-speed railway has 23 stations, namely Beijing South Railway Station (BJ, 1), Langfang Railway Station (LF, 2), Tianjin Railway Station (TJ, 3), Cangzhou West Railway Station (CZ, 4), Dezhou East Railway Station (DZ, 5), Jinan West Railway Station (JN, 6), Tai'an Station (TA, 7), Qufu East Station (QF, 8), Tengzhou East Station (TZ, 9), Zaozhuang Station (ZZ, 10), Xuzhou East Station (XZ, 11), Suzhou East Station (SZ, 12), Bengbu South Station (BB, 13), Dingyuan Station (DY, 14), Chuzhou Station (CHZ, 15), Nanjing South Station (NJ, 16), Zhenjiang South Station (ZJ, 17), Danyang North Station (DAY, 18), Changzhou North Station (CHAZ, 19), Wuxi East Railway Station (WX, 20), Suzhou North Railway Station (SUZ, 21), Kunshan South Railway Station (KS, 22), and Shanghai Hongqiao Railway Station (SH, 23). The lengths of the 22 sections are 59 km, 72 km,

88 km, 108 km, 79 km, 56 km, 71 km, 58 km, 36 km, 65 km, 75 km, 77 km, 53 km, 62 km, 64 km, 64 km, 25 km, 32 km, 57 km, 26 km, 41 km, and 50 km, respectively. There are five original departure stations, namely Beijing South Railway Station, Tianjin West Railway Station, Jinan West Railway Station, Nanjing South Railway Station, and Shanghai Hongqiao Railway Station. The other parameters are set as follows.

(1) Section running time. The running time is calculated based on the average running speed of the train (300 km/h).
(2) Minimum dwell time. The dwell time of the high-speed railway is usually set in a range from 2 to 5 min. For major stations with capacity to depart and arrive trains, the dwell time can be increased to 10 min.
(3) Headway. This section mainly controls the minimum headway between adjacent trains occupying the same resource, which can be up to 5 min in peak period.
(4) Train capacity and marshaling. The Beijing-Shanghai high-speed trains are organized into 16 carriages with a seating of 1004 people. The attendance rate can be set to 1 under emergency conditions.
(5) The hardware environment. Intel Core 2 Duo E7500 and 2 GB memory computer.

2. Emergency scenario settings

Assuming that equipment failure occurs in the Wuxi-Kunshan section (the high-speed railway) due to bad weather, the train can only run at a limited speed, which reduces the daily carrying capacity of the section to 50 trains and is expected to last for 5 days. Due to the long-term impact of the emergency and the severe loss of the line capacity, the train routes set should be searched first, and then, the line plan is generated according to the passenger demand.

The OD passenger demand of the Beijing-Shanghai high-speed is shown in Table 5.5. High-grade passengers (52%) tend to choose stop schedule NSS and SMS, medium-grade passengers (28%) tend to SMS and SSS, and low-grade passengers (28%) tend to SSS and SAS.

3. Train routes generation

The remained carrying capacity of the existing general speed railway and the intercity railway, which is not affected by the emergency, is about $\widetilde{N}_{JY} = (24, 25, 26)$ and $\widetilde{N}_{CJ} = (46, 50, 53)$, respectively. The distance restriction Rs_s of the departure and terminal station is set to 300 km, because the total length of the Shanghai-Nanning intercity railway is about 300 km. The minimum capacity restriction Rs_c and the maximum length restriction of rerouting route Rs_l are set to 3 km and 600 km, respectively.

The set of departure and terminal station is generated firstly.

In Fig. 5.8, three stations, Wuxi East Railway Station (WX), Changzhou North Station (CHAZ), and Nanjing South Station (NJ), not only satisfy the restriction Rs_s with Suzhou North Railway Station (SUZ), but also have the capacity to cross lines.

Table 5.5 Passenger travel demand for Beijing-Shanghai high-speed railway (HSR)

Station	LF	TJ	CZ	DZ	JN	TA	QF	TZ	ZZ	XZ	SZ	BB	DY	CHZ	NJ	ZJ	DAY	CHAZ	WX	SUZ	KS	SH
BJ	135	792	297	213	741	306	429	96	99	438	81	132	118	63	783	84	209	120	114	105	99	1263
LF		141	93	60	66	15	34	60	25	30	21	23	23	21	84	13	68	11	24	17	13	87
TJ			225	246	366	105	222	132	57	360	45	168	98	39	219	48	106	51	54	42	45	297
CZ				42	75	18	18	45	12	36	9	21	22	6	72	12	37	21	21	18	15	87
DZ					147	54	66	39	45	42	24	33	12	0	78	21	20	27	27	24	24	96
JN						207	210	187	192	171	30	51	96	21	339	54	61	72	75	69	75	390
TA							51	36	48	123	15	21	55	0	90	18	36	27	27	27	24	96
QF								42	36	42	12	18	16	6	111	30	40	39	39	36	42	138
TZ									38	63	51	32	21	0	86	68	87	23	45	39	21	94
ZZ										72	15	57	9	6	48	51	24	36	36	45	81	102
XZ											75	147	115	105	297	87	137	108	147	147	258	357
SZ												99	82	132	141	45	61	51	51	30	48	183
BB													56	108	162	114	112	120	123	126	120	396
DY														49	175	142	81	67	53	88	32	109
CHZ															72	30	42	39	84	96	60	240
NJ																102	237	111	114	171	108	1281
ZJ																	106	120	114	168	150	183
DAY																		103	88	77	136	154
CHAZ																			81	300	120	255
WX																				450	261	429
SUZ																					288	435
KS																						315

Therefore, $seq_s = \{WX, CHAZ, NJ\}$. Similarly, for Kunshan South Railway Station (KS), $seq_e = \{KS, SH\}$.

The time densities and the capacities of the rerouting sections are calculated and shown in Table 5.6.

The rerouted paths are searched according to the density of capacities of routes, as shown in Table 5.7, where D represents the sum of density of rerouting path capacities, C represents the intercity railway, and E represents the existing ordinary speed railway.

Analyzing the ten paths in Table 5.7, the path 8 and 9 starting from Wuxi East Railway Station (WX), which is the nearest station to Suzhou North Railway Station (SUZ), are obviously inferior to the previous seven paths because of the long length and high time densities of the capacities. Path 1 and path 4 (bold) are finally selected due to the short running distance. Furthermore, the results of the searching route will be changed because the adjustment of the restrictions. For example, when $Rs_s = 100$ km, $Rs_l = 300$ km, then $seq_s = \{WX, CHAZ\}$, $seq_e = \{KS\}$, and the path 1 and 4 are directly obtained.

Therefore, there are three paths that can make full use of other lines' capacity on the railway network to meet the passenger flow demand under emergency conditions. The first route is the Beijing-Shanghai high-speed railway. The second route is composed

Table 5.6 Time densities and time densities of all sections' capacity

Number	Sections	Length (km)	Average speed (km/h)	*Tes*	*Des*
1	NJ-ZJ (Intercity)	71	(250,265,300)	186.33	0.005367
2	ZJ-CHAZ (Intercity)	72	(250,265,300)	183.75	0.005442
3	CHAZ-WX (Intercity)	39	(250,265,300)	339.23	0.002948
4	WX-SUZ (Intercity)	42	(250,265,300)	315.00	0.003175
5	SUZ-KS (Intercity)	34	(250,265,300)	389.11	0.002570
6	KS-SH (Intercity)	50	(250,265,300)	264.60	0.003779
7	ZJ-CHAZ (Existing)	72	(160,170,200)	53.47	0.018702
8	CHAZ-WX (Existing)	39	(160,170,200)	98.71	0.010131
9	WX-SUZ (Existing)	42	(160,170,200)	91.66	0.010910
10	SUZ-KS (Existing)	35	(160,170,200)	110.00	0.009091
11	KS-SH (Existing)	49	(160,170,200)	78.57	0.012728
12	WX-CX (Existing)	59	(160,170,200)	65.25	0.015326
13	CX-HZ (Existing)	126	(160,170,200)	30.00	0.033333
14	HZ-SH (Intercity)	160	(250,265,300)	82.68	0.012095
15	HZ-SH (Existing)	188	(160,170,200)	20.47	0.048852
16	NJ-WH (Existing)	125	(160,170,200)	30.80	0.032468
17	WH-CX (Existing)	178	(160,170,200)	21.62	0.046253
18	NJ-HF (Intercity)	166	(250,265,300)	79.69	0.012549
19	HF-WH (Existing)	144	(160,170,200)	26.73	0.037411

Table 5.7 Rerouting path sets

Number	Rerouting path	Length	*D*
1	**CHAZ(C)-WX(C)-SUZ(C)-KS**	**115**	**0.0086**
2	CHAZ(C)-WX(C)-SUZ(C)-KS(C)-SH	165	0.0123
3	NJ(C)-ZJ(C)-CHAZ(C)-WX(C)-SUZ(C)-KS	166	0.0195
4	**CHAZ(E)-WX(E)-SUZ(C)-KS**	**115**	**0.0236**
5	CHAZ(C)-WX(C)-SUZ(E)-KS(E)-SH	165	0.0279
6	CHAZ(E)-WX(E)-SUZ(E)-KS(E)-SH	165	0.0428
7	NJ(C)-ZJ(E)- CHAZ(E)-WX(E)-SUZ(E)-KS	144	0.0476
8	WX(E)-CX(E)-HZ(C)-SH	381	0.0607
9	WX(E)-CX(E)-HZ(E)-SH	409	0.0975
10	NJ(E)-WH(E)-CX(E)-HZ(C)-SH	589	0.1241

of the section from Beijing to Changzhou (Beijing-Shanghai high-speed railway), the section from Changzhou to Kunshan (Shanghai-Nanjing intercity railway), and the section from Kunshan to Shanghai (Beijing-Shanghai high-speed railway). The third path is similar to the second one, except that the existing oridinary speed railway is chosen at the section from Changzhou to Wuxi.

4. Line planning

According to the characteristics of the problem, the stop schedule is set as follows: $NSS = 9, \quad SMS = 36, \quad SSS = 75, \quad SAS = 6$, so the dimension of the plant growth node is 126. The iteration procedure of the solution is shown in Fig. 5.9. After more than 103 iterations, the optimal solution converges to 42 671.38 with a calculation time of 316 s. The line plan is shown in Fig. 5.10, where station 1 (BJ), station 3 (TJ), station 6 (JN), station 16 (NJ), and station 23 (SH) are the major stations that can be as the original and terminal stations. In addition, the three stop patterns in Fig. 5.10 are represented by dots, triangles, and squares, respectively. The total service frequency is 143, which meets the transport demand of all passengers.

5.4 Chapter Summary

Firstly, a bi-level programming model is established in this chapter. The upper level (stop schedule and service frequency optimization) aims to minimize the total operation cost and the number of the unserved passengers, while the lower level (passenger flow allocation) takes the maximum passenger service and the minimum travel time of all passengers as the optimization target. Then, according to the characteristics of the bi-level programming for the line planning under emergency conditions, the hybrid intelligent algorithm for plant multi-direction growth simulation algorithm combined with branch-and-bound method is proposed. Finally, the methods are applied to the

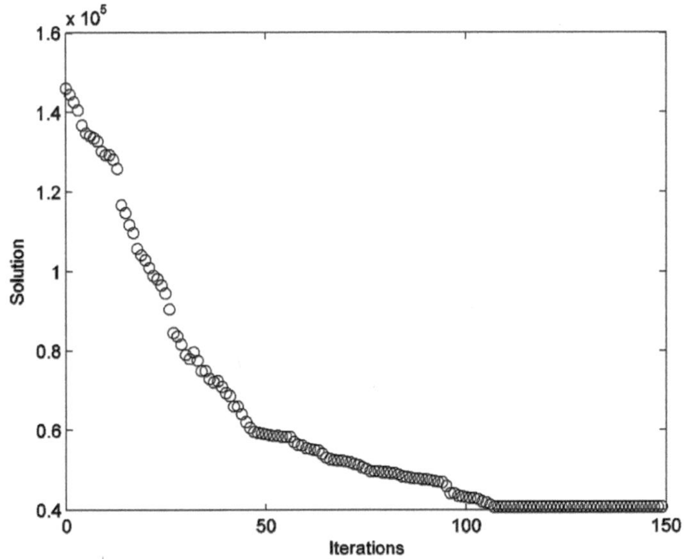

Fig. 5.9 Iteration procedure of the solution

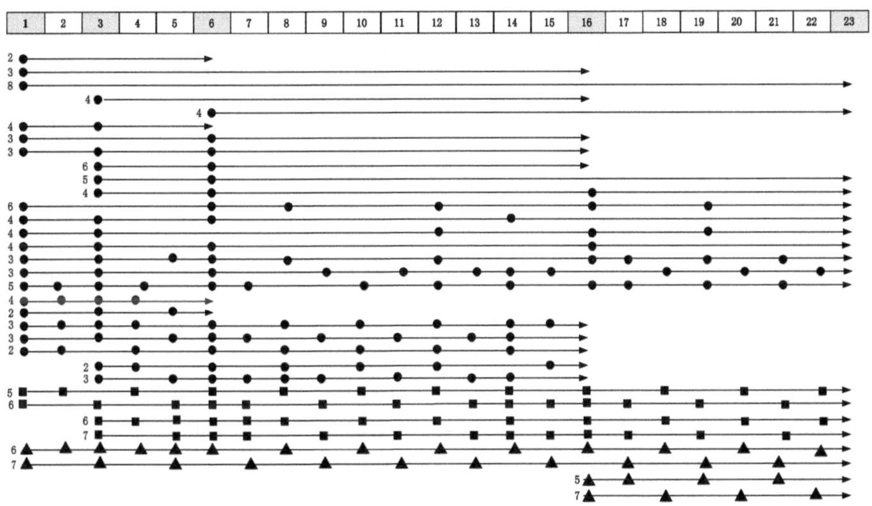

Fig. 5.10 Optimal line plan

Taiwan high-speed railway and Beijing-Shanghai high-speed railway, and the feasibility and effectiveness of the model and algorithm to solve the problem of high-speed railway line planning under emergency are proved. The results obtained can be the foundation of the next chapter.

References

1. Lianbo D (2007) Study on the optimal problems of passenger train plan for dedicated passenger traffic line. Central South University, Changsha (in Chinese)
2. Feng S, Wenliang Z, Yan C et al (2008) Optimization study on passenger train plans with elastic demands. J China Railw Soc 3:150 (in Chinese)
3. Jian H (2013) Optimization of train operation plan for network of dedicated passenger lines based on passenger flow dynamic adjustment. Southwest Jiaotong University, Chengdu (in Chinese)
4. Ruihua Xu, Xiaolei Z (2005) Study on train plans optimization for passenger traffic special line. J Tongji Univ (Nat Sci) 12:1608–1703 (in Chinese)
5. Zha W, Fu Z (2000) Research on the optimization method of through passenger train plan. J China Railw Soc 5:1–5 (in Chinese)
6. Xin Du, Yongtao N, Baoming H et al (2010) Train service planning for passenger dedicated railway line based on analyzing importance of nodes. J Beijing Jiaotong Univ 34(6):5–10 (in Chinese)
7. Zhang Y, Ren M, Du W (1998) Optimization of high-speed train operation. J Southwest Jiaotong Univ (Nat Sci) 4:400-404 (in Chinese)
8. Xiaoqiu J, Xiaoyu G (2011) Research on a model and its algorithm about the line planning on the network for passenger. J Syst Eng 26(2):216–221 (in Chinese)
9. Shouqing D, Haifeng Y, Qunren L (2012) Research on genetic optimization of operation scheme for PDL trains based on network flow distribution. China Railw Sci 33(4):105–111 (in Chinese)
10. Chang YH, Yeh CH, Shen CC (2000) A multiobjective model for passenger train services planning: application to Taiwan's high–speed rail line. Transp Res Part B 34(2):91–106
11. Goossens JW, van Hoesel S, Kroon L (2004) A branch-and-cut approach for solving railway line-planning problems. Transp Sci 38(3):379–393
12. Goossens JW, Van Hoesel S, Kroon L (2014) Optimizing halting station of passenger railway lines [EB/OL] (2014-09-01) [2016-03-08]. https://arno.unimaas.nl/ show.cgi?fid=803
13. Goossens JW, van Hoesel S, Kroon L (2006) On solving multi-type railway line planning problems. Eur J Oper Res 168(2):403–424
14. Chakroborty P, Wivedi T (2000) Optimal route network design for transit systems using genetic algorithms. Eng Optim 34(1):83–100
15. Laporte G, Mesa JA, Perea F (2000) A game theoretic framework for the robust railway transit network design problem. Transp Res Part B 44(4):447–459
16. Schŏebel A, Schwarze S, A game-theoretic approach to line planning [EB/OL] (2011–09–01) [2016–03–08]. https://drops.dagstuhl.de/opus/volltexte/2006/688/pdf/06002.SchoebelAnita
17. Wang L, Jia LM (2011) A two-layer optimization model for high-speed railway line planning. J Zhejiang Univ Sci 12(12):902–913
18. Fliege J, Vicente LN (2006) A multicriteria approach to bilevel optimization. J Optim Theory Appl 131(2):209–225
19. Glackin JA (2006) Solving bilevel linear programming using multiple objective linear programming. Rensselaer Polytechnic Institute, Troy, New York
20. Audet C, Hansen P, Jaumard B (1997) Links between linear bilevel and mixed 0–1 programming problem. J Optim Theory Appl 93:273–300
21. Jian D (2009) Some ALgorithm of bilevel programming problems. Jilin University, Jinlin (in Chinese)
22. Yongchao G (2007) Study on performance of intelligent optimization algorithm and search space. Shandong University, Jinan (in Chinese)
23. Tong Li (2004) The research on bilevel integer programming algorithm based on plant growth simulation. Tianjin University, Tianjin (in Chinese)
24. Tong Li, Zhongtuo W (2010) Plant growth simulation and the thinking in knowledge innovation. J Manage Sci China 13(3):87–96 (in Chinese)
25. Guney K, Durmus A, Basbug S (2009) A plant growth simulation algorithm for pattern nulling of linear antenna arrays by amplitude control. Progr Electromagn Res B 17:69–84

26. Chun W, Haozhong C (2007) A plant growth simulation algorithm and its application in power transmission network planning. Autom Electr Power Syst 31(7):24–28 (in Chinese)
27. Litu Y, Kai W, Xuncheng H et al (2008) Application of trees growth simulation algorithm to solve reactive power optimization problem. J Zhengzhou Univ (Eng Sci) 29(2):69–72 (in Chinese)
28. Wenxing Z (2000) A filled function method for integer programming. Acta Math Applicatae Sin 23(4):486–487 (in Chinese)
29. Meng Z, Hu Q, Yang X (2002) A method of non-linear penalty functino for solving integer programming and mixed integer programming. Control Decis 17(3):310–314 (in Chinese)
30. Jian X, Zhihong Z (2006) An algorithm for solving the global optimization of nonlinear integer programming. J Shijiazhuang Univ 8(6):49–53 (in Chinese)
31. Luo W, Yu J, Huang J (2008) Bionic algorithm for solving nonlinear integer programming. Comput Eng Appl 44(7):57–59, 68 (in Chinese)
32. Wang H (2006) Study on the train scheme of passenger transport special line. China Academy of Railway Sciences (in Chinese)

Chapter 6
Train Timetable Rescheduling for High-Speed Railway Under Emergency Conditions

The high-speed railway train timetable rescheduling refers to the process of restoring the train orderly operation as soon as possible by rescheduling the train schedule when the train actual timetable deviates from the train scheduled timetable and the train disordered. Adjustment of high-speed railway operation plan under emergency conditions is a very complicated and important task. Establishing a complete set of relevant theories, methods, and strategies of transport organization will play a guiding role in the operation of high-speed railway trains under emergency conditions.

At present, there are a lot of literatures on train timetable rescheduling. Study on train timetable adjustment algorithm is mainly divided into two categories: the method based on operations research optimization theory and the method based on artificial intelligence.

The method based on operations research optimization theory mainly uses mathematical programming to study the optimization rescheduling algorithm. This method is relatively mature and has many research results. Its disadvantage is that in the case of a large number of stations and trains, the real-time performance is relatively poor, dispatch experience of dispatchers is difficult to effectively use. The hybrid 0–1 linear optimization model of single-line section train timetable rescheduling is constructed. This model comprehensively considers the adjustment of train operation plan and the utilization of station arrival and departure tracks. In view of the adjustment of train operation plan to the NPC problem, an effective large-scale system decomposition algorithm, namely dynamic area local optimization algorithm, is proposed. This algorithm optimizes local problems by constructing the train state space tree and applying the branch-and-bound method [1]. A linear programming model of train operation adjustment in single-line section is established to select the best crossing line and passing station as the target. The branch-and-bound method in operation research optimization theory is used to solve the model [2]. The adjustment of train operation is treated as a complex job shop problem. An optimization model is established with the target of minimizing total delay time and using the branch-and-bound method to solve the problem [3]. A planning model of train operation adjustment is

L. Jia et al., *High-Speed Railway Operation Under Emergent Conditions*,
Advances in High-speed Rail Technology,
https://doi.org/10.1007/978-3-662-63033-4_6

established with the target of minimizing the total delay time of trains and solved by the exhaustive method [4]. The planning model takes into account the impact of fuzzy passenger demand on the preparation of single-line railway train diagrams and using the branch-and-bound method to solve the model [5]. Using the analytic hierarchy process (AHP) principle in operation research to compare and analyze the main influencing factors in the train operation adjustment. Construct a judgment matrix and use the simple weighted average method to obtain the relative proportion of the three types of running trains. Finally, the existing cross-line trains are used as reference objects to obtain the weight of other passenger dedicated trains in the operation adjustment objective function [6]. A linear programming model of "constraint optional" for the operation adjustment problem of double-track railway trains is established and solved by dual algorithm [7]. With on the definition of train priority, node, alternative arc, and alternative equivalent parameters, a train operation adjustment model based on the alternative graph is established. Use the local search algorithm to solve the model [8]. A "Constrained Programming Model" for the train operation adjustment in the double-track section is established, with the goal of minimizing the total delay time. The branch-and-bound method is used to solve the problem [9]. Establish a high-speed train operation process model by using the convert maximal algebraic method. Different timetable rescheduling are formed by updating the state matrix in the train operation process model. All the adjustment schemes are input into the optimization model of train operation adjustment scheme which is based on order optimization method, and the train operation adjustment scheme with less total delay time is obtained through simulation calculation [10]. Taking directed arc and directed train as the main line, this paper constructs the mathematical model for the optimization of train operation diagram based on railway network. A decomposition algorithm is proposed to solve the sub-problems of the original problem and to gradually obtain the overall solution of the network train operation diagram [11]. A complex railway network train operation adjustment model is established with the optimization target of minimizing the train travel time. The order optimization theory and method are introduced to solve the problem [12]. A linear programming model of train operation adjustment in single-line section is established with the goal of maximizing the total benefits of train operation. Use Lagrange−heuristic method to solve the problem [13].

The method based on artificial intelligence is to use expert system, neural network, genetic algorithm, Petri net theory, and other methods to optimize traffic scheduling. These methods combine the rich empirical knowledge of the dispatcher and solve the large-scale combination problem in train scheduling optimization through the heuristic search mechanism of artificial intelligence. The research on the optimization of the scheduling algorithm has made great progress. In order to reduce train running time and improve train running ability, a two-layer model of train adjustment and optimization is established. The optimization model design of the first layer determines the appropriate number of stops and trains. The second layer is designed to reasonably arrange the inter-station time of train operation to achieve low energy consumption. This problem is solved by using genetic algorithm. Finally, the effectiveness of the double-layer train adjustment optimization model and algorithm is proved through the case study of the high-efficiency and energy-saving train running on part of the long-distance line [14]. In this paper, it considers the flexibility of train scheduling and uses genetic algorithm to solve the problem [15]. The train operation adjustment optimization model is solved by combining ant colony algorithm and maximal algebra. Author merges train operation adjustment problems into large-scale job shop scheduling problems. By means of the introduced train route matrix and the sequential matrix, the optimization model of train operation adjustment is constructed. Aiming at the characteristics of large model solution space and complex constraints, an ant colony algorithm is proposed to optimize the train layout sequence. Then use maximal algebra method to solve the method of scheduling train arrival and departure time [16]. Because the study of train operation adjustment by computer is far from satisfying the practical requirements, the theory of discrete event dynamic system is put forward. Author establishes a new method of train adjusting event-driven state space model. A hierarchical decision making and rolling optimization method are used to design a general algorithm suitable for train operation planning and adjustment of various railway sections in China [17]. Based on the analysis of the hierarchical rolling adjustment algorithm for the dynamic system model of discrete events, the fuzzy theory is combined. In order to improve the algorithm, the key links such as layer processing and conflict resolution are fuzzy treated [18]. According to the principle and basic method of train operation adjustment, the mathematical model of train operation adjustment in double-line section is designed. Using the three-group cooperative particle swarm algorithm can improve the time efficiency of the model. It can also play a good role in assisting decision making for railway dispatching adjustment [19]. According to the characteristics of train operation, reducing the total delay time of train group and improving the absorption level of adjacent trains to passenger flow are taken as performance indexes. The train operation adjustment problem is optimized based on the selective genetic algorithm through analyzing the optimization model of multi-track train operation control [20]. Combined with the current railway transport organization, the train operation adjustment problem is solved by searching the optimal layout sequence of trains to be adjusted in the feasible solution space. A hybrid 0–1 linear programming model with the optimization goal of minimum train delay rate is established. Tabu search algorithm is used to adjust the train layout sequence, and the validity and reliability of the model and algorithm are verified [21]. Use the train order entropy to describe the difference between the train's adjustment order and the running order in the basic diagram. Based on this, a train operation adjustment model based on the minimum train order entropy is established. When solving the model, the train set to be adjusted is divided into sub-train sets with different adjustment priorities. Then, according to the priority order, solve the optimal scheme in each sub-train set. Finally, merge the sub-train set operation schemes to obtain the final solution [22]. This paper proposes an artificial intelligent train operation adjustment algorithm based on train operation state derivation diagram. Decompose the train operation adjustment problem into a single train line calculation problem. Realize the representation of process knowledge in train operation adjustment through constructing the diagram of train operation state. And the efficiency of the algorithm is optimized by introducing occupation record table [23]. The heuristic algorithm is used to solve the train operation adjustment graph theory model. Its essence is an improvement of the greedy algorithm [24]. The adjustment model of two-line segment trains is established with the aim of minimizing the total delay time. Use the fuzzy Petri net technique to solve this model [25]. Design a parallel network algorithm under network conditions and multiple train routes. This method first proposes the hierarchical node representation of the railway network structure and the serial event representation of the train schedule. Merge and group train event sequences into stations and zones. This paper proposes an adjustment method of train operation plan, which unifies the distribution of train points, stratifies the train priority, and transfers the train event state in parallel trigger section [26]. The genetic algorithm is used to solve the optimal model of train operation adjustment [27]. Based on the knowledge expression of rough set theory and the extraction process of rough set decision rules, this paper proposes a corresponding evaluation system of driving command decision. In this way, the train operation plan can be adjusted [28]. Establish a hybrid integer programming model for the adjustment of double-track segment trains. The optimization goal is to minimize the total cost of train delays. Finally, design a genetic algorithm to solve this problem [29]. Establish the train operation adjustment model of two-line dispatch section. When solving the model, use the theory of large systems to classify the trains hierarchically, thus to decompose the original problem to be solved into several sub-problems. After this, design particle swarm optimization to solve the model [30]. Analyze train delay, adjustment strategy, and adjustment process, and construct a train operation adjustment model. The dual heuristic dynamic programming algorithm is used to solve the train operation adjustment model [31].

In this chapter, the adjustment of train operation plan is different from the adjustment of train operation plan under normal conditions. We have comprehensively considered goals such as the low total train delay time, the low number of severely delayed trains, and the stability of the operation plan. And trains have strong ability to deal with secondary interference when meet emergencies. This chapter focuses on the model of train operation adjustment and its solving algorithm. Then, establish a bi-level programming model of traffic scheduling optimization and train operation adjustment under emergency conditions. At the same time, considering the characteristics of different target orientations in the event of an emergency, two train operation adjustment models with different objects are established. Design algorithms to solve the model respectively. Then, in order to prove the feasibility of the two algorithms which is proposed in this chapter, the emergency synthesis scenario is set up respectively. Finally, verify the validity of the train operation adjustment under emergency conditions which is proposed in this chapter.

6.1 The Strategies of High-Speed Railway Under Emergency Conditions

High-speed railway has the characteristics of high speed and high density. Once the emergency happens, the loss is serious. Thus, high-speed railway has a relatively high demand for traffic safety system. Not only to ensure the high security of equipment such as lines, locomotives, traction power supplies, and communication signals, but also to take emergency measures in time to deal with sudden natural disasters (such as strong winds, heavy rain, heavy snow, earthquakes, and other disasters). It is the premise to ensure the efficient and accurate implementation of emergency measures to formulate comprehensive high-speed railway operation rules under emergency conditions. The operation rules of high-speed railway need to be formulated according to the nature and level of emergency. Usually, it includes rules like changing operating modes, implementing early warning, speed-limiting operations, or suspension of traffic. The rules of high-speed railway under emergency conditions in China have their own characteristics. In this section, according to the railway technical management regulations (part of high-speed railway), the operation rules of high-speed railway in China are listed under such emergency conditions as high wind, heavy rain, snow and ice, foreign matter invasion limit, and earthquake. Secondly, according to the special operating environment of Lanzhou-Xinjiang high-speed railway line, the paper analyzes the operation rules of Lanzhou-Xinjiang high-speed railway line in windy weather. Because of the different natural environments, geographical conditions, and operating methods, there are also large differences in the rules of high-speed railways under emergency conditions in foreign countries. Taking Shinkansen in Japan as an example, this paper analyzes the train running strategies of Shinkansen in Japan under such emergency conditions as strong wind, heavy rain and flood, blizzard, earthquake, and high temperature [32]. It is the key to improve the safety of high-speed railway transportation in China to further improve the operation rules of high-speed railway under emergency conditions.

6.1.1 The Strategies of High-Speed Railway Under Emergency Conditions

In order to avoid the impact of emergencies on China's railway operation safety, according to the types of emergencies, China's railway technical management regulations (part of high-speed railway) have formulated operation rules in disaster environments. When occurrence of the emergency condition of high-speed railway, the operation of railway should be carried out in strict accordance with its requirements.

1. Strong wind

Strong wind is one of the major meteorological disasters that affects the safety of railway transportation. It not only affects the stability of train operation, but also may

Table 6.1 Train running strategies under strong wind in high-speed railway sections in China

Surrounding wind speed (m/s)	The train running strategies
Under 15	Normal speed
15–20	Running speed no more than 300 km/h
20–25	Running speed no more than 200 km/h
25–30	Running speed no more than 120 km/h
More than 30	High-speed trains cannot enter the wind zone

Table 6.2 Train running strategies in windy weather at high-speed railway stations in China

Surrounding wind speed (m/s)	The running strategies
Under 15	Running speed no more than 80 km/h
More than 15	Running speed no more than 45 km/h

damage railway equipment, lead to the breakdown of the railway network, and even cause casualties.

The train running strategies of China's high-speed railways in windy weather are shown in Table 6.1.

When the Central Line of the track is 1750 mm away from the edge of the platform, the main line and the arrival and departure tracks are used to handle the passing of high-speed trains. The operation strategies are shown in Table 6.2.

Under special circumstances, as the Lanzhou-Xinjiang Passenger Dedicated Line is a high-speed railway located in China's plateau, high altitude, and alpine regions, the traffic rules of the Lanzhou-Xinjiang Passenger Dedicated Line have their own characteristics under emergency conditions. In Lanzhou-Xinjiang Passenger Dedicated Line which has no windshield section, trains need to follow the railway technical management regulations (high-speed railway section), when meeting the windy weather. In windshield section and the section from LiaodunBei-HongcengNan, train has to limit its speed in windy weather. As shown in Table 6.3.

2. Heavy rain

Heavy rain often causes traffic disruption and endangers railway transport safety in severe cases.

The train running strategies of China's high-speed railway under heavy rain conditions are shown in Table 6.4.

3. Snow and ice disasters

In the case of snow and ice disaster, the safety and stability of railway traction power supply equipment, catenary and the normal switch will be affected. The braking

Table. 6.3 Train running strategies in the Lanzhou-Xinjiang passenger dedicated line under windy weather

Line	Surrounding wind speed/(m/s)	No windshield	Windshield	Section from LiaodunBei-HongcengNan
Lanzhou-Xinjiang passenger dedicated line	Under 20	Follow the railway technical management regulations (high-speed railway section)	Normal speed	Normal speed
	20–25		Running speed no more than 200 km/h	Running speed no more than 160 km/h
	25–30		Running speed no more than 160 km/h	Running speed no more than 120 km/h
	30–35		Running speed no more than 80 km/h	Running speed no more than 80 km/h
	More than 35		High-speed trains cannot enter the wind zone	EMU trains cannot enter the wind zone

Table 6.4 Train running strategies of high-speed railway in China under heavy rain conditions

Rainfall of 1 h/(mm)	The train running strategies	Remove speed limitation
45 ~ 60	Train speed limit 120 km / h	When the one hour rainfall drops to 20 mm or less and lasts for more than 30 min, the speed limit can be gradually lifted
More than 60	Train speed limit 45 km/h	

effect is not good, which has a negative impact on the transportation efficiency and safety.

The train running strategies of China's high-speed railway under snow and ice weather are shown in Table 6.5.

4. Limitation of foreign objects incursion

The train dispatcher should immediately notify the train driver that the train has entered the alarm location and have not passed the alarm location within the section to stop immediately after receiving the warning message of the foreign objects incursion limitation system. No more trains run into this section. Meanwhile, report to the Dispatch officer on duty (associated officer on duty). Dispatch officer on duty (associated officer on duty) should immediately notify the equipment management unit to rush to the site for inspection and processing.

Before the inspector of the equipment management unit arrives at the alarm point, the train dispatcher checks the situation through the video monitoring system. If there is any abnormality or it cannot be confirmed, the operation of the train can only be

Table 6.5 Train running strategies of China's high-speed railway under snow and ice weather

Snow location	Snow and ice conditions	Track type	The train running strategies	Remark
Train section	Medium snow, snow cover	Ballastless tracks	Speed limit 250 km/h and below	When snow thickness of sleeper in ballastless track is more than 100 mm, the speed limit is 200 km/h and below
		Ballast tracks	Speed limit 200 km/h and below	
	Heavy snow	Ballastless tracks	Speed limit 200 km/h and below	
		Ballast tracks	Speed limit 160 km/h and below	When snow thickness of sleeper in ballast track is more than 50 mm, the speed limit is 160 km/h and below
EMU bogie	Ice (When train speed limit is required)	Ballastless tracks	Speed limit 250 km/h and below	
		Ballast tracks	Speed limit 200 km/h and below	
Catenary wire	Ice (When the pantograph cannot get the current smoothly)	Ballastless and ballast tracks	Speed limit 160 km/h and below	

Note The definition of medium snow, heavy snow and blizzard is subject to the announcement or observation of the meteorological department

organized after it has been checked and handled by the equipment management unit. If there is no abnormality, the dispatcher orally notifies the driver to resume operation one by one. The driver should drive the train with the speed of stopping at an obstacle. EMU trains do not exceed 40 km / h, and other trains do not exceed 20 km/h.

5. Earthquake

When the train dispatcher receives the earthquake monitoring alarm message of the earthquake monitoring subsystem or receives the on-site earthquake report, it shall immediately turn off the relevant signals (during station control, inform station watchman to turn off the signal). At the same time, notify the relevant trains to stop. Train drivers organize train attendants to take emergency measures according to the actual situation. The train dispatcher shall immediately report to the duty director (associated officer on duty) of the dispatching office and notify the equipment management unit of the public works, electricity, power supply, communication, and house construction to inspect. After the equipment management unit checks and processes, organize the driving according to the driving restrictions registered by the equipment management unit.

6. Others

In case of fire, bad weather, it is difficult to identify the signal. If train failure and other conditions happen, it should be based on the actual situation of the site to stop.

6.1.2 Train Running Strategies of Japanese Shinkansen High-Speed Railway Under Emergency Conditions

Japan is a disaster-prone country. Typhoons, rainstorms, heavy snow, earthquakes, and other natural disasters are frequent. However, since the Shinkansen operation in October 1964, no passenger has been killed or injured. The minimum train section is 5 min. Its railway traffic regulations are relatively complete.

1. Strong wind

Under extreme conditions, the natural wind direction, wind speed, vehicle weight, shape, track subgrade structure, and train operating speed all affect the overturning extreme wind speed of the train. These endanger the safety of trains.

Strong winds can also damage catenary.

Table 6.6 shows the train running strategies of high-speed trains in Tohoku, Joetsu, and Nagano Shinkansen under strong wind conditions.

Table 6.6 Train running strategies of high-speed trains in Tohoku, Joetsu, and Nagano Shinkansen under strong wind conditions

Wind speed/(m/s)	The general section	Set a certain standard wind wall section
20–25	The train speed limit is 160 km/h	No speed limit
25–30	The train speed limit is 70 km/h, and the train can be stopped according to the specific situation	The train speed limit is 160 km/h
30–35	Out of service	The train speed limit is 70 km/h, and the train can be stopped according to the specific situation
More than 35	Out of service	Out of service

Note Wind speed refers to instantaneous wind speed

2. Heavy Rains and Floods

Heavy rains and floods have caused disasters such as flooded railway lines, unstable roadbeds, debris flows, and landslides.

Table 6.7 shows the train running strategies of the Tokaido Shinkansen high-speed train under heavy rain and flood conditions.

3. Blizzard

Japan is one of the snowy countries, and blizzards are more harmful to railway operation safety. For example, snowdrifts formed by blizzards can affect driving safety when they are too high. Snow will also affect the normal operation of turnouts, which may cause damage to power supply equipment.

The train running strategies of JR East Japan's high-speed trains under blizzard conditions are shown in Table 6.8.

4. Earthquake

The earthquake has a severe effect on the railway subgrade, especially the bridges and tunnels in the quake zone.

(1) Train running strategies under earthquake conditions

Table 6.9 shows the train running strategies of the Sanyo Shinkansen high-speed train in Japan under earthquake conditions.

(2) Train running strategies after stopping under earthquake conditions.

5. High Temperature

The high temperature reduces the safety of the long seamless rail, and it also changes in the rail lock temperature or decreases track bed resistance.

There are two types of Japanese Shinkansen train running strategies at high temperatures: One is the general section and the other is the ballastless bridge. According to the measured transverse resistance of road bed, each kind of road is controlled by rail temperature.

(1) General section

Table 6.11 shows the operating rules of high-speed trains on the Shinkansen in Japan under high temperature conditions.

(2) Ballastless bridge

Table 6.12 shows the operating rules of the high-speed trains on the Japan Shinkansen ballastless bridge at high temperatures.

6. Others

Fires, train failures, and catenary failures should be stopped in accordance with the actual situation at the site.

Table 6.7 Train running strategies of the Tokaido Shinkansen high-speed train under heavy rain and flood conditions

Operation limitation			24 h cumulative continuous rainfall/mm	1 h Rainfall/mm	Continuous rainfall + 1 h rainfall/mm/mm	Rainfall report	Note
Alert	Type 3		100–110	25	–	1 time/h	
	Type 2		120–130	30	110 + 20	2 times/h	Inspection every 3–4 h
	Type 1		140	35	120 + 25		Inspection every 2 h
Speed limitation	170 km/h	Area B	–	40	140 + 30 or 160 + 2	2 times/h	Real-time ground inspection, add appropriate inspection
		Area A	–	45	150 + 30 or 180 + 2		
	70 km/h	Area B	–	45	152 + 30 or 180 + 2		
Stop running	General section		–	50	150 + 40	6 times/h	During continuous rainfall, emergency inspections in Area B are needed. If there is a concentrated rainstorm, pay attention to the emergency inspection of the location
	Viaduct slag-free bridge		–	70	150 + 60		

Note (1) The third kind of alert refers to the regular inspection and alert in a predetermined section and a designated place to pay attention to during equipment maintenance
(2) The second type of alert refers to periodic inspections and vigilance around geotechnical structures and tunnel openings other than the third type of alert objects
(3) The first type of alert refers to periodic inspections and inspections at pre-designated sections other than the second type of objects or at places deemed likely to be affected by disasters
(4) Caution: The rainfall reaches the promulgated standard, there is almost no possibility of disaster, and some precursors of disaster can be predicted, and vigilance is needed
(5) Speed limit operation: Rainfall has reached the promulgated standard. Experience has shown that there are no disasters, no abnormal rainfall, and the possibility of minor disasters. It is necessary to consider speed limit operation
(6) Stop operation: The rainfall has reached the promulgated standard, and there is a possibility of disaster, and the operation needs to be stopped
(7) Area B: the inspection section with continuous rainfall of more than 150 mm and the rainfall of 40 mm; other areas A
(8) The inspection section when the rainfall reaches 50 mm is called the "point of attention"

Table 6.8 Train running strategies of JR East Japan's high-speed trains under blizzard conditions

Snow depth (on rail surface)/cm	Train running strategies
9–17	Train speed limit 245 km/h
17–19	Train speed limit 210 km/h
19–22	Train speed limit 160 km/h
22–30	Train speed limit 110 km/h
30 以上	Traffic termination

6.2 Train Dispatching Programming Model Under Emergency Conditions

According to the description of the organization problem of train operation under emergency conditions in chap. 2. This chapter studies the combined problem of traffic scheduling optimization and train operation adjustment. This section analyzes the factors influencing the implementation of train rescheduling. Consider rescheduling strategies such as train rerouting, train reconnections, and train canceling. The optimization model of train rescheduling under emergency condition is established. Then analyze the uncertain factors in the model. According to the different preferences of decision-makers, the expected value model and the opportunity constraint model of train rescheduling optimization are established.

6.2.1 Uncertain Two-Layer Programming

In the real-world problems, there are a large number of multi-layer or two-layer optimization problems. Multi-layer programming provides a mathematical modeling basis for the study of hierarchical decision-making problems. In general, each layer in a multi-layer program has its own decision variables and optimization objectives. Its essence is that upper-level decision-makers influence the decision-making behavior of lower-level decision-makers through decision making. But it is not directly involved in the decisions of the lower levels. Lower decision-makers use upper decisions as constraints. And they have full authority to freely decide how to optimize their own goals. At the same time, the results are fed back to the upper layer. The feedback of decision result between different layers is an important mechanism for the realization of multi-layer programming. If a decision problem has only two layers, it becomes a two-layer program. This is a widely used form of multi-level planning. With the development of uncertainty theory, in multi-level programming, the constraints and variables and parameters in the objective function of each layer are usually accompanied by inherent random or fuzzy uncertainty. The transmission and feedback information between layers may also carry a certain degree of uncertainty. So uncertain multi-level planning came into being. According to different measures

Table 6.9 Train running strategies of the Sanyo Shinkansen high-speed train in Japan under earthquake conditions

Sensor maximum value /Gal	Judgment of seismic intensity	Operating rules		Emergency inspection		Speed up		
		Stop	Speed limitation	Ground inspection	Added inspection	30 km/h	70 km/h	70 km/h
40–80	Under 3		The speed limit is 70 km/h within the monitoring range of the sensor. Before the magnitude of the earthquake was clear. The speed limit was 30 km/h. The special case has a speed limit of 30 km/h		Nothing; Special case is the same as the “speed limit” section		Electrical Equipment	Equipment
	More than 4		The speed limit is 30 km/h within the monitoring range of the sensor. If there are equipment and electrical personnel to increase the speed limit of 70 km / h. The special case has a speed limit of 30 km/h		Special case is the same as the “speed limit” section	Electrical Equipment	Electrical Equipment	Equipment

(continued)

Table 6.9 (continued)

Sensor maximum value /Gal	Judgment of seismic intensity	Operating rules		Emergency inspection		Speed up		
		Stop	Speed limitation	Ground inspection	Added inspection	30 km/h	70 km/h	70 km/h
80–120	Under 3		Same as above		Same as above	Electrical Equipment	Electrical Equipment	Equipment

(continued)

Table 6.9 (continued)

Sensor maximum value /Gal	Judgment of seismic intensity	Operating rules		Emergency inspection		Speed up		
		Stop	Speed limitation	Ground inspection	Added inspection	30 km/h	70 km/h	70 km/h
	4	Within the scope of shock sensor supervision	The speed limit is 30 km/h within the monitoring range of the sensor. If there are electrical personnel to increase the speed limit of 70 km / h. The special case has a speed limit of 30 km/h	Specific locations in parking zones	Same as above	Electrical equipment	Electrical equipment	Equipment
	More than 5	Same as above	Same as above	Parking section	Same as above	Electrical Equipment	Electrical Equipment	Equipment

(continued)

Table 6.9 (continued)

Sensor maximum value /Gal	Judgment of seismic intensity	Operating rules		Emergency inspection		Speed up		
		Stop	Speed limitation	Ground inspection	Added inspection	30 km/h	70 km/h	70 km/h
		Same as above	Same as above	Same as above	Same as above	Electrical Equipment	Electrical Equipment	Equipment

Note (1) "Electrical equipment" means that equipment and electrical personnel are available to increase speed
(2) "Special cases" refer to one of the following situations:
① an earthquake occurs when the continuous rainfall reaches 120 mm or more;
② an earthquake occurs after sunset (including dense fog), except for the "*" line;
③ the temperature rises, the orbit An earthquake occurred when the temperature reached above 50 °C.

Table 6.10 Train running strategies of the Tohoku, Joetsu, and Nagano Shinkansen high-speed trains after the earthquake stops

Sensor maximum value/Gal	Operating rules		Train running strategies	
	Stop	Speed limitation	Walking inspection	Added inspection
Under 80	Restoring operation according to dispatchers		Nothing	Nothing
80–120		The speed limit is 30 km/h within 12 km of both ends of the sensor points exceeding 80 Gal. If there are equipment and electrical personnel to increase the speed limit of 70 km/h. However, the speed limit is 30 km/h in the following cases: (1) An earthquake occurs when the rail temperature is above 55 °C in a ballasted track; (2) Specially specified sections in the event of an earthquake after sunset (including dense fog and heavy snow) (except for ground inspections)	Nothing	Carry out inspections on speed limit zones

(continued)

of uncertain variables, Liu established the expected value bi-layer programming model and the opportunity constraint bi-layer programming model [33].

1. Classical Two-Layer Programming Model

The universal mathematical expression of the two-layer programming model is as follows.

$$\max_{x} F(x, y)$$

Table 6.10 (continued)

Sensor maximum value/Gal	Operating rules		Train running strategies	
	Stop	Speed limitation	Walking inspection	Added inspection
More than 120	Above 120 Gal, between the sensor and the adjacent sensor	The speed limit between the ground sensor and the neighboring sensor over 120 Gal is 70 km/h after the ground inspection, but the speed limit is 30 km/h in the following cases: An earthquake occurs when the rail temperature of the ballasted track is above 55 °C	Walking inspections of parking areas	Carry out inspections on speed limit zones

Table 6.11 Operating rules of high-speed trains on the Shinkansen in Japan under high temperature conditions

Track temperature/°C	Measurement of transverse resistance of crossing bed/(N/each pillow)	Train running strategies
More than 64	Under 12,748	Stop running
60–64	Under 8826	Stop running
	Under 8826	Drive slowly at 70 km/h
58–60	Under 8826	Determine whether to slow down at 70 km/h based on special inspections
	Under 8826	Temperature observation
53–58	Under 8826	Special inspection
	Under 8826	Temperature observation
48–53	Under 8826	Temperature observation
45–48	Under 8826	Special inspection in area A

$$\text{s.t. } G(x, y) \leq 0 \tag{6.1}$$

$y = y(x)$ is the solution of the following program.

$$\max_{x} f(x, y)$$
$$\text{s.t. } g(x, y) \leq 0 \tag{6.2}$$

Table 6.12 High-speed train running strategies on ballastless bridge under high temperature in Shinkansen

Track temperature/°C	Measurement of transverse resistance of crossing bed/(N/each pillow)	Train running strategies
More than 64	Under 12,748	Stop running
60–64	Under 10,787	Stop running
	Under 10,787	Drive slowly at 70 km/h
58–60	Under 10,787	Determine whether to slow down at 70 km/h based on special inspections
	Under 10,787	Temperature observation
53–58	Under 10,787	Special inspection
	Under 10,787	Temperature observation
48–53	Under 10,787	Temperature observation
45–48	Under 10,787	Special inspection in area B

Note (1) special inspection means that when the rail temperature reaches the specified level, the workers from the work area shall send personnel to the possible dangerous places (such as the narrow road shoulder, the transition section of railway and bridge, etc.), enter the first protective grid, and inspect the direction and level of the line outside the adjacent protective grid
(2) A, B section means that track segments with track joints where the longitudinal stability factors are different

The two-level planning model is composed of the upper-level planning model (6.1) and the lower-level planning model (6.2). The solution x of the upper programming problem is used as the input of the lower programming. It provides constraints for lower-level decisions. The solution y of the lower-level planning problem feed back to the upper-level planning. It is a basis for upper-level optimization.

2. Expected Two-Layer Programming Model

If there are some uncertain parameters in the upper or lower planning, the above classic model will become as follows.

$$\begin{aligned} &\max_x F(x, y, \xi) \\ &\text{s.t. } G(x, y, \xi) \le 0 \end{aligned} \tag{6.3}$$

$y = y(x)$ is the solution of the following program.

$$\begin{aligned} &\max_x f(x, y, \xi) \\ &\text{s.t. } g(x, y, \xi) \le 0 \end{aligned} \tag{6.4}$$

ξ is a fuzzy, random, or multi-dimensional uncertain variable.

If the feasible solution of the above plan is defined by expectation constraints, it means that the objective function and constraints containing uncertain variables in the original plan are treated as expectations. Then the expected value two-layer plan can be obtained. Its form is as follows.

$$\begin{aligned} &\max_x E[F(x, y, \xi)] \\ &\text{s.t. } E[G(x, y, \xi) \le 0] \end{aligned} \tag{6.5}$$

$y = y(x)$ is the solution of the following program.

$$\begin{aligned} &\max_x E[f(x, y, \xi)] \\ &\text{s.t. } E\big[g(x, y, \xi) \le 0\big] \end{aligned} \tag{6.6}$$

3. Opportunity Constrained Two-Layer Programming Model

If decision-makers want to maximize the optimistic returns under the opportunity constraint, they can establish an opportunity constrained two-layer programming model with optimistic value.

$$\begin{aligned} &\max_x \max_{\overline{F}} \overline{F} \\ &\text{s.t.} \\ &\quad \text{Ch}\big[F(x, y, \xi) \ge \overline{F}\big] \ge \beta \\ &\quad \text{Ch}[G(x, y, \xi) \le 0] \ge \alpha \end{aligned} \tag{6.7}$$

$y = y(x)$ is the solution of the following programs.

$$\begin{aligned} &\max_x \max_{\overline{f}} \overline{f} \\ &\text{s.t..} \\ &\quad \text{Ch}\big[f(x, y, \xi) \ge \overline{f}\big] \ge \eta \\ &\quad \text{Ch}[g(x, y, \xi) \le 0] \ge \delta \end{aligned} \tag{6.8}$$

$\alpha, \beta, \eta, \delta$ are the confidence levels given by the decision-maker.

If decision-makers want to maximize the pessimistic returns under the opportunity constraint, they can establish an opportunity constrained two-layer programming model with pessimistic value:

$$\begin{aligned} &\max_x \min_{\overline{F}} \overline{F} \\ &\text{s.t.} \\ &\quad \text{Ch}\big[F(x, y, \xi) \le \overline{F}\big] \ge \beta \end{aligned}$$

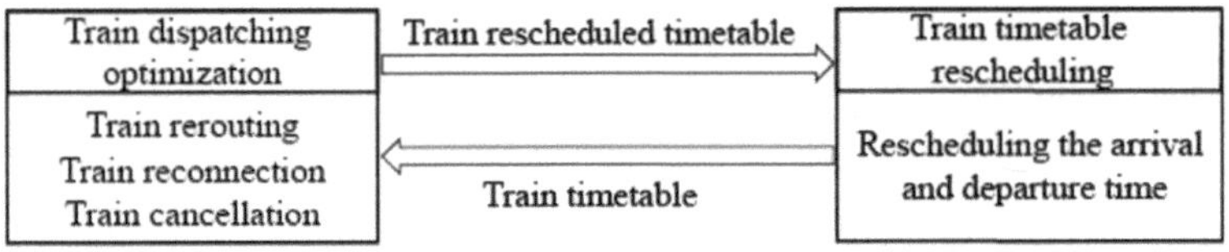

Fig. 6.1 Two-layer planning of train organization under emergency conditions

$$\mathrm{Ch}[G(x, y, \xi) \le 0] \ge \alpha \tag{6.9}$$

$y = y(x)$ is the solution of the following program:

$$\begin{aligned} &\max_{x} \max_{\overline{f}} \overline{f} \\ &\text{s.t.} \\ &\quad \mathrm{Ch}\left[f(x, y, \xi) \le \overline{f}\right] \ge \eta \\ &\quad \mathrm{Ch}[g(x, y, \xi) \le 0] \ge \delta \end{aligned} \tag{6.10}$$

This uncertain two-layer programming model provides a modeling basis for two-layer programming problems in uncertain environments. However, in the real environment, due to the different system structure parameters and the environment, it is necessary to improve the characteristics of the problem. The two-layer planning in this chapter includes two levels which are train dispatching optimization and train timetable rescheduling. The structure is shown in Fig. 6.1. Uncertain variables are involved in the planning of the upper and lower levels. This chapter will make a detailed analysis of this two-layer programming.

(1) Train dispatching optimization.

Train dispatching optimization is the upper-layer planning in the two-layer planning. It refers to the emergency conditions, the affected line of the train whether to take train detour, train reconnection, train stop, and other scheduling strategies to optimize ①. This reduces dependence on the capacity of the affected lines. It also takes full advantage of the capabilities of other lines on the network to complete the process of passenger transportation as much as possible. Its optimization results provide the information such as the running route, number of trains, and stop patterns of the trains which are needed for the lower-layer planning. At the same time, it obtains the optimization results of the lower-layer planning as the input of the train dispatching optimization.

(2) Train timetable rescheduling.

Train timetable rescheduling is the lower-layer planning in the two-layer planning. It refers to the process of reducing train delays and resuming driving plan by adopting strategies such as reducing train section running time, train dwell time, and changing train crossing mode after giving the optimization results of the upper layer under the emergency condition. And it provides the adjusted train working diagram for the upper-layer planning.

6.2.2 Analysis of Influencing Factors

1. Analysis of Influencing Factors of Train Detour

After emergency, how to make full use of the capacity of the railway network to make up for the loss of capacity of the affected lines is the key to improve passenger service levels. Therefore, detouring to other routes (train detours) is one of the frequent scheduling strategies used by dispatchers. The main influencing factors are as follows.

(1) Train detour is based on railway network conditions. The impact range and length of time of different types and levels of emergencies on the basic railway network are very different. For example, line failure or train failure generally affects only one line, and the repair time is faster. However, if a large-scale natural disaster occurs, such as a blizzard or an earthquake, it often affects several lines in a large area. After emergency, the change of the railway network structure is the basis of implementing the train detour strategy.

(2) The choice of detours. According to the Chap. 4, after emergencies, multiple routes will be generated according to the actual situation of the railway network. How to choose an appropriate detour route for each train is the key to optimize traffic scheduling.

(3) Train distribution on detour routes. The spatial and temporal distribution of trains on detours determines the remaining capacity on the track and the time for detours to get on and off the train. It influences the selection of detours by trains at a micro level.

(4) The original stop pattern for detour trains. After train has detoured to other routes, it still needs to provide services for passengers who get on and off the train according to the train level and stop mode. Therefore, the station should try to include the established station or be in the same city as the established station.

(5) The original train operation plan of the detour train. The final adjustment goal of trains detoured to other routes is still to try to run according to the train diagram. The impact of train detour on passenger travel should be minimized.

(6) Ability time density of train detours. According to Chap. 4, ability time density is a measure of the reliability of train operations on detours. It is more meaningful than simply considering the length of the detour route.

2. Analysis of Influencing Factors of Train Reconnection

Train reconnection refers to the change from 8-group trains scheduled to 16-group trains due to natural disasters, railway accidents, and other emergencies or major holidays, when the line's capacity decreases or passenger flow significantly increases. Or to combine two or more trains into a scheduling strategy. This section mainly considers merging the two planned trains to operate under emergency conditions in order to reduce the use of line capacity. The influencing factors of train reconnection are as follows.

(1) Static attributes of the train. The static attributes of the train include the type of train, the number of trains, the length of the train, and so on. In general, only two trains in the same type of short group can be reconnected. This is the basic attribute that determines whether two trains can be reconnected.

(2) Train operation plan. In general, two short trains with similar stopping plans and similar arrival and departure times can be reconnected. While decreasing capacity utilization, it reduces train delay. Therefore, the scheduled timetable is also the basis of the train reconnection strategy.

(3) The arrival and departure tracks of the station. After train is reconnected, the length of the train should be less than the length of the arrival and departure tracks which are in the station. This ensures that the throat and turnout capacity will not be continuously occupied when the train stops, and it is convenient for passengers to get on and off the train normally.

(4) Line carrying capacity. Under emergency conditions, both the section carrying capacity and the station carrying capacity may suffer different degrees of losses. This is the main reason for adopting the train reconnection strategy.

(5) Station marshaling line. The marshaling line is the arrival and departure tracks which is used for reconnection and disconnection of trains. When the reconnection and marshaling operations of the station are frequent, the marshaling line can be used to handle this operation. At other times, marshaling line generally uses as the arrival and departure tracks [34]. The difference between marshaling line and arrival and departure tracks is that in addition to the length requirements, it also needs a signal device to ensure the safe of train reconnection and marshaled.

3. Analysis of Influencing Factors of Train Canceling

Train canceling is the rescheduling strategy of the scheduled timetable, when occurring emergencies such as natural disasters or railway accidents, which have resulted in a significant reduction in passenger flow or line passing capacity. This section mainly considers the canceled strategy for trains when the line passing capacity is reduced. The influencing factors of train canceling are as follows.

(1) Train priority. Train canceling is the worst influence of the three traffic scheduling strategies. In the same situation, the stop of long-distance, high-class trains leads to bring more losses. Therefore, train priority is an important factor to determine whether a train is out of service.

(2) Train operation plan. Similar to the analysis of influencing factors of train reconnection, how to determine whether a train is out of service and make the adjusted train run according to the train diagram as far as possible is the key to train stop strategy optimization.

(3) Line carrying capacity. Similar to the analysis of influencing factors of train reconnection, the loss of line carrying capacity is also the basis of train out-of-service strategy optimization.

(4) Passengers. Unlike train detours and train reconnections, train stop can severely affect passengers. It makes passengers impossible to reach their destination, and affected passengers have to transfer to other modes of transportation, or to other trains. Therefore, the convenience of passenger transfer should be considered when canceling the train.

6.2.3 Construction of Train Dispatching Optimization Model Under Emergency Condition

1. Relevant parameters

Train dispatching optimization problem is based on railway network. To maintain the consistency of this book, the network model in this section is the same as defined in the previous chapters.

(1) Set of detour routes

Using the route search model based on capacity time density in Chap. 4, we can obtain multiple alternate routes. Written as Route $= \{\text{rte}_i | i = 1, 2, \ldots, N_{\text{Route}}\}$, *Route* is a set of alternate routes; N_{Route} is the size of Route, which means the number of feasible detours in Route. rte_i is route i sorted by capacity time density, written as $\text{rte}_i = \{v_s, v_1, \ldots, v_t\}$, which represents a route from v_s to v_t. It should be noted that multiple routes in the Route often occupy some nodes or arcs in the railway network at the same time. The capacity of these nodes and arcs will be allocated to multiple routes accordingly. In Fig. 6.2, there are three routes for the starting points v_1^1 and v_6^1:

(1) $\text{rte}_1 = \{v_1^1, v_2^1, v_3^1, v_4^1, v_5^1, v_6^1\}$ This is the original route on the line. As the arc section (v_3^1, v_4^1) is damaged, a detour needs to be found.

(2) $\text{rte}_2 = \{v_1^1, v_2^1, v_2^2, v_3^2, v_4^2, v_5^2, v_5^1, v_6^1\}$

(3) $\text{rte}_3 = \{v_1^1, v_1^2, v_2^2, v_3^2, v_4^2, v_5^2, v_5^1, v_6^1\}$

It can be seen that the arc sections (v_2^2, v_3^2), (v_3^2, v_4^2), (v_4^2, v_5^2) are occupied by the routes rte_2 and rte_3 at the same time. The arc section (v_5^1, v_6^1) is occupied by three routes at the same time.

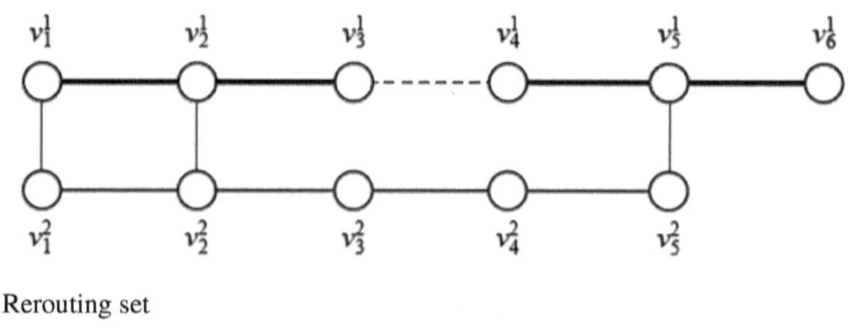

Fig. 6.2 Rerouting set

(2) Penalty

All of the three scheduling strategies have an impact on passenger travel. Although train detour changes the route of passengers and increases the travel time of passengers, it can effectively utilize the capacity of the network to complete the passenger transportation service. Therefore, train detour is a common dispatching strategy under emergency conditions. Compared with the other two dispatching strategies, the penalty cost is smaller.

Although the train reconnection will not change the route of the train, when the two trains reconnect at the middle station, the front train needs to wait for the arrival of the rear train, which will also increase the travel time of passengers. In addition, train reconnection involves complex train entry and exit dispatching, which requires the cooperation of communication signal systems and traction power supply systems. It is riskier than train detour strategy.

Train canceling cannot carry passengers. In addition to the economic loss caused by ticket refunds, it also reduces the service satisfaction and social benefits of passengers. Therefore, the penalty cost of train canceling dispatching strategy is high.

(3) Other relevant parameters

Other relevant parameters are as follows:

$\mathrm{Inid}_{i,j}$ Scheduled departure time of train i at station j;
$\mathrm{Inia}_{i,j}$ Scheduled arrival time of train i at station j;
$x_{i,j}$ The actual departure time of train i at station j, which is the calculation result of the lower-layer planning;
$y_{i,j}$ The actual arrival time of train i at station j, which is the calculation result of the lower-layer planning;
c_i^{delay} Train i unit delay cost;
c_i^{run} Train i unit mileage operating cost;
g_i Train i marshaling type; only when the train is a short formation (eight formations), it is possible to reconnect with another short formation train;
c_i^{cancel} Penalty costs arising from the suspension of train i;
c_i^{reroute} Penalty costs for trains i detours to other routes;
$c_{i,j}^{\mathrm{merge}}$ Penalty costs for reconnection of train i and train j;
ΔC_e In time period T, the remaining passing capacity of arc e on the railway network;
δ The number of trains that need to be scheduled due to the capacity loss of this line;
Trn The set of affected trains in the time period T affected by the emergency:
Stn The set of affected stations in the time period T affected by the emergency:
len_e The length of the arc e;

2. Decision variables

The decision variables of the train dispatching optimization model under emergency conditions are as follows.

(1) Train rerouting variables.

$$\mathrm{rrt}_{i,\mathrm{rte}} = \begin{cases} 1 & \text{if train } i \text{ detours to the path rte} \\ 0 & \text{others} \end{cases}$$

(2) Train reconnection variables.

$$\mathrm{meg}_{i,j} = \begin{cases} 1 \text{ if train } i \text{ is reconnected by train } j \\ 0 \text{ others} \end{cases}$$

If train *i* is reconnected by train *j*, it can be considered that these two trains logically become one train and have the same train number as train *j*.

(3) train canceling variables.

$$\mathrm{cal}_i = \begin{cases} 1 \text{ if train } i \text{ is out of service} \\ 0 \text{ others} \end{cases}$$

3. Objective function

The objective function of the traffic dispatching optimization model is mainly concerned with the punishment of train detour, train reconnection, and train canceling.

(1) Penalty of train rerouting.

$$Z_{\mathrm{rrt}}^{\mathrm{p}} = \min \sum_{i \in \mathrm{Trn}} c_i^{\mathrm{reroute}} \mathrm{rrt}_{i,\mathrm{rte}} \tag{6.11}$$

$$Z_{\mathrm{rrt}}^{\mathrm{r}} = \min \sum_{i \in \mathrm{Trn}} c_i^{\mathrm{run}} \mathrm{rrt}_{i,\mathrm{rte}} \sum_{e \in \mathrm{rte}} \mathrm{len}_e \tag{6.12}$$

The train detour punishment is made up of two parts. $Z_{\mathrm{rrt}}^{\mathrm{p}}$ represents the train detour penalty cost because of the impact of social benefits. $Z_{\mathrm{rrt}}^{\mathrm{r}}$ represents the train operating cost on the detour route. The latter is designed to allow trains to choose relatively short detours.

(2) Penalty of train reconnection.

$$Z_{\mathrm{meg}}^{\mathrm{p}} = \min \sum_{i \in \mathrm{Trn}} c_{i,j}^{\mathrm{merge}} \sum_{j \in \mathrm{Trn}} \mathrm{meg}_{i,j} \tag{6.13}$$

$$Z_{\mathrm{meg}}^{\mathrm{t}} = \min \sum_{i \in \mathrm{Trn}} \sum_{j \in \mathrm{Trn}} \sum_{v \in \mathrm{Stn}} \left(\left| \mathrm{Inid}_{i,v} - \mathrm{Inid}_{j,v} \right| + \left| \mathrm{Inia}_{i,v} - \mathrm{Inia}_{j,v} \right| \right) \mathrm{meg}_{i,j} \tag{6.14}$$

The train reconnection punishment is made up of two parts. $Z_{\text{meg}}^{\text{p}}$ represents the train reconnection penalty cost. $Z_{\text{meg}}^{\text{t}}$ represents the difference in arrival and departure time between the two reconnected trains. The purpose of this target selection is to choose two trains with small differences in arrival and departure times to reconnect in order to reduce the impact on passenger travel time. The loss of other time because of train reconnection is controlled by the delay time in the lower-layer planning.

(3) Punishment of train canceling.

$$Z_{\text{cal}}^{\text{p}} = \min \sum_{i \in \text{Trn}} c_i^{\text{cancel}} \text{cal}_i \tag{6.15}$$

Only the social benefits penalty of train canceling is considered.

4. Feedback of operation adjustment results.

$$Z_{\text{delay}}^{\text{low}} \tag{6.16}$$

$Z_{\text{delay}}^{\text{low}}$ represents the optimization results of lower-layer planning. It includes the punishment of train delay, the punishment of station delay, and the number of severely delayed trains. The feedback of the optimization results of the lower-layer planning is also an important part to measure the optimization results of the upper-layer planning, in order to achieve the overall optimization of the upper and lower layers.

5. Constraints

The constraints of the traffic dispatching model under emergency conditions are as follows.

(1) Uniqueness constraint

(1) Rerouting trains can only detour on one route.

$$\sum_{rte \in \text{Route}} \text{rrt}_{i,\text{rte}} \leq 1 \; i \in \text{Trn} \tag{6.17}$$

(2) Canceling trains cannot detour to other routes.

$$\text{cal}_i + \sum_{rte \in \text{Route}} \text{rrt}_i \leq 1 \; i \in \text{Trn} \tag{6.18}$$

(3) If a train meets the conditions of reconnection, it can only reconnect with another one train.

$$\sum_{i=j} \text{meg}_{i,j} = 0 \; i, j \in \text{Trn} \tag{6.19}$$

$$\text{meg}_{i,j} + \text{meg}_{j,i,} \leq 1 \; i, j \in \text{Trn} \tag{6.20}$$

(4) The reconnected trains cannot be canceled.

$$\sum_{j \in \mathrm{Trn}} \mathrm{meg}_{i,j} + \mathrm{cal}_i \le 1 \; i \in \mathrm{Trn} \tag{6.21}$$

$$\sum_{i \in \mathrm{Trn}} \mathrm{meg}_{i,j} + \mathrm{cal}_j \le 1 \; j \in \mathrm{Trn} \tag{6.22}$$

(5) The reconnected trains cannot be rerouted.

$$\sum_{j \in \mathrm{Trn}} \mathrm{meg}_{i,j} + \sum_{\mathrm{rte} \in \mathrm{Route}} \mathrm{rrt}_{i,\mathrm{rte}} \le 1 \le 1 \; i \in \mathrm{Trn} \tag{6.23}$$

$$\sum_{i \in \mathrm{Trn}} \mathrm{meg}_{i,j} + \sum_{\mathrm{rte} \in \mathrm{Route}} \mathrm{rrt}_{i,\mathrm{rte}} \le 1 \; j \in \mathrm{Trn} \tag{6.24}$$

(2) Capability constraints.

(1) The total number of trains that need to be rerouted, reconnected, and canceled cannot be less than the number of trains that need to be dispatched because of the capacity loss.

$$\sum_{i \in \mathrm{Trn}} \sum_{j \in \mathrm{Trn}} \mathrm{meg}_{i,j} + \sum_{i \in \mathrm{Trn}} \mathrm{cal}_i + \sum_{i \in \mathrm{Trn}} \sum_{rte \in \mathrm{Route}} \mathrm{rrt}_{i,\mathrm{rte}} \ge \delta \tag{6.25}$$

(2) The remaining capacity of all sections on the railway network cannot be less than the sum of the number of rerouted trains.

$$\sum_{i \in \mathrm{Trn}} \sum_{\mathrm{rte} \in \mathrm{Route}} \mathrm{rrt}_{i,\mathrm{rte}} \le \Delta C_e \quad e \in \mathrm{Trn} \quad \mathrm{rte} \in \mathrm{Route} \tag{6.26}$$

(3) The remaining capacity of all stations on the railway network cannot be less than the sum of the number of rerouted trains.

$$\sum_{i \in \mathrm{Trn}} \sum_{rte \in \mathrm{Route}} \mathrm{rrt}_{i,\mathrm{rte}} \le \Delta C_v \quad v \in \mathrm{Trn} \quad \mathrm{rte} \in \mathrm{Route} \tag{6.27}$$

(3) Train rerouting constraints

The set of the graph fixed stop of train i is denoted as Sts_i, which is as follows.

$$\mathrm{Sts}_i = \left\{\mathrm{sts}_{i,1}, \mathrm{sts}_{i,2}, \ldots, \mathrm{sts}_{i,N_{\mathrm{stsn}}}\right\} = \left\{\mathrm{sts}_{i,n} | n = 1, 2, \ldots, N_{\mathrm{stsn}}\right\} \tag{6.28}$$

N_{sts} is the number of the stations where trains need to be stopped.

The new set of stops after the change of route of train i is denoted as $\mathrm{Sts}n_i$, which is as follows.

$$\mathrm{Stsn}_i = \left\{\mathrm{stsn}_{i,1}, \mathrm{stsn}_{i,2}, \ldots, \mathrm{stsn}_{i,N_{\mathrm{stsn}}}\right\} = \left\{\mathrm{stsn}_{i,n} | n = 1, 2, \ldots, N_{\mathrm{stsn}}\right\} \quad (6.29)$$

N_{stsn} is the number of stations where the train stop after the train changes its route.

When the train detours to other routes, it still needs to meet the passenger's boarding and alighting requirements. Therefore, the stations where the train stops on the changed route should include the original stop stations. There are the following constraints.

$$\mathrm{Stsn}_i \supseteq \mathrm{Sts}_i \; i \in \mathrm{Trn} \quad (6.30)$$

For all stop station $\mathrm{Sts}_{i,k}$ not included in Stsn_i, their hub is the same as $\mathrm{Stsn}_{i,k}$, which is as follows.

$$Sn_{\mathrm{Sts}_{i,k}} = Sn_{\mathrm{Stsn}_{i,k}} \; \forall \mathrm{Sts}_{i,k} \notin \mathrm{Stsn}_i \; i \in \mathrm{Trn} \quad (6.31)$$

Sn_v is the hub of station v.

(4) Train reconnection constraints

(1) Station restrictions for passengers getting on and off.

If train i and train j are reconnected and they share the train number of train j. In order to meet the passenger's boarding and alighting requirements on train i, the stop station set of train j needs to include the stop station set of train i.

$$\begin{aligned} \mathrm{Sts}_j = \left\{\mathrm{sts}_{j,1}, \mathrm{sts}_{j,2}, \ldots, \mathrm{sts}_{j,N_{\mathrm{stsn}}}\right\} \supseteq Sts_i = \left\{\mathrm{sts}_{i,1}, \mathrm{sts}_{i,2}, \ldots, \mathrm{sts}_{i,N_{\mathrm{stsn}}}\right\} \\ i \in \mathrm{Trn} \; \mathrm{j} \in \mathrm{Trn} \end{aligned} \quad (6.32)$$

If a stop $\mathrm{sts}_{i,k}$ of train i is not in the scheduled stop station set of train j, the new stop sequence of train j is required to include $\mathrm{sts}_{i,k}$.

$$\mathrm{Stsn}_j = \mathrm{Sts}_j \cup \left\{\mathrm{sts}_{i,k} \middle| \forall \mathrm{Sts}_{i,k} \notin \mathrm{Sts}_j, k = 1, 2, \ldots, N_{\mathrm{sts}_i}\right\} i \in \mathrm{Trn} \quad j \in \mathrm{Trn} \quad (6.33)$$

(2) Constraint of type of reconnected train.

If train i is reconnected by train j, train i and train j are required to be of the same type, and both of them are short group trains.

$$\mathrm{Ty}_i = \mathrm{Ty}_j \quad (6.34)$$

Ty_i is the type of train i.

$$\mathrm{Len}_i = \mathrm{Len}_j = \mathrm{Len}_G \quad (6.35)$$

Len_i is the length of train i, Len_G is the length of the eight marshaled trains.

6.2.4 Analysis of Uncertain Factors in Train Dispatching Optimization Model

1. Analysis of uncertain factors in objective function.

(1) Penalty for train rerouting

c_i^{reroute} is the penalty for train i rerouting to other routes. The determination of this penalty cost is related to the type and level of the emergency. Due to the different levels of emergencies, the degree of passenger acceptance of train detour is also different. For example, under the influence of a small-scale and short time emergency, the trains on this line have not been greatly disturbed. Passengers do not want the trains they take to detour to other routes. When a large-scale, long-term natural disaster occurs, or a serious railway accident, the loss of line capacity is large and the line repair takes a long time. Passenger's goal is to be able to reach their destinations. At this point, their acceptance of the train's detour will increase. Therefore, as the type and level of emergencies change, c_i^{reroute} should also change within a range.

In addition, the penalty of train detour mainly reflects the impact of the strategy on passenger satisfaction and the loss of social benefits. And these are all human psychological feelings. It is related to factors such as the purpose of travel, occupation, consumption level, and even the economic development level of the whole society. It is often a vague concept in the minds of the public, and it is difficult to describe it with precise values. Therefore, the c_i^{reroute} is expressed as a fuzzy number, which can more truly reflect people's acceptance of train detour under emergency conditions. So the objective function (6.11) can be transformed into as follows.

$$Z_{\text{rrt}}^{\text{p}} = \min \sum_{i \in \text{Trn}} \tilde{c}_i^{\text{reroute}} \text{rrt}_{i,\text{rte}} \tag{6.36}$$

(2) Train operating costs on detours

c_i^{run} is the operating cost per unit mileage of train *i*. This cost is related to the characteristics of the line, the type of train, the power (fuel) consumption of the locomotive, etc.[35–37]. Due to the differences in the basic characteristics of the lines (such as bridges, tunnels, culverts) and the types of trains such as diesel locomotives, electric locomotives, and EMUs, the operating cost unit mileage of trains in different sections changes within a range. It is difficult to express with a certain value and more consistent with the characteristics of fuzzy variables. So the objective function (6.12) can be transformed into as follows.

$$Z_{\text{rrt}}^{r} = \min \sum_{i \in \text{Trn}} \tilde{c}_i^{\text{run}} \text{rrt}_{i,\text{rte}} \sum_{e \in \text{Trn}} \text{len}_e \tag{6.37}$$

(3) Penalty of train reconnection

$c_{i,j}^{\text{merge}}$ represents the penalty for train *i* being reconnected by train *j*. Its value is also related to the type and level of the emergency. And it is more suitable for the expression of the fuzzy variables. So the objective function (6.13) can be transformed into as follows.

$$Z_{\text{meg}}^{\text{p}} = \min \sum_{i \in \text{Trn}} \tilde{c}_{i,j}^{\text{merge}} \sum_{j \in \text{Trn}} \text{meg}_{i,j} \tag{6.38}$$

(4) Penalty of train canceling

In the same way, the penalty cost c_i^{cancel} for canceling train i is also suitable for expressing fuzzy variables. So the objective function (6.15) can be transformed into as follows.

$$Z_{\text{cal}}^{\text{p}} = \min \sum_{i \in \text{Trn}} \tilde{c}_{i,j}^{\text{cancle}} \text{cal}_i \tag{6.39}$$

2. Analysis of uncertain factors in constraints

(1) Carrying capacity of sections

According to the Chap. 3, the carrying capacity of sections changes with the speed limit value. It has both fuzzy and random properties. Chapter 3 has described the derivation process of interval passing capacity in detail, which will not be described here.

(2) Train reception and departure capacity of the station

The capacity of reception and departure trains in a station is determined by the number of available arrival and departure tracks, the time occupied by arrival and departure tracks, and the operation time of throat turnout. The less time of station operation which occupies the arrival and departure tracks, and the less operation time of throat turnout, the stronger capacity of receiving and dispatching trains in a station. The time occupied by the arrival and departure tracks and the operation time of the throat turnout are determined by the train speed, equipment usage, and manual operation proficiency. They have a certain range of changes. And they also have ambiguity. In addition, under emergency conditions, the capacity of receiving and dispatching trains in a station is similar to the through capacity. With the change of the impact of the emergency and the maintenance and restoration work, it also has a certain randomness. Therefore, the description by fuzzy random variable is more consistent with the actual operating environment.

So the capacity constraints in the model are transformed into:

(1) Constraints of loss capacity.

$$\sum_{i \in \text{Trn}} \sum_{j \in \text{Trn}} \text{meg}_{i,j} + \sum_{i \in \text{Trn}} \text{cal}_i + \sum_{i \in \text{Trn}} \sum_{\text{rte} \in \text{Route}} \text{rrt}_{i,rte} \geq \hat{\delta} \tag{6.40}$$

(2) Constraints of remaining through capacity on routes.

$$\sum_{i \in \text{Trn}} \sum_{\text{rte} \in \text{Route}} \text{rrt}_{i,rte} \leq \Delta \hat{C}_e \quad e \in \text{rte} \quad \text{rte} \in \text{Route} \tag{6.41}$$

(3) Constraints of remaining capacity of reception and departure trains in a station on detour routes.

$$\sum_{i \in \text{Trn}} \sum_{\text{rte} \in \text{Route}} \text{rrt}_{i,\text{rte}} \leq \Delta \hat{C}_v \quad v \in \text{rte} \quad \text{rte} \in \text{Route} \tag{6.42}$$

6.2.5 *Train Dispatching Optimization Model Under Uncertain Environment*

According to the previous analysis, the expected value model and the opportunity constraint model are two widely used methods in uncertain planning. Their core ideas are to transform problems with random or fuzzy parameters into deterministic programming by using the correlation measure of uncertain variables. So in this section, the expected value model and the opportunity constraint model of the traffic dispatching optimization will be established separately.

1. Expected value model of traffic dispatching optimization

By taking the expected value of the uncertain variable in the original problem, the expected value model of traffic scheduling optimization can be obtained. The specific description is as follows.

(1) Objective function

$$\begin{aligned}
\min Z = {} & w_1 E\left[Z^{\text{p}}_{\text{rrt}}\right] + w_2 E\left[Z^{r}_{\text{rrt}}\right] \\
& + w_3 E\left[Z^{\text{p}}_{\text{meg}}\right] + w_4 E\left[Z^{\text{p}}_{\text{cal}}\right] \\
& + w_5 Z^{t}_{\text{meg}} + w_6 Z^{\text{low}}_{\text{deley}} \\
= {} & w_1 E\left[\sum_{i \in \text{Trn}} \tilde{c}_i^{\text{reroute}} \text{rrt}_{i,rte}\right] + w_2 E\left[\sum_{i \in Trn} \tilde{c}_i^{\text{run}} rrt_{i,\text{rte}} \sum_{e \in \text{rte}} \text{len}_e\right] \\
& + w_3 E\left[\sum_{i \in \text{Trn}} \tilde{c}_{i,j}^{\text{merge}} \sum_{j \in Trn} \text{meg}_{i,j}\right] \\
& + w_4 E\left[\sum_{i \in \text{Trn}} \tilde{c}_i^{\text{cancle}} \text{cal}_i\right] \\
& + w_5 \sum_{i \in \text{Trn}} \sum_{j \in \text{Trn}} \sum_{v \in \text{Stn}} \left(\left|\text{Inid}_{i,v} - \text{Inid}_{j,v}\right| + \left|\text{Inia}_{i,v} - \text{Inia}_{j,v}\right|\right) \\
& \text{meg}_{i,j} + w_6 Z^{\text{low}}_{\text{deley}}
\end{aligned}$$

$$= w_1 \sum_{i\in \mathrm{Trn}} E\left[\tilde{c}_i^{\mathrm{reroute}}\right]\mathrm{rrt}_{i,rte} + w_2 \sum_{i\in \mathrm{Trn}} E\left[\tilde{c}_i^{\mathrm{run}}\right]\mathrm{rrt}_{i,rte} \sum_{e\in rte} \mathrm{len}_e$$

$$+ w_3 \sum_{i\in \mathrm{Trn}} E\left[\tilde{c}_{i,j}^{\mathrm{merge}}\right] \sum_{j\in Trn} \mathrm{meg}_{i,j}$$

$$+ w_4 \sum_{i\in \mathrm{Trn}} E\left[\tilde{c}_i^{\mathrm{cancle}}\right]\mathrm{cal}_i$$

$$+ w_5 \sum_{i\in \mathrm{Trn}} \sum_{j\in \mathrm{Trn}} \sum_{v\in \mathrm{Stn}} \left(\left|\mathrm{Inid}_{i,v} - \mathrm{Inid}_{j,v}\right| + \left|\mathrm{Inia}_{i,v} - \mathrm{Inia}_{j,v}\right|\right)$$

$$\mathrm{meg}_{i,j} + w_6 Z_{\mathrm{deley}}^{\mathrm{low}} \tag{6.43}$$

w_i is the weight of the i-th sub-objective.

(2) Constraints

(1) Constraints of reduced capacity.

$$\sum_{i\in \mathrm{Trn}} \sum_{j\subset \mathrm{Trn}} \mathrm{meg}_{i,j} + \sum_{i\subset \mathrm{Trn}} \mathrm{cal}_i + \sum_{i\subset \mathrm{Trn}} \sum_{\mathrm{rte}\in \mathrm{Route}} \mathrm{rrt}_{i,\mathrm{rte}} \geq E\left[\hat{\delta}\right] \tag{6.44}$$

(2) Constraints of remaining carrying capacity on routes.

$$\sum_{i\in \mathrm{Trn}} \sum_{\mathrm{rte}\in \mathrm{Route}} \mathrm{rrt}_{i,\mathrm{rte}} \leq E\left[\Delta \hat{C}_e\right] e \in \mathrm{rte}, \mathrm{rte} \in \mathrm{Route} \tag{6.45}$$

(3) Constraints of remaining capacity of reception and departure trains in a station on reroutes.

$$\sum_{i\in \mathrm{Trn}} \sum_{\mathrm{rte}\in \mathrm{Route}} \mathrm{rrt}_{i,rte} \leq E\left[\Delta \hat{C}_v\right] \quad v \in \mathrm{rte} \quad \mathrm{rte} \in \mathrm{Route} \tag{6.46}$$

Other deterministic constraints remain unchanged.

For the expected values of fuzzy variables and fuzzy random variables, some scholars have proposed different definitions. This chapter adopts the definition in literature [33]. The description is as follows.

In fuzzy theory, $\mathrm{Pos}\{A\}$ represents the probability measure of the occurrence of event A, $\mathrm{Nec}\{A\}$ represents the necessity measure of the occurrence of event A, $\mathrm{Cr}\{A\}$ represents the credibility measure of the occurrence of event A, Ac represents the complement of event A. The relationship among the three is as follows.

$$\mathrm{Nec}\{A\} = 1 - \mathrm{Pos}\{A\}, \mathrm{Cr}\{A\} = \frac{1}{2}(\mathrm{Pos}\{A\} + \mathrm{Nec}\{A\})$$

If ξ is a fuzzy variable, the expectation of ξ is as follows.

$$E[\xi] = \int_{0}^{+\infty} \mathrm{Cr}\{\xi \geq r\}\mathrm{d}r - \int_{-\infty}^{0} \mathrm{Cr}\{\xi \leq r\}\mathrm{d}r \tag{6.47}$$

If ξ is a fuzzy random variable, the expectation of ξ is as follows.

$$E[\xi] = \int_{0}^{+\infty} \mathrm{Pos}\{\omega \in \Omega | E[\xi(\omega)] \geq r\}\mathrm{d}r - \int_{-\infty}^{0} \mathrm{Pos}\{\omega \in \Omega | E[\xi(\omega)] \leq r\}\mathrm{d}r \tag{6.48}$$

2. Opportunity constraint model of traffic dispatching optimization

Sometimes people do not always care about maximizing expected benefits or minimizing expected costs. But they are more concentrated on the opportunities which are from the risk of driving adjustments. Opportunity-constrained programming controls the reliability of the system in a confidence level manner. According to different targets, it can be divided into pessimistic value model, optimistic value model, and Hurwicz standard model.

(1) Pessimistic value model of traffic dispatching optimization

If using the pessimistic value and opportunity measure of uncertain variables, we can obtain the pessimistic value model of traffic dispatching optimization. The description is as follows.

Objective function:

$$\min \min_{\overline{Z}} \overline{Z} \tag{6.49}$$

Constraints:

$$\mathrm{Cr}\left\{ w_1 Z_{\mathrm{rrt}}^{\mathrm{p}} + w_2 Z_{\mathrm{rrt}}^{\mathrm{p}} + w_3 Z_{\mathrm{meg}}^{\mathrm{p}} + w_4 Z_{\mathrm{cal}}^{\mathrm{p}} + w_5 Z_{\mathrm{meg}}^{\mathrm{t}} + w_6 Z_{\mathrm{deley}}^{\mathrm{low}} \geq \overline{Z} \right\} \geq \alpha \tag{6.50}$$

$$\mathrm{Ch}\left\{ \sum_{i \in \mathrm{Trn}} \sum_{j \in \mathrm{Trn}} \mathrm{meg}_{i,j} + \sum_{i \in \mathrm{Trn}} \mathrm{cal}_i + \sum_{i \in \mathrm{Trn}} \sum_{\mathrm{rte} \in \mathrm{Route}} \mathrm{rrt}_{i,\mathrm{rte}} \geq \hat{\delta} \right\} \geq \beta \tag{6.51}$$

$$\mathrm{Ch}\left\{ \sum_{i \in \mathrm{Trn}} \sum_{\mathrm{rte} \in \mathrm{Route}} \mathrm{rrt}_{i,\mathrm{rte}} \leq \Delta \hat{C}_e \right\} \geq \delta \tag{6.52}$$

$$\mathrm{Cr}\left\{ \sum_{i \in \mathrm{Trn}} \sum_{\mathrm{rte} \in \mathrm{Route}} \mathrm{rrt}_{i,\mathrm{rte}} \leq \Delta \hat{C}_v \right\} \geq \eta \tag{6.53}$$

Other deterministic constraints remain unchanged.

In this way, the objective function is transformed into pessimistic values of minimizing various punishments. Capacity constraints also translate into opportunities at a certain level of confidence (α, β, δ, η).

(2) Optimistic value model of traffic dispatching optimization

If using the optimistic value and opportunity measure of uncertain variables, we can obtain the optimistic value model of traffic dispatching optimization. The description is as follows.

Objective function:

$$\min \max_{\overline{Z}} \overline{Z} \tag{6.54}$$

Constraints:

$$\mathrm{Cr}\left\{w_1 Z_{\mathrm{rrt}}^{\mathrm{p}} + w_2 Z_{\mathrm{rrt}}^{\mathrm{r}} + w_3 Z_{\mathrm{meg}}^{\mathrm{p}} + w_4 Z_{\mathrm{cal}}^{\mathrm{p}} + w_5 Z_{\mathrm{meg}}^{\mathrm{t}} + w_6 Z_{\mathrm{deley}}^{\mathrm{low}} \geq \overline{Z}\right\} \geq \alpha \tag{6.55}$$

$$\mathrm{Ch}\left\{\sum_{i\in\mathrm{Trn}}\sum_{j\in\mathrm{Trn}} \mathrm{mcg}_{i,j} + \sum_{i\in\mathrm{Trn}} \mathrm{cal}_i \mid \sum_{i\in\mathrm{Trn}}\sum_{\mathrm{rte}\in\mathrm{Route}} \mathrm{rrt}_{i,\mathrm{rte}} \geq \hat{\delta}\right\} \geq \beta \tag{6.56}$$

$$\mathrm{Ch}\left\{\sum_{i\in\mathrm{Trn}}\sum_{\mathrm{rte}\in\mathrm{Route}} \mathrm{rrt}_{i,\mathrm{rte}} \leq \Delta\hat{C}_e\right\} \geq \delta \tag{6.57}$$

$$\mathrm{Cr}\left\{\sum_{i\in\mathrm{Trn}}\sum_{\mathrm{rte}\in\mathrm{Route}} \mathrm{rrt}_{i,\mathrm{rte}} \leq \Delta\hat{C}_v\right\} \geq \eta \tag{6.58}$$

Other deterministic constraints remain unchanged.

(3) Hurwicz standard model

If both pessimism and optimism are considered, the parameter $\mu \in (0, 1)$ can be introduced to indicate the decision -maker's tendency to choose between pessimism and optimism. Establish the Hurwicz standard model. The description is as follows.

Objective function:

$$\min(\mu \max_{\overline{M}} \overline{M} + (1-\mu) \min_{\overline{Z}} \overline{Z}) \tag{6.59}$$

Constraints:

$$\mathrm{Cr}\left\{w_1 Z_{\mathrm{rrt}}^{\mathrm{p}} + w_2 Z_{\mathrm{rrt}}^{\mathrm{r}} + w_3 Z_{\mathrm{meg}}^{\mathrm{p}} + w_4 Z_{\mathrm{cal}}^{\mathrm{p}} + w_5 Z_{\mathrm{meg}}^{\mathrm{t}} + w_6 Z_{\mathrm{deley}}^{\mathrm{low}} \leq \overline{M}\right\} \geq \alpha \tag{6.60}$$

$$\mathrm{Cr}\left\{w_1 Z_{\mathrm{rrt}}^{\mathrm{p}} + w_2 Z_{\mathrm{rrt}}^{\mathrm{r}} + w_3 Z_{\mathrm{meg}}^{\mathrm{p}} + w_4 Z_{\mathrm{cal}}^{\mathrm{p}} + w_5 Z_{\mathrm{meg}}^{\mathrm{t}} + w_6 Z_{\mathrm{deley}}^{\mathrm{low}} \leq \overline{Z}\right\} \geq \alpha \tag{6.61}$$

$$\mathrm{Ch}\left\{\sum_{i\in \mathrm{Tm}}\sum_{j\in \mathrm{Tm}} \mathrm{meg}_{i,j} + \sum_{i\in \mathrm{Trn}} \mathrm{cal}_i + \sum_{i\in \mathrm{Trn}}\sum_{\mathrm{rte}\in \mathrm{Route}} \mathrm{rrt}_{i,\mathrm{rte}} \geq \delta\right\} \geq \beta \tag{6.62}$$

$$\mathrm{Ch}\left\{\sum_{i\in \mathrm{Trn}}\sum_{\mathrm{rte}\in \mathrm{Route}} \mathrm{rrt}_{i,\mathrm{rte}} \leq \Delta\tilde{C}_e\right\} \geq \delta \tag{6.63}$$

$$\mathrm{Cr}\left\{\sum_{i\in \mathrm{Trn}}\sum_{\mathrm{rte}\in \mathrm{Route}} \mathrm{rrt}_{i,\mathrm{rte}} \leq \Delta\tilde{C}_v\right\} \geq \eta \tag{6.64}$$

Other deterministic constraints remain unchanged.

Through the establishment of the above model, the original problem can be transformed into deterministic planning.

See Sects. 6.4 and 6.5 for the solution algorithm.

6.3 Train Timetable Rescheduling Model Under Emergency Conditions

After the completion of the upper-layer planning, the information of the train's route, stop, and speed is determined. The lower-layer planning focuses on the adjustment of train schedules. And it will feedback the adjustment results to the upper planning. Thus, the train scheduling optimization and the feedback rolling optimization of train operation adjustment are realized. Finally, we can get the railway operation organization scheme under emergency conditions.

6.3.1 Analysis of Influencing Factors

The research content of this layer is to reschedule the arrival and departure time of the train according to the principles and strategies of transportation organization under the emergency when the train operation plan is determined. The main influencing factors are as follows.

1. Based on railway network

Transportation organization optimization under emergency conditions is based on railway network conditions. As the lower level of the optimized train operation adjustment is no exception. Therefore, the object of adjustment of the lower layer is the train on all routes in the upper optimization.

2. Train delay tolerance

Under emergency conditions, due to the low train speed limit or even the breaking of the track, the network topology changes are caused. It has a large capacity loss. It often causes all trains to be delayed within the time period affected by the emergency. It does not make sense to discuss the number of delayed trains at this time. In this case, a delay tolerance level can be set to identify trains that are seriously. Under emergency conditions, it is more meaningful to reduce the number of severely delayed trains.

3. Train timetable

The train timetable describes the train's operation in the section and the time of arrival or departure of the station. This is the basis of train operation adjustment. Whether the train runs on the section or stops at the station for passenger operations such as getting on and off services, loading water, waiting, and so on, it can be regarded as the occupation of the line equipment by the train. Therefore, literature [38] can be adopted. Train operations are regarded as discrete events in time. When a train passes through a station without stopping, the duration of the event can be considered to be zero.

4. Large hub

Large hubs are generally located in densely populated, economically developed cities. In the whole transportation system, it assumes missions for the majority of passenger travel. Delays at these hubs can have a worse impact than at some of the smaller stations. Therefore, it is necessary to minimize the delay time of large hubs under emergency conditions.

5. Uncertain environment

Similar to the upper-layer planning, the adjustment of train operation under emergency conditions also involves a large number of uncertain variables. The objective function and constraints also have fuzziness or randomness. This makes the model solution more complicated.

6.3.2 Construction of Train Timetable Rescheduling Model Under Emergency Condition

1. Relevant parameters

The relevant parameters involved in the train timetable rescheduling model are as follows.

p: A train operation (discrete event), depending on the location of the train operation, can be expressed as section operation or station operation;

inis_p: The original start time of operation p, which can be expressed as the original start time of the section operation or the original start time of the station operation (arrival time).

inie_p: The original end time of operation p can be expressed as the original arrival time. It also means the original end time of station operations, which is also the original start time of the next section operation;

zt_p: Minimum train operating time on p. If p is station operation, then it represents the minimum station operation time. If p is an interval operation, then it represents the minimum operation time of the section.

w_i: The degree of delay tolerance of train i

$$\lambda_{p,j} = \begin{cases} 1 \text{ If operation } p \text{ takes up the route, } p \in \text{Itv}, j \in \text{Rte}; \\ 0 \text{ Others} \end{cases}$$

$$h_p = \begin{cases} 1 \text{ If operation } p \text{ is stopping operation of original station, } p \in \text{Itv}; \\ 0 \text{ other} \end{cases}$$

c_i^{tmd}: Penalty for unit delay of train i;

c_k^{stnd}: Penalty for unit delay of hub stations k;

M: A very large integer.

In addition, Table 6.13 describes some set definitions. For the convenience of expression, station and the section are uniformly referred to as the section. A section may represent a station or a section in which the train runs.

Table 6.13 explains the definition of the set in the model as follows:

(1) For convenience of understanding, p is used to represent the identity of elements in all sets related to train operation (such as: Itv, Itv_i, $\text{Itv}_i^{\text{stn}}$, $\text{Itv}_i^{\text{sec}}$ and Itv_k.

(2) In all the sets related to train operations, $p + 1$ represents the first operation after p on the j route, $\tilde{p}$ means all operations after p.

2. Decision variables

The decision variables involved in the train timetable rescheduling model are as follows.

Table 6.13 Set definitions in the model

Parameters	Description	Size	Index
Trn	Trains set	T	i
Rte	The set of all affected routes, including original and detour routes	R	j
Seg	The set of all sections, that is, the sum of stations and sections	S	k
$\text{Stn} \subseteq \text{Seg}$	Station set	N	n
$\text{Sec} \subseteq \text{Seg}$	Section set	S–N	–
Itv	The set of all train operations, the sum of station operations and section operations	–	P
$\text{Itv}_i \subseteq \text{Itv}$	The set of original operations of train i	–	–
$\text{Itv}_i^{\text{stn}} \subseteq \text{Itv}_i$	The set of original station operations of train i	–	–
$\text{Itv}_k \subseteq \text{Itv}$	The set of original operations of section k	–	–
$\text{Itv}_i^{\text{sec}} \subseteq \text{Itv}_i$	The set of original section operations of train i	–	–
Trk_k	The set of arrival and departure tracks of station k	L_k	l

(1) Related variables of operation time

s_p: Rescheduled start time for operation p.
e_p: Rescheduled end time for operation p.
d_p: The delay time of job p, that is, the difference between the original end time or original start time of the adjusted operation p.

(2) additional variables

$$b_i = \begin{cases} 1 \text{ if the delay time over the delay tolerance } w_i \\ \quad \text{when train } i \text{ arrives at the last station, } i \in \text{Trn} \\ 0 \text{ other} \end{cases}$$

$$\eta_{p,l} = \begin{cases} 1 \text{ if station operation } p \text{ occupies the arrival and departure track } l, \\ \quad p \in \text{Itv}_k, l \in \text{Trk}_k, k \in \text{Stn} \\ 0 \text{ other} \end{cases}$$

$$\alpha_{p,\tilde{p}} = \begin{cases} 1 \text{ Operation } p \text{ is earlier than operation } \tilde{p}, p, \tilde{p} \in \text{Itv}_k, k \in \text{Seg} \\ 0 \text{ other} \end{cases}$$

$$\beta_{p,\tilde{p}} = \begin{cases} 1 \text{ Operation } p \text{ is latter than operation } \tilde{p}, p, \tilde{p} \in \text{Itv}_k, k \in \text{Seg} \\ 0 \text{ other} \end{cases}$$

3. Objective function

The objective function of the train operation adjustment model under emergency conditions mainly considers the delayed punishment, the stability of the train operation plan and the number of seriously delayed trains.

(1) Delay penalty

$$z_{\text{delay}}^{\text{train}} = \min \sum_{i \in \text{Trn}} \left(c_i^{\text{trnd}} \cdot \left(\sum_{p \in Itv_i^{\text{sec}}} (d_p \cdot h_{p+1}) + \sum_{p \in Itv_i^{\text{stn}}} (d_p \cdot h_p) \right) \right) \tag{6.65}$$

$$z_{\text{delay}}^{\text{stn}} = \min \sum_{k \in \text{Stn}} (c_k^{\text{stnd}} \cdot \sum_{p \in \text{Itv}_k} d_p) \tag{6.66}$$

$Z_{\text{delay}}^{\text{train}}$ represents the penalty for delayed arrival and departure of trains at stations. The delay of different class trains will affect passengers to different degrees. Therefore, the objective function designed in this section is not simply to sum up the train delay time, but to give different punishments to different types of train delay per unit time. In Eq. (6.65), $\sum_{p \in \text{Itv}_i^{\text{sec}}} (d_p \cdot h_{p+1})$ is the penalty for delayed arrival of train i, $\sum_{p \in \text{Itv}_i^{\text{stn}}} (d_p \cdot h_p)$ is the penalty for delayed departure of train i.

In addition, the following two points should be noted in Eq. (6.65).

(1) For trains detoured to other routes, the stop may not be the original station but other stations located in the same transportation hub as the original station.

(2) The delay statistics of the reconnected trains are only conducted for the reconnected trains, and the arrival and departure time of the reconnected trains is taken as the arrival and departure time of the subsequent trains.

$Z_{\text{delay}}^{\text{stn}}$ represents the penalty for delayed arrival of trains at different grades stations. In China's railway transportation, the number of passengers on and off at a few large hub stations such as Beijing and Shanghai accounts for a large proportion in the total railway passenger transport. Some economically developed medium-sized cities are also hubs for tourists. For a large number of low-grade stations, the number of passengers on and off the is relatively small. So there are different delayed penalties for different levels of stations, in order to minimize train delay at higher-level stations. It avoids the impact on a large number of passenger travel.

(2) Stability of train operation plan

(1) The definition of stability of train operation plan

In recent years, academic community has paid more and more attention to the stability analysis of train operation plans. There are several methods for studying the stability of train operation planning. Most of these methods are based on Petri net theory or extremely algebraic methods [39–41]. The research on the stability of train operation plan mainly focuses on the following aspects.

① When trains are disturbed during operation, the time required for all trains to resume running according to the train diagram, such as literature [39].
② After the train is disturbed during operation, what is the adjustable margin, that is, what is the adjustable time prepared for the timetable, such as literature [40]
③ Assume that the train was disturbed during the operation, and the statistics of the delay propagation caused by the initial delay [42]. Literature [42] designed a simulation method and conducted statistics.

In addition, literature [41, 43–51] studied the method of evaluating train operation diagram. But the most of the research done so far is mostly on the evaluation level. Literature [52] defines the concept of stability of train operation diagram. The stability of train operation plan is defined qualitatively. Literature [53] describes the stability of the train operation plan from three aspects: conflict, structure, and clustering. However, when evaluating whether the train operation plan is stable, a reference value is required without exception. In the actual evaluation, the reference value is difficult to give. Literature [54] stipulates that under the system ideal state, the system sensitivity and restoring force of the train operation diagram under the assumption that the train operation does not deviate from the threshold are called the stability of the train operation diagram. Some indexes for evaluating the stability of train diagram are given. In this section, based on the current research status of stability of train operation diagram, the concept of stability of train operation plan under emergency conditions is proposed.

Define 6.1 train operation plan stability

Stability of train operation plan is the ability of train operation plan to restore the train to the original plan after the train operation is disturbed.

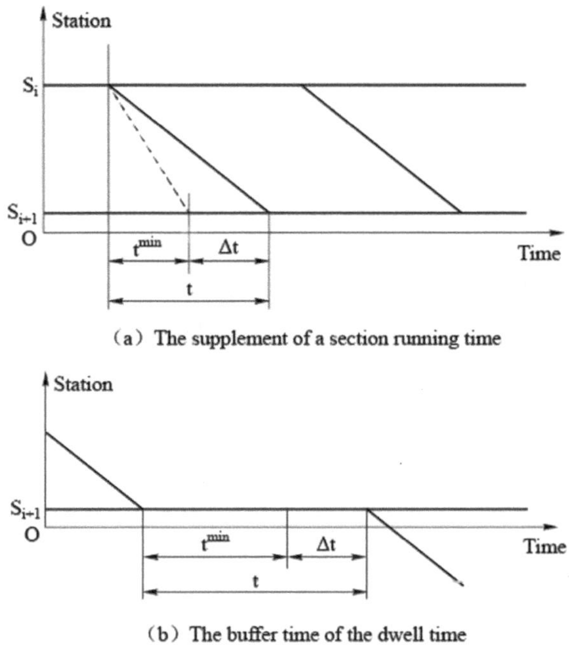

Fig. 6.3 Buffer time of the section running and dwell

(2) Quantification of train operation plan stability

There are two main factors affecting the stability of the train operation plan. One is the buffer time of the train section running operation and station dwelling operation, as shown in Fig. 6.3.

The other is the probability of interference in train operation and stop operation. The probability of an emergency occurring in different sections is different. In order to introduce the stability of the train operation plan into the train operation adjustment model as the optimization objective, it must be quantified. The following is the quantitative treatment of train operation plan stability [55].

Firstly, research the relevant conception of train section operation. Suppose that there are M trains on the studied track j with $N + 1$ stations, and N intervals in this section. Define the operation end time matrix of the adjusted train on route j as $E = \{e_p\}_{M\times N}$, start time matrix is $S = \{s_p\}_{M\times N}$.

When $\lambda_{p,j} = 1, p \in \mathrm{Itv}_i^{sec}$, the runing time matrix of M train running in N sections, $T^{j,r}$ is as follows.

$$T^{j,x} = \{t_p^{j,x}\}_{M\times N} = \{S_p - e_p\}_{M\times N} \tag{6.67}$$

The free-flow running time matrix of M train running in N sections, $\left(T^{j,r}\right)^{\min}$ is as follows.

$$\left(T^{j,r}\right)^{\min} = \left\{\left(t_p^{j,r}\right)^{\min}\right\}_{M\times N} \tag{6.68}$$

Interval running time margin matrix, $\Delta T^{j,r}$ is as follows.

$$\Delta T^{j,r} = \left\{\Delta t_p^{j,r}\right\}_{M\times N} = \left\{t_p^{j,r} - \left(t_p^{j,r}\right)^{\min}\right\}_{M\times N} \tag{6.69}$$

Train operation adjustability matrix, $A^{j,r}$ is as follows.

$$A^{j,r} = \left\{\alpha_p^{j,r}\right\}_{M\times N} = \left\{\Delta t_p^{j,r}/t_p^{j,r}\right\}_{M\times N} \tag{6.70}$$

Because the probability of the occurrence of the interference caused by the emergency conditions is different in each section, it must be considered when studying the dynamic performance indicators of the train operation plan. The probability vector generated by interference in each interval $R^{j,r}$ is as follows.

$$R^{j,r} = \left\{r_k^{j,r}\right\}_N = \left[r_1^{j,r}, r_2^{j,r}, \ldots, r_N^{j,r}\right] \tag{6.71}$$

Relative adjustability of train operation, $\left(A^{j,r}\right)^{\mathrm{Re}}$ is as follows.

$$\left(A^{j,r}\right)^{\mathrm{Re}} = \left\{\left(\alpha_p^{j,r}\right)^{\mathrm{Re}}\right\}_{M\times N} = \left\{\alpha_p^{j,r}/r_k^{j,r}\right\}_{M\times N} \tag{6.72}$$

$r\left(\left(A^{j,r}\right)^{\mathrm{Re}}\right)$ is rank of the matrix $\left(A^{j,r}\right)^{\mathrm{Re}}$.

Discussion:

① If $r\left(\left(A^{j,r}\right)^{\mathrm{Re}}\right) < M$, then the train's relative adjustability is strong. The $r\left(\left(A^{j,r}\right)^{\mathrm{Re}}\right)$ is smaller, the more balanced the running margin of the train is relative to the probability of interference, the greater the relative adjustability of the train operating plan.

② The variance of the relative adjustability of a train in all sections reveals the allocation of train operating capacity in all sections. Similarly, the variance of the relative adjustability of all trains on the same interval reveals the distribution of the service capacity of the interval on each train. The smaller the value, the more balanced the service capacity of the section is on different trains. The quantified value of the stability of the train operation plan for the interval operation plan S^r is as follows.

$$\begin{aligned} S^r = 1/\Big\{ & \alpha^{j,r}\left[\mathrm{dev}\left(\left(a_p^{j,r}\right)^{\mathrm{Re}} | p=1\right), \mathrm{dev}\left(\left(a_p^{j,r}\right)^{\mathrm{Re}} | p=2\right), \ldots, \mathrm{dev}\left(\left(a_p^{j,r}\right)^{\mathrm{Re}} | p=N\right)\right]^T \\ & + \beta^{j,r}\left[\mathrm{dev}\left(\left(a_p^{j,r}\right)^{\mathrm{Re}} | p=1\right), dev\left(\left(a_p^{j,r}\right)^{\mathrm{Re}} | p=2\right), \ldots, \mathrm{dev}\left(\left(a_p^{j,r}\right)^{\mathrm{Re}} | p=M\right)\right]^T \Big\} \\ & a^{j,r} = \left(a_1^{j,r}, a_2^{j,r}, \ldots, a_N^{j,r}\right), \beta^{j,r} = \left(\beta_1^{j,r}, \beta_2^{j,r}, \ldots, \beta_M^{j,r}\right) \end{aligned} \tag{6.73}$$

Among them, the vector, $\alpha^{j,r}$ represents the weight of each section, and the vector, $\beta^{j,r}$ represents the weight of each train.

Similarly, when $h_p = 1, p \in \text{Itv}_i^{\text{stn}}$, M train dwell time matrix at N stations $T^{j,w}$ is as follows.

$$T^{j,\mathrm{w}} = \left\{t_p^{j,w}\right\}_{M\times N} = \left\{e_p - s_p\right\}_{M\times N} \tag{6.74}$$

The minimum dwell time matrix of M train at N stations $\left(T^{j,w}\right)^{\min}$ is as follows.

$$\left(T^{j,\mathrm{w}}\right)^{\min} = \left\{\left(t_p^{j,w}\right)^{\min}\right\}_{M\times N} \tag{6.75}$$

Trains dwell time margin matrix $\Delta T^{j,w}$ is as follows.

$$\Delta T^{j,\mathrm{w}} = \left\{\Delta t_p^{j,w}\right\}_{M\times N} = \left\{t_p^{j,w} - \left(t_p^{j,w}\right)^{\min}\right\}_{M\times N} \tag{6.76}$$

Train stop adjustability matrix $A^{j,w}$ is as follows.

$$A^{j,\mathrm{w}} = \left\{a_p^{j,w}\right\}_{M\times N} = \left\{\Delta t_p^{j,w} / t_p^{j,w}\right\}_{M\times N} \tag{6.77}$$

Similarly, the probability of interference from emergency conditions at different stations is different. The probability vector generated by the interference located at each station $R^{j,w}$ is as follows.

$$R^{j,\mathrm{w}} = \left\{r_k^{j,w}\right\}_N = \left[r_1^{j,w}, r_2^{j,w}, \ldots, r_N^{j,w}\right] \tag{6.78}$$

Train dwell relative adjustability matrix $\left(A^{j,w}\right)^{\mathrm{Re}}$ is as follows.

$$\left(A^{j,w}\right)^{\mathrm{Re}} = \left\{\left(a_p^{j,w}\right)^{\mathrm{Re}}\right\}_{M\times N} = \left\{a_p^{j,w} / r_k^{j,w}\right\}_{M\times N} \tag{6.79}$$

$r\left(\left(A^{j,w}\right)^{\mathrm{Re}}\right)$ is rank of the matrix $\left(A^{j,w}\right)^{\mathrm{Re}}$.

The quantized stability value of the train stop pattern of the train operation plan S^w is as follows.

$$\begin{aligned}
S^w = 1/\Big\{&\alpha^{j,w}\Big[\mathrm{dev}\Big(\left(a_p^{j,w}\right)^{\mathrm{Re}}\,|p=1\Big), \mathrm{dev}\Big(\left(a_p^{j,w}\right)^{\mathrm{Re}}\,|p=2\Big), \ldots, \mathrm{dev}\Big(\left(a_p^{j,w}\right)^{\mathrm{Re}}\,|p=N\Big)\Big]^T \\
&+ \beta^{j,w}\Big[\mathrm{dev}\Big(\left(a_p^{j,w}\right)^{\mathrm{Re}}\,|p=1\Big), \mathrm{dev}\Big(\left(a_p^{j,w}\right)^{\mathrm{Re}}\,|p=2\Big), \ldots, \mathrm{dev}\Big(\left(a_p^{j,w}\right)^{\mathrm{Re}}\,|p=M\Big)\Big]^T\Big\} \\
\alpha^{j,w} = &\left(\alpha_1^{j,w}, \alpha_2^{j,w}, \ldots, \alpha_N^{j,w}\right), \beta^{j,w} = \left(\beta_1^{j,w}, \beta_2^{j,w}, \ldots, \beta_M^{j,w}\right)
\end{aligned} \tag{6.80}$$

Among them, the vector, $\alpha^{j,w}$ represents the weight of each section, and the vector, $\beta^{j,w}$ represents the weight of each train.

In summary, the stability quantification value S of the train operation plan is as follows.

$$S = S^r \times S^w \tag{6.81}$$

(3) Number of severely delayed trains

$$z_{\text{delay}}^{\text{trncount}} = \min \sum_{i \in \text{Trn}} b_i \tag{6.82}$$

The disrupted scenario studied in this chapter has the characteristics of large loss of line capacity and long impact time of emergency. Almost all trains will be delayed during the emergency period. Therefore, it is more meaningful to aim at reducing train delays.

4. Constraints

(1) Constraints of operation time

(1) Train minimum operation time constraints

$$e_p \geq s_p + zt_p p \in \text{Itv} \tag{6.83}$$

If p is the operation in a station, Eq. (6.83) represents the minimum operation time constraint of train station. If p is section operation, Eq. (6.83) represents the minimum operation time constraint of train section.

(2) Train earliest departure time constraints

$$e_p \geq \text{ inis }_p p \in Itv, h_p = 1 \tag{6.84}$$

It is stipulated that the train departure time cannot be earlier than the scheduled departure time when the train operation is adjusted.

(3) Train delay constraint

$$e_p - \text{inie}_p = d_p, \, p \in \text{Itv} \tag{6.85}$$

(4) Train operation continuation constraint

$$e_p = s_{p+1} \quad p \in \text{Itv } p \neq \text{last}(\text{Itv}_i) \quad i \in \text{Trn} \tag{6.86}$$

Among them, $\text{last}(\text{Itv}_i)$ represents the last scheduled operation. According to the book's definition of train operations, the end time of the previous operation is equal to the start time of the next operation.

(2) Station capacity constraints

(1) Station arrival and departure track constraints

$$\sum_{l \in \text{Trk}_k} \eta_{p,l} = 1 \quad p \in \text{Itv}_k \quad k \in \text{Stn} \tag{6.87}$$

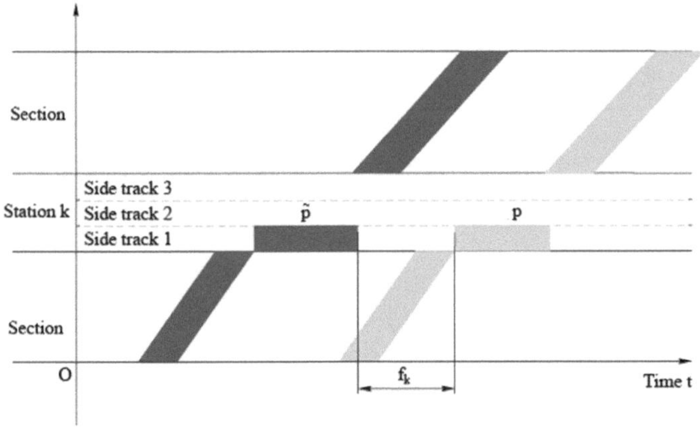

Fig. 6.4 Time interval between the arrival and departure trains at the station

When the train is operating at the station, it can only occupy one arrival and departure track in the station.

(2) Constraint of two trains occupying one arrival and departure tracks

Except for reconnection of trains, it is not considered that two trains occupy one arrival and departure tracks at the same time. For each station, if there are two train operations occupying one arrival and departure tracks at the same time, at least one of α and β is 1. And meet the train interval f_k of the arrival and departure tracks in the station.

$$\eta_{p,l}+\eta_{\tilde{p},l} - 1 \leq \alpha_{p,\tilde{p}} + \beta_{p,\tilde{p}} \quad p, \tilde{p} \in \mathrm{Itv}_k \quad p < \tilde{p}\, l \in \mathrm{Itv}_k \quad k \in \mathrm{Stn} \tag{6.88}$$

If two trains enter the same arrival and departure tracks in the station consecutively, the time interval between the arrival and departing trains at the station is shown in Fig. 6.4.

At this time, the following constraints must be satisfied.

$$s_{\tilde{p}} - e_p \geq f_k \alpha_{p,\tilde{p}} - M\left(1 - \alpha_{p,\tilde{p}}\right) \quad p, \tilde{p} \in \mathrm{Itv}_k \quad k \in \mathrm{Stn} \tag{6.89}$$

$$s_p - e_{\tilde{p}} \geq f_k \beta_{p,\tilde{p}} - M\left(1 - \alpha_{p,\tilde{p}}\right) \quad p, \tilde{p} \in \mathrm{Itv}_k \quad k \in \mathrm{Stn} \tag{6.90}$$

(3) Constraints on the departure-arrival time interval between departure and arrival of trains in the same direction

In stations where simultaneous arrival and departure of trains in the same direction is prohibited, the interval time between the departure of one train from the station and the arrival of another train in the same direction at the station is called the interval time between departure and arrival of trains in the same direction at different times, denoted as fd_k. Its traditional description is shown in Fig. 6.5. According to the model proposed in this chapter, it can be expressed as Fig. 6.6.

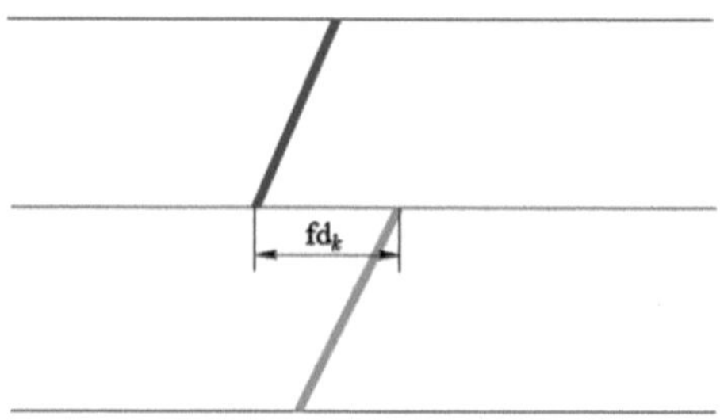

Fig. 6.5 Time interval between departure and arrival of trains in the same direction (traditional description)

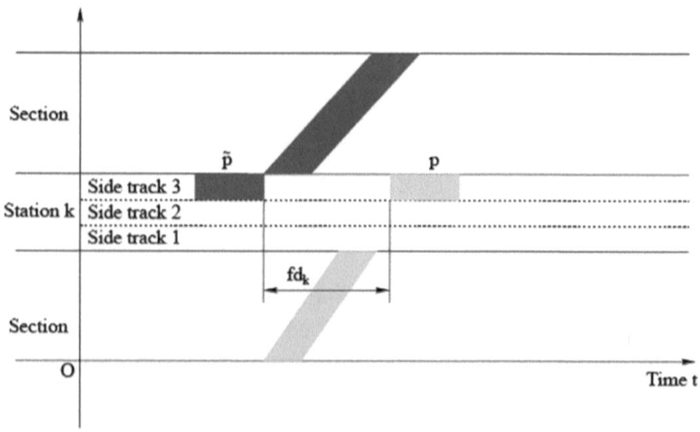

Fig. 6.6 Time interval between departure and arrival of trains in the same direction

The constraints can be described as follows.

$$s_p - e_{\tilde{p}} \geq \mathrm{fd}_k \beta_{p,\tilde{p}} - M\left(1 - \beta_{p,\tilde{p}}\right) \quad p, \tilde{p} \in \mathrm{Itv}_k \quad k \in \mathrm{Stn} \tag{6.91}$$

(4) Constraints on the arrival-departure time interval between of trains in the same direction

In stations where simultaneous arrival and departure of trains in the same direction is prohibited, the interval time between the arrival of one train from the station and the departure of another train in the same direction at the station is called the interval time between arrival and departure of trains in the same direction at different times, denoted as df_k. Its traditional description is shown in Fig. 6.7. According to the model proposed in this chapter, it can be expressed as Fig. 6.8.

The constraints can be described as follows.

$$e_{\tilde{p}} - s_p \geq \mathrm{df}_k \alpha_{p,\tilde{p}} - M\left(1 - \alpha_{p,\tilde{p}}\right) \quad p, \tilde{p} \in \mathrm{Itv}_k \quad k \in \mathrm{Stn} \tag{6.92}$$

Fig. 6.7 Time interval between arrival and departure of trains in the same direction (traditional description)

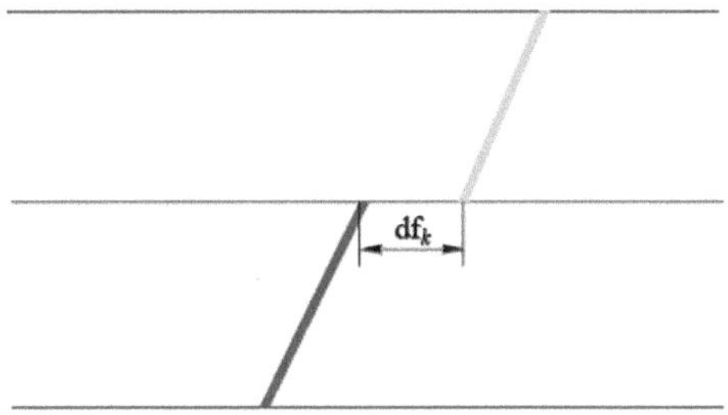

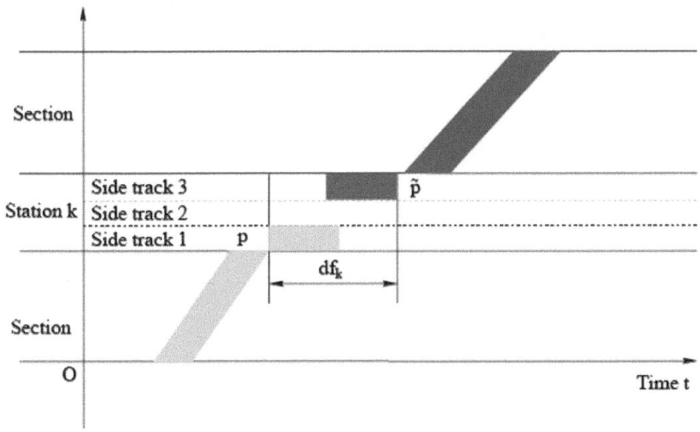

Fig. 6.8 Time interval between arrival and departure of trains in the same direction

(5) Interval time constraints at different direction of arrival.

In a single-track section, when two trains from opposite directions meet at the station, the minimum interval time between the arrival of the train in a certain direction at the station and the arrival or passage of the train in the opposite direction at the station is called the interval time of different arrival, denoted as fds_k. When a train stops and another train passes, its traditional description is shown in Fig. 6.9. According to the model proposed in this chapter, it can be expressed as Fig. 6.10.

When both trains stop, the traditional description is shown in Fig. 6.11. According to the model proposed in this chapter, it can be expressed as Fig. 6.12.

The constraints can be described as follows.

Fig. 6.9 Time interval constraints (when a train stops and another train through, traditional description)

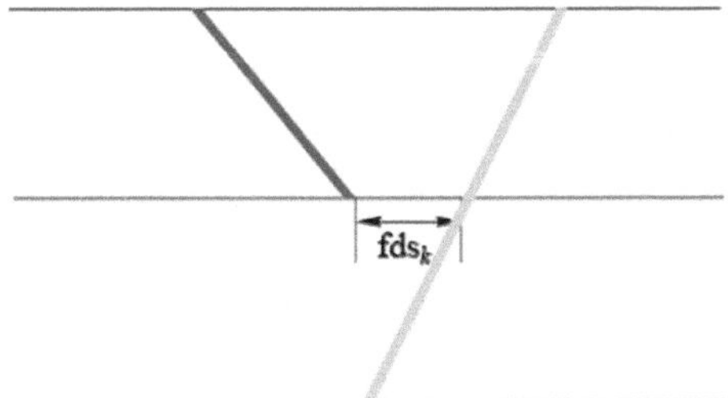

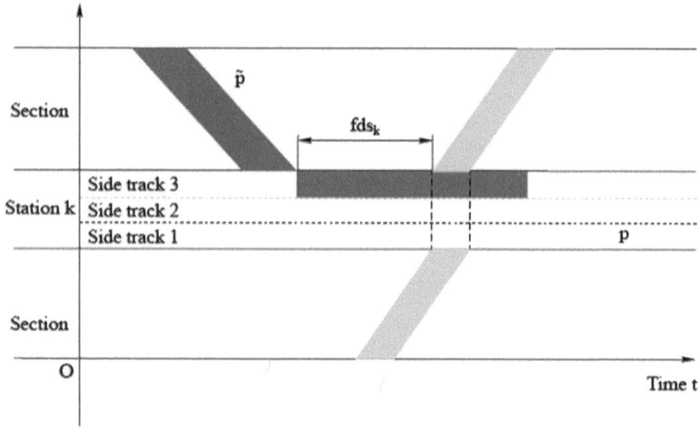

Fig. 6.10 Time interval constraints at different times of arrival (when a train stops and another train passes)

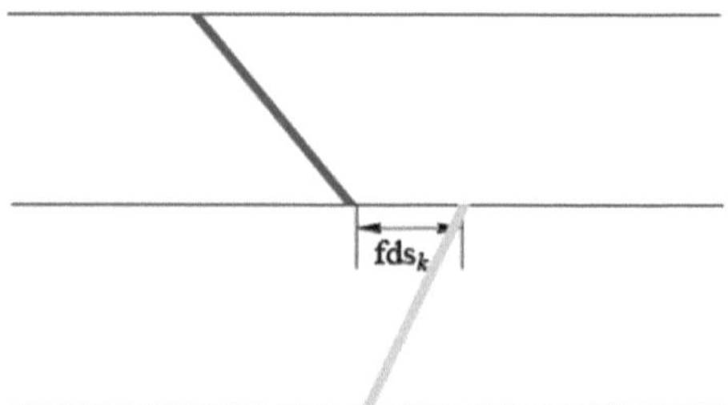

Fig. 6.11 Time interval between two arrival trains in different directions (when both trains stop, traditional description)

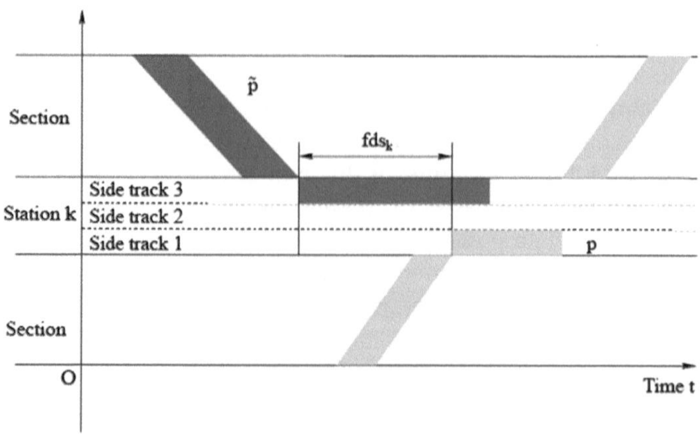

Fig. 6.12 Time interval between two arrival trains in different directions (when both trains stop)

$$s_p - e_{\tilde{p}} \geq \text{fds}_k \beta_{p,\tilde{p}} - M\left(1 - \beta_{p,\tilde{p}}\right) \quad p, \tilde{p} \in \text{Itv}_k \quad k \in \text{Seg} \tag{6.93}$$

(6) Time interval constraints between the meeting of trains.

In a single-track section, the minimum interval between the arrival or passage of a train at a station and the departure of another trains from the station to the opposite section is sent to the same section is called the interval time between the meeting of trains, denoted as fdh_k. When a train stops and another train passes, its traditional description is shown in Fig. 6.13. According to the model proposed in this chapter, it can be expressed as Fig. 6.14.

When both trains stop, the traditional description is shown in Fig. 6.15. According to the model proposed in this chapter, it can be expressed as Fig. 6.16.

The constraints can be described as

$$s_p - e_{\tilde{p}} \geq \text{fdh}_k \beta_{p,\tilde{p}} - M\left(1 - \beta_{p,\tilde{p}}\right) \quad p, \tilde{p} \in \text{Itv}_k \quad k \in \text{Sec} \tag{6.94}$$

(3) Interval constraints

(1) Interval occupation constraints.

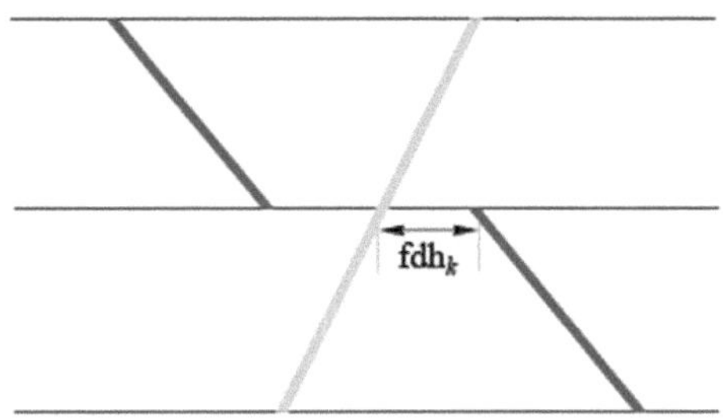

Fig. 6.13 Time interval constraint between the meeting trains (when a train stops and another train passes, traditional description)

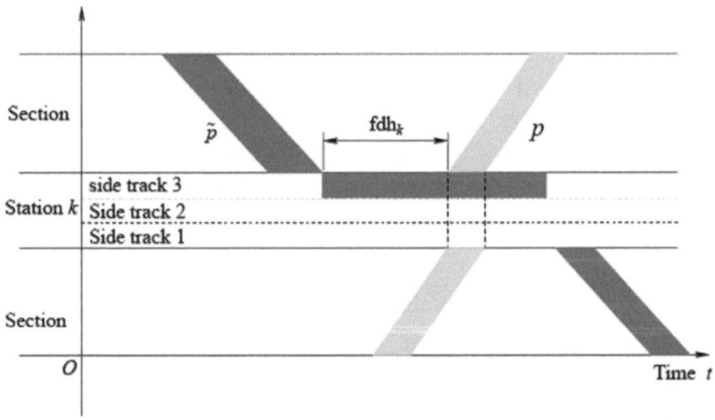

Fig. 6.14 Time interval constraint between the meeting of trains (when a train stops and another train passes)

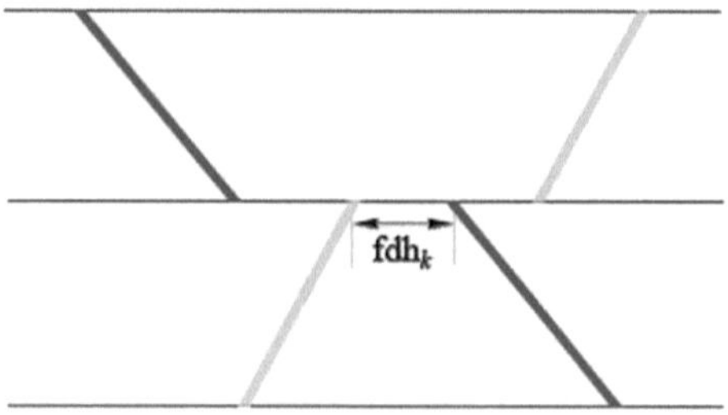

Fig. 6.15 Time interval constraint between the meeting of trains (when both trains stop, traditional description)

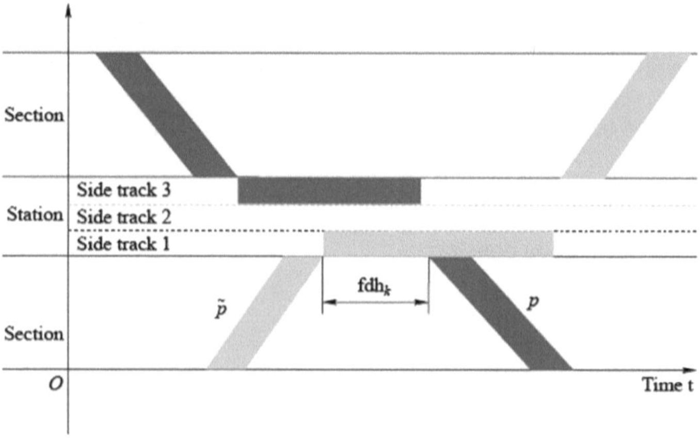

Fig. 6.16 Time interval constraint between the meeting of trains (when both trains stop)

$$\alpha_{p,\tilde{p}} + \beta_{p,\tilde{p}} = 1 \quad p, \tilde{p} \in \mathrm{Itv}_k \quad k \in \mathrm{Sec} \tag{6.95}$$

When the train runs in the interval, it is impossible for the two operations of the train to occupy the same block section at the same time

(2) Headway constraint

There are more than two trains running in the same direction in a station interval under fixed block condition. Or more than two trains are running in the target range control mode under moving block conditions. The minimum interval between two running trains is called the interval time between two trains, denoted as hw_k. According to whether the tracked train stops, it can be divided into the following four categories.

① Both trains departure after dwelling at the station.

The traditional description of time interval between two trains in same direction is shown in Fig. 6.17. According to the model proposed in this chapter, it can be expressed as Fig. 6.18.

The constraints can be described as follows.

$$s_p - s_{\tilde{p}} \geq \mathrm{hw}_k \tag{6.96}$$

Fig. 6.17 Time interval between two trains in same direction (when both trains stop at a station, traditional description)

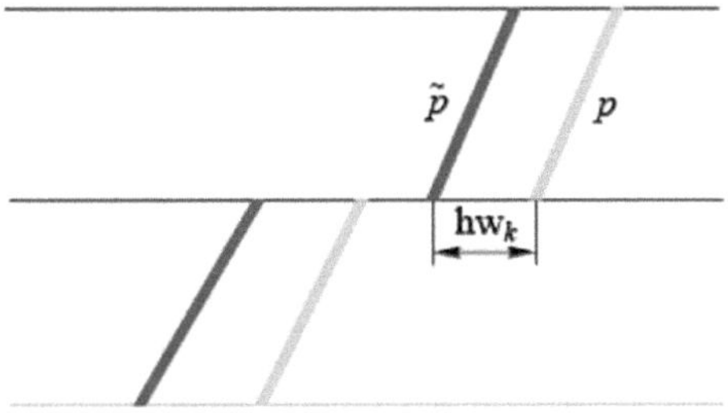

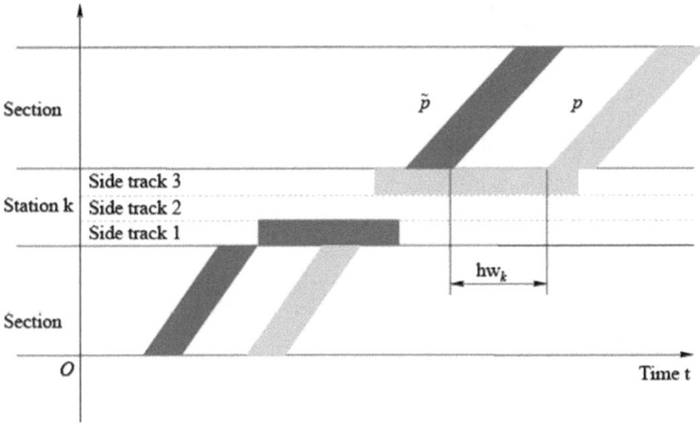

Fig. 6.18 Time interval between two trains in same direction (when both trains stop at a station)

② Both trains pass directly at the station.

The traditional description is shown in Fig. 6.19. According to the model proposed in this chapter, it can be expressed as Fig. 6.20.

The constraints can be described as follows.

$$s_p - s_{\tilde{p}} \geq hw_k \tag{6.97}$$

Fig. 6.19 Time interval at a station between adjacent trains (when both trains pass at the station, traditional description)

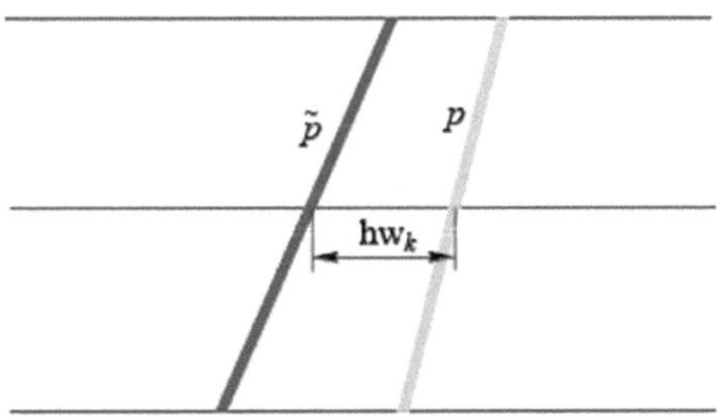

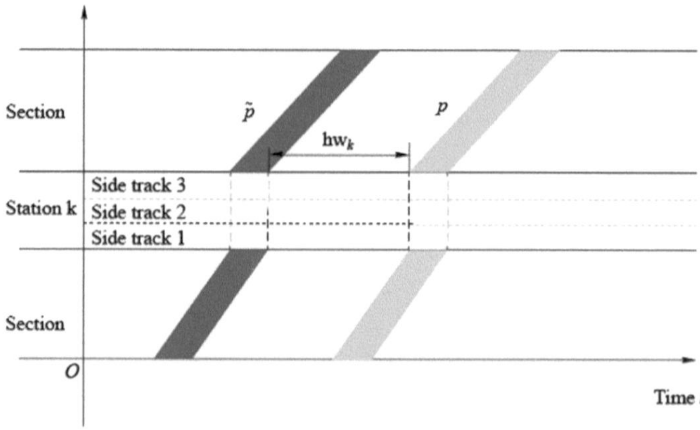

Fig. 6.20 Time interval at a station between adjacent trains (when both trains pass at the station)

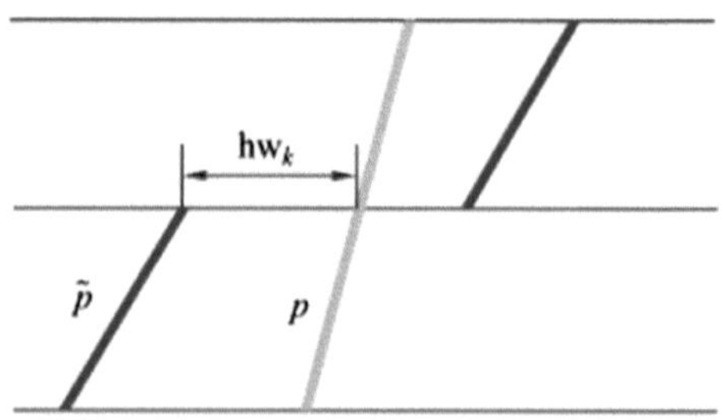

Fig. 6.21 Time interval between arrival and through (traditional description)

③ The interval time between arrival and passage.

The minimum interval between the arrival of the first train at the station and the passing of the second train in the same direction is called the interval time between arrival and passing. The traditional description is shown in Fig. 6.21. According to the model proposed in this chapter, it can be expressed as Fig. 6.22.

The constraints can be described as follows.

$$e_p - e_{\tilde{p}} \geq \mathrm{hw}_k \tag{6.98}$$

④ The time interval between through and departure.

The minimum interval between the passing of the first train at the station and the passing of the second train in the same direction is called the interval time between passing and departure. The traditional description is shown in Fig. 6.21. According to the model proposed in this chapter, it can be expressed as Fig. 6.22.

The constraints can be described as follows.

$$s_p - s_{\tilde{p}} \geq \mathrm{hw}_k \tag{6.99}$$

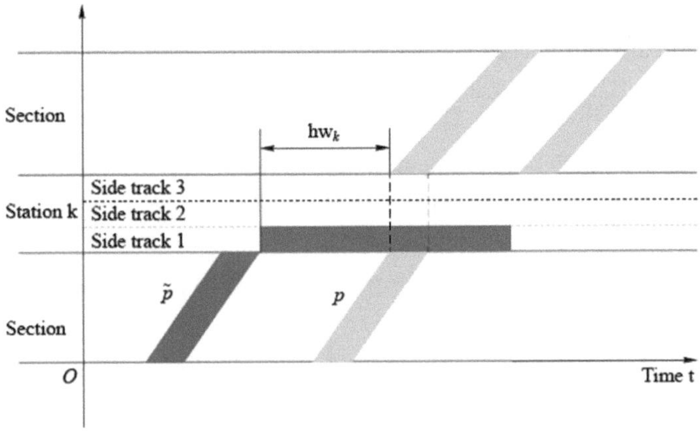

Fig. 6.22 Time interval between arrival and passage

Fig. 6.23 Time interval between adjacent trains (former train passes and latter train stops at the station, traditional description)

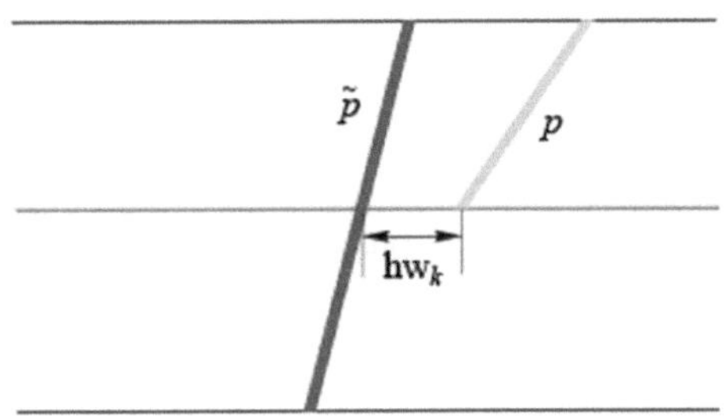

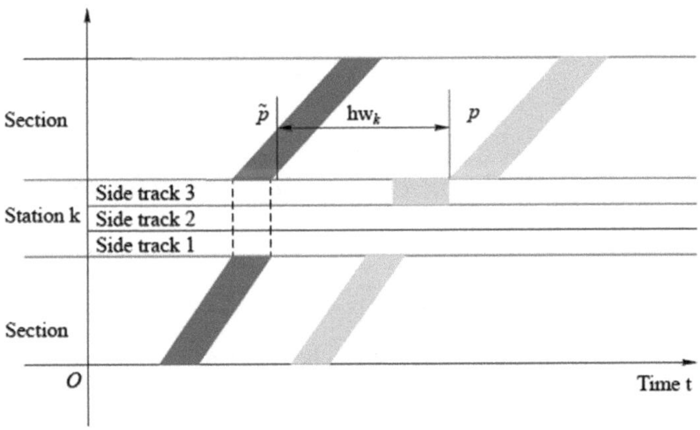

Fig. 6.24 Time interval between adjacent trains (former train passes and latter train stops at the station)

After the train operation is adjusted, new overtaking may occur, so the parameters, $\alpha_{p,\tilde{p}}$ and $\beta_{p,\tilde{p}}$ are introduced. The above constraints are uniformly described as

$$s_{\tilde{p}} - s_p \geq \mathrm{hw}_k \alpha_{p,\tilde{p}} - M\left(1 - \alpha_{p,\tilde{p}}\right) \quad p, \tilde{p} \in \mathrm{Itv}_k \quad k \in \mathrm{Sec} \tag{6.100}$$

$$s_p - s_{\tilde{p}} \geq \mathrm{hw}_k \beta_{p,\tilde{p}} - M\left(1 - \beta_{p,\tilde{p}}\right) \quad p, \tilde{p} \in \mathrm{Itv}_k \quad k \in \mathrm{Sec} \tag{6.101}$$

$$e_{\tilde{p}} - e_p \geq \mathrm{hw}_k \alpha_{p,\tilde{p}} - M\left(1 - \alpha_{p,\tilde{p}}\right) \quad p, \tilde{p} \in \mathrm{Itv}_k \quad k \in \mathrm{Sec} \tag{6.102}$$

$$e_p - e_{\tilde{p}} \geq \mathrm{hw}_k \beta_{p,\tilde{p}} - M\left(1 - \beta_{p,\tilde{p}}\right) \quad p, \tilde{p} \in \mathrm{Itv}_k \quad k \in \mathrm{Sec} \tag{6.103}$$

(4) Constraints of seriously delayed trains

$$d_{\mathrm{last}(\mathrm{Itv}_i)} - w_i \leq M b_i \tag{6.104}$$

This constraint is related to the objective function of the number of severely delayed trains. When the delay time of the train arriving at the terminal exceeds the delay tolerance, the value of b_i is 1.

(5) Overtaking constraints

Low-priority trains can be overtaken by at most three higher-priority trains at the same station.

$$\sum_{\tilde{p} \in \mathrm{Itv}_k} \beta_{p,\tilde{p}} \leq 3 \quad \forall p \in \mathrm{Itv}_k \quad k \in \mathrm{Seg} \tag{6.105}$$

6.3.3 Analysis on the Characteristics of Train Timetable Rescheduling Model

1. Express train tracking effectively

Generally, the traditional train timetable rescheduling model represents the tracking train interval time as the constraint with absolute value. For example, $|p - \tilde{p}| \geq I$ represents the tracking train interval time between any two trains must be larger than I. However, in the real operation environment, dozens or even hundreds of trains are often involved, so it is difficult to determine the sequence of any two operations of any two trains. In particular, the adjustment of train operation under emergency conditions will involve trains on multiple tracks. These trains may also be tracked in a certain section or station, which further increases the complexity of train tracking constraints. Headway time constraints is an important parameter to ensure the safe operation of trains. This constraint with absolute value is very difficult to simplify in the actual calculation. Therefore, this kind of traditional operation adjustment model is often difficult to solve. It can only be solved by some heuristic algorithms such as genetic algorithm and particle swarm optimization. In addition, the process of solving is to first ignore various constraints in the model, then adjust the solutions that do not meet the constraints, and then solve again. Although this is a feasible method in theory, the situation involving multiple tracks and a large number of trains in the actual operating environment will often cause the program to enter an endless loop and fail to find a solution that meets the constraints. Or it takes a lot of time to generate a feasible solution, and the model solution is inefficient. In this section, the $\alpha_{p,\tilde{p}}$ variable clearly expresses the sequence of scheduled train operations. The $\beta_{p,\tilde{p}}$ variable describes the sequence of train operations after adjustment. This design makes the constraint of tracking train interval time no longer has the trouble of absolute value. It is transformed into a linear constraint, which reduces the difficulty of solving the model. In addition, the discrete event p of the train itself does not distinguish the route. Therefore, it is also applicable to the situation of multiple routes in an emergency. Furthermore, the design of the $\beta_{p,\tilde{p}}$ variable makes it clear whether the train is crossing or not. The physical meaning of the model is clearer. By setting the $\beta_{p,\tilde{p}}$, we can flexibly control the crossing mode between trains. Therefore, the train operation adjustment model based on discrete events not only simplifies the constraint expression of the traditional model, but also embodies the principles and strategies of train operation adjustment.

2. Effectively simplify capacity constraints

The constraint of the traditional train timetable rescheduling model on the station capacity stipulates that the number of trains dwelling at the station at any time shall not be greater than the number of its arrival and departure tracks. This means that it needs to constrain the number of trains at stations at any moment. The constraint is generally expressed as an inequality with continuous time t. For example, $\sum_{k=1}^{n} [\mu(x_{i,k}, t) - \mu(y_{i,k}, t)] \leq Nl_i$, and $\mu(x, t) = \begin{cases} 1 & x \leq t \\ 0 & x > t \end{cases}$, Nl_i is the number of arrival and departure tracks in the station. $x_{i,k}$, $y_{i,k}$ are respectively the arrival and departure time of train k at station i. But continuous time t brings great difficulties to the solution of the model. Traditional optimization methods cannot deal with this constraint. Therefore, the solution of the model can only adopt the same treatment as the tracking train interval time constraint, or simply ignore this constraint. Define various types of train operations as time-continuous discrete events p. It has both spatial and temporal properties of resource occupancy. Therefore, a linear combination of discrete event p (see Sect. 6.3.2, "station capacity constraint") can be used to realize the double constraint of space and time of station arrival and departure tracks. It eliminates the trouble of continuous time t in the traditional model and greatly simplifies the solution of the model.

6.3.4 Analysis of Uncertainty Factors in the Train Operation Adjustment Model

1. Analysis of uncertainty factors in the objective function

(1) train delay penalty

c_i^{trnd} represents the penalty for unit delay of train *i*. c_i^{trnd} is related to the type of emergency, level of emergency, train type, and the number of passengers carried by the train and other factors. When the emergency level is low, the impact scope is relatively small, and the duration is relatively short, passengers will have a strong aversion to train delay. The unit delay penalty should be set to a larger value. If the level of the emergency is high, the impact range is relatively large and the duration is long, then the passenger's expectations will change to reach the destination as soon as possible, and they will become less sensitive to the delay time of the train. The unit delay penalty should be set to a small value. In addition, if the train's priority is high, the running distance is long and the number of passengers is large, c_i^{trnd} should be set to a larger value. If the train's priority is low, the running distance is short and the number of passengers is small, c_i^{trnd} should be set to a small value.

In addition, this parameter is introduced to describe the impact of train delays on passenger travel, rather than simply adding up the delay time. Passenger's feeling of train delay is a kind of human psychological activity, which is related to people's occupation, consumption level, quality of life and social, economic, and cultural development level. In the public mind, it is often a vague concept. It is difficult to express in precise numerical terms. Therefore, c_i^{trnd} will be expressed as a fuzzy number, which can more truly reflect people's punishment for different types of train delays under emergency conditions. So Eq. (6.65) can be converted into as follows.

$$z_{\text{delay}}^{\text{train}} = \min \sum_{i \in \text{Trn}} \left(\tilde{c}_i^{\text{trnd}} \bullet \left(\sum_{p \in \text{Itv}_i^{\text{sec}}} (d_p \bullet h_{p+1}) + \sum_{p \in \text{Itv}_i^{\text{stn}}} (d_p \bullet h_p) \right) \right) \tag{6.106}$$

(2) station delay penalty

c_k^{stnd} represents the penalty for unit delay of hub station k. c_k^{stnd} is related to the type of emergency, level of emergency, station level, and the number of passengers passing through the station and other factors. Similar to the penalty for unit delay of train, when the level of emergency is low, the impact range is relatively small, the duration is relatively short, and the station grade is high, a larger value of c_k^{stnd} should be set. Instead, a smaller value of c_k^{stnd} should be set. In addition, this parameter is introduced to describe the impact of train delays on passenger travel. Due to the great difference in the number of passengers on and off at different levels of stations, their importance in the railway transport system is different. As some large hub stations are often located in densely populated cities with larger station yard size and better infrastructure, the passenger service capacity of these stations is far greater than that of ordinary stations. As a result, passengers have higher expectations for these large stations, where delays tend to have a more severe impact. Therefore, c_k^{stnd} will be expressed as a fuzzy number, which can more truly reflect people's punishment for different types of station delays under emergency conditions. So Eq. (6.65) can be converted into as follows.

$$z_{\text{delay}}^{\text{stn}} = \min \sum_{k \in \text{Stn}} \left(\tilde{c}_k^{\text{stnd}} \bullet \sum_{p \in \text{Itv}_k} d_p \right) \tag{6.107}$$

2. Analysis of uncertainty factors in constraints

(1) Train operation time

zt_p represents the section running time of the train or the time of station operation. The section running time of the train is composed of the pure running time of the interval and the additional starting and stopping time. The section length, line characteristics, train type and train speed determine the zt_p size. In the actual operating environment, especially under emergency conditions, due to the different characteristics of the bridge, tunnel, culvert, and other lines, the speed of the train is often not a fixed value, but changes within a certain range. In addition, the natural climate, driver's driving behavior and driving skill level, and the status of related equipment such as electricity and signals for railway operations will also increase the uncertainty of the train's operating hours. Train station operation time generally includes the pure operation time of the station, the time for passengers to get on and off the train, the preparation time of the driver and the steward, and other additional time. Factors such as station operation type, station level, and train type jointly determine the station operation time. In actual operating environment, the natural climate, driver's driving behavior and driving skill level, and the status of related equipment such as signals and turnout will also increase the uncertainty of the station operating hours. Therefore, zt_p will be expressed as a fuzzy number, which can more truly reflect the station operation time. So Eq. (6.83) can be converted into as follows.

$$e_p \geq s_p + \widetilde{\text{zt}}_p, \quad p \in \text{Itv} \tag{6.108}$$

(2) interval time between arrival and departure trains

The interval time between arrival and departure trains at the station is related to the length of arrival and departure tracks, the speed of trains entering and leaving the station, the length of trains, and other factors. The natural climate, the driver's driving behavior and the state of the equipment associated with the station receiving and delivering the train also add to the uncertainty. Therefore, in this section, the interval time of various receiving and dispatching trains is expressed as a fuzzy parameter, and Eqs. (6.89)–(6.94) can be converted into as follows.

$$s_{\tilde{p}} - e_p \geq \tilde{f}_k \alpha_{p,\tilde{p}} - M\left(1 - \alpha_{p,\tilde{p}}\right) \quad p, \tilde{p} \in \text{Itv}_k \quad k \in \text{Stn} \tag{6.109}$$

$$s_p - e_{\tilde{p}} \geq \tilde{f}_k \beta_{p,\tilde{p}} - M(1 - \beta_{p,\tilde{p}}) \quad p, \tilde{p} \in \mathrm{Itv}_k \quad k \in \mathrm{Stn} \tag{6.110}$$

$$s_p - e_{\tilde{p}} \geq \widetilde{\mathrm{fd}}_k \beta_{p,\tilde{p}} - M(1 - \beta_{p,\tilde{p}}) \quad p, \tilde{p} \in \mathrm{Itv}_k \quad k \in \mathrm{Stn} \tag{6.111}$$

$$e_{\tilde{p}} - s_p \geq \widetilde{\mathrm{df}}_k \alpha_{p,\tilde{p}} - M(1 - \alpha_{p,\tilde{p}}) \quad p, \tilde{p} \in \mathrm{Itv}_k \quad k \in \mathrm{Stn} \tag{6.112}$$

$$s_p - e_{\tilde{p}} \geq \widetilde{\mathrm{fds}}_k \beta_{p,\tilde{p}} - M(1 - \beta_{p,\tilde{p}}) \quad p, \tilde{p} \in \mathrm{Itv}_k \quad k \in \mathrm{Seg} \tag{6.113}$$

$$s_p - e_{\tilde{p}} \geq \widetilde{\mathrm{fdh}}_k \beta_{p,\tilde{p}} - M(1 - \beta_{p,\tilde{p}}) \quad p, \tilde{p} \in \mathrm{Itv}_k \quad k \in \mathrm{Sec} \tag{6.114}$$

(3) Tracking train interval

hw_k represents the tracking train interval in section k. According to the analysis of the tracking train interval in Chap. 3, the tracking train interval under the condition of quasi-moving block is determined by several factors, such as t_{action}(the time when the braking command is issued to the train control vehicle equipment to receive it and the response time when the driver takes the braking measures), t_{delay}(delay time for starting the train braking system), a (train brake deceleration), v (the average speed of the train). Train tracking interval under fixed block condition is determined by train length, block length and average train speed. Due to the influence of line conditions, drivers' behavior, equipment status, external environment and many other factors, hw_k has a certain degree of fuzziness. Equations (6.100)–(6.103) can be converted into as follows.

$$s_{\tilde{p}} - s_p \geq \widetilde{\mathrm{hw}}_k \alpha_{p,\tilde{p}} - M(1 - \alpha_{p,\tilde{p}}) \quad p, \tilde{p} \in \mathrm{Itv}_k \quad k \in \mathrm{Sec} \tag{6.115}$$

$$s_p - s_{\tilde{p}} \geq \widetilde{\mathrm{hw}}_k \beta_{p,\tilde{p}} - M(1 - \beta_{p,\tilde{p}}) \quad p, \tilde{p} \in \mathrm{Itv}_k \quad k \in \mathrm{Sec} \tag{6.116}$$

$$e_{\tilde{p}} - e_p \geq \widetilde{\mathrm{hw}}_k \alpha_{p,\tilde{p}} - M(1 - \alpha_{p,\tilde{p}}) \quad p, \tilde{p} \in \mathrm{Itv}_k \quad k \in \mathrm{Sec} \tag{6.117}$$

$$e_p - e_{\tilde{p}} \geq \widetilde{\mathrm{hw}}_k \beta_{p,\tilde{p}} - M(1 - \beta_{p,\tilde{p}}) \quad p, \tilde{p} \in \mathrm{Itv}_k \quad k \in \mathrm{Sec} \tag{6.118}$$

(4) Tolerance of delayed trains

w_i Indicates the delay tolerance of train i. According to the above analysis, it is more meaningful to reduce the number of severely delayed trains than to directly set the goal of reducing the number of delayed trains under emergency conditions. The value of delay tolerance is related to the type and level of the emergency and other attributes. In addition, passengers' acceptance of train delays is a vague feeling in people's hearts, which is difficult to express in precise number. Therefore, the expression of w_i as a fuzzy number can more truly reflect passengers' acceptance of different types of train delays under emergency conditions. Therefore, the Eq. (6.104) can be converted into as follows.

$$d_{\text{last}(\text{Itv}_i)} - \tilde{w}_i \leq Mb_i \tag{6.119}$$

6.4 Fuzzy Train Timetable Rescheduling Based on Tolerance Method

Based on the analysis of the uncertainties of the train operation adjustment model in Sect. 6.3.4, the tolerance method is designed in this section to improve the objective function of the model under the condition that the constraint conditions are not relaxed or less relaxed.

6.4.1 Fuzzy Linear Programming

Fuzzy linear programming can be divided into fuzzy target resource type, right-end term coefficient fuzzy type, price system fuzzy type, and full coefficient fuzzy type according to the location of fuzzy parameters. In this chapter, the Zimmermann symmetric model is used to solve the problem [56].

A fuzzy linear programming with full coefficients can be expressed as follows.

$$\begin{aligned} &\max z = \tilde{c}^T x \\ &\text{s.t. } \tilde{A}x \leq \tilde{b} \\ &\quad x \geq 0 \end{aligned} \tag{6.120}$$

$\tilde{\boldsymbol{c}}$,$\tilde{\boldsymbol{A}}$,$\tilde{\boldsymbol{b}}$ is a fuzzy number. The linear programming with fuzzy numbers for constraints can be expressed as [57].

$$\begin{aligned} &\max z = c^T x \\ &\text{s.t. } Ax \leq b \\ &\quad x \geq 0 \end{aligned} \tag{6.121}$$

Among them $\boldsymbol{A} = (a_{ij})_{m \times n}$, **b**、**c** and **x** have corresponding dimensions. And assuming that $r(\boldsymbol{A}) = m$. Suppose the maximum tolerance of the i-th resource constraint is p_i, and its membership function is defined as follows.

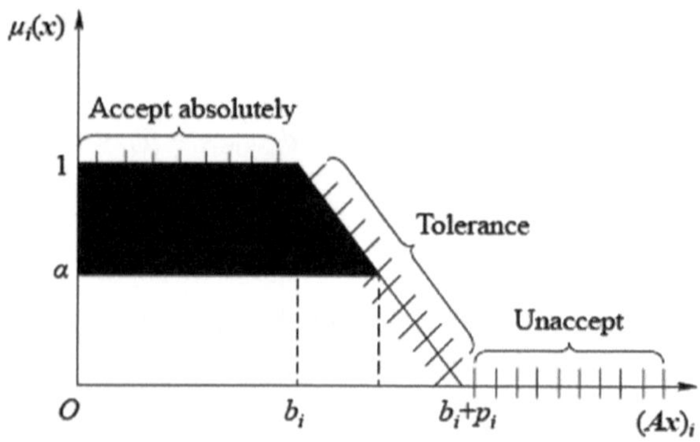

Fig. 6.25 Membership function

$$\mu_i(x) = \begin{cases} 1 & (Ax)_i < b_i \\ 1 - \frac{(Ax)_i - b_i}{p_i} & b_i \le (Ax)_i \le b_i + p_i \\ 0 & b_i + p_i < (Ax)_i \end{cases} \tag{6.122}$$

The membership function is shown in Fig. 6.25.

Based on the idea of the Zimmermann symmetric model, the fuzziness of the constraint conditions inevitably leads to the fuzziness of the objective function [58]. Using the solution method of the linear programming problem, we can get the solution Z_1 when the constraint condition is not relaxed and the solution Z_0 when the relaxation is large. The scalability of the target can be expressed as Eq. (6.123), and the tolerance index refers to d_0.

$$Z = cx \gtrsim Z_1 \tag{6.123}$$

$$d_0 = Z_0 - Z_1 \tag{6.124}$$

$$Z_1 = \max_{x \in X_L} z(x) \tag{6.125}$$

$$Z_0 = \max_{x \in X_U} z(x) \tag{6.126}$$

$$X_L = \{x \in R | a_{i1}x_1 + a_{i2}x_2 + \cdots + a_{in}x_n \le b_i\} \tag{6.127}$$

$$X_U = \{x \in R | a_{i1}x_1 + a_{i2}x_2 + \cdots + a_{in}x_n \le b_i + p_i\} \tag{6.128}$$

The membership function at this time is defined as follows.

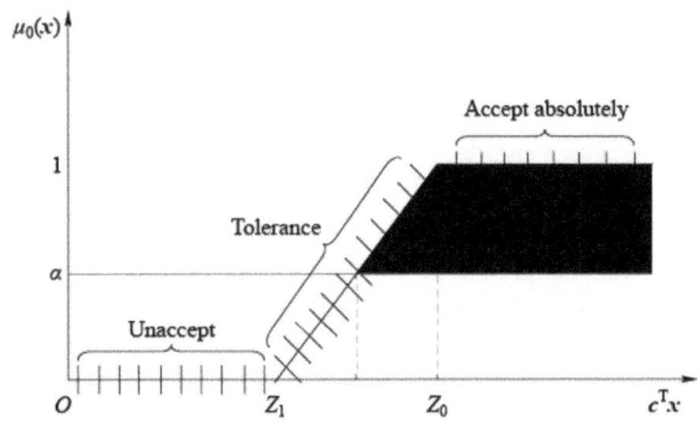

Fig. 6.26 Membership function of objective function

$$\mu_0(x) = \begin{cases} 1 & c^T x > Z_0 \\ 1 - \frac{Z_0 - c^T x}{d_0} & Z_1 \le c^T x \le Z_0 \\ 0 & c^T x < Z \end{cases} \tag{6.129}$$

The membership function is shown in Fig. 6.26. The cut set of resource constraints is defined as follows.

$$X_\alpha = \left\{ x \in R_+^n | \mu_i(x) \ge \alpha, \forall i = 1, 2, \ldots, m \right\} \tag{6.130}$$

Among them, μ_1,μ_2,...,μ_m are memberships of resource constraints. At this time, the optimal decision is as follows.

$$\boldsymbol{x}^* = \arg(\max_{\boldsymbol{x}} \mu_{\mathrm{D}}(\boldsymbol{x})) \tag{6.131}$$

In fact, due to $\mu_{\mathrm{D}}(\boldsymbol{x}) = \min\{\mu_0(\boldsymbol{x}), \mu_1(\boldsymbol{x}), \ldots, \mu_m(\boldsymbol{x})\}$, the problem can be transformed into the following linear programming [59].

$$\max \alpha \tag{6.132}$$

$$\text{s.t. } \mu_0(\boldsymbol{x}) \ge \alpha \tag{6.133}$$

$$\mu_i(\boldsymbol{x}) \ge \alpha \quad \forall i = 1, 2, \ldots, m \tag{6.134}$$

$$\alpha \in [0, 1] \quad x \ge 0 \tag{6.135}$$

Expand the objective function and the membership function of resource constraints as described above, and then obtain the clear linear programming (LP) described below.

$$\max \alpha \tag{6.136}$$

$$\text{s.t. } \boldsymbol{c}^{\mathrm{T}}\boldsymbol{x} \geq Z_1 + \alpha(Z_0 - Z_1) \tag{6.137}$$

$$(\boldsymbol{Ax})_i \leq b_i + (1 - \alpha)p_i \ \forall i = 1, 2, \ldots, m \tag{6.138}$$

$$\alpha \in [0, 1] \quad x \geq 0 \tag{6.139}$$

The optimal solution is the optimal decision ($\boldsymbol{x}^*$) to reach the maximum degree of membership (α^*), which is recorded as:

$$S^* = \{(x^*, \alpha^*)|\alpha \in [0, 1], x^* \text{ is the solution of LP}\} \tag{6.140}$$

6.4.2 *Fuzzy Adjustment Model of Train Timetable Based on Tolerance Method*

The above method of fuzzy linear programming is to improve the objective function greatly under the condition that the constraint conditions are relaxed slightly. The adjustment model of train operation is a multi-objective mixed-integer programming with 0–1 variable, and its target price coefficient and the right end coefficient of the constraint conditions are vague. Therefore, symmetry ideology will be used to solve the problem.

The steps of modeling of fuzzy train operation adjustment based on tolerance method are as follows.

(1) Determine the tolerance of the ambiguity coefficient in the system.

According to the decision-maker's preference or the stability requirements of the system, the tolerance of the fuzzy coefficient in the adjustment model of original train operation plan is set. The variables are described as follows:

$H_{c_i^{\mathrm{trnd}}}$—Tolerance of unit delay time penalty for train i;

$H_{c_k^{\mathrm{stnd}}}$—Tolerance of unit delay time penalty for hub station k;

H_{zt_p}—Tolerance of train running time in section or station operating time;

H_{f_k}—Tolerance of time interval between train reception and departure at the station;

H_{fd_k}—Tolerance of time interval between two trains in the same direction departure and reception at station not at the same time;

H_{df_k}—Tolerance of time interval between two trains in the same direction reception and departure at station not at the same time;

H_{fds_k}—Tolerance of time interval between two opposing trains arriving at station not at the same time;

H_{fdh_k}—Tolerance of time interval for two meeting trains at station;

H_{hw_k}—Tolerance of time interval between trains spaced by automatic block signals in section k;

H_{w_i}—Tolerance of the tolerance of train i delays.

(2) Solve the clear model without loose constraints.

Without loose constraints, the adjustment model of train operation is transformed into a mixed-integer programming with 0–1 variables and continuous variables. Then, the traditional branch-and-bound method or cutting plane method can be used to solve the model. For details, please refer to Sect. 5.2. In this section, the delay time penalty and the number of severely delayed trains are used as the objective function. At this time, the model is described as follows.

Objective function:

$$\begin{aligned}\min Z^* &= w_1 z_{\text{delay}}^{\text{train}} + w_2 z_{\text{delay}}^{\text{stn}} + w_3 z_{\text{delay}}^{\text{trncount}} \\ &= w_1 \sum_{i \in \text{Trn}} \left(c_i^{\text{trnd}} \bullet \left(\sum_{p \in \text{Itv}_i^{\text{sec}}} (d_p \bullet h_{p+1}) + \sum_{p \in \text{Itv}_i^{\text{stn}}} (d_p \bullet h_p) \right) \right) \\ &+ w_2 \sum_{k \in \text{Stn}} \left(c_k^{\text{stnd}} \bullet \sum_{p \in \text{Itv}_k} d_p \right) + w_3 \sum_{i \in \text{Trn}} b_i \end{aligned} \tag{6.141}$$

Among them, wi is the weight of the i-th sub-goal.

Constraints with ambiguity:

$$g_1 = e_p - s_p \geq \text{zt}_p \quad p \in \text{Itv} \tag{6.142}$$

$$g_2 = Mb_i - d_{\text{last}(\text{Itv}_i)} \geq -w_i \tag{6.143}$$

$$g_3 = s_{\tilde{p}} - e_p \geq f_k \alpha_{p,\tilde{p}} - M(1 - \alpha_{p,\tilde{p}}) \quad p, \tilde{p} \in \text{Itv}_k \quad k \in \text{Stn} \tag{6.144}$$

$$g_4 = s_p - e_{\tilde{p}} \geq f_k \beta_{p,\tilde{p}} - M(1 - \beta_{p,\tilde{p}}) \quad p, \tilde{p} \in \text{Itv}_k \quad k \in \text{Stn} \tag{6.145}$$

$$g_5 = s_p - e_{\tilde{p}} \geq f d_k \beta_{p,\tilde{p}} - M\,(1 - \beta_{p,\tilde{p}}) \quad k \in \text{Stn} \tag{6.146}$$

$$g_6 = e_{\tilde{p}} - s_p \geq \text{df}_k \alpha_{p,\tilde{p}} - M(1 - \alpha_{p,\tilde{p}}) \quad p, \tilde{p} \in \text{Itv}_k \quad k \in \text{Stn} \tag{6.147}$$

$$g_7 = s_p - e_{\tilde{p}} \geq \text{fds}_k \beta_{p,\tilde{p}} - M(1 - \beta_{p,\tilde{p}}) \quad p, \tilde{p} \in \text{Itv}_k \quad k \in \text{Seg} \tag{6.148}$$

$$p, \tilde{p} \in \text{Itv}_k \quad k \in \text{Sec} \tag{6.149}$$

$$g_9 = s_{\tilde{p}} - s_p \geq \text{hw}_k \alpha_{p,\tilde{p}} - M(1 - \alpha_{p,\tilde{p}}) \quad p, \tilde{p} \in \text{Itv}_k \quad k \in \text{Sec} \tag{6.150}$$

$$g_{10} = s_p - s_{\tilde{p}} \geq \text{hw}_k \beta_{p,\tilde{p}} - M(1 - \beta_{p,\tilde{p}}) \quad p, \tilde{p} \in \text{Itv}_k \quad k \in \text{Sec} \tag{6.151}$$

$$g_{11} = e_{\tilde{p}} - e_p \geq \text{hw}_k \alpha_{p,\tilde{p}} - M(1 - \alpha_{p,\tilde{p}}) \quad p, \tilde{p} \in \text{Itv}_k \quad k \in \text{Sec} \tag{6.152}$$

$$g_{12} = e_p - e_{\tilde{p}} \geq \text{hw}_k \beta_{p,\tilde{p}} - M(1 - \beta_{p,\tilde{p}}) \quad p, \tilde{p} \in \text{Itv}_k \quad k \in \text{Sec} \tag{6.153}$$

Certain constraints: $g_8 = s_p - e_{\tilde{p}} \geq \text{fdh}_k \beta_{p,\tilde{p}} - M(1 - \beta_{p,\tilde{p}})$

$$\eta_{p,l} + \eta_{\tilde{p},l} - 1 \leq \alpha_{p,\tilde{p}} + \beta_{p,\tilde{p}} \quad p, \tilde{p} \in \text{Itv}_k \quad p < \tilde{p} \quad l \in \text{Trk}_k \quad k \in \text{Stn} \tag{6.154}$$

$$e_p \geq \text{inis}_p \quad p \in \text{Itv} \quad h_p = 1 \tag{6.155}$$

$$e_p - \text{inie}_p = d_p \quad p \in \text{Itv} \tag{6.156}$$

$$e_p = s_{p+1} \quad p \in \text{Itv}_i \quad p \neq \text{last}(\text{Itv}_i) \quad i \in \text{Trn} \tag{6.157}$$

$$\sum_{l \in \text{Trk}_k} \eta_{p,l} = 1 \quad p \in \text{Itv}_k \quad k \in \text{Stn} \tag{6.158}$$

$$\alpha_{p,\tilde{p}} + \beta_{p,\tilde{p}} = 1 \quad p, \tilde{p} \in \text{Itv}_k \quad k \in \text{Sec} \tag{6.159}$$

$$\sum_{\tilde{p} \in \text{Itv}_k} \beta_{p,\tilde{p}} \leq 3 \quad \forall p \in \text{Itv}_k \quad k \in \text{Seg}$$

(3) Solve the certain model under loose constraints.

Note that delay time penalty in the objective function is also ambiguous. Therefore, the penalty coefficient in the model should be reduced by the corresponding tolerance.

Objective function is as follows.

$$\begin{aligned} \min Z^0 = {} & w_1 \sum_{i \in \text{Trn}} \left((c_i^{\text{trnd}} - H_{c_i^{\text{trnd}}}) \bullet \left(\sum_{p \in \text{Itv}_i^{\text{sec}}} (d_p \bullet h_{p+1}) + \sum_{p \in \text{Itv}_i^{\text{stn}}} (d_p \bullet h_p) \right) \right) \\ & + w_2 \sum_{k \in \text{Stn}} \left((c_k^{\text{stnd}} - H_{c_k^{\text{stnd}}}) \bullet \sum_{p \in \text{Itv}_k} d_p \right) + w_3 \sum_{i \in \text{Trn}} b_i \end{aligned} \tag{6.160}$$

Constraints with ambiguity:

$$g_1 = e_p - s_p \geq \text{zt}_p - H_{\text{zt}_p}, \ p \in \text{Itv} \tag{6.161}$$

$$g_2 = Mb_i - d_{\text{last}(\text{Itv}_i)} \geq -w_i - H_{w_i} \tag{6.162}$$

$$g_3 = s_{\tilde{p}} - e_p \geq (f_k - H_{f_k})\alpha_{p,\tilde{p}} - M(1 - \alpha_{p,\tilde{p}})\ p,\ \tilde{p} \in \mathrm{Itv}_k\ k \in \mathrm{Stn} \tag{6.163}$$

$$g_4 = s_p - e_{\tilde{p}} \geq (f_k - H_{f_k})\beta_{p,\tilde{p}} - M(1 - \beta_{p,\tilde{p}})\ p,\ \tilde{p} \in \mathrm{Itv}_k\ k \in \mathrm{Stn} \tag{6.164}$$

$$g_5 = s_p - e_{\tilde{p}} \geq (\mathrm{fd}_k - H_{\mathrm{fd}_k})\beta_{p,\tilde{p}} - M(1 - \beta_{p,\tilde{p}})\ p,\ \tilde{p} \in \mathrm{Itv}_k\ k \in \mathrm{Stn} \tag{6.165}$$

$$g_6 = e_{\tilde{p}} - s_p \geq (\mathrm{df}_k - H_{\mathrm{df}_k})\alpha_{p,\tilde{p}} - M(1 - \alpha_{p,\tilde{p}})\ p,\ \tilde{p} \in \mathrm{Itv}_k\ k \in \mathrm{Stn} \tag{6.166}$$

$$g_7 = s_p - e_{\tilde{p}} \geq (\mathrm{fds}_k - H_{\mathrm{fds}_k})\beta_{p,\tilde{p}} - M(1 - \beta_{p,\tilde{p}})\ p,\ \tilde{p} \in \mathrm{Itv}_k\ k \in \mathrm{Seg} \tag{6.167}$$

$$g_8 = s_p - e_{\tilde{p}} \geq (\mathrm{fdh}_k - H_{\mathrm{fdh}_k})\beta_{p,\tilde{p}} - M(1 - \beta_{p,\tilde{p}})\ p,\ \tilde{p} \in \mathrm{Itv}_k\ k \in \mathrm{Sec} \tag{6.168}$$

$$g_9 = s_{\tilde{p}} - s_p \geq (\mathrm{hw}_k - H_{\mathrm{hw}_k})\alpha_{p,\tilde{p}} - M(1 - \alpha_{p,\tilde{p}})\ p,\ \tilde{p} \in \mathrm{Itv}_k\ k \in \mathrm{Sec} \tag{6.169}$$

$$g_{10} = s_p - s_{\tilde{p}} \geq (\mathrm{hw}_k - H_{\mathrm{hw}_k})\beta_{p,\tilde{p}} - M(1 - \beta_{p,\tilde{p}})\ p,\ \tilde{p} \in \mathrm{Itv}_k\ k \in \mathrm{Sec} \tag{6.170}$$

$$g_{11} = e_{\tilde{p}} - e_p \geq (\mathrm{hw}_k - H_{\mathrm{hw}_k})\alpha_{p,\tilde{p}} - M(1 - \alpha_{p,\tilde{p}})\ p,\ \tilde{p} \in \mathrm{Itv}_k\ k \in \mathrm{Sec} \tag{6.171}$$

$$g_{12} = e_p - e_{\tilde{p}} \geq (\mathrm{hw}_k - H_{\mathrm{hw}_k})\beta_{p,\tilde{p}} - M(1 - \beta_{p,\tilde{p}})\ p,\ \tilde{p} \in \mathrm{Itv}_k\ k \in \mathrm{Sec} \tag{6.172}$$

(4) Calculate the objective function and membership function with fuzzy attribute constraints, and then establish a fuzzy operation adjustment model based on tolerance method.

First, calculate the membership function of the objective function as follows.

$$\mu_0 = \begin{cases} 1 & Z < Z^* \\ 1 - \frac{Z - Z^*}{Z^0 - Z^*} & Z^* \leq Z \leq Z^0 \\ 0 & Z > Z^0 \end{cases} \tag{6.173}$$

Its membership function is shown in Fig. 6.27.

The membership function of train dwelling time constraints is:

$$\mu_1 = \begin{cases} 1 & g_1 > \mathrm{zt}_p \\ 1 - \frac{\mathrm{zt}_p - g_1}{H_{\mathrm{zt}_p}} & \mathrm{zt}_p - H_{\mathrm{zt}_p} \leq g_1 \leq \mathrm{zt}_p \\ 0 & g_1 < \mathrm{zt}_p - H_{\mathrm{zt}_p} \end{cases} \tag{6.174}$$

Its membership function is shown in Fig. 6.28.

The forms of the membership functions of other constraints are similar to the formula (6.174). The membership function of tolerance of train delays constraints is

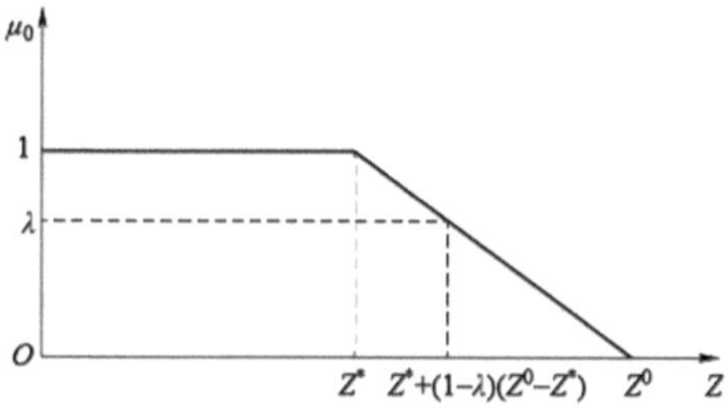

Fig. 6.27 Membership function of objective function

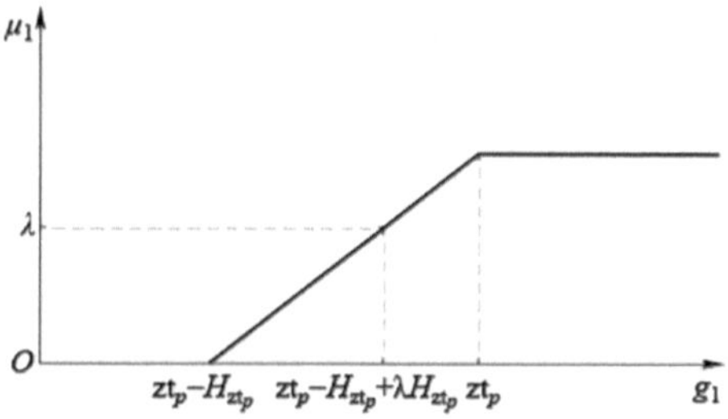

Fig. 6.28 Membership function of train dwelling time constraints

as follows.

$$\mu_2 = \begin{cases} 1 & g_2 > -w \\ 1 - \dfrac{-w_i - g_2}{H_{w_i}} & -w_i - H_{w_i} \le g_2 \le -w_i \\ 0 & g_2 < -w_i - H_{w_i} \end{cases} \tag{6.175}$$

The membership function of train arrival and departure at the station constraints is as follows.

$$\mu_3 = \begin{cases} 1 & g_3 > f_k \\ 1 - \dfrac{f_k - g_3}{H_{f_k}} & f_k - H_{f_k} \le g_3 \le f_k \\ 0 & g_3 < f_k - H_{f_k} \end{cases} \tag{6.176}$$

$$\mu_4 = \begin{cases} 1 & g_4 > f_k \\ 1 - \dfrac{f_k - g_4}{H_{f_k}} & f_k - H_{f_k} \le g_4 \le f_k \\ 0 & g_4 < f_k - H_{f_k} \end{cases} \tag{6.177}$$

$$\mu_5 = \begin{cases} 1 & g_5 > \mathrm{fd}_k \\ 1 - \dfrac{\mathrm{fd}_k - g_5}{H_{fd_k}} & \mathrm{fd}_k - H_{\mathrm{fd}_k} \le g_5 \le \mathrm{fd}_k \\ 0 & g_5 < \mathrm{fd}_k - H_{\mathrm{fd}_k} \end{cases} \tag{6.178}$$

$$\mu_6 = \begin{cases} 1 & g_6 > \mathrm{df}_k \\ 1 - \dfrac{\mathrm{df}_k - g_3}{H_{\mathrm{df}_k}} & df_k - H_{df_k} \le g_6 \le \mathrm{df}_k \\ 0 & g_6 < \mathrm{df}_k - H_{\mathrm{df}_k} \end{cases} \tag{6.179}$$

$$\mu_7 = \begin{cases} 1 & g_7 > \mathrm{fds}_k \\ 1 - \dfrac{\mathrm{fds}_k - g_7}{H_{fds_k}} & \mathrm{fds}_k - H_{\mathrm{fds}_k} \le g_7 \le \mathrm{fds}_k \\ 0 & g_7 < \mathrm{fds}_k - H_{\mathrm{fds}_k} \end{cases} \tag{6.180}$$

$$\mu_8 = \begin{cases} 1 & g_8 > \mathrm{fdh}_k \\ 1 - \dfrac{\mathrm{fdh}_k - g_8}{H_{fdh_k}} & \mathrm{fdh}_k - H_{\mathrm{fdh}_k} \le g_8 \le \mathrm{fdh}_k \\ 0 & g_8 < \mathrm{fdh}_k - H_{\mathrm{fdh}_k} \end{cases} \tag{6.181}$$

$$\mu_9 = \begin{cases} 1 & g_9 > \mathrm{hw}_k \\ 1 - \dfrac{\mathrm{hw}_k - g_9}{H_{hw_k}} & \mathrm{hw}_k - H_{\mathrm{hw}_k} \le g_9 \le \mathrm{hw}_k \\ 0 & g_9 < \mathrm{hw}_k - H_{\mathrm{hw}_k} \end{cases} \tag{6.182}$$

The form of the membership function of constraints g_{10}, g_{11}, g_{12} is the same as that of Eq. (6.182). When the membership function of the constraint condition and the objective function are recorded as λ, the following determined plan is obtained.

Objective function:

$$\mathrm{Max}\lambda \tag{6.183}$$

Constraint conditions:

Objective constraints of the original problem:

$$w_1 \sum_{i \in \mathrm{Trn}} \left(c_i^{\mathrm{trnd}} \bullet \left(\sum_{p \in \mathrm{Itv}_i^{\mathrm{sec}}} (d_p \bullet h_{p+1}) + \sum_{p \in \mathrm{Itv}_i^{\mathrm{stn}}} (d_p \bullet h_p) \right) \right)$$
$$+ w_2 \sum_{k \in \mathrm{Stn}} \left(c_k^{\mathrm{stnd}} \bullet \sum_{p \in \mathrm{Itv}_k} d_p \right)$$

$$+ w_3 \sum_{i \in \mathrm{Trn}} b_i \le Z^* + (1 - \lambda)(Z^0 - Z^*) \tag{6.184}$$

Other constraints:

$$e_p - s_p \ge \mathrm{zt}_p + (\lambda - 1) H_{\mathrm{zt}_p} \;\; p \in \mathrm{Itv} \tag{6.185}$$

$$M b_i - d_{\mathrm{last}(\mathrm{Itv}_i)} \ge -w_i + (\lambda - 1) H_{w_i} \tag{6.186}$$

$$s_{\tilde{p}} - e_p \ge (f_k + (\lambda - 1) H_{f_k}) \alpha_{p,\tilde{p}} - M(1 - \alpha_{p,\tilde{p}}) \;\; p, \tilde{p} \in \mathrm{Itv}_k \;\; k \in \mathrm{Stn} \tag{6.187}$$

$$s_p - e_{\tilde{p}} \ge (f_k + (\lambda - 1) H_{f_k}) \beta_{p,\tilde{p}} - M(1 - \beta_{p,\tilde{p}}) \;\; p, \tilde{p} \in \mathrm{Itv}_k \;\; k \in \mathrm{Stn} \tag{6.188}$$

$$s_p - e_{\tilde{p}} \ge (\mathrm{fd}_k + (\lambda - 1) H_{\mathrm{fd}_k}) \beta_{p,\tilde{p}} - M(1 - \beta_{p,\tilde{p}}) \;\; p, \tilde{p} \in \mathrm{Itv}_k \;\; k \in \mathrm{Stn} \tag{6.189}$$

$$e_{\tilde{p}} - s_p \ge (\mathrm{df}_k + (\lambda - 1) H_{\mathrm{df}_k}) \alpha_{p,\tilde{p}} - M(1 - \alpha_{p,\tilde{p}}) \;\; p, \tilde{p} \in \mathrm{Itv}_k \;\; k \in \mathrm{Stn} \tag{6.190}$$

$$s_p - e_{\tilde{p}} \ge (\mathrm{fds}_k + (\lambda - 1) H_{\mathrm{fds}_k}) \beta_{p,\tilde{p}} - M(1 - \beta_{p,\tilde{p}}) \;\; p, \tilde{p} \in \mathrm{Itv}_k \;\; k \in \mathrm{Seg} \tag{6.191}$$

$$s_p - e_{\tilde{p}} \ge (\mathrm{fdh}_k + (\lambda - 1) H_{\mathrm{fdh}_k}) \beta_{p,\tilde{p}} - M(1 - \beta_{p,\tilde{p}}) \;\; p, \tilde{p} \in \mathrm{Itv}_k \;\; k \in \mathrm{Sec} \tag{6.192}$$

$$s_{\tilde{p}} - s_p \ge (\mathrm{hw}_k + (\lambda - 1) H_{\mathrm{hw}_k}) \alpha_{p,\tilde{p}} - M(1 - \alpha_{p,\tilde{p}}) \;\; p, \tilde{p} \in \mathrm{Itv}_k \;\; k \in \mathrm{Sec} \tag{6.193}$$

$$s_p - s_{\tilde{p}} \ge (\mathrm{hw}_k + (\lambda - 1) H_{\mathrm{hw}_k}) \beta_{p,\tilde{p}} - M(1 - \beta_{p,\tilde{p}}) \;\; p, \tilde{p} \in \mathrm{Itv}_k \;\; k \in \mathrm{Sec} \tag{6.194}$$

$$e_{\tilde{p}} - e_p \ge (\mathrm{hw}_k + (\lambda - 1) H_{\mathrm{hw}_k}) \alpha_{p,\tilde{p}} - M(1 - \alpha_{p,\tilde{p}}) \;\; p, \tilde{p} \in \mathrm{Itv}_k \;\; k \in \mathrm{Sec} \tag{6.195}$$

$$e_p - e_{\tilde{p}} \ge (\mathrm{hw}_k + (\lambda - 1) H_{\mathrm{hw}_k}) \beta_{p,\tilde{p}} - M(1 - \beta_{p,\tilde{p}}) \;\; p, \tilde{p} \in \mathrm{Itv}_k \;\; k \in \mathrm{Sec} \tag{6.196}$$

Other certain constraints are unchanged. Since M is an integer with a very large value, Eqs. (6.187)–(6.196) can be transformed into the following linear constraints.

$$s_{\tilde{p}} - e_p \ge f_k + (\lambda - 1) H_{f_k} - M(1 - \alpha_{p,\tilde{p}}) \;\; p, \tilde{p} \in \mathrm{Itv}_k \;\; k \in \mathrm{Stn} \tag{6.197}$$

$$s_p - e_{\tilde{p}} \ge f_k + (\lambda - 1) H_{f_k} - M(1 - \beta_{p,\tilde{p}}) \;\; p, \tilde{p} \in \mathrm{Itv}_k \;\; k \in \mathrm{Stn} \tag{6.198}$$

$$s_p - e_{\tilde{p}} \geq \text{fd}_k + (\lambda - 1)H_{\text{fd}_k} - M(1 - \beta_{p,\tilde{p}})\ p, \tilde{p} \in \text{Itv}_k\ k \in \text{Stn} \tag{6.199}$$

$$e_{\tilde{p}} - s_p \geq \text{df}_k + (\lambda - 1)H_{\text{df}_k} - M(1 - \alpha_{p,\tilde{p}})\ p, \tilde{p} \in \text{Itv}_k\ k \in \text{Stn} \tag{6.200}$$

$$s_p - e_{\tilde{p}} \geq \text{fds}_k + (\lambda - 1)H_{\text{fds}_k} - M(1 - \beta_{p,\tilde{p}})\ p, \tilde{p} \in \text{Itv}_k\ k \in \text{Seg} \tag{6.201}$$

$$s_p - e_{\tilde{p}} \geq \text{fdh}_k + (\lambda - 1)H_{\text{fdh}_k} - M(1 - \beta_{p,\tilde{p}})\ p, \tilde{p} \in \text{Itv}_k\ k \in \text{Sec} \tag{6.202}$$

$$s_{\tilde{p}} - s_p \geq \text{hw}_k + (\lambda - 1)H_{\text{hw}_k} - M(1 - \alpha_{p,\tilde{p}})\ p, \tilde{p} \in \text{Itv}_k\ k \in \text{Sec} \tag{6.203}$$

$$s_p - s_{\tilde{p}} \geq \text{hw}_k + (\lambda - 1)H_{\text{hw}_k} - M(1 - \beta_{p,\tilde{p}})\ p, \tilde{p} \in \text{Itv}_k\ k \in \text{Sec} \tag{6.204}$$

$$e_{\tilde{p}} - e_p \geq \text{hw}_k + (\lambda - 1)H_{\text{hw}_k} - M(1 - \alpha_{p,\tilde{p}})\ p, \tilde{p} \in \text{Itv}_k\ k \in \text{Sec} \tag{6.205}$$

$$e_p - e_{\tilde{p}} \geq \text{hw}_k + (\lambda - 1)H_{\text{hw}_k} - M(1 - \beta_{p,\tilde{p}})\ p, \tilde{p} \in \text{Itv}_k\ k \in \text{Sec} \tag{6.206}$$

After the transformation of the above steps, the original train timetable rescheduling problem with fuzzy parameters is transformed into a certain mixed-integer linear programming. The tolerance principle of fuzzy train timetable rescheduling model is shown in Fig. 6.29. Find the maximum value of membership λ, so that the objective function can be greatly improved under the condition that the constraints are less loose.

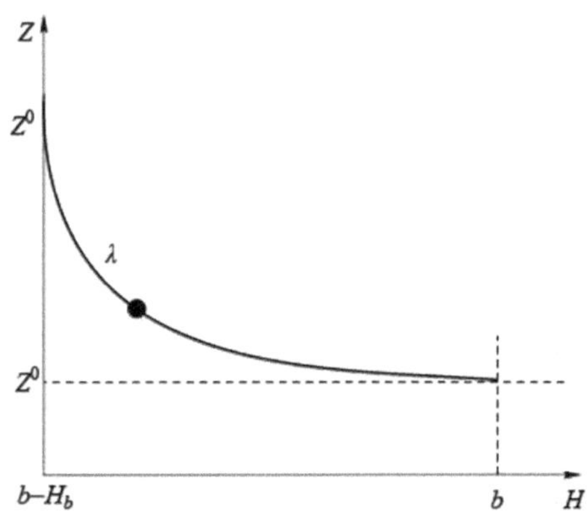

Fig. 6.29 Tolerance principle of fuzzy train timetable rescheduling model

6.4.3 Bi-Level Programming Algorithm for Train Organization Under Emergency Conditions

After the traffic scheduling rules are given, after processing the uncertain parameters in the previous section, the lower-level programming is transformed into a mixed-integer programming for train operation adjustment. Therefore, in this section, a hybrid intelligent algorithm for plant multi-directional growth simulation (embedded with branch-and-bound method) is used to solve the bi-level programming problem of train organization under emergency conditions.

1. **Design of decision variables**

In the original programming, 0–1 variables were designed to describe the train scheduling strategy. Although this design method can clearly describe the relationship between trains and the relationship between trains and detour routes, it will greatly increase the dimension of the variables and generate a large number of uniqueness constraints, which makes it easy to generate numerous non-feasible solutions during the iteration process. This shortcoming will affect solution efficiency, and sometimes even the model cannot get the optimal solution.

Therefore, during the problem solving process, the affected trains are designed as a set of variables, and various integer values of the variables represent the various strategies that the trains can take, which greatly reduces the dimension of the variables and satisfy the uniqueness constraint. The detailed description of the design of decision variables is as follows.

Assuming that there are m trains affected and n routes that can be detoured under emergency conditions, the design variable $\boldsymbol{x} = (x_1, x_2, \ldots, x_m)$, and x is an integer variable. When train i adopts the original route strategy,$x_i = 0$; when train i adopts the merge strategy with train j,$x_i = j, i \neq j$; when train i adopts the detour route strategy k,$x_i = m+k$; and when train i adopts the withdrawal strategy,$x_i = m+n+1$.

In this design, the dimension of the variable is m, and the feasible solution set of the variable is an integer from 0 to $m + n + 1$, that is, $x_i \in [0, m + n + 1]$. In comparison, the dimension of variables in the original problem is $m+mn+m(m-1)$, which is much larger than the number after the change.

2. **Solving process**

The specific steps to solve the bi-level programming model of the train organization under emergency conditions are as follows.

Step1 Enter the number of affected trains, detours, types of trains, number of train compositions, train schedules and other information, design a decision variable **x**, and determine the correspondence between the value of the variable and the strategy it adopts.

Step2 Select the initial solution of the decision variable $\boldsymbol{x}^0$, let $\boldsymbol{x}_{\min} = \boldsymbol{x}^0$, calculate the objective function value of $\boldsymbol{x}^0$, and let $f_{\min} = f(\boldsymbol{x}^0)$. At this time, the initial base point $\boldsymbol{x}^0$ can be regarded as a node with growth ability,$\boldsymbol{x}^G = \boldsymbol{x}^0$; let F be the maximum value that can receive light. The number of iterations $k = 0$; the solution space set $S = \{\boldsymbol{x}^0\}$, and the number of solutions is $\Omega(S) = 1$.

It is worth noting that the composition of $f(\boldsymbol{x}^0)$ not only includes the penalty costs of various scheduling strategies, but also includes feedback of the adjustment results of the lower-level planning under $\boldsymbol{x}^0$ strategy, so this step needs to call Step7.

Step3 Generate new branches.

Step3.1 Vertical growth direction.

First set the angle between the new branches to 90°, and use $\boldsymbol{x}^G$ to make straight line segments parallel to each number axis, take $\boldsymbol{x}^G$ as the center, and intercept the node S'_{i_1,j_1} on the trunk in step λ according to the positive and negative directions of the coordinate axis($\mathrm{i}_1 = 1, 2, \ldots, 2n$, $j_1 = 1, 2, \ldots, m_{i_1}$).

Step3.2 Random growth direction.

Three branches are randomly selected from n directions, a new trunk is formed in a direction symmetrical to the three branches, and then $\boldsymbol{x}^G$ is taken as the center, and the node s'_{i_{2n+1},j_1}, $j_1 = 1, 2, \ldots, m_{i_{2n+1}}$ on the trunk is also intercepted with the step λ, and the process is repeated n times, generating n new growth directions.

Since the value of the decision variable in this problem represents different scheduling strategies but there is no necessary connection between the strategies, in order to ensure the diversity of the solution space, the step size λ in the above steps can be randomly selected in the feasible region $(0, m + n]$.

When train i adopts the strategy of merging with train j, $x_i = j$, and $x_j = i$ is required. Therefore, when the arbitrary solution s_i is taken in $\lfloor 1, m \rfloor$, that is $s_i = j, i \neq j, j \in [1, m]$, it is necessary to judge whether the stopping schedule of train i and train j are the same, and whether the train types meet the multi-locomotive constraints. If they are all satisfied, then $s_j = i$ is directly set; if they are not satisfied, the step size λ is re-selected to grow a new branch.

In addition, determine whether each solution meets the capacity constraints. If it is not satisfied, choose a new direction or step size to grow a new branch.

At this point, a new node set S' is generated.

Step4. Update the optimal node.

First, find the function value of new growth point according to Step 7, and let $s'_{p,q}$ be the node with the smallest function value, that is, $f^*(s'_{p,q}) = \min\{f(s'_{i_1,j_1}), i_1 = 1, 2, \cdots, 3n, j_1 = 1, 2, \cdots, m_{i_1}\}$.

If $f^*(s'_{p,q}) < f_{\min}$,let $\boldsymbol{x}_{\min} = s'_{p,q}$,$f_{\min} = f^*(s'_{p,q})$.

If $f^*(s'_{p,q}) = f_{\min}$,let $\boldsymbol{x}_{\min} := \boldsymbol{x}_{\min} \cup \{s'_{p,q}\}$.

Step5. Calculate the growth probability of a node.

Let $S := S \cup S'$,$F = f_{\min} + \sum (f(s) - f_{\min})/\Omega(S)$.

For each $s \in S$,

if $F \leq f(s)$,let $P_s = 0$,$S := S \backslash \{s\}$;

if $F > f(s)$,let $P_s = (F - f(s))/\left(\sum_{s \in S} (F - f(s))\right)$.

Step6. Select a new base point.

Generate a random number $\eta \in (0, 1)$, and select the node st as the new base point which satisfies $\sum_{s}^{t-1} P_s < \eta < \sum_{s}^{t} P_s$, $t \in [1, \Omega(S)]$. Let $\boldsymbol{x}^G = s_t$,$k = k + 1$,repeat Step3.

Step7. Use branch and bound method to adjust train arrival and departure times.

Traditional branch-and-bound methods often need to traverse all possible branches to obtain the global optimal solution. However, in fact, train operation adjustment problems, especially under emergency conditions, often involve a large number of decision variables, and the constraints are complex. Therefore, the branch can be selected according to the characteristics of the train operation adjustment problem, instead of the traditional method of traversing the branch irregularly, thereby improving the problem solving efficiency. The specific steps are as follows.

UB is the upper bound of the optimization result, NF is the index set of the problem to be solved, $S(P_i)$ is the feasible set of each branch problem Pi of the mixed-integer programming, P0 represents the original train operation adjustment problem, $S(\overline{P}_i)$ is the feasible set of relaxation problem $\overline{P}_i$, $\overline{\boldsymbol{x}}$ is a feasible solution for $S(P_i)$.

Step7.1. Let $\text{NF} = \{0\}, \overline{\boldsymbol{x}} = \emptyset, \text{UB} = +\infty$.

Step7.2. Select the subscript $\upsilon \in \text{NF}$, and solve the relaxation problem $\overline{P}_\upsilon$. Let the optimal solution be $\boldsymbol{x}^\upsilon$ and the optimal value be f_υ. If $f_\upsilon \geq \text{UB}$, let $\text{NF} := \text{NF}\backslash\{v\}$,go to Step7.5. Otherwise, go to Step7.3.

Step7.3. If $f_\upsilon < \text{UB}$ and $\boldsymbol{x}^\upsilon \in S(P_0)$, let $\text{UB} = f_\upsilon$,$\overline{x} = x^\upsilon$,$\text{NF}\backslash\{v\}$, go to Step7.5. Otherwise, go to Step7.4.

Step7.4. If $f_\upsilon < \text{UB}$ and $\boldsymbol{x}^\upsilon \notin S(P_0)$, the problem needs to be branched. Since the most critical integer variable in the train operation adjustment problem is the additional variables $\alpha_{p,\tilde{p}}$ and $\beta_{p,\tilde{p}}$ that represent the crossover relationship between trains, the operation with the earliest start time in the high-level train is selected to branch, and $S(P_\upsilon)$ is decomposed into two subsets $S(P_{\upsilon_1})$ and $S(P_{\upsilon_2})$. Let $\text{NF} := \text{NF}\backslash\{v\} \cup \{v_1, v_2\}$, go to Step7.2.

The above method replaces the traditional method of traversing branches randomly, which not only speeds up the problem solving speed, but also meets the adjustment principle that low-level trains cannot cross high-level trains.

Step7.5. If all the branch sets to be solved (NF) are traversed, the calculation is terminated, and $\overline{\boldsymbol{x}}$ is the optimal solution for the lower-level planning, otherwise it proceeds to Step 7.2.

Step8. Iteration terminates.

According to the scale of the problem, when the number of iterations for which the optimal solution remains unchanged reaches a certain number, a model is set to terminate and an optimal rescheduled timetable is output.

The algorithm flowchart is shown in Fig. 6.30.

6.4.4 *Case Study*

According to the description in the above section, the train scheduling adjustment under emergency conditions is a multi-stage closed-loop uncertain optimization process including bi-level planning triggered by an emergency. The key relationship is closed-loop feedback optimization and fuzzy random model of bi-level planning, and any of them is a complex problem.

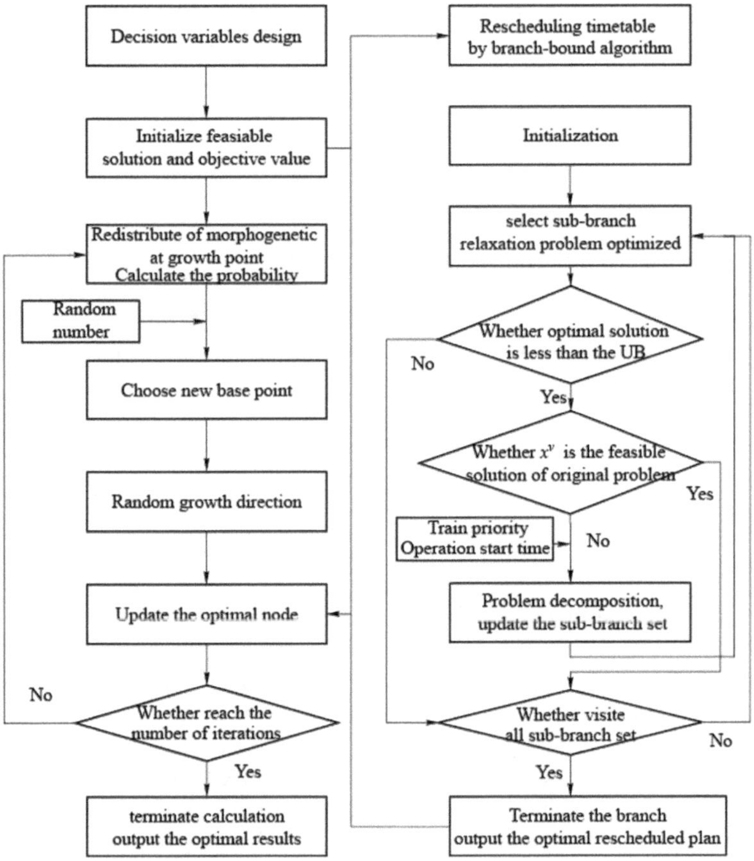

Fig. 6.30 Algorithm flowchart

In order to prove the feasibility of the related theories and methods proposed in this book, this section will carry out experimental analysis in a step-by-step manner.

First, verify the validity of the bi-level programming model without considering fuzzy random factors; then, without considering the upper-level planning, verify the effectiveness of the fuzzy train operation adjustment model based on the tolerance method; then, on the premise that the above two experiments are feasible and the solution efficiency is high, a comprehensive scenario of the emergency is set up to realize the entire multi-stage closed-loop fuzzy random optimization process including the bi-level programming triggered by the emergency. Finally, prove the validity and feasibility of the research contents in this section.

1. Experimental setup

In this section, the data from the Shanghai-Nanjing section of the Beijing-Shanghai high-speed rail is intercepted for experiments. The Shanghai-Nanjing section is the busiest section of the Beijing-Shanghai high-speed rail. The two nearest railway lines

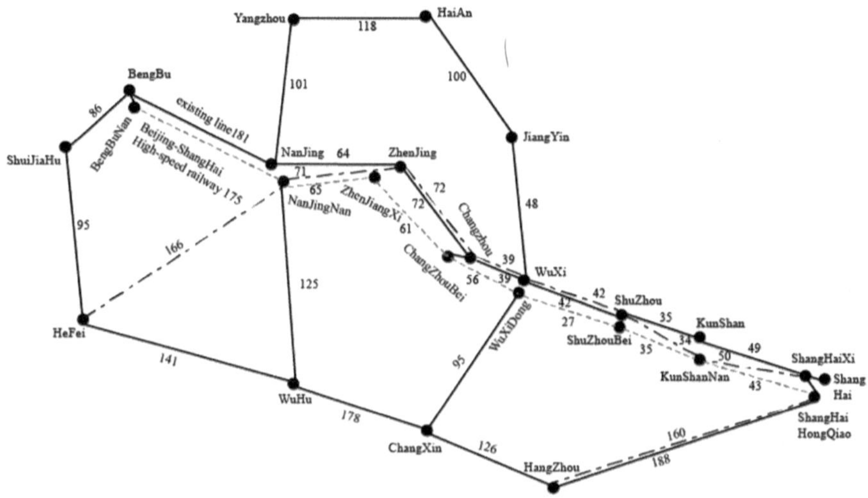

Fig. 6.31 Physical railway network around the Shanghai-Nanjing section

are the Shanghai-Nanjing intercity railway and the Beijing-Shanghai General speed railway.

Figure 6.31 shows a physical network around the Shanghai-Nanjing section. As the three basic line networks used for empirical analysis in this section, it consists of three types of railway lines, namely the existing General speed railway (solid line), the intercity railway (dotted line), and the new high-speed railway (dotted line).

As the Beijing-Shanghai high-speed railway operates on two lines, this section only considers the transportation organization from Nanjing to Shanghai, and does not consider the adjustment of operations of station. The train timetable of Shanghai-Nanjing section can be seen in Fig. 6.32.

There are seven stations on the Shanghai-Nanjing section of the Beijing-Shanghai high-speed railway, including NanJingNan Station, ZhenJiangXi Station, ChangZhouBei Station, WuXiDong Station, SuZhouBei Station, KunShanNan Station, and Shanghai Hongqiao Station. The length of the six sections is 65,110, 61,050, 56,400, 26810, 31,350, and 43,570 m.

According to the train schedule data in the literature [15], the operating start and end time of the Shanghai-Nanjing section is from 6:30 to 23:30. A total of 60 trains were operated during this period, including 14 high-speed trains and 46 medium-speed trains. Among them, G and DJ represent high-speed trains, D represents medium and high speed trains, and K represents medium- and high-speed trains operating across lines. The operating speed range of L and DJ trains is set between 380 and 360 km/h, the operating speed range of G trains is set between 330 and 350 km/h, and the operating speed range of D trains is set between 250 and 300 km/h. The other operating parameters are set as follows.

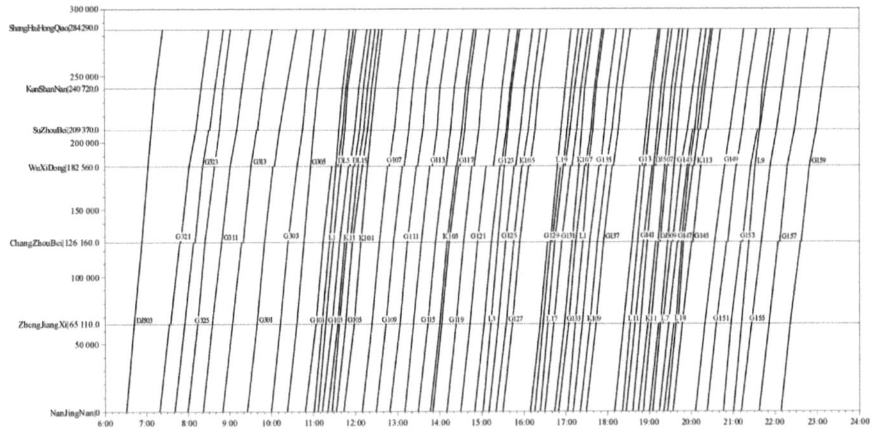

Fig. 6.32 Train timetable of Shanghai-Nanjing section

(1) Section running time: discounted according to the running speed of the train.

(2) Minimum dwell time: The dwell time of high-speed trains is generally short, usually 1 to 2 min, but in the case of large train disturbances, the dwell time can be shortened, so it is uniformly set to 1 min.

(3) Time interval between trains spaced by automatic block signals: In the actual train running process, the time interval between trains includes many different situations. This section mainly controls the minimum interval time of adjacent operations occupying the same resource and uniformly sets it to 2.5–3 min.

(4) The latest train arrival time: According to the train timetable (see Fig. 6.32), the latest train arrival time is less than 24 o'clock.

(5) Station capacity constraints: The number of arrival and departure lines at NanJingNan Station and Shanghai Hongqiao Station is 28 and 32, respectively, and the number of arrival and departure lines at other stations is 6.

(6) The hardware environment of the experiment: a computer with an Intel Core 2 Duo E7500 and 2 GB of memory.

2. **Example Analysis of Bi-Level Programming Model**

The purpose of this case study is to verify the effectiveness of the bi-level programming model and algorithm. Assume that when the Shanghai-Nanjing section of the Beijing-Shanghai high-speed railway appears catastrophic weather, train operation adjustment experiments are performed under different speed limit conditions of 50–300 km/h.

The lower-level planning is carried out first, and the adjustment results are shown in Table 6.14. Among them, η_1 represents the number of trains delayed by more than 30 min, η_2 is the number of trains delayed by more than 60 min, η_3 is the number of trains exceeding the latest arrival time, and η_4 is the number of delayed trains on the line.

Table 6.14 Adjustment results of lower-level planning

Speed limit section	speed limit value/(km/h)	η_1	η_2	η_3	η_4	total delay time/min	calculation time/s
First section	300	0	0	0	1	3.81	4.83
	200	0	0	0	25	98.21	5.24
	150	0	0	0	47	384.81	18.69
	100	0	0	0	52	1158.81	9.22
	80	23	0	0	52	1743.31	6.63
	50	**40**	**20**	**1**	**52**	**3501.31**	**6.69**
First and second sections	300	0	0	0	4	7	5.22
	200	0	0	0	47	383.13	8.13
	150	0	0	0	52	1125.28	6.19
	100	60	0	0	52	2640.14	5.28
	80	**18**	**42**	**1**	**52**	**3776.15**	**5.81**
	50	**0**	**60**	**3**	**52**	**7180.28**	**5.67**
All sections	300	0	0	0	15	54.05	5.13
	200	11	0	0	52	1359.95	3.7
	150	51	9	0	52	3065.95	3.72
	100	**0**	**60**	**2**	**52**	**6476.95**	**4.27**
	80	**0**	**60**	**4**	**52**	**9035.95**	**3.67**
	50	**0**	**60**	**11**	**52**	**16716.15**	**4.14**

As can be seen from Table 6.14, when the speed limit intervals is the first interval or the first and second intervals, and the speed limit value is higher than 100 km/h, η_1 is 0. The comparison of speed limits in different sections shows that the more the number of speed limit sections, the lower the speed limit value. This shows that the higher the level of the emergency, the more serious the impact on the train, and the greater the degree of delay of the train.

In addition, when the speed limit value of the train is very low, the value of η_3 is greater than 0, which means that some trains cannot complete all operations within the operating time. At this time, just adjusting the schedule of the lower-level plan can no longer provide a suitable train operation scheme, but requires a bi-level planning of train scheduling optimization and train operation adjustment.

Based on the bold data in Table 6.14, a two-level planning is performed. Assuming that the impact of the event is large, the existing Beijing-Shanghai General speed railway and Shanghai-Nanjing intercity railway can be used for train detours. The remaining capacity of them is 3 trains and 2 trains, respectively.

In order to describe clearly, the number of trains that need to adopt the scheduling strategy in the Shanghai-Nanjing section is iterated from 1 to 13. Each iteration will generate a new train diagram. Due to space limitations, only the results summarized according to various indicators will be summarized, see Table 6.15.

Table 6.15 Bi-level planning performance

Number of dispatched trains	Speed limit section	Speed limit value/(km/h)	Multi-locomotive train	Reroute train	Canceling train	1	2	3	4	Total delay time/min	Calculation time/s
$\delta = 1, \ldots, 5$	First section	50		G157, G159, K113, K105, K103		5	0	0	0	3217.6	15.27
	First and second sections	80		G153, G155, G157, G159, K113		4	1	0	8	3484.91	15.87
		50		G153, G155, G157, G159, K113		0	5	0	8	6608.88	16.64
	All sections	100		G151, G153, G155, G157, G159		0	55	0	47	5971.05	14.85
		80		G153, L9, G155, G157, G159		0	55	0	47	8300.7	14.88

(continued)

Table 6.15 (continued)

Number of dispatched trains	Speed limit section	Speed limit value/(km/h)	Multi-locomotive train	Reroute train	Canceling train	1	2	3	4	Total delay time/min	Calculation time/s
		50		G153, L9, G155, G157, G159		0	55	6	47	15,467.03	16.32
$\delta = 6$	First section	50	G141-DJ507	G157, G159, K113, K105, K103		34	20	0	49	3160.98	16.25
	First and second sections	80	G141-DJ507	G153, G155, G157, G159, K113		14	41	0	47	3424.73	15.17
		50	G141, DJ507	G153, G155, G157, G159, K113		0	54	0	47	6491.93	14.37
	All sections	100	G141, DJ507	G151, G153, G155, G157, G159		0	54	0	46	5866.53	15.66

(continued)

Table 6.15 (continued)

Number of dispatched trains	Speed limit section	Speed limit value/(km/h)	Multi-locomotive train	Reroute train	Canceling train	1	2	3	4	Total delay time/min	Calculation time/s
		80	G141, DJ507	G153, L9, G155, G157, G159		0	54	0	46	8153.53	13.26
		50	G141, DJ507	G153, L9, G155, G157, G159		0	54	5	46	15,211.93	14.29
$\delta=7$	First section	50	DJ505, G103, G141, DJ507	G157, G159, K113, K115, K103		35	18	0	48	3098.01	16.33
	First and second sections	80	DJ505, G103, G141, DJ507	G153, G155, G157, G159, K113		13	40	0	46	3362.55	15.81
		50	DJ505, G103, G141, DJ507	G153, G155, G157, G159, K113		0	53	0	46	6366.15	15.73

(continued)

Table 6.15 (continued)

Number of dispatched trains	Speed limit section	Speed limit value/(km/h)	Multi-locomotive train	Reroute train	Canceling train	1	2	3	4	Total delay time/min	Calculation time/s
	All sections	100	DJ505, G103, G141, DJ507	G151, G153, G155, G157, G159		0	53	0	45	5750.48	14.39
		80	DJ505, G103, G141, DJ507	G153, L9, G155, G157, G159		0	53	0	45	7994.83	15.88
		50	G141, DJ507	G151, G153, L9, G155, G157	G159	0	53	4	45	14,978.83	13.24
$\delta=8$	First section	50	DJ505, G103, G141, DJ507, G145, G143	G157, G159, K109, K115, K101		5	7	0	7	3037.26	15.58
	First and second sections	80	DJ505, G103, G141, DJ507, G145, G143	G153, G155, G157, G159, K113		2	0	0	5	3302.73	15.71

(continued)

Table 6.15 (continued)

Number of dispatched trains	Speed limit section	Speed limit value/(km/h)	Multi-locomotive train	Reroute train	Canceling train	1	2	3	4	Total delay time/min	Calculation time/s
		50	DJ505, G103, G141, DJ507, G145, G143	G153, G155, G157, G159, K113		0	2	0	5	6250.73	15.23
	All sections	100	DJ505, G103, G141, DJ507	G151, G153, G155, G157, G159	K113	0	2	0	5	5648.96	13.55
		80	DJ505, G103, G141, DJ507	G153, L9, G155, G157, G159	G151	0	2	0	4	7850.66	14.93
		50	DJ505, G103, G141, DJ507	G151, G153, L9, G155, G157	G159	0	2	3	4	14,650.2	16.65
$\delta=9$	First section	50	DJ505, G103, G141, DJ507, G145, G143, G129, G131	G157, G159, K101, K115, K103		4	7	0	6	2983.88	15.35

(continued)

Table 6.15 (continued)

Number of dispatched trains	Speed limit section	Speed limit value/(km/h)	Multi-locomotive train	Reroute train	Canceling train	1	2	3	4	Total delay time/min	Calculation time/s
	First and second sections	80	DJ505, G103, G141, DJ507, G145, G143	G153, G155, G157, G159, K103	K111	3	8	0	5	3227.95	16.46
		50	DJ505, G103, G141, DJ507, G145, G143	G153, G155, G157, G159, K103	K105	0	1	0	5	6128.98	13.85
	All sections	100	DJ505, G103, G141, DJ507, G145, G143	G151, G153, G155, G157, G159	K113	0	1	0	4	5552.45	14.98
		80	DJ505, G103, G141, DJ507	G153, L9, G155, G157, G159	G151, K113	0	1	0	4	7706.5	13.61
		50	DJ505, G103, G141, DJ507	G151, L9, G155, G157, G159	G149, G153	0	1	2	3	14,233.3	14.82

(continued)

Table 6.15 (continued)

Number of dispatched trains	Speed limit section	Speed limit value/(km/h)	Multi-locomotive train	Reroute train	Canceling train	1	2	3	4	Total delay time/min	Calculation time/s
	First section	50	DJ505, G103, G141, DJ507, G145, G143, G129, G131	G157, G159, K101, K115, K103	K113	3	7	0	6	2923.53	14.65
δ=10	First and second sections	80	DJ505, G103, G141, DJ507, G145, G143	G153, G155, G157, G159, K103	K111, K109	3	7	0	5	3164.76	14.82
		50	DJ505, G103, G141, DJ507, G145-G143	G153, G155, G157, G159, K103	K105, K111	0	0	0	5	5996.68	15.40
	All sections	100	DJ505, G103, G141, DJ507, G145, G143	G151, L9, G155, G157, G159	G153, K113	0	0	0	3	5434.83	13.88
		80	DJ505, G103, G141, DJ507	G151, G159, L9, G155, G157	G153, G149, K113	0	0	0	3	7562.33	13.19

(continued)

Table 6.15 (continued)

Number of dispatched trains	Speed limit section	Speed limit value/(km/h)	Multi-locomotive train	Reroute train	Canceling train	1	2	3	4	Total delay time/min	Calculation time/s
		50	DJ505, G103, G141, DJ507	G149, G151, G153, L9, G157	G155, G159, K113	0	0	1	3	13,959.0	13.83
$\delta=11$	First section	50	DJ505, G103, G141, DJ507, G145, G143, G129, G131	G157, G159, K105, K115, K103	K113, K111	2	7	0	6	2866.5	14.70
	First and second sections	80	DJ505, G103, G141, DJ507, G145-G143	G153, G155, G157, G159, K115	K111, K109, K113	12	37	0	45	3106.78	13.38
		50	DJ505, G103, G141, DJ507, G145, G143	G153, G155, G157, G159, K101	K105, K111, K113	0	49	0	45	5876.9	14.21
	All sections	100	DJ505, G103, G141, DJ507, G145, G143	G151, L9, G155, G157, G159	G153, K113, K115	0	49	0	42	5332.31	15.73

(continued)

Table 6.15 (continued)

Number of dispatched trains	Speed limit section	Speed limit value/(km/h)	Multi-locomotive train	Reroute train	Canceling train	1	2	3	4	Total delay time/min	Calculation time/s
		80	DJ505, G103, G141, DJ507, G145, G143	G149, G155, G159, L9, G157	K113, G153, G151	0	49	0	43	7423.16	13.49
		50	DJ505, G103, G141, DJ507, G145, G143	G151, G153, L9, G155, G157	G159, G149, K113	0	49	1	43	13,697.9	15.50
$\delta=12$	First section	50	DJ505, G103, G141, DJ507, G145, G143, G129, G131	G157, G159, K105, K115, K103	K113, K111, K109	1	7	0	6	2807.6	15.27
	First and second sections	80	DJ505, G103, G141, DJ507, G145, G143	G153, G155, G157, G159, K101	K111, K109, K113, K115	2	6	0	5	3036.18	14.53
		50	DJ505, G103, G141, DJ507, G145, G143	G153, G155, G157, G159, K103	K105, K109, K111, K113	0	8	0	5	5763.36	14.54

(continued)

Table 6.15 (continued)

Number of dispatched trains	Speed limit section	Speed limit value/(km/h)	Multi-locomotive train	Reroute train	Canceling train	1	2	3	4	Total delay time/min	Calculation time/s
	All sections	100	DJ505, G103, G141, DJ507, G145-G143	G151, L9, G155, G157, G159	G153, K113, K115, K109	0	8	0	2	5224.8	13.71
		80	DJ505, G103, G141, DJ507, G145, G143	G149, G155, G159, L9, G157	K101, K113, G153, G151	0	48	0	42	7265.58	13.48
		50	DJ505, G103, G141, DJ507, G145, G143	G149, G151, G153, L9, G157	K105, K113, G155, G159	0	8	0	2	13,413.8	14.51
$\delta=13$	First section	50	DJ505, G103, G141, DJ507, G145, G143, G129, G131	G157, G159, K105, K115, K103	K113, K111, K109, K107	1	6	0	6	2741.58	14.85
	First and second sections	80	DJ505, G103, G141, DJ507, G145, G143	G153, G155, G157, G159, K103	K111, K109, K113, K115, K101	2	5	0	5	2972.0	15.53

(continued)

Table 6.15 (continued)

Number of dispatched trains	Speed limit section	Speed limit value/(km/h)	Multi-locomotive train	Reroute train	Canceling train	1	2	3	4	Total delay time/min	Calculation time/s
		50	DJ505, G103, G141, DJ507, G145, G143	G153, G155, G157, G159, K103	K105, K109, K111, K113, K115	0	7	0	5	5647.45	15.14
	All sections	100	DJ505, G103, G141, DJ507, G145, G143	G151, G153, L9, G155, G157	K115, K109, K113, K101, G159	0	7	0	3	5109.86	15.85
$\delta=13$	All sections	80	DJ505, G103, G141, DJ507, G145, G143	G149, G151, L9, G157, G159	K115, K101, K113, G153, G155	0	7	0	2	7120.41	15.27
		50	DJ505, G103, G141, DJ507, G145, G143	G151, G153, L9, G155, G157	K115, K105, K113, G149, G159	0	7	0	2	120.918.6	15.22

It can be known from Table 6.15 that when iterating to 12, the adjustment results of all limited speed cases are acceptable. In the same situation, the train detour strategy has the highest priority, followed by the train reconnection strategy, and the train shutdown strategy has the lowest priority, which reflects the adjustment principle of train dispatching under emergency conditions.

Trains starting with K (K101, K103, K105, K107, K109, K111, K113, and K115) are bypassed or stopped because they are cross-line trains on the Beijing-Shanghai high-speed railway, and the train rank is lower. Under the conditions of speed and dispatching capacity, the collection of bypass trains and out-of-service trains has changed (e.g., when the speed limit value in all sections is 80 km/h and the number of dispatched trains is less than 10, L9 takes the bypass strategy instead of G151).

This is because the upper-level planning not only considers the level of the train itself, but also considers the level of the emergency (speed limit value) and the impact of trains on the operating environment, so that all trains can complete all operations within the operating time.

In addition, according to the calculation time statistics, it can be seen that the characteristics of the algorithm (a hybrid intelligent algorithm for plant multi-directional growth simulation) that adapts to the bi-level planning problem of train scheduling optimization and train operation adjustment under emergency conditions. It can get rich driving organization at a faster speed in realistic and complex environments.

The algorithm also proves the feasibility of the bi-level programming model for train scheduling optimization and train operation adjustment and provides a basis for further comprehensive experiments under fuzzy random environments.

3. Example analysis of fuzzy train operation adjustment model based on tolerance method

In the upper-level planning, if the fuzzy random parameters are expected or related, they can be converted into a deterministic model for solving. Therefore, the purpose of this case study is to verify the effectiveness of the train operation adjustment model based on the tolerance method in an uncertain environment. In order to verify that the tolerance method can ideally deal with uncertain parameters and obtain a satisfactory optimal solution when the train is subject to large disturbances (speed limit) and minor disturbances (a train is late), this group of experiments will be performed in six groups.

(1) The delay of the single train and single section

Assume that train G103 is 20 min late in the NanJingNan-ZhenJiangXi section. First, do not relax the constraints, that is, the minimum speed of the train and the maximum time of each operation. The result is: 3 trains are delayed 10–20 min, 1 trains are delayed 20–30 min, the total delayed time is 47.5 min, the objective function value is 368,784, and the calculation time is 2.38 s.

Then relax the constraints to the maximum, that is, the maximum operating speed of the train and the minimum of each operation time. The result is: 1 trains are delayed 0–10 min, 2 trains are delayed 10–20 min, the total delayed time is 31.5 min, the objective function value is 350,449, and the calculation time is 2.49 s.

Comparing the two results, it can be found that when the constraints are relaxed to the maximum, the adjustment results have been greatly improved, but at the same time, the running risk and operating cost of the train have been greatly increased. The maximum membership value is 0.86, which means that the speed of high-speed trains and medium–high-speed trains is 365 km/h and 335 km/h, and the interval between tracking trains is 2.9 min. There are 1 train with a delay of 0–10 min, and 2 trains with a delay of 10–20 min. The total delay time is 33.55 min, and the calculation time is 5.78 s.

The adjustment strategy is to extend the dwelling time of K115 and K101 at ChangZhouBei Station and to be crossed by L15 and G105 respectively, reflecting the principle of priority of high-level trains. Due to space limitations, this book only draws the adjustment results of the fuzzy train operation adjustment model based on tolerance method (Hereinafter referred to as the tolerance model), see Fig. 6.33. For detailed train schedules of the above experiments, see Tables A.1 to A.3 in Appendix A.

(2) the delay of two trains and multiple section

Assume that G305 is 30 min late in the NanJingNan-ZhenJiangXi section, and G103 is delayed in the ZhenJiangXi-ChangZhouBei section, and that it is 20 min late due to the train's own factors.

The experimental process is similar to the delay of the single train and single section. The results when the constraints are not relaxed are: There are 2 trains that are late within 0–10 min, 3 trains that are late within 10–20 min, and 3 trains that are late within 20–30 min, the total delay time is 122.43 min, the objective function value is 860,720, and the calculation time is 3.48 s.

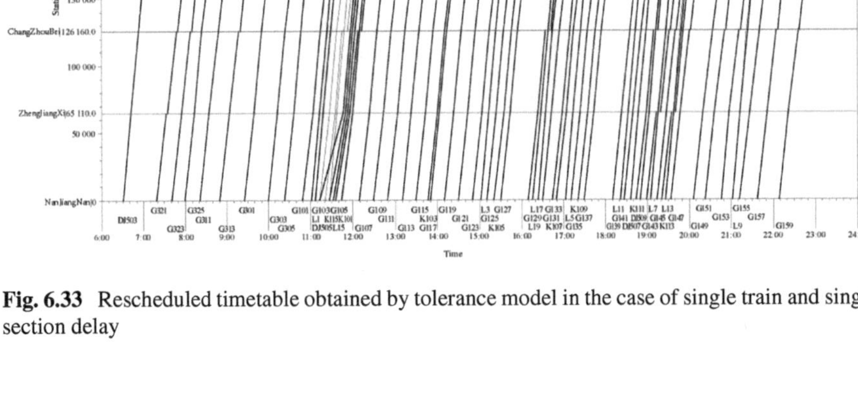

Fig. 6.33 Rescheduled timetable obtained by tolerance model in the case of single train and single section delay

The results when the constraints are relaxed to the maximum are: 2 trains within 0 to 10 min late, 2 trains within 10 to 20 min late, 3 trains between 20 to 30 min late, the total delay time is 109.18 min, the objective function value is 848 871, and the calculation time is 3.36 s.

The above results are substituted into the tolerance model, and the maximum membership value obtained is 0.894 25, that is, the running speed of the high-speed train and the medium–high-speed train at this time are 363 and 334 km/h, and the interval between the tracking trains is 2.95 min. There are 2 trains that are late within 0–10 min, 2 trains that are late within 10–20 min, and 3 trains that are late 20–30 min. The total delay time is 114.29 min, and the calculation time is 5.95 s.

The adjustment strategy was that K115 was overtaken by trains L15, K101, and G015 at ZhenJiangXi Station and ChangZhouBei Station, respectively. The tolerance model adjustment results are shown in Fig. 6.34. The detailed train schedules for the above experiments are shown in Tables A.4–A.6 in Appendix A.

(3) Single section speed limit

It is assumed that the speed limit of the NanJingNan-ZhenJiangXi section is affected by natural disasters, and the train running speed is 50–60 km/h. The results when the constraints are not relaxed are: 19 trains within 40–50 min late, 30 trains within 50–60 min late, 11 trains over 60 min late, total delay time is 3 269.32 min, the objective function value is 9294,339, and the calculation time is 3.36 s.

After relaxing the constraints, set the train running speed to 60 km / h, track the train interval to 2.5 min, and stop the station to 1 min. The result is: There are 20 trains that are late within 30 to 40 min, 37 trains within 40 to 50 min late, 3 trains within 50 to 60 min late, and no trains over 60 min late. The total delay time is

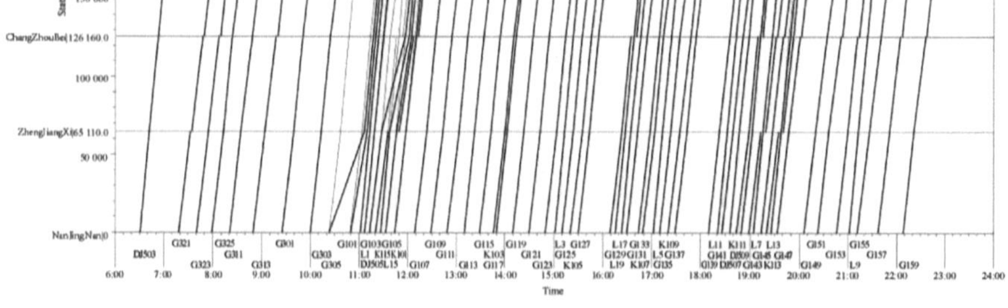

Fig. 6.34 Rescheduled timetable obtained by tolerance model in the case of two trains and multiple sections delay

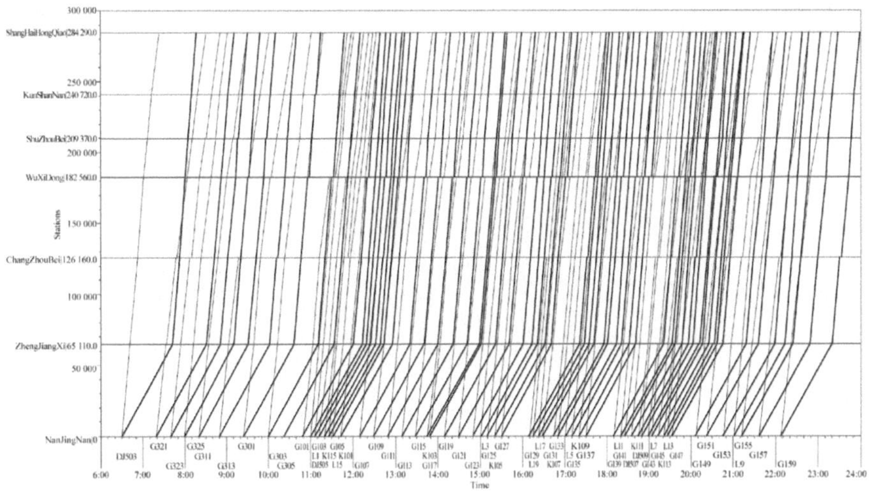

Fig. 6.35 Rescheduled timetable obtained by the tolerance model in the case of single section speed limitation

2486.32 min, and the objective function value is 8577577. The calculation time is 3.39 s.

The above results are substituted into the tolerance model, and the maximum membership value obtained is 0.706. At this time, the train running speed is 53 km / h, the interval between tracking trains is 2.85 min, and there are 5 trains within 30 to 40 min late, 39 trains within 40 to 50 min late, 16 trains within 50 to 60 min late, and no trains longer than 60 min late. The total delay time is 2753.17 min, and the calculation time is 4.38 s. The adjustment result of the tolerance model at this time is shown in Fig. 6.35. Because a large number of train adjustments are involved under speed limit conditions, detailed timetables under speed limit conditions are omitted.

(4) full-range speed limit

It is assumed that the speed limit is set in all sections of the Shanghai-Nanjing section, and the train running speed is 150–170 km / h. The results when the constraints are not relaxed are: 2 trains within 30–40 min late, 18 trains within 40–50 min late, 34 trains within 50–60 min late, 2 trains over 60 min late, the total delay time is 3,052.95 min, the objective function value is 7,969,213, and the calculation time is 3.38 s.

The results obtained after relaxing the constraints are: 5 trains within 20–30 min late, 38 trains within 30–40 min late, 17 trains within 40–50 min late, no trains within 50 to 60 min late and over 60 min late, the total delay time is 2248.26 min, the objective function value is 7473473, and the calculation time is 3.32 s.

Substituting the above results into the tolerance model, the maximum membership value was found to be 0. 689 117, at which time the train running speed was 156 km / h, and the interval between tracking trains was 2.80 min. There are 19 trains within

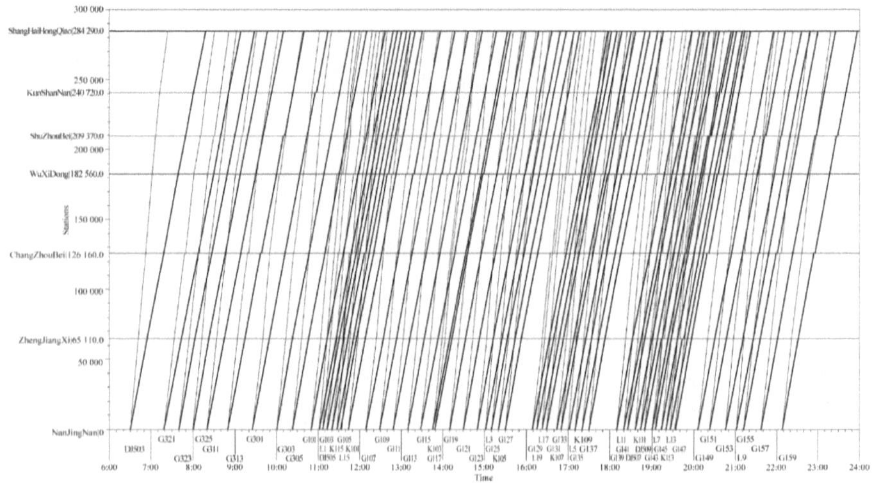

Fig. 6.36 Rescheduled timetable obtained by the tolerance model in the case of all sections speed limitation

30 to 40 min late, 25 trains within 40 to 50 min late, 16 trains within 50 to 60 min late, no trains late over 60 min, the total delay time is 2655.16 min, and the calculation time is 4.13 s. The adjustment result of the tolerance model at this time is shown in Fig. 6.36.

(5) single section speed limitation and single train delay

Assume that the speed limit in the NanJingNan-ZhenJiangXi section is 50–60 km / h, and the remaining sections can run normally. At the same time, the G103 train is 30 min later in the ZhenJiangXi-ChangZhouBei section on the basis of the delay in the first section, and other parameters are the same as the previous experiment.

The results when the constraints are not relaxed are: 19 trains are delayed within 40–50 min, 24 trains are delayed 50–50 min, 17 trains are delayed more than 60 min, and the total delay time is 3,305.19 min. The objective function value is 9367,637, and the calculation time is 8.05 s.

When the constraints are relaxed to the maximum, the result is: There are 3 trains within 20–30 min late, 34 trains within 30–40 min late, 17 trains within 40–50 min late, 1 train within 50–60 min late, 5 trains late more than 60 min, and the total delay time is 2557.32 min. The objective function value is 8671, 072, and the calculation time is 8.31 s.

Substituting the above results into the tolerance model, the maximum membership value is 0.740 856, that is, the speed of the train in the speed limit intervals is 52 km / h, the speed of the high-speed train and medium–high-speed train is 368 km/h and 339 km / h, and the interval time for tracking trains is 2.87 min. There are 17 trains within 30–40 min late, 23 trains within 40–50 min, 15 trains within 50–60 min late, and 5 trains late over 60 min. The total delay time is 2823.34 min, and the calculation time is 10.48 s.

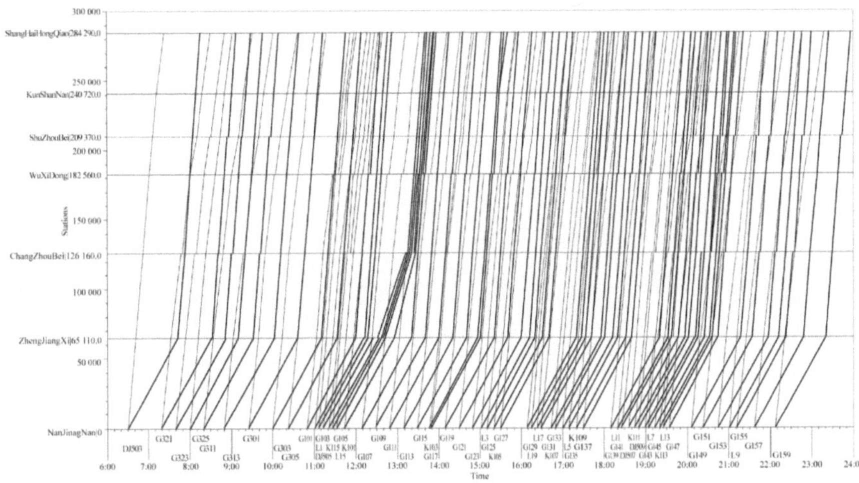

Fig. 6.37 Rescheduled timetable obtained by the tolerance model in the case of single section speed limitation and single train delay

The adjustment strategy is that K115 train was crossed by L15, K101, and G105 at ChangZhouBei Station, and then K101 was crossed by G105 at WuXiDong Station. The tolerance model adjustment results are shown in Fig. 6.37. The train schedules for the above experiments are shown in Tables A.7–A.9 in Appendix A.

(6) All section speed limitation and multiple trains delay

It is assumed that the speed limit is set in all sections of the Shanghai-Nanjing section, and the train runs at a speed of 150–170 km/h. In addition, the train G305 is 30 min late in the NanJingNan-ZhenJiangXi section, and G103 is 15 min late in the ZhenJiangXi-ChangZhouBei section. The other parameters are the same as above.

The result when the constraints are not relaxed is: 1 train is delayed within 30–40 min, 17 trains are delayed within 40–50 min, 27 trains are delayed within 50–60 min, and 15 trains are delayed more than 60 min. The total delay time is 3197.86 min, the objective function value is 8 086 249, and the calculation time is 8.47 s.

When the constraint conditions are relaxed to the maximum, the result is: There are 5 trains within 20–30 min late, 35 trains within 30–40 min late, 14 trains within 40–50 min late, 3 trains within 50–60 min late, and 3 trains late more than 60 min. The total delay time is 2397.15 min, the objective function value is 7588, 658, and the calculation time is 8.48 s.

Substituting the above results into the tolerance model, the maximum membership value was found to be 0.792877, that is, the speed of the train in the speed limit zone was 155 km/h, and the headway between adjacent trains was 2.89 min. There are 17 trains within 30–40 min late, 23 trains within 40–50 min late, 18 trains within 50–60 min late, and 2 trains over 60 min late. The whole trains are late. The total delay time is 2704.43 min, and the calculation time is 10.89 s.

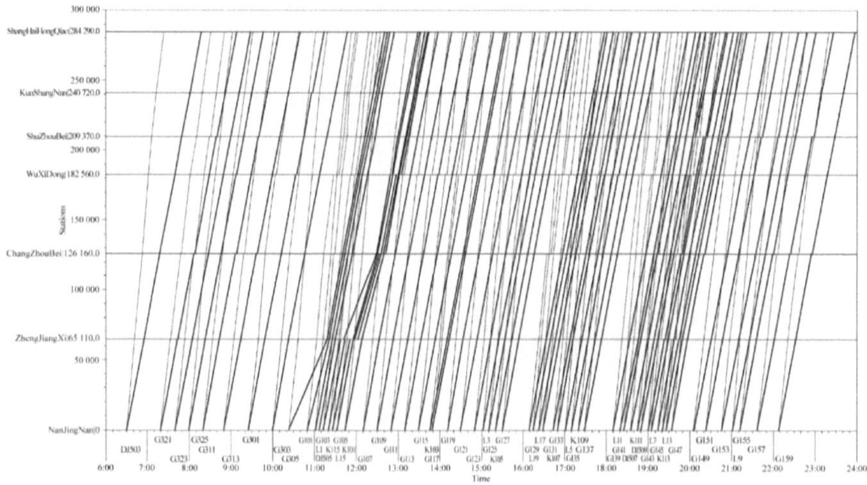

Fig. 6.38 Rescheduled timetable obtained by the tolerance model in the case of all sections speed limitation and multiple trains delay

The rescheduling strategy is: the L1 train canceled the plan to cross the G101 train at WuXiDong Station, thereby reducing the dwell time of the G101 train; and the K115 train changed its stop pattern: It was crossed by L15 and K101 at ZhenJiangXi Station, and then was crossed by G105 train at ChangZhouBei Station. The adjustment result obtained by its tolerance model is shown in Fig. 6.38. The train schedule of the above experiment is shown in Tables A.10 – A.12 of Appendix A.

Table 6.16 is a summary of the above experimental results. Among them, Case1 ~ Case6 respectively represent the above six groups of experiments; C1, C2, and C3 represent non-relaxation constraints, relaxation constraints, and uncertain constraints; R1, R2, R3, R4, R5, R6, and R7 represent the number of trains that are 0–10 min late, 10–20 min late, 20–30 min late, 30–40 min late, 40–50 min late, 50–60 min late, and more than 60 min late respectively; S is the value of the objective function; Z_{delay} represents the total delay time, the unit is min; N is the number of new overtaking trains or canceled overtaking trains in the adjustment strategy; T represents the calculation time of the simulation experiment, the unit is second.

From the above experiments, it can be known that the tolerance method can handle uncertain parameters well regardless of minor disturbances (such as a single train being delayed) or a long-term and large-scale impact (full-range speed limit during the entire operating period). This method has greatly improved the objective function under the conditions of less relaxing constraints, maintaining low operating costs, and high safety. In particular, it makes full use of the principle that high-priority trains can overtake low-priority trains, achieving more flexible dispatching strategy, reducing the number of delayed trains and the delay time.

In addition, because the solving algorithm selects branches based on the characteristics of the train operation adjustment problem, it speeds up the solution of the

Table 6.16 Adjustment results of different experiments

		R1	R2	R3	R4	R5	R6	R7	S	Z_{delay}/min	N	T/s
Case1	C1	0	3	1	0	0	0	0	368,784	47.5	2	2.38
	C2	1	2	0	0	0	0	0	350,449	31.5	2	2.49
	C3	1	2	0	0	0	0	0	0.863 92	33.5	1	5.78
Case2	C1	2	3	3	0	0	0	0	860,720	122.43	2	3.48
	C2	2	2	3	0	0	0	0	848,871	109.18	2	3.36
	C3	2	2	3	0	0	0	0	0.89425	114.29	2	5.95
Case3	C1	0	0	0	0	19	30	11	9,294,339	3269.32	–	3.36
	C2	0	0	0	20	37	3	0	8,577,661	2486.32	–	3.39
	C3	0	0	0	5	39	16	0	0.70606	2753.17	–	4.38
Case4	C1	0	0	0	2	18	34	6	7,969,213	3052.95	–	3.38
	C2	0	0	5	38	17	0	0	7,473,849	2248.26	–	3.32
	C3	0	0	0	19	25	16	0	0.689117	2655.16	–	4.13
Case5	C1	0	0	0	0	19	24	17	9,367,637	3305.19	2	8.05
	C2	0	0	3	34	17	1	5	8,671,072	2557.32	3	8.31
	C3	0	0	0	17	23	15	5	0.740,856	2823.34	3	10.48
Case6	C1	0	0	0	1	17	27	15	8,086,249	3197.86	3	8.47
	C2	0	0	5	35	14	3	3	7,588,658	2397.15	4	8.48
	C3	0	0	0	17	23	15	5	0.792877	2704.43	4	10.89

problem and makes it possible to implement a multi-stage closed-loop fuzzy random optimization process including bi-level programming under contingency conditions in a limited time.

4. Comprehensive case analysis

The smooth progress of the above experiments provides the basis for the study of the comprehensive examples below. The comprehensive experiments under fuzzy random conditions will be performed below to further prove the effectiveness of related theories and algorithms.

(1) Disturbance scenarios settings

Assume that at 8 a.m. on a certain day, due to bad weather, the speed limit of ZhenJiangXi-WuXiDong section of the Beijing-Shanghai high-speed railway is 160 km / h, and SuZhouBei-KunShanNan section is broken due to track failure. It is estimated that it will take 3 h to resume driving, and six high-speed trains will be involved later, namely G325, G311, G313, G301, G303, and G305. The other parameter settings are consistent with the parameter settings in the "Experimental setup" section.

In the above-mentioned emergency scenario, due to the long duration of the emergency and the speed limit or interruption in multiple sections, the railway network structure was partially changed, and the planned transportation plan could not be completed, which was in line with the characteristics of the traffic scheduling problem. Therefore, the entire set of models and methods proposed in this article can be used for solving.

(2) Calculation of carrying capacity in the section

According to the development status of emergencies, the section carrying capacity change process is divided into three phases according to hourly intervals. It is expected that the train in the first stage will maintain a speed limit of 160 km/h; the probability of the train turning to the normal state in the second stage is 0.1, the probability of maintaining 160 km/h is 0.4, and the probability of turning to a speed limit of 70 km / h is 0.5; the probability of the train turning to the normal state in the third stage is 0.4, and the probability of maintaining 160 km/h is 0.5, and the probability of turning to a speed limit of 70 km/h is 0.1.

According to the calculation method of the railway section passing capacity under the emergency conditions in Chap. 3, the expected value of the comprehensive carrying capacity is:

$$
\begin{aligned}
E(\hat{N}_{\text{sec}}) = & \frac{75 + 79 \times 2 + 83}{4} \times 0.04 + \frac{69 + 74 \times 2 + 79}{4} \\
& \times 0.21 + \frac{65 + 70 \times 2 + 76}{4} \times 0.21 \\
& + \frac{63 + 69 \times 2 + 75}{4} \times 0.20 + \frac{59 + 65 \times 2 + 72}{4} \\
& \times 0.29 + \frac{55 + 61 \times 2 + 69}{4} \times 0.05 \\
= & 3.16 + 15.54 + 14.75 + 13.8 + 18.92 + 3.08 \\
= & 69.25
\end{aligned}
$$

According to the calculation process, it is known that the section carrying capacity calculation method based on the fuzzy Markov chain is adapted to that the ability value itself has the characteristics of fuzzy and random dynamic changes under special conditions. In addition, the final expression of the passage capacity of the section retains the above characteristics, which is more in line with the actual operating environment of the following vehicles under special conditions.

(3) Generation of rerouting routes

This case will generate a alternate route set based on the route search method (based on capability time density) proposed in Sect. 4.3. Analyzing the capacity of the speed limit section, it was found that $\hat{N}_{\text{sec}}$ after the speed limit was still much larger than the number of trains that needed to be adjusted, so only the short-circuited section (SuZhouBei-KunShanNan) was searched for the alternate route set.

Since other sections and stations are not affected by emergencies, the remaining capacity of other lines can be directly determined. Assume that the remaining capacity of the existing normal speed railway and intercity railway in the train network diagram

Table 6.17 Rerouting set

Number	Rerouting	Length/km	T
1	Changzhou (C)-Wuxi (C)-Suzhou (C)-KunShanNan	115	0.0086
2	Nanjing (C)-Zhenjiang (C)-Changzhou (C)-Wuxi (C)-Suzhou (C)-KunShanNan	166	0.0195
3	Changzhou (E)-Wuxi (E)-Suzhou (C)-KunShanNan	115	0.0236
4	Nanjing (C)-Zhenjiang (E)-Changzhou (E)-Wuxi (E)-Suzhou (E)-KunShanNan	144	0.0476
5	Wuxi (E)-Changxing (E)-Hangzhou (C)-Shanghai	381	0.0607
6	Wuxi (E)-Changxing (E)-Hangzhou (E)-Shanghai	409	0.0975
7	Nanjing (E)-Wuhu (E)-Changxing (E)-Hangzhou (C)-Shanghai	589	0.1241

is about $\tilde{N}_{JY}$ =(20, 23, 24) and $\tilde{N}_{CJ}$ =(46, 50, 53). Since the total length of the Shanghai-Nanjing section is about 300 km, the distance limit Rs_s between origin station and destination station is set to 300 km; the minimum capacity limit Rs_c is 3, and the maximum length of the alternate route Rs_l is 600 km.

First, a set of origin station and destination station for detours is generated.

According to Fig. 6.31, it can be seen that the stations with cross-line capability whose distance from the SuZhouBei Station meets the limit Rs_s are WuXiDong Station, ChangZhouBei Station, and NanJingNan Station. You can get the starting station set seq_s ={WuXiDong Station, ChangZhouBei Station, NanJingNan Station} in the route search. Similarly, if you search for stations with cross-line capability whose distance from the KunShanNan Station meets the limit Rs_s, you can get the ending station set seqe = {KunShanNan Station, Shanghai West Station, Shanghai Hongqiao Station}.

Then search the detours according to the capacity time density. The results are shown in Table 6.17, where T is the sum of the inverse of the circuitous capacity time density, C represents the intercity railway, and E represents the remaining normal speed railway.

Analyzing the results in Table 6.17, although WuXiDong Station is the closest station to SuZhouBei Station, when this station is selected as the starting station, the alternative route is long and the time reliability is low, so Route 5 and Route 6 are significantly worse than the other routes.

As the remaining capacity of any of the routes in Table 6.17 can satisfy the detour of the affected trains, and Route 1 and Route 3 are components of Route 2 and Route 4 respectively, the routes with short length, strong time reliability and few crossing lines should be selected, that is, routes 1 and 3.

In addition, you can adjust the search limit and repeat the above steps. For example, if you set Rs_s to 100 km and Rs_l to 300 km, the starting station set in the route search becomes seqs = {WuXiDong, ChangZhouBei}. At this time, you can directly obtain route 1 and route 3.

(4) Bi-level planning for train organization

This section consists of three parts. First, adjust train operation under speed limit conditions. Since carrying capacity of the ZhenJiangXi-Wuxi section (after the speed limit) is still far greater than the number of trains that need to be adjusted, the lower-level plan is implemented firstly.

Then, on the basis of the new schedule, search the train affected by the disconnection, and perform the scheduling optimization and train operation adjustment double-layer planning to obtain the adjusted schedule on detours.

Finally, the train returns to the original route and executes the lower-level planning to obtain the final adjustment plan. It is worth noting here that due to the dynamic changes in train running status and emergencies, the adjustment objects of the first part and the second part are often different.

(1) Timetable rescheduling under speed limit conditions.

When the speed limit value is 160 km/h, the average speed of the train changes to (130, 140, 155 km/h), the penalty c_i^{trnd} for the delayed unit of the train is (1, 1.1, 1.2), and the tolerance degree w_i for late is (28, 30, 32 min).

Since there is no original station and terminal station in this section, the late penalty of the station can be ignored, and the optimization target weights are $w_1 = 1, w_2 = 0, w_3 = 1\ 000$ respectively. The other parameter settings are the same as the "Experimental setup".

First of all, do not relax the constraints, that is, the average train speed is 130 km/h, the delay of the train is 1.2, and the tolerance of the delay is 28 min. The implementation of the lower-level planning results are as follows: 2 trains are delayed within 20–30 min, and 4 trains are delayed within 30–40 min. The total delay time is 177.48 min, and the objective function value is 68658.

After relaxing the constraints, set the train's speed as 155 km/h, the train's late penalty as 1, and the tolerance for late delay as 32 min. The results are: 2 trains are delayed within 10–20 min, 4 trains are delayed within 20–30 min, and no train is delayed more than 30 min. The total delays are 125.18 min, and the objective function value is 16026.

The above results are substituted into the tolerance model, and the maximum membership value obtained is 0.715. At this time, the average train's speed is 137 km/h, the train late penalty is 1.36, and the late tolerance is 29.13 min. Six trains are delayed within 20 to 30 min, and no train is delayed more than 30 min. The total delays are 170.12 min.

In the results of the tolerance model, although the overall delays have not been greatly improved, the adjusted trains have not exceeded the tolerance level for delays under the condition that the average speed is close to the minimum value, but the objective function has been greatly improved. The calculation time of the above three models is less than 0.01 s. The adjustment result of its tolerance model is shown in Fig. 6.39. For the schedule of the above experiments, see Tables A.13–A.15 in Appendix A.

(2) Optimization of dispatching strategies and train operation.

After the speed limit adjustment, G303 and G305 entered the section from SuZhouBei Station to KunShanNan Station later than 11:00, so the trains that are rescheduled at this time are G323, G325, G311, G313, and G301.

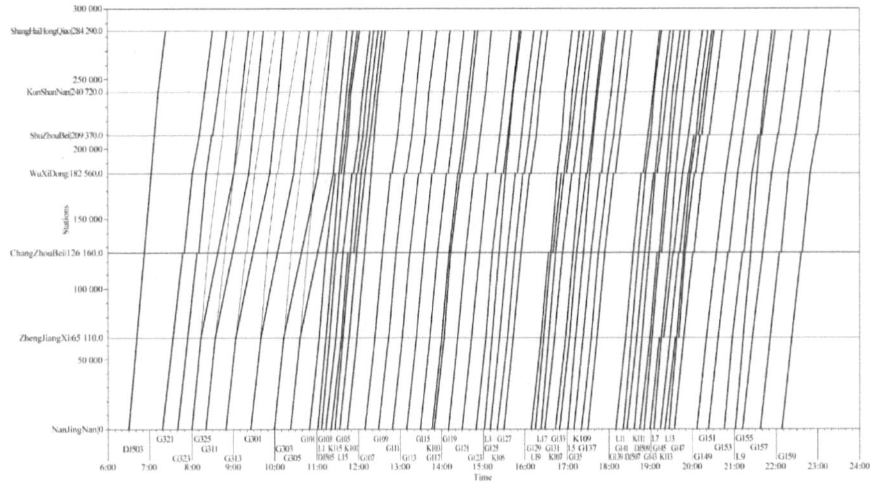

Fig. 6.39 Rescheduled timetable obtained by the tolerance model in case of two sections speed limitation

Due to the track interruption, only the adjustment of the lower-level timetable cannot provide a feasible solution, and a two-level planning under the guidance of traffic scheduling optimization is required. In the experiment, a pessimistic value-tolerance model is used to solve. The specific steps are as follows.

First, design decision variables. According to the above, the two detour routes are numbered as Route 1 (Changzhou(E)-Wuxi(E)-Suzhou(C)-KunShanNan) and Route 2 (Changzhou(C)-Wuxi (C)-Suzhou(C)-KunShanNan), involving five high-speed trains, and the stopping schedule of the alternative route can be regarded as the same as the original route. So let $\boldsymbol{x} = (x_1, x_2, x_3, x_4, x_5)$ be the decision variable and x_i be the scheduling strategy of the i-th train.

When train i adopts the original route operation strategy, $x_i = 0$; when train i adopts the merge strategy with train j,$x_i = j,\ i \neq j$; when train i adopts the detour strategy k, $x_i = 5 + k, k = 1, 2$; when train i adopts a shutdown strategy, $x_i = 8$.

Since the sum of the capacity time density of Route 1 is much larger than Route 2, the cost of detour to Route 1 is less. Suppose the penalty fee for the train detour to route 1 is $\tilde{C}^{\text{reroute},1} = (20{,}000, 22{,}000, 24{,}000)$, the penalty fee for detour to route 2 is $\tilde{C}^{\text{reroute},2} = (22{,}000, 24{,}000, 26{,}000)$, the penalty fee for train reconnection is $\tilde{C}^{\text{merge}} = (40{,}000, 42{,}000, 44{,}000)$, and the penalty fee for train outage is $\tilde{C}^{\text{cancle}} = (80{,}000, 82{,}000, 84{,}000)$.

In addition, according to the recorded historical operation data, the operating costs per kilometer of the intercity railway and the existing normal speed railway are $c^{\text{run,c}} = (30{,}32{,}34)$ and $c^{\text{run,j}} = (9, 9.2, 9.4)$ respectively. The unit delay penalty for inter-line trains is $c^{\text{Kd}} = (1, 1.1, 1.2)$ and the unit delay penalty for trains on this line is $c^{\text{Bd}} = (2, 2.4, 2.8)$.

The scheduled train operation plans of the Shanghai-Nanjing intercity railway and the existing Beijing-Shanghai normal speed railway are shown in Fig. 6.40 and Fig. 6.41, which involve 15 high-speed trains (G), 4 express trains (T), and 3 fast trains (K).

The DPGSA solution was used to obtain the relevant parameters and optimization target (Z) values under different credibility measures α, as shown in Table 6.18.

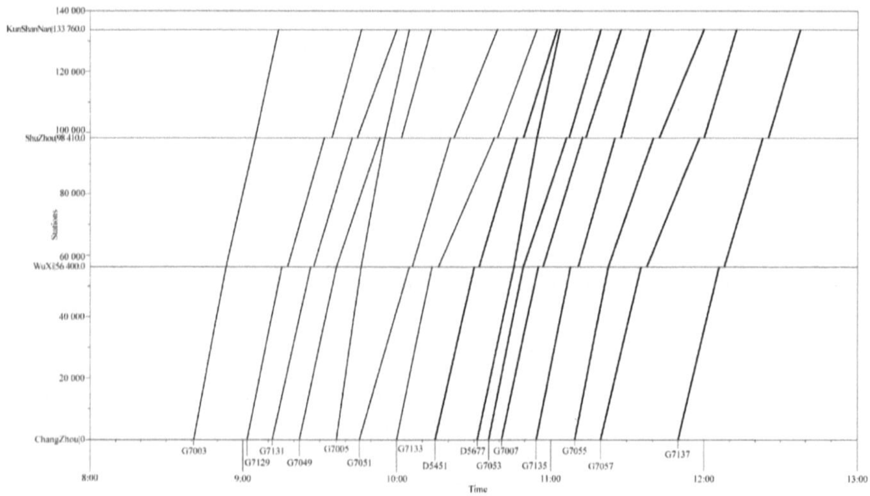

Fig. 6.40 Scheduled timetable of Shanghai-Nanjing high-speed railway

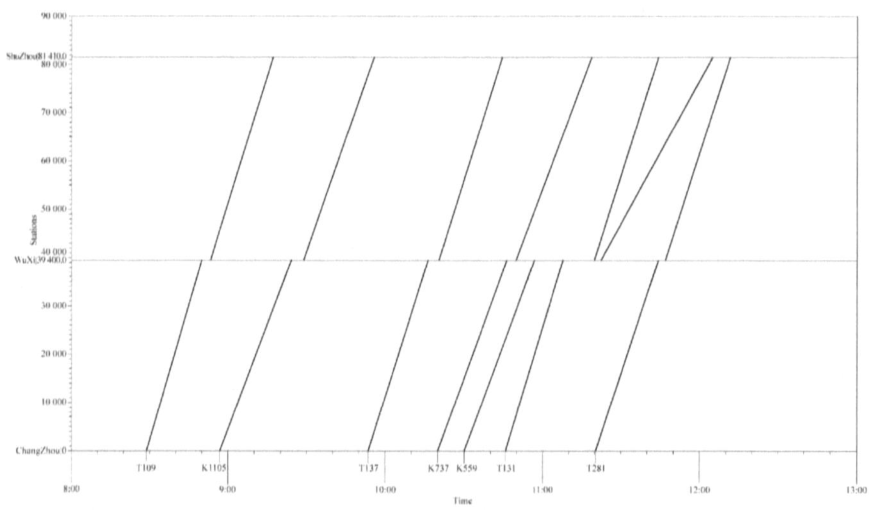

Fig. 6.41 Scheduled timetable of Beijing-Shanghai existing railway

Table 6.18 Parameters and optimization target values under different credibility measures

α	$\tilde{C}^{\text{reroute},1}$	$\tilde{C}^{\text{reroute},2}$	$\tilde{C}^{\text{merge}}$	$\tilde{C}^{\text{cancle}}$	$c^{\text{run},c}$	$c^{\text{run},j}$	c^{Kd}	c^{Bd}	Z
0.1	20,400	22,400	40,400	80,400	30.4	9.04	1.02	2.08	260,834
0.2	20,800	22,800	40,800	80,800	30.8	9.08	1.04	2.16	266,418
0.3	21,200	23,200	41,200	81,200	31.2	9.12	1.06	2.24	272,002
0.4	21,600	23,600	41,600	81,600	31.6	9.16	1.08	2.32	277,586
0.5	22,000	24,000	42,000	82,000	32	9.2	1.1	2.4	282,747.2
0.6	22,400	24,400	42,400	82,400	32.4	9.24	1.12	2.48	287,850.8
0.7	22,800	24,800	42,800	82,800	32.8	9.28	1.14	2.56	292,954.5
0.8	23,200	25,200	43,200	83,200	33.2	9.32	1.16	2.64	298,058.1
0.9	23,600	25,600	43,600	83,600	33.6	9.36	1.18	2.72	303,161.8
1.0	24,000	26,000	44,000	84,000	34	9.4	1.2	2.8	308,265.4

Table 6.19 Specific objective function values related to c_1,c_2

c1	c2			
	1.9	2.0	2.1	2.2
1.8	1.317E−05	9.046E−06	2.041E−06	2.658E−05
1.9	1.097E−05	9.046E−06	1.024E−06	2.264E−06
2.0	9.623E−06	1.084E−05	0.039E−06	1.058E−06
2.1	1.258E−05	1.256E−05	1.106E−06	6.264E−06
2.2	1.091E−05	1.063E−05	1.856E−06	1.045E−05

Only the adjustment results when the values of α are equal to 0.3 or 0.7 are listed. When α=0.3, all trains run on the alternate route 2 (Shanghai-Nanjing intercity railway), see Fig. 6.42. When α=0.7, train G313 detours on route 1 (Beijing-Shanghai existing normal speed railway), and other trains still detour on route 2, see Fig. 6.43 and Fig. 6.44.

Train G313 departs from Changzhou Station at 9:30 to detour the existing normal speed railway. After arriving at Suzhou Station at 9:59, it departs at 10:00 to depart from Shanghai-Nanjing intercity railway and finally arrives at KunShanNan Station at 10:09.

Figure 6.45 shows the iterative process of the optimization target under different credibility measures. The detailed timetable of the above experiment is shown in Tables A.16 and A.17 of Appendix A.

(3) The train returns to its original route.

After the train has completed the detours, it will return to the Beijing-Shanghai high-speed railway from KunShanNan Station.

When α = 0.3, that is, all trains run on route 2, the adjustment results are shown in Fig. 6.46. At this time, the detour of the train makes the train G7005 on the Shanghai-

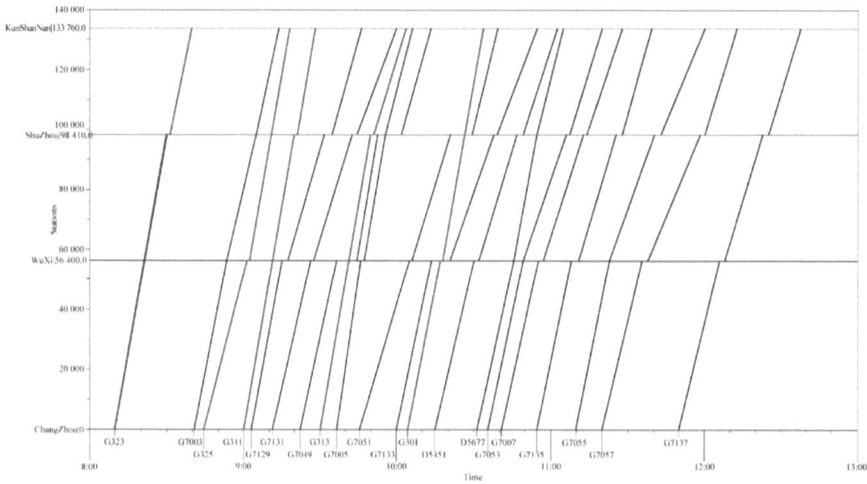

Fig. 6.42 Scheduled timetable of Shanghai-Nanjing railway ($\alpha = 0.3$)

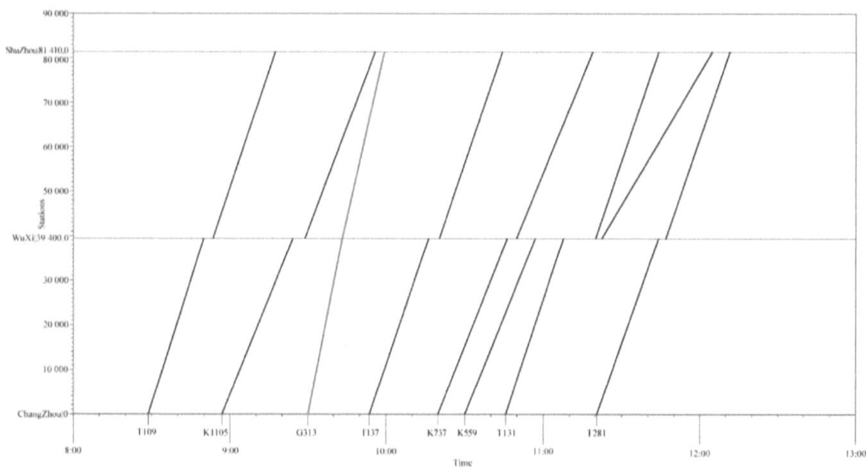

Fig. 6.43 Scheduled timetable of Beijing-Shanghai existing railway ($\alpha = 0.7$)

Nanjing intercity railway delayed. The final total delay time is 96 min. Among them, there are 1 trains within 0 to 10 min late, and 6 trains within 10 to 20 min late.

When $\alpha = 0.7$, the adjustment result is similar to that shown in Fig. 6.46. The total delays are 101 min. Among them, there are 5 trains within 10 to 20 min late, and 1 train within 20 to 30 min late. At that time, although the total delays increased, and trains delayed more than 20 min appeared, trains on other routes were not affected by the detour.

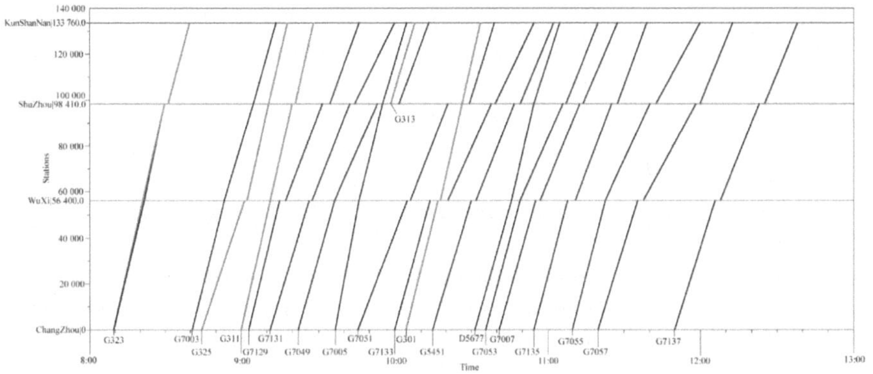

Fig. 6.44 Scheduled timetable of rerouting train in Shanghai-Nanjing railway ($\alpha = 0.7$)

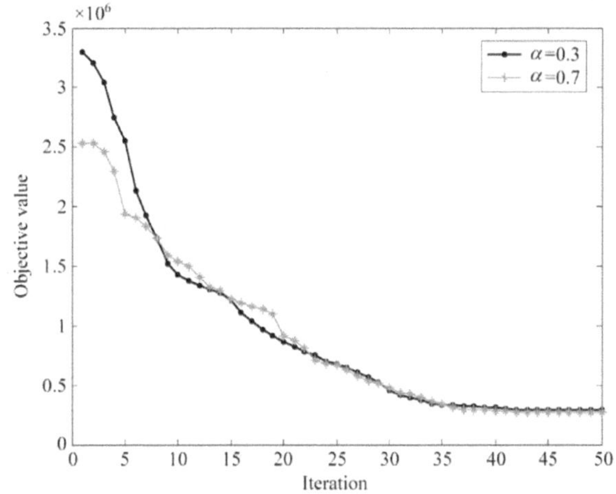

Fig. 6.45 Iteration of optimization target under different credibility measures

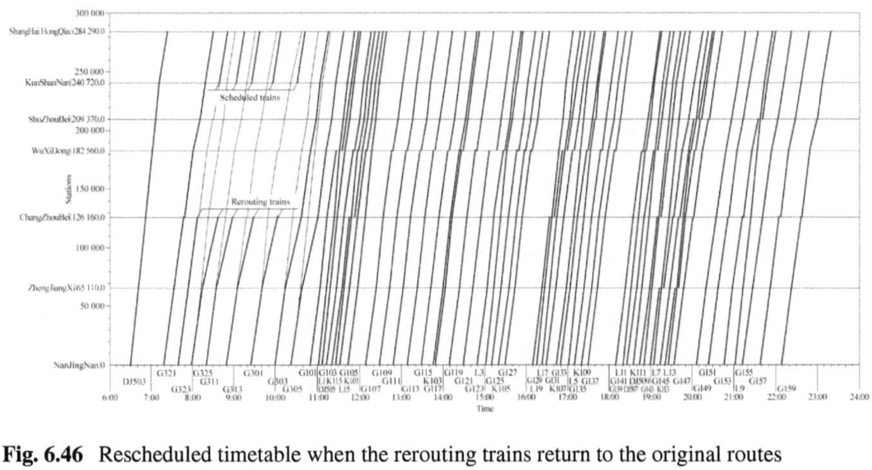

Fig. 6.46 Rescheduled timetable when the rerouting trains return to the original routes

The optimization of train operation organization under emergency conditions is a multi-stage dynamic adjustment process. Different from the previous practice of adjusting only the affected sections, we first implement the train operation adjustment under the speed limit. Based on the adjustment, find the trains affected by the disconnection, implement the two-layer optimization of traffic scheduling optimization and train operation adjustment. Finally, complete the entire process of returning the train to the original route, providing a comprehensive solution for decision-makers.

In this example, the number of affected trains is far less than the remaining capacity provided on the detour, so the upper-layer optimization results only use the train detours strategy. In addition, decision-makers can choose different adjustment schemes based on credibility measures of uncertain parameters.

6.5 Train Timetable Rescheduling Model Based on Convergent Fuzzy Particle Swarm Optimization

In this section, authors propose convergent fuzzy particle swarm optimization, another efficient algorithm, for solving train timetable rescheduling problem.

6.5.1 Objective Function Design

The train timetable rescheduling model belongs to the NP-hard problem. This section considers a different objective function from the previous section to design a solution algorithm to provide more reference for the train timetable rescheduling model.

The general penalty of train delays is closely related to the time of train delays. It not only reflects the total delay time of all trains to be adjusted, but also reflects the importance of different grades of trains. It is the primary goal of train timetable rescheduling.

The stability of the train operation plan reflects the ability of the train operation plan to repair itself after being disturbed, which reveals the ability to resist sudden interference. The emergency caused the state of the line section to be unstable. An emergency that occurs in a certain line section or station will affect the adjacent sections and stations. When trains run in these sections or work at these stations, they will be subject to these secondary disturbances caused by emergencies.

Therefore, in order to cope with these secondary disturbances, the train operation adjustment should optimize the stability of the train operation plan on the basis of minimizing the general penalty value of the train's delay as much as possible, so that it has a strong adaptability in the execution process.

In summary, the train operation adjustment model takes the generalized penalty of train delay and the stability of the train operation plan as the objective functions. For

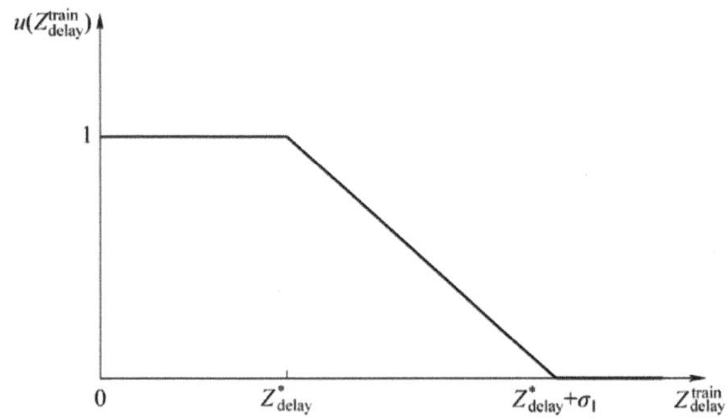

Fig. 6.47 Fuzzy membership function of generalized penalty satisfaction for delayed trains

this kind of multi-objective programming problem, this section designs a satisfaction measure.

Assume that the optimal target value or approximate optimal value obtained by separately optimizing the first target (generalized penalty for delayed trains) is Z_{delay}^{*}. Based on this, a fuzzy membership function that satisfies the reality is constructed.

$$u_1 = \begin{cases} 1 & Z_{delay}^{train} \le Z_{delay}^{*} \\ 1 - (Z_{delay}^{train} - Z_{delay}^{*})/\sigma_1 & Z_{delay}^{*} < Z_{delay}^{train} < Z_{delay}^{*} + \sigma_1 \\ 0 & Z_{delay}^{train} \ge Z_{delay}^{*} + \sigma_1 \end{cases} \tag{6.207}$$

The corresponding satisfaction fuzzy membership function curve is shown in Fig. 6.47.

Among them,σ_1 is determined by the decision-maker based on the fuzzy expectations of the relative importance of the target Z_{delay}^{train} to reach the ideal value Z_{delay}^{*}, and the value of σ can make the following problems feasible.

$$Z_{delay}^{train} = Z_{delay}^{*} + \sigma_1 \tag{6.208}$$

Similarly, the optimal target value or approximate optimal value obtained by separately optimizing the second objective (stability of the train operation plan) is S^*. Based on this, a satisfaction membership function consistent with the actual situation is constructed.

$$u_2 = \begin{cases} 1 & S \ge S^* \\ 1 - (S^* - S)/\sigma_2 & S^* - \sigma_2 < S < S^* \\ 0 & S^* \le S^* - \sigma_2 \end{cases} \tag{6.209}$$

The corresponding fuzzy membership function curve of the stability satisfaction of the train operation plan is shown in Fig. 6.48.

Among them,σ_2 is determined by the decision-maker based on the fuzzy expectations of the relative importance of the target S to reach the ideal value S^*, and the value of σ_2 can make the following problems feasible.

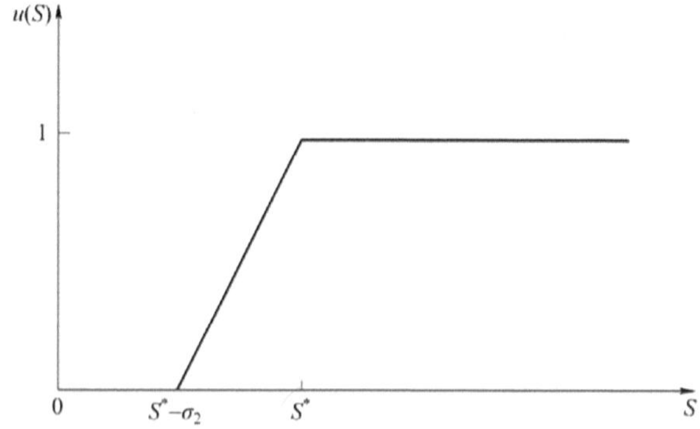

Fig. 6.48 Fuzzy membership function of train operation plan stability satisfaction

$$S = S^* - \sigma_2 \tag{6.210}$$

The triangle fuzzy membership function is used to define the satisfaction of the generalized penalty of train delay and the satisfaction of the stability of the train operation plan. Of course, you can also use the Bell function, Gaussian function, Sigmoid function, and Trapezoid function to define according to the actual situation.

For the synthesis of two satisfaction measures, there are many methods, such as taking small, taking large, algebraic product, algebraic sum, weighted sum, and so on. Different integrated approaches reflect the importance that decision-makers attach to different goals. This section proposes the following comprehensive approach.

$$\max Z = \theta \max\{\mu_1, \mu_2\} + (1 - \theta) \min\{\mu_1, \mu_2\}, \theta \in [0, 1] \tag{6.211}$$

Among them,θ is the optimistic and cautious tendency parameter of decision-makers. If the value of θ is large, the weight of the satisfaction factor in the comprehensive target is larger, and the weight of the pessimistic factor is correspondingly reduced, indicating that the decision-maker is cautiously optimistic. On the contrary, policymakers are optimistic and cautious. By setting the value of θ, the optimization intention and preferences of decision-makers in decision making can be accurately expressed.

6.5.2 Convergent Fuzzy Particle Swarm Optimization

Particle swarm optimization [60] is an evolutionary computing technology, first proposed by Kennedy and Eberhart in 1995. This algorithm originates from the study of the behavior of bird predation. It is an optimization tool based on iterative technology, which is mostly used to solve nonlinear and multi-extreme complex optimization problems. This optimization method is simple and easy to understand, and easy to model and implement [60–62].

Particle swarm optimization (PSO) has been widely studied in academic circles since it was proposed in 1995. In literature [63], a fuzzy particle swarm optimization (FPSO) algorithm was proposed, but the choice of membership function in the particle motion equation was not discussed in depth. Literature [64] proposed an instinct-based satisfaction particle swarm algorithm. Literature [65] analyzed several variants of the particle swarm algorithm and evaluated the reachability of these variants. Literature [66] proposed a full-information particle swarm algorithm. Literature [67] made improvements to particle swarm optimization. Literature [68] studied the fuzzy adaptive particle swarm algorithm. Reference [69] studied the use of fuzzy adaptive particle swarm algorithm to solve the problem of unit commission calculation. References [70, 71] studied the generation of fuzzy rules and the selection of membership functions using hybrid particle swarm optimization. Literature [72] studied the weighted fuzzy reasoning based on particle swarm optimization. The above literatures propose a variety of particle swarm optimization algorithms, but none have comprehensively studied the efficiency and accuracy of the algorithm through experiments.

Particle swarm algorithm is inspired by the swarm intelligence phenomenon that occurs during the foraging of birds and fish in nature [60]. Suppose particle i contains two N-dimensional vectors, one is a Boolean vector $\mathbf{x}_i$, which represents a candidate solution, or the state of particle i, and the other is a real number vector $\mathbf{v}_i$, called the velocity of particle i. In particle swarm algorithm, the velocity vector represents the particle's speed and direction in each dimension of the solution.

Let $K(i)$ represent the neighbor of particle i, and let $\mathbf{p}_i$ represent the optimal position that particle i passes. During each iteration, each particle i adjusts its state and flying speed according to the following equation.

$$\boldsymbol{v}_{i+1} = \omega \boldsymbol{v}_i + c_1 r_1 (\boldsymbol{p}_i - \boldsymbol{x}_i) + c_2 r_2 (\boldsymbol{p}_g - \boldsymbol{x}_i) \tag{6.212}$$

$$\boldsymbol{x}_{i+1} = \boldsymbol{x}_i + \boldsymbol{v}_{i+1} \tag{6.213}$$

Among them,ω is called the inertia factor, and the value is between [0, 1], which usually decreases gradually as the iteration progresses [73]. c_1 and c_2 are two constants. Generally, $c_1 + c_2 = 4$ is taken to indicate the degree of influence of a particle group on a single particle [74].r_1 and r_2 are two random numbers between 0 and 1. $\mathbf{p}_g$ represents the optimal position experienced by all particles in the particle population. In this section, a convergence fuzzy particle swarm algorithm is introduced to solve the train operation adjustment model established in Sect. 6.3.2.

1. Fuzzy particle swarm optimization

Literature [63] first proposed the fuzzy particle swarm algorithm in 2005, which is a generalization of the basic particle swarm algorithm. The difference from the basic particle swarm algorithm is that it is not just that the optimal particle can affect other neighboring particles, but that several particles can affect other neighboring particles.

The degree to which each particle is affected by other neighboring particles depends on the membership of these particles, and the amount of attraction depends on the corresponding fuzzy variables of the particles. This makes full use of the advantages of each particle around the particle, and the algorithm has higher accuracy.

Membership is a fuzzy variable. Bell, Gaussian, Sigmoid, Triangle, and Trapezoid membership functions are several common membership functions (MF). Their function curves are shown in Fig. 6.49 [75].

Each function expression is as follows.

$$\mathrm{Bell}(x; c, \sigma) = 1/(1 + ((x - c)/\sigma)^2) \quad \sigma = 1, c = -5 \tag{6.214}$$

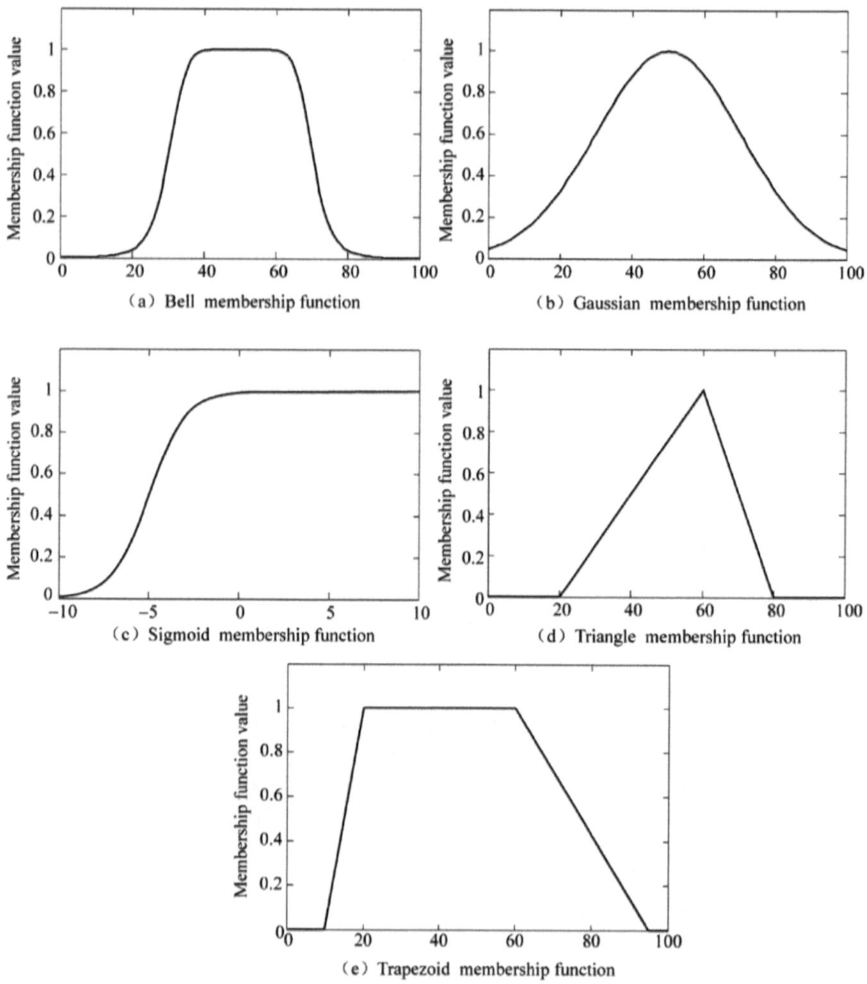

Fig. 6.49 Membership functions

$$\text{Gaussian}(x; c, \sigma) = \exp(-1/2((x-c)/\sigma)^2) \quad \sigma = 20, c = 50 \tag{6.215}$$

$$\text{Sigmoid}(x; c, \sigma) = 1/(1 + \exp(-\sigma(x-c))) \quad \sigma = 20, c = 50 \tag{6.216}$$

$$\text{Triangle}(x; a, b, c) = \max(\min((x-a)/(b-a) \quad (c-x)/(c-b)), 0),$$
$$a = 20, b = 60, c = 80 \tag{6.217}$$

$$\text{Trapezoid}(x; a, b, c, d) = \max(\min((x-a)/(b-a) \quad 1, (d-x)/(d-c)), 0)$$
$$a < b \le c < d, a = 10, b = 20, c = 60, d = 95 \tag{6.218}$$

A constant k is artificially determined, which can be changed according to the accuracy requirements of problem solving. k represents the number of neighboring particles existing around a particle that can affect its motion. Let h be one of the k-optimal particles, and $f(\boldsymbol{p}_g)$ represents the fitness value of the optimal particle.

Then, if the membership function is based on the Bell membership function, then the membership function is defined as

$$\phi(h) = 1/(1 + ((f(\boldsymbol{p}_h) - f(\boldsymbol{p}_g))/\beta)^2) \tag{6.219}$$

If the membership function is based on the Gaussian membership function, then the membership function is defined as

$$\phi(h) = \exp(-1/2(f(\boldsymbol{p}_h) - f(\boldsymbol{p}_g)/\beta)^2) \tag{6.220}$$

If the membership function is based on the Sigmoid membership function, then the membership function is defined as

$$\phi(h) = 1/(1 + \exp(-\beta(f(\boldsymbol{p}_h) - f(\boldsymbol{p}_g)))) \tag{6.221}$$

If the membership function is based on the Triangle membership function, then the membership function is defined as

$$\phi(h) = \max\left(\min\left(\beta\frac{f(\boldsymbol{p}_h) - f(\boldsymbol{p}_g^0)}{f(\boldsymbol{p}_g) - f(\boldsymbol{p}_g^0)}, \beta\frac{f(\boldsymbol{p}_g^1) - f(\boldsymbol{p}_h)}{f(\boldsymbol{p}_g^1) - f(\boldsymbol{p}_g)}\right), 0\right) \tag{6.222}$$

If the membership function is based on the Trapezoid membership function, then the membership function is defined as

$$\phi(h) = \max\left(\min\left(\beta\frac{f(\boldsymbol{p}_h) - f(\boldsymbol{p}_g^0)}{f(\boldsymbol{p}_g) - f(\boldsymbol{p}_g^0)}, 1, \beta\frac{f(\boldsymbol{p}_g^1) - f(\boldsymbol{p}_h)}{f(\boldsymbol{p}_g^1) - f(\boldsymbol{p}_g)}\right), 0\right) \tag{6.223}$$

Because $f(\boldsymbol{p}_h) \leq f(\boldsymbol{p}_g)$,$\phi(h)$ is a decreasing function. When $f(\boldsymbol{p}_h) = f(\boldsymbol{p}_g)$, the function value is 1. As the gap between $f(\boldsymbol{p}_h)$ and $f(\boldsymbol{p}_g)$ becomes larger, the function value approaches zero. In order to avoid excessive effects from the fitness function, define $\beta = f(\boldsymbol{p}_g)/\ell$, where ℓ is a parameter given by the user. For a certain $f(\boldsymbol{p}_h)$, the larger ℓ is, the smaller $\phi(h)$ is.$f(\boldsymbol{p}_g^0)$ and $f(\boldsymbol{p}_g^1)$ are the two boundary values of Triangle membership function and Trapezoid membership function, respectively. In the fuzzy particle swarm algorithm, the velocity equation is defined as

$$\boldsymbol{v}_{i+1} = \omega \boldsymbol{v}_i + c_1 r_1 (\boldsymbol{p}_i - \boldsymbol{x}_i) + \sum_{h \in B(i,k)} \phi(h) c_2 r_2 (\boldsymbol{p}_h - \boldsymbol{x}_i) \tag{6.224}$$

where $B(i, k)$ represents the set of k particles adjacent to particle i. Each particle is affected by its own optimal position $\boldsymbol{p}_i$ and the surrounding k-optimal particles. The degree to which each adjacent particle affects it depends on the value of $\phi(h)$. It can be inferred that when the value of k is set to 1, the fuzzy particle swarm algorithm is transformed into a basic particle swarm algorithm.

2. Introduction of convergence factors

Clerc strictly proved that the particle swarm algorithm must converge after introducing a convergence factor [61]. In the particle motion equation, c_1 and c_2 are two constants, usually set to 2 in the basic particle swarm algorithm. In Clerc's model, φ is defined as the sum of c_1 and c_2. The convergence factor k can be defined as

$$k = 2/\left|2 - \varphi - \sqrt{|\varphi^2 - 4\varphi|}\right| \tag{6.225}$$

Therefore, the motion equation of the convergent particle swarm optimization (convergent PSO, CPSO) algorithm is

$$\boldsymbol{v}_{i+1} = k[\omega \boldsymbol{v}_i + c_1 r_1 (\boldsymbol{p}_i - \boldsymbol{x}_i)] + c_2 r_2 (\boldsymbol{p}_g - \boldsymbol{x}_i)] \tag{6.226}$$

This model can be seen as a generalization of the basic particle swarm algorithm. When $\varphi = 4$, $k = 1$. The algorithm degenerates into a basic particle swarm algorithm ,and it is difficult to converge. According to Clerc's conclusion [61], when $\varphi > 4$, the algorithm has strong convergence characteristics, and the convergence characteristics are almost linear.

The disadvantage of the basic particle swarm algorithm is that it is difficult to converge. After introducing a convergence factor, it guarantees a fast convergence speed [76, 77]. However, the accuracy of the algorithm is still difficult to guarantee.

Therefore, the concept of membership function in fuzzy particle swarm algorithm was introduced into the model, and a new particle swarm algorithm—convergent fuzzy particle swarm optimization (CFPSO) algorithm was proposed [78].

The difference between this algorithm and fuzzy particle swarm algorithm is that a convergence factor is added to the particle's motion equation. Its motion equation is shown in Eq. (6.227).

$$\boldsymbol{v}_{i+1} = k[\omega \boldsymbol{v}_i + c_1 r_1(\boldsymbol{p}_i - \boldsymbol{x}_i)] + \sum_{h \in B(i,k)} \phi(h) c_2 r_2(\boldsymbol{p}_h - \boldsymbol{x}_i)] \tag{6.227}$$

3. Parameter design and calculation result analysis

(1) Convergence factor determination

According to Clerc's model,φ is equal to $c_1 + c_2$, and the convergence factor k is defined as $k = 2/\left|2 - \varphi - \sqrt{|\varphi^2 - 4\varphi|}\right|$. So the value of c_1 and c_2 is the key to the problem.

In order to determine the optimal values of c1 and c2, the Rastrigin function shown in formula (6.228) is used as a test function for numerical calculation experiments. The particle swarm is designed to be 100, each particle has 20 dimensions, and the loop iterates 100 times. As can be seen from Fig. 6.50, as c_1 and c_2 approach 2.0 and 2.1, respectively, the objective function value becomes smaller and smaller. Table 6.14 shows the specific objective function values. When c_1 and c_2 take 2.0 and 2.1 respectively, the objective function value is the smallest.

According to the conclusion that the algorithm has better convergence performance when Clerc's $\varphi > 4$, $\varphi = 2.0 + 2.1 = 4.1 > 4$, so $c_1 = 2.0$ and $c_2 = 2.1$. At this time, $k = 0.73$.

(2) Selection of membership function

From the four commonly used fuzzy membership functions of Bell, Gaussian, Sigmoid, and Triangle membership functions, select a suitable function for the

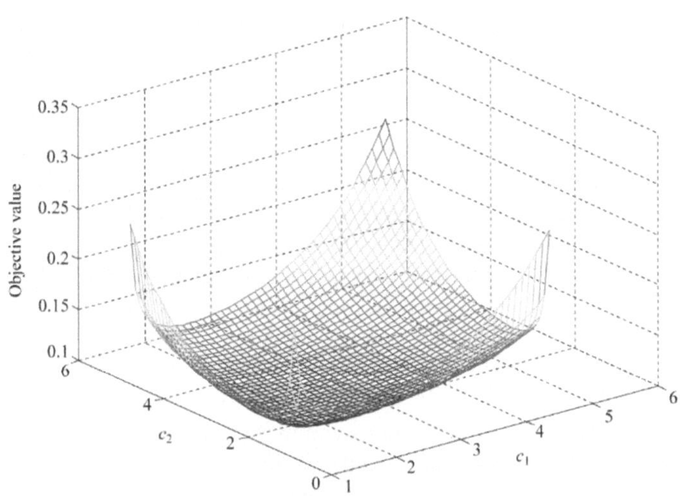

Fig. 6.50 Relation surface of objective function value and c_1, c_2

Table. 6.20 Relationship between objective function value and membership function

Standard function	Membership function	Optimal solution	Average solution	Number of times the target is reached
f_1	Gaussian	5.1287E−11	3.4687E−07	58
	Bell	8.2256E−11	1.6721E−07	55
	Sigmoid	6.1259E−11	4.2684E−07	53
	Triangle	2.4511E−08	8.5784E−04	30

particle swarm motion equation. Using f_1 as the standard function and four functions as the membership functions, 100 experiments were run each. The size of the particle swarm is 100. Each particle contains 20 dimensions. The target value of the standard function is 1.000E−07. The maximum velocity v_{max} of the particles is limited to 10.

Table 6.20 lists the optimal and average solutions of the standard functions calculated using various membership functions, and the number of times the target value is reached. Using the Triangle membership function, the target value can be reached 30 times, which is less than using the other three membership functions.

The calculation results show that although the Triangle membership function is highly efficient, the accuracy of the calculation is the lowest among the four membership functions. In addition, it can be seen that the optimal and average solutions of the standard equations obtained by using the Bell membership function are better than the other three membership functions, and it reaches the target value more times, so the Bell membership function has the highest accuracy.

Figure 6.51 shows the relationship between the number of times that the optimal value is reached and the time consumed when experimentally calculating with different membership functions.

It can be seen from Fig. 6.51 that the Triangle membership function takes the least time to reach the objective function value of the same number of times.

Therefore, although the Triangle membership function has the lowest calculation accuracy among the four membership functions, it has the highest efficiency. This is because the calculation complexity of the Triangle membership function is much lower than that of several other membership functions, so you can find a balance between the calculation efficiency and the calculation accuracy by replacing the membership function in the particle swarm motion equation.

When comparing the performance of the algorithm with experimental data in this section, the Triangle membership function is used in the FPSO and CFPSO algorithms.

(3) Calculation results and analysis

This section uses four standard functions commonly used in academia as test functions, each running 100 times to analyze the performance of the new algorithm. The size of the particle swarm is 100, and the dimensions of the particles are 20, 40, and 60, respectively.

The f_1 function is a generalized Rastrigin function, as shown in Eq. (6.228).

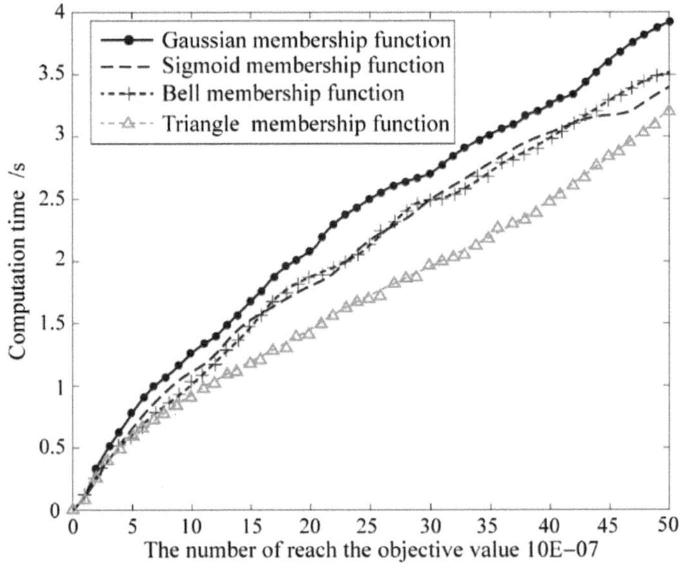

Fig. 6.51 Relationship between the number of times reaching the objective value using different membership functions the elapsed time

$$f_1(x) = \sum_{i=1}^{n} (x_i^2 - 10\cos(2\pi x_i) + 10), x_i \in [-5.12, 5.12] \tag{6.228}$$

The f_2 function is a generalized Griewank function, as shown in Eq. (6.229).

$$f_2(x) = \frac{1}{4\,000}\sum_{i=1}^{n} x_i^2 - \prod_{i=1}^{n}\cos\left(\frac{x_i}{\sqrt{i}}\right) + 1, x_i \in [-600, 600] \tag{6.229}$$

The f3 function is a generalized Rosenbrock function, as shown in Eq. (6.230).

$$f_3(x) = \sum_{i=1}^{n} (100(x_i^2 - x_{i+1})^2 + (1 - x_i)^2), x_i \in [-30, 30] \tag{6.230}$$

The f4 function is a generalized Ackley function, as shown in Eq. (6.231).

$$f_4(x) = -20 \times \exp\left(-0.2\sqrt{\frac{1}{n}\sum_{i=1}^{n} x_i^2}\right) - \exp\left(\frac{1}{n}\sum_{i=1}^{n}\cos(2\pi x_i)\right) + 20 + e, x_i \in [-30, 30] \tag{6.231}$$

Table 6.21 Experimental data of four standard functions

Test function	Algorithm	Optimal solution	Average solution	Number of times the objective function value is reached
f_1	PSO	2.639E−11	3.172E−02	26
	FPSO	1.449E−11	8.256E−03	53
	CFPSO	1.825E−11	9.672E−03	50
f_2	PSO	2.664E−14	9.963E−08	76
	FPSO	5.467E−15	8.124E−08	83
	CFPSO	7.879E−15	9.549E−08	80
f_3	PSO	4.531E−11	8.854E−06	72
	FPSO	6.712E−12	3.645E−06	80
	CFPSO	7.564E−12	6.152E−06	78
f_4	PSO	7.125E−11	2.457E−04	50
	FPSO	6.587E−12	2.365E−05	66
	CFPSO	9.145E−12	3.264E−05	63

Table 6.21 lists the results of four standard functions calculated by three algorithms. The optimal or average solution of the four functions calculated by the FPSO algorithm is the smallest of the three algorithms.

Therefore, of the three algorithms, the FPSO algorithm has the highest accuracy. In the case of running this experiment 100 times, the number of times that the CFPSO algorithm is used to achieve the objective function value is more than that of the PSO algorithm, but slightly less than the FPSO algorithm.

As shown in Fig. 6.52, as the number of times the objective function value is reached increases, the calculation time consumed also increases. The time consumed by the FPSO algorithm is very close to the time consumed by the basic PSO algorithm. So choosing Triangle membership function as the fuzzy membership function has little effect on the calculation efficiency of the algorithm.

The CFPSO algorithm consumes less time than the FPSO algorithm because the convergence factor in the algorithm works. In addition, when the number of times to reach the objective function value is small, the CFPSO algorithm takes more time than the basic PSO algorithm. This is because it takes some time to calculate the value of the convergence factor.

The CFPSO algorithm inherits the advantages of high calculation accuracy of the FPSO algorithm, taking into account the calculation efficiency, and has the best algorithm performance.

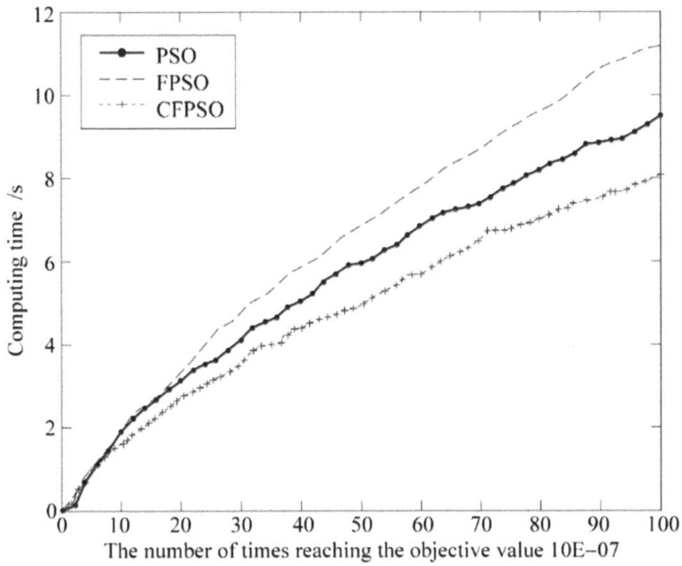

Fig. 6.52 Relationship between the number of times reaching the objective value and the time consuming

6.5.3 Implementation of Train Timetable Rescheduling Based on CFPSO Algorithm

1. Particle Swarm Design

Aiming at the problem of train timetable rescheduling in this chapter, the size of the particle swarm is designed to be 20–50, which can ensure the speed of iterative calculation of train operation adjustment and prevent the result of train timetable rescheduling from falling into a local optimum [79].

2. Design of position vector

According to the train timetable rescheduling model established in Sect. 6.3, the position vector of the particle swarm for this model is designed. The circuitous route set includes several routes. For each route p, the number of trains assigned to it is n_p and the number of stations it passes is NS_p, then there are $NS_p - 1$ sections on the route.

Because a train's arrival and departure time at each station correspond to two decision variables, then there are $2 \times n_p \times NS_p$ decision variables for n_p trains at NS_p stations, and because the route set contains L routes, the number of decision variables is $\sum_{p=1}^{L} (2 \times n_p \times NS_p)$.

In addition, design 0–1 variable $x_p^{k,i}$. When $x_p^{k,i} = 1$, it means that the route designated by the i-th train of the k-th kind is p, the number of train types is TN, there are N^k trains for a certain type of train, and because the detours include L routes, so the number of 0–1 variable is $L \times \mathrm{TN} \times N^k$.

Then the position information of each particle contains $\sum_{p=1}^{L} (2 \times n_p \times \mathrm{NS}_p) + L \times \mathrm{TN} \times N^k$ variables, which correspond to the arrival and departure time of all trains at each station on the L tracks, and the decision variables of the trains, namely

$$
\begin{aligned}
&\boldsymbol{P}_i = (a_1^{1,1}, a_2^{1,1}, \ldots, a_{\mathrm{NS}_1}^{1,1}, a_1^{1,2}, a_2^{1,2}, \ldots, a_{\mathrm{NS}_1}^{1,2}, \ldots, a_1^{1,n_1}, a_2^{1,n_1}, \ldots, a_{\mathrm{NS}_1}^{1,n_1},\\
&a_1^{2,1}, a_2^{2,1}, \ldots, a_{\mathrm{NS}_2}^{2,1}, a_1^{2,2}, a_2^{2,2}, \ldots, a_{\mathrm{NS}_2}^{2,2}, \ldots, a_1^{2,n_2}, a_2^{2,n_2}, \ldots, a_{\mathrm{NS}_2}^{2,n_2},\\
&\ldots,\\
&d_1^{L,1}, d_2^{L,1}, \ldots, d_{\mathrm{NS}_L}^{L,1}, d_1^{L,2}, d_2^{L,2}, \ldots, d_{\mathrm{NS}_L}^{L,2}, \ldots, d_1^{L,n_L}, d_2^{2,n_L}, \ldots, d_{\mathrm{NS}_L}^{2,n_L},\\
&d_1^{1,1}, d_2^{1,1}, \ldots, d_{\mathrm{NS}_1}^{1,1}, d_1^{1,2}, d_2^{1,2}, \ldots, d_{\mathrm{NS}_1}^{1,2}, \ldots, d_1^{1,n_1}, d_2^{1,n_1}, \ldots, d_{\mathrm{NS}_1}^{1,n_1},\\
&d_1^{2,1}, d_2^{2,1}, \ldots, d_{\mathrm{NS}_2}^{2,1}, d_1^{2,2}, d_2^{2,2}, \ldots, d_{\mathrm{NS}_2}^{2,2}, \ldots, d_1^{2,n_2}, d_2^{2,n_2}, \ldots, d_{\mathrm{NS}_2}^{2,n_2},\\
&\ldots,\\
&d_1^{L,1}, d_2^{L,1}, \ldots, d_{\mathrm{NS}_L}^{L,1}, d_1^{L,2}, d_2^{L,2}, \ldots, d_{\mathrm{NS}_L}^{L,2}, \ldots, d_1^{L,n_L}, d_2^{2,n_L}, \ldots, d_{\mathrm{NS}_L}^{2,n_L}, \qquad (6.232)\\
&x_1^{1,1}, x_2^{1,1}, \cdots, x_L^{1,1}, x_1^{1,2}, x_2^{1,2}, \cdots, x_L^{1,2}, \cdots, x_1^{1,N^1}, x_2^{1,N^1}, \cdots, x_L^{1,N^1},\\
&x_1^{2,1}, x_2^{2,1}, \cdots, x_L^{2,1}, x_1^{2,2}, x_2^{2,2}, \cdots, x_L^{2,2}, \cdots, x_1^{2,N^2}, x_2^{2,N^2}, \cdots, x_L^{2,N^2},\\
&\cdots,\\
&x_1^{\mathrm{TN},1}, x_2^{\mathrm{TN},1}, \cdots, x_L^{\mathrm{TN},1}, x_1^{\mathrm{TN},2}, x_2^{\mathrm{TN},2}, \cdots, x_L^{\mathrm{TN},2}, \cdots, x_1^{\mathrm{TN},N^{\mathrm{TN}}}, x_2^{\mathrm{TN},N^{\mathrm{TN}}}, \cdots, x_L^{\mathrm{TN},N^{\mathrm{TN}}})\\
&i = 1, 2, \cdots, \mathrm{popsize}
\end{aligned}
$$

Particle swarm initialization is actually assigning a value to each variable of the position vector of each particle. At the time of initialization, the arrival time and departure time of the train at the railway station should be considered, rather than randomly generated randomly. That is, the value of the train's arrival and departure variables at the station are within a certain range.

The initial value of $x_j^{k,i}$, which is a 0–1 variable, is determined by generating a random number. That is, a random variable rand is generated. If its value is greater than a certain threshold μ, set $x_j^{k,i} = 1$, otherwise, set $x_j^{k,i} = 0$. Then according to the value of $x_j^{k,i}$, determine n_p^k, and then determine n_p.

There are two cases to determine the value of the train's arrival and departure variables at the station. One is that after some trains are optimized in the upper-level planning, the newly selected routes are the same as those of the original operation plan. At this time, the arrival and departure time of the train at each station can be designed with reference to the arrival and departure time in the original operation plan.

The other is that the route of some trains has changed. The arrival and departure time of each train should be comprehensively considered according to the original arrival and departure time of the train in the original operating plan, the technical operation time of the train at the station, and the minimum running time of various types of trains in various sections.

3. Design of speed vector

Design the fuzzy particle velocity according to the method described in Sect. 6.5.2.

4. Calculate stopping conditions

There are two strategies for stopping the calculation during the design of the particle swarm optimization algorithm for train operation adjustment. A method for measuring the number of iterations is to set a maximum number of iterations of the particle swarm algorithm based on the cycle number control strategy of a general evolutionary algorithm. When the number of iterations is reached, the calculation stops.

The other is the method of setting the objective function, that is, setting an objective function value as the optimization criterion for iterative calculation. When the value of the decision variable makes the objective function value reach this criterion, the calculation is ended.

The first strategy is the most commonly used strategy, and its design is quite simple. When the number of iterations is set reasonably, it can obtain a better calculation result. The disadvantage is that it cannot save calculation time. For a problem with certain complexity, the required calculation time is basically the same.

The advantage of the second strategy is that it can control the index of the objective function value. It can save calculation time to some extent when dealing with some simple optimization problems, and its disadvantages are also obvious.

First of all, for some optimization problems, such as train operation adjustment problems, especially under emergency conditions, it is difficult to predict the optimization criteria of the objective function value.

Furthermore, when the value of the objective function is set too harshly, the calculation is in an infinite loop, which may lead to a situation without a solution. Therefore, the iterative calculation in this section stops adopting the first strategy.

In fact, when the loop operation reaches a certain number of times, the objective function value is basically close to the optimal value, and the subsequent calculations have little effect on the optimization of the objective function value.

Therefore, based on a large number of numerical experiments, the upper limit of the number of iterations is set to 200, which not only ensures that the calculation results are as close to optimal as possible, but also ensures the calculation efficiency, which is consistent with the real-time requirements of on-site production command.

5. Calculation framework

The calculation steps for train operation adjustment are as follows.

Step1 Initialize the particle swarm population. Set the size of the particle swarm population as N_{swarm}, the position vector of each particle is composed of

$\sum_{p=1}^{L} (2 \times n_p \times \mathrm{NS}_p) + L \times \mathrm{TN} \times N^k$ variables. Set the initial value of each variable based on the train timetable and the delay situation caused by the emergency conditions.

Step2 Calculate the objective function value Z according to formula (6.211) and the position vector of each particle in the population. Record the optimal position of each particle, and record the number of the particle that minimizes the objective function value.

Step3 According to formula (6.227), calculate the flying speed of each particle.

Step4 Calculate the new position of each particle according to the position calculation formula. If the required accuracy or number of iterations is reached, go to Step 5; otherwise, go to Step 2.

Step5 According to the optimal position of the optimal particle in the population, the arrival and departure times of each train at the station after the train diagram adjustment are given.

6. The performance of train operation adjustment principles in solving under emergency conditions

According to the principle of train operation adjustment, the method of adjusting the arrival and departure time of the train at the station is used in the algorithm design. A particle contains a number of variables, which represent the train routing decision and the time of arrival and departure of each train on each track. The grade of each train can be obtained by checking k in n_p^k.

Regarding the principle of "first to high-priority trains, and low-priority trains to follow," check whether the arrival and departure time of low-priority trains is set earlier than that of high-level trains under the conflicting conditions of train running time in particles (i.e., the difference between the arrival and departure times of the two trains at the station is less than the minimum arrival and departure time interval). If this is the case, adjust the variables representing the arrival and departure times of the two trains at the station among the particle variables.

Regarding the principle of "the trains of the same class, punctual trains have absolute priority, and late trains follow," the principle is to check whether there are conflicts between trains at the same time and to prioritize the arrangement of late trains. If this exists the case, adjust the variables of train arrival and departure time.

Regarding the principle of "all trains of the same class are sorted according to the earliest possible departure time," when setting the departure time of particles at the station, try to approach the earliest possible time (by the arrival time of the train at the station plus the minimum stay time at the station).

For the principle of "partial trains with special requirements, proper priority can be given," directly changing the group to which the train belongs, that is, directly changing its priority.

Regarding the principle that "low-priority trains cannot overtake high-priority trains," check whether there is a low-priority train in the front station whose arrival and departure time are later than that of the high-priority train, and the arrival and departure time of the rear station are earlier than the high-priority train. If so, reassign the variables at the time of arrival and departure at the rear station.

For the "trains of the same priority, can overtake" and "high-priority trains can cross the low-priority trains" two principles, reflected in the setting of trains at the station sooner or later, you do not deal with this situation.

7. Steps for Solving Bi-level Programming Model of Train Organization under Emergency Conditions

The steps for solving the bi-level programming model of the train organization under emergency conditions are as follows.

Step1 Initialize parameters, including various parameter coefficients, the overall satisfaction value of the lower-level planning, and the total number of iterations.

Step2 Set the objective function membership value of the upper-level plan as $1-\alpha$, use it as the lower limit of the objective function membership value, and solve the upper-level plan (train scheduling optimization model) to get n_p^k.

Step3 Set n_p^k as a constraint of the lower-level optimization model (train operation adjustment model) and solve it, the specific steps are as follows.

Step3.1 For the train allocation information of each route, the dispatching section of the trains is determined.

Step3.2 Search for trains in all dispatching sections during the adjustment period to generate a train set that contains all trains need to be adjusted.

Step3.3 Read the train diagram data of the rescheduled train set.

Step3.4 The rescheduled train set is divided into several train groups, and the adjustment priority level of each train group is determined.

Step3.5 The train operation plan in each section is adjusted to solve the train operation adjustment problem.

Step3.6 The obtained result is as a train operation plan.

Step4 Determine whether the objective function value of the lower-level plan (train operation adjustment model) has reached the set value, and if so, terminated; otherwise, increase the total number of iterations by 1, and go to Step 5.

Step5 Determine whether the total number of iterations has reached the set value or not, and if so, the procedure is terminated; otherwise, reset the membership function value $1-\alpha$ of the objective function of the upper-layer plan, and obtain a new set of solutions n_p^k, then go to Step 3.

6.5.4 Case Study

The proposed models are applied to the railway network composed of the Beijing-Shanghai high-speed railway and surrounding railways. The case study illustrates how to use the convergence fuzzy particle swarm algorithm to realize train operation adjustment.

1. Disturbance scenario assumptions

At present, China's railway network is composed of three types of railway lines: existing ordinary speed railways, existing intercity railways, and new high-speed railways. Existing intercity railways include Beijing-Tianjin Intercity Railway, Hefei-Nanjing Intercity Railway, Shijiazhuang-Taiyuan Intercity Railway, etc., and new high-speed railways include Wuhan-Guangzhou high-speed railway and Beijing-Shanghai high-speed railway.

For empirical analysis, this section draws a physical railway network to the east of railway line (Jining-Datong-Taiyuan-Luoyang-Xiangfan-Liuzhou), north of railway line (liuzhou-Hengyang-Zhuzhou-Yingtan-Jinhua), and south of railway line (Jining-ZhangJiakou-Huludao) as the basic line network for empirical analysis, as shown in Fig. 6.53.

Assume that at 8 o'clock × × × year × × × month × × × day, line failures occur on the Beijing-Shanghai high-speed railway DK703 + 903 and DK856 + 321, resulting in an average speed limit of 160 km/h for Xuzhou East-Suzhou East high-speed railway. The up and down of the high-speed railway between South and Dingyuan is interrupted, and it is estimated that it will take 4–8 h to resume traffic. The physical railway network around the accident site is shown in Fig. 6.54.

Assuming that the upstream direction line is not affected, only the train operation adjustment problem in the downstream direction is studied. In order to avoid causing interference to the up train, this example does not consider the use of the up direction line to organize reverse traffic when generating a detour. The research content does not involve the internal route arrangement and operation of the station, so the internal operation organization of the station is not studied in the case study.

Under the condition of this emergency, both the upside and downside lines of the high-speed railway line are interrupted, and the train operation cannot be organized in the reverse direction. It takes a long time to restore the line, so we must find a new downstream route between Xuzhou and Nanjing, and organize trains to pass this bottleneck section in time. Therefore, the following problems need to be addressed.

(1) Find new routes between Xuzhou and Nanjing.

(2) Carry out train route allocation.

(3) Adjust the train scheduled timetable. The time range for train adjustment is from 8 to 12 o'clock. The space range is the new train running route. The involved trains are all trains on the above route from 8 to 12 o'clock.

Note: The solid line in the figure is the existing railway, the dashed line is the new high-speed railway, and the dotted line indicates the existing intercity railway line, the same applies as below.

2. Calculation of the capacity of the influence area

According to the capacity estimation of the railway section under emergency conditions in Chap. 3, the capacity of the Xuzhou East-Suzhou East section is 115 columns (the detailed process is omitted) in the time period calculated in this example.

Since the other sections and stations are not affected by the emergency, you can directly determine the downward remaining capacity of the Xuzhou-Bengbu, Bengbu-Nanjing, Bengbu-Shuijiahu, Shuijiahu-Hefei, Hefei-Nanjing (intercity) sections. They are 20, 20, 12, 12, 62 columns, respectively. Bengbu South Station,

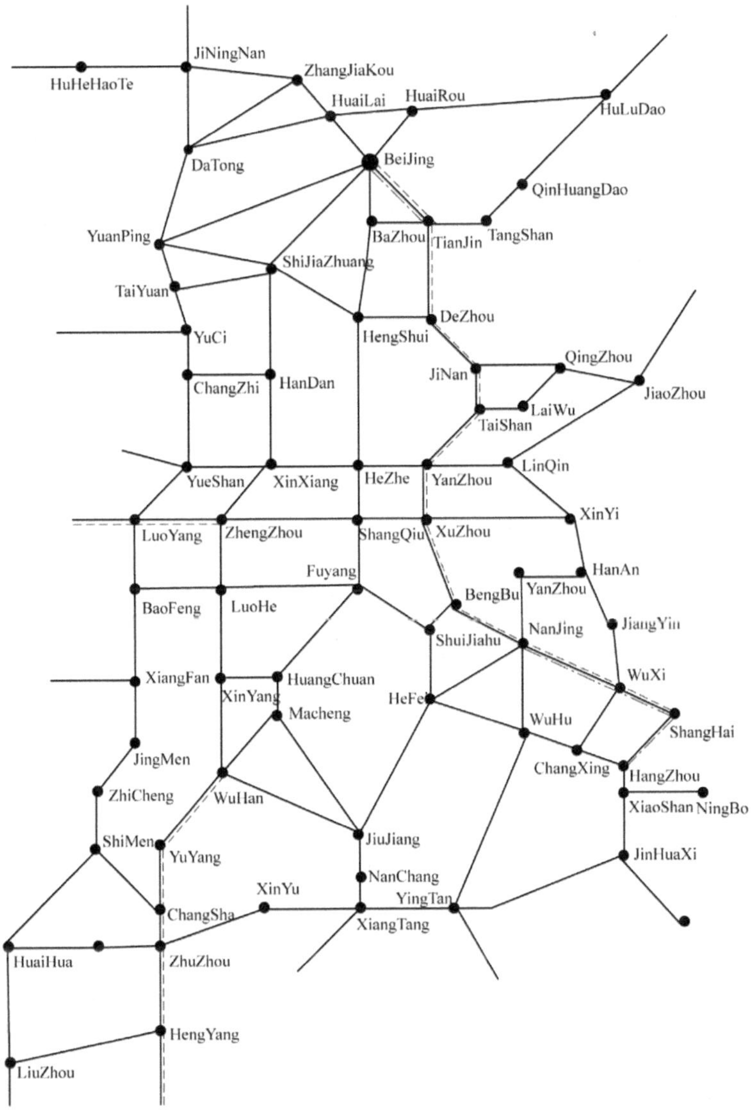

Fig. 6.53 Physical network of some national railway

Xuzhou Station, Bengbu Station, Nanjing Station, Shuijiahu Station, and Hefei Station (intercity) have a remaining capacity of 70, 32, 38, 30, 20, 70 in the downward direction.

3. Generation of train rerouting routes

This example will generate a set of reroutes based on the route search method based on sufficient capacity proposed in Chap. 4. First determine the origin station and

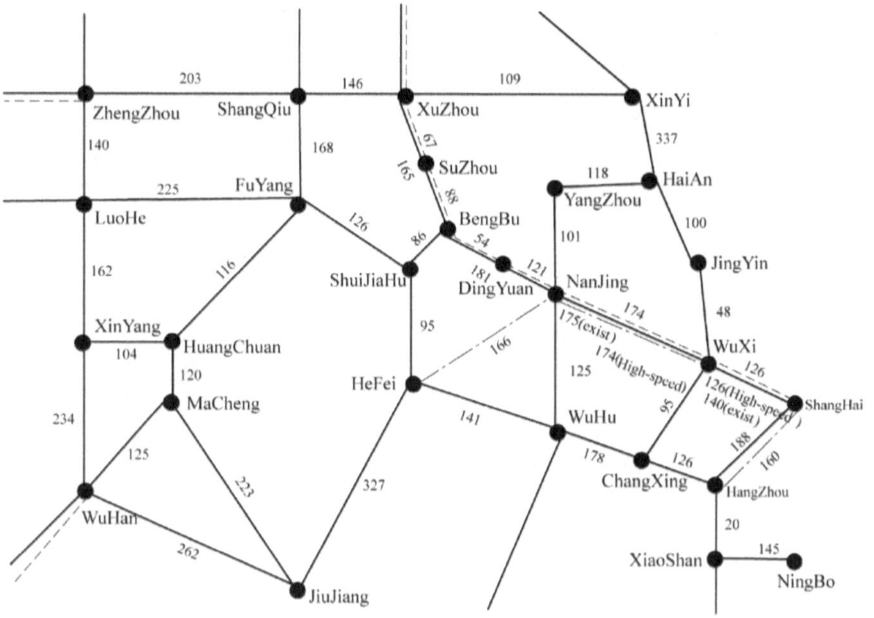

Fig. 6.54 Physical network of the railway around the accident site

destination station of the route generation. The accident affected two sections, namely Xuzhou East-Bengbu South and Bengbu South-NanJingNan. Then it is easy to know that the original station of the route is Xuzhou Station, and the terminal station is Nanjing Station.

As can be seen from Fig. 6.54, in addition to the Beijing-Shanghai high-speed railway, there are also existing Beijing-Shanghai ordinary speed railways connecting Xuzhou hub to Nanjing hub.

According to design data, it can be seen that Xuzhou and Nanjing hubs have Beijing-Shanghai high-speed railway departure and end stations and can handle train's departure and arrival operations. Moreover, Xuzhou East Station and NanJingNan Station are separately connected with Xuzhou and Nanjing stations passing by the existing ordinary speed railway line, so the Xuzhou and Nanjing hubs have the ability to cross other lines.

Bengbu South Station has the capability of departure and arrival, but can only handle train connections and stand-off operations. It cannot handle train departure and arrival operations, but there is also a connected line between Bengbu South Station and Bengbu Station. Therefore, Bengbu Station also has the ability to cross other lines.

The above physical network is the foundation of the route generation.

According to the method described in Sect. 4.3, the design sufficient capacity value is 40 (estimated by the spatial and temporal characteristics of the emergency), and a list of detour routes can be obtained, as shown in Table 6.22.

Table 6.22 List of Xuzhou-Nanjing Detour Route Sets (adequate capacity plan)

Number	Route	Length/km
1	Xuzhou(via high-speed railway)-Bengbu (via existing General speed railway)-Nanjing	336
2	Xuzhou(via existing General speed railway) - Bengbu (via existing General speed railway)-Nanjing	346
3	Xuzhou (via high-speed railway) - Bengbu - Shuijiahu - Hefei (via intercity railway)-Nanjing	502

It can be seen that the routes are generated according to this method, and the capacity problem of the route is considered, that is, the distribution of train running routes can be realized on these routes, and there are fewer sections involved, which is of great reference value for solving the train operation route allocation problem.

4. Relevant data description

(1) Trains involved in train organization

In the case of this emergency, according to the set of circuitous routes shown in Table 6.22, in this example, the train involved in the adjustment of the train operation between the Xuzhou hub and the Nanjing hub is divided into two categories (except for high-speed trains, other train data are collected from the national railway train timetable on August 1, 2010).

(1) Those who entered the sections before 8 o'clock include:

① express train: T114/1, T54, T284/1, T166/3, T115/5, T65, T131/4/1, T118/5, T140/37;
② fast train: K190/87, K8471, K374/1, K559/8.

This type of train is only considered during train operation adjustment and does not participate in train route allocation.

(2) Trains arriving at Xuzhou and entering the track from 8:00 to 12:00 include:

① High-speed trains: G301, G303, G305, G101, G103, G105, G107, G109, G111, G113 (data taken from the research results of key technologies and system project optimization design of high-speed train operation organization scheme);
② Drive multiple unit: D88/5;
③ Express trains and fast trains: K58/5、K518/5、K101/4/1;
④ General speed trains: 1230/27;
⑤ Slow trains: 10,135, 10,625, 11,301, 23,005, 11,305;
⑥ Irregular train: None;
⑦ Other train: None.

For this type of train, first run route allocation, and then comprehensively consider the information of type (1) train to make train operation adjustments.

(2) Established train operation plan on the track

The train operation plan is established on the track.

(3) Train operating parameters

(1) Minimal operating time of various trains running on existing railway lines.

According to the literature [80] on the definition of the operating speed of various trains and the line conditions of the Beijing-Shanghai high-speed railway and the existing Beijing-Shanghai railway, in this example, when adjusting the train operation, the following rules are obeyed.

On existing General speed railway lines, the high-speed trains, the high-speed railway train (D-head), and intercity trains (C-head) must not exceed 200 km/h, the speed of direct and express passenger trains (Z-head and T-head) must not exceed 160 km/h, the speed of fast passenger trains (K-shaped head) must not exceed 120 km/h, the speed of ordinary passenger trains must not exceed 100 km/h, and the speed of freight trains must not exceed 100 km/h. See Table D.1 in Appendix D for the minimum pure operating times of the train in the Xuzhou-Nanjing section.

(2) Pure running times in high speed railways under different speed limit conditions.

The pure running times of high-speed railway trains are related to speed limits. Under different speed-limiting conditions, the pure running times of high-speed trains in the Xuzhou East-NanJingNan section are listed in Table D.2 of Appendix D.

(3) Train start and stop additional time.

The length of the additional time for train start and stop mainly depends on the type of locomotive, the weight of the train, the maximum speed of the train in the section, and the cross-section and plane conditions of the inbound and outbound lines.

General regulations: The additional time for passenger trains to start is 1 to 2 min, the additional time for parking is 1 min; the additional time for cargo trains to start is 2 to 3 min, and the additional time for parking is 1 to 2 min [43]. In view of this, in this example, all passenger trains take 1 min for the start and stop additional time, and cargo trains take 2 min for the start and stop additional time.

(4) Headway times between trains using automatic block signals.

The headway between adjacent trains on high-speed railways is 3 min, and the headway between adjacent trains on general speed railways is 6 min.

(5) Train dwell time.

In principle, the dwell time of trains in the adjustment plan is not less than the dwell time in the scheduled operation plan. However, in the case of a severely delayed train, the train dwell time can be appropriately compressed. The dwell time for a passenger boarding is not less than 1 min.

5. The solution to the train organization problem in this example

(1) Train rerouting problem

Take the three routes in Table 6.22 as the target routes for train running routes, as shown in Fig. 6.55. Routes 1, 2, and 3 are the new three routes determined. In this

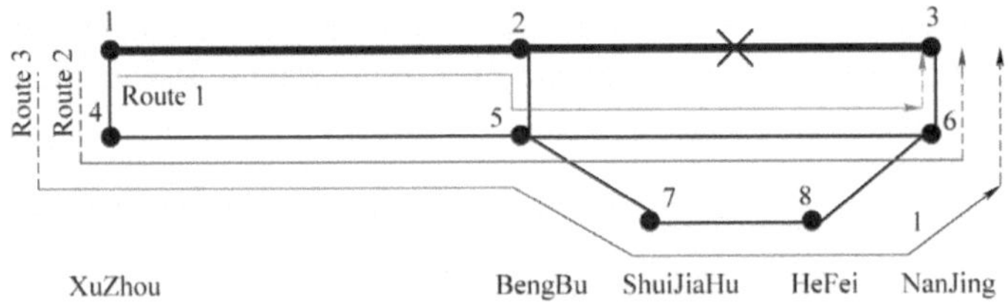

Fig. 6.55 Xuzhou-Nanjing train routes under emergency conditions

local network, there are eight node stations involved and six sections. This example uses LINGO11 to solve the train dispatch optimization model

The solution of the train dispatch optimization model is:$n_1^1 = 10, n_2^1 = 0, n_2^2 = 1$, $n_2^3 = 3$, $n_2^4 = 1$, $n_2^5 = 5$, $n_3^1 = n_3^2 = n_3^3 = n_3^4 = n_3^5 = 0$。

In addition, it is not difficult to find that all trains are assigned to Route 1 and Route 2, and there are no trains assigned to Route 3, which is related to the large capacity value (40 columns) set when the route is generated. There are actually 20 trains that need to be allocated. It is necessary to set the sufficient capacity value too large in practical applications, because it is difficult to accurately calculate the number of trains to bc allocated, and in addition, a part of capacity needs to be reserved to deal with uncertain conditions on the site.

(2) Rescheduling train timetable

According to the demonstration of the convergent of the fuzzy particle swarm algorithm described in Sect. 6.5.2 ("Convergent Fuzzy Particle Swarm Optimization"), train operation adjustments are performed on the trains assigned to Route 1 and Route 2. The adjustment period is from 8 to 12 o'clock. The train timetable after 12:00 is a pre-schedule in this example, as shown in bold font in Tables B.3 and B.4 of Appendix B. After 12 o'clock, other trains will appear on the two routes, and the operating plan needs to be readjusted under the actual train running at 12 o'clock and the emergency situation at that time. In this example, the train operation adjustment model was solved using PowerBuilder 9.0, and the database platform was Oracle 10.0.

The maximum allowable delay time per train is set to be 600 s, and the satisfactory delay time is as 480 s. Then, the generalized penalty value for the delay of the train is

$$Z_{\text{delay}}^* = [6 \times 480 \times 10 \times 3 + 6 \times 480 \times (1.5 \times 1 + 1.3 \times 3 + 1.1 \times 1 + 0.8 \times 5)] \times 2 = 233\,280$$

$$Z_{\text{delay}}^* + \sigma_1 = [6 \times 600 \times 10 \times 3 + 6 \times 600 \times (1.5 \times 1 + 1.3 \times 3 + 1.1 \times 1 + 0.8 \times 5)] \times 2 = 291\,600$$

When the delay penalty value of the train is greater than or equal to 291,600, the total satisfaction is 0; when the total late time of the train is less than 233,280, the total satisfaction is 1.

Set $\sigma_1 = 58\ 320$.

Then, $u_1 = 1 - (Z_{\text{delay}}^{\text{train}} - Z_{\text{delay}}^{*})/\sigma_1 = 1 - (Z_{\text{delay}}^{\text{train}} - 233\ 280)/58\ 320$.

For stability satisfaction, $S^* = 1, \sigma_2 = 0.8$, then $u_2 = 1 - (S^* - S)/\sigma_2 = 1 - (1 - S)/0.8$.

When the stability of the train operation plan is 1, the stability satisfaction is 1; when the stability of the train operation plan is less than 0.2, the stability satisfaction is 0.

When calculating the stability of the train operation plan, the section weight $\boldsymbol{\alpha}^{\mathrm{r}}$, the station weight $\boldsymbol{\alpha}^{\mathrm{w}}$, the probability of trains disrupted in sections $\mathrm{PR}^{\mathrm{p,r}}$ and the probability of trains disrupted at stations $\mathrm{PR}^{\mathrm{p,w}}$, and the train weight of each section on the track 1 and track 2 are set. Because the weight of the section is closely related to the length of the section, the weight of the section is defined in this example as the ratio of the length of the section to the length of the entire line; see Table E.1 in Appendix E.

Because the weight of the station on the route is related to the type of line where the station is located and the level of the station, the weight of each station defined in this example is shown in Table E.2 of Appendix E.

Due to the limitation of the actual situation, in this example, no probability data on the impact of the existing Beijing-Shanghai speed railway on train operation was collected, and the Beijing-Shanghai high-speed railway has not yet been put into operation, so this example also lacks data on the probability of the impact of unexpected events on train operation.

Therefore, in this example, the probability that the train is subjected to secondary disturbances in each section and station is 0.01. The weights of different trains are consistent with the penalty coefficients for delayed trains.

A total of 80 iterative calculations were performed in this example, and the calculation results are shown in Fig. 6.56. It can be seen from Fig. 6.56 that the total late time satisfaction value has an upward trend during the iterative calculation process, reaching a maximum value of 0.888 at the 71st generation. At this time, the general late penalty value is 239,812.

Starting from the 72nd generation, the total late satisfaction value oscillated near the optimal value, and the further optimization of the general late penalty value cannot be realized. In the iterative calculation process, the stability satisfaction value of the train timetable first increased and then gradually decreased, reaching a maximum value of 0.872 in the 41st generation.

It can be seen that after the 42nd generation, total late satisfaction and stability satisfaction are mutually exclusive, and an increase in the satisfaction value of one party will inevitably lead to a decrease in the satisfaction value of the other party.

The comprehensive satisfaction value of total late satisfaction and stability satisfaction is considered comprehensively, and it reaches the optimal value when iteration to 42 generations, and the value is 0.866. At this time, the total late satisfaction

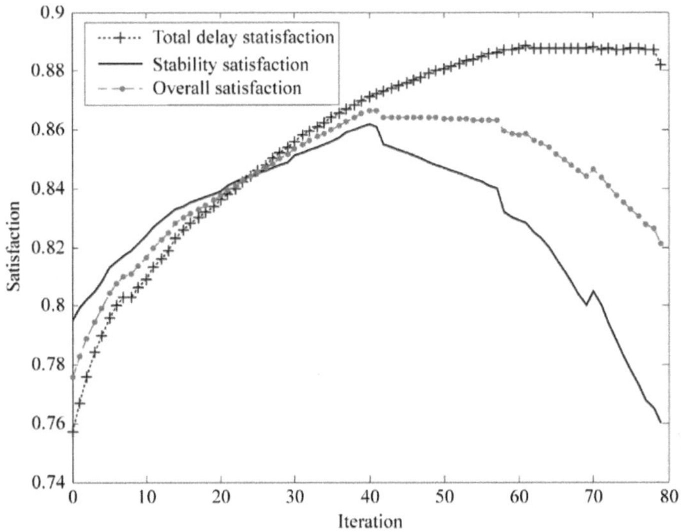

Fig. 6.56 Satisfaction value of train rescheduling model

value is 0.878 and the stability satisfaction value is 0.861. At this time, the operation adjustment result obtained is optimal.

The calculation result at this time is restored to the train operation plan, see Tables B.3 and B.4 in Appendix B, and the train diagram is shown in Figs. C.2 and C.3 in Appendix C. For the sake of intuition, Figure C.3 plots the running lines of high-speed trains on the section where Track 1 and Track 2 overlap, and is shown in the same figure as the running lines of other trains.

From Fig. C.2 in Appendix C, it can be seen that after the high-speed train is adjusted for operation, the travel speed of the train has decreased, causing a certain degree of delay. No crossings occurred between high-speed trains.

It is not difficult to find from Appendix C, Fig. C.3, that T118, T284/1, T65, T131/4/1, T559/8, T166/3, and T140/37 do not conflict with high-speed trains because it enters track 2 at an earlier time. Therefore, the running time of the above trains need not be adjusted. Because T54 has the same speed as high-speed trains in the Bengbu-Nanjing section, there is no running conflict with high-speed trains. 10135, 10625, 11301, 23005, and 11305 are freight trains. Due to the late entry time and the low speed of the train, there is no conflict with other trains, and their operation plans have not changed.

When other trains conflict with high-speed trains, there are two ways to resolve them. First, organize trains to run on time, that is, try to intersperse other trains between high-speed train lines, including T114, D88/5, and D88/5 has always been running on time.

Second, according to the minimum headway of the train at the station, the trains are arranged to run at a later time, including K58/5, K518/5, K190/87, K8471, K371/4, K104/1, 1230/27. Among them, K190/87 was adjusted to resume operation at Nanjing Station; K518/5 adjusted to arrive at Nanjing Station on time. However, due to the conflict between the departure time of K518/5 and G109, the train departed 2 min later; K104/1 organizes the acceleration of on-site technical operations at Yongning and Forest Farm stations and reduces the dwell time.

In summary, the train operation plan adjustment scheme obtained by the algorithm designed in this section meets the needs of the following train operation adjustments in the event of an emergency. It not only fully reduces the general penalty value of the delayed operation of the train, but also makes the layout of the train operation line in the train diagram more balanced, that is, improves the stability of the train operation plan.

From the above analysis, it can be seen that the adjustment plan of the train operation obtained by the algorithm designed in this section is very reasonable. This method not only considers the late penalty of the train, but also fully considers the stability of the train operation plan, and provides conditions for dealing with uncertain secondary interference caused by unexpected events.

(3) Analysis of calculation results

(1) It can be seen from the solution results of train's route allocation that the membership of the fuzzy coefficient is positively related to the objective function value. After the fuzzy membership function is used to blur the cost coefficient to a certain extent, the decision plan will change; when the idealization of the cost coefficient reaches a certain level, the decision plan will no longer change, and the total cost of the train will not change significantly.

Therefore, in actual production, the cost coefficient is fuzzy, which is as close as possible to the actual production on site, which not only meets the requirements of the possibility of fuzziness value, but also reduces the total train cost. Using the methods provided in this section, several options with reference value are calculated for decision-makers to choose.

(2) By considering the comprehensive optimization goal of the total delay satisfaction and stability satisfaction, there is a certain gap between the solution result and the solution result that only considers the total delay time as the optimization goal in the total delay time. The reason is to take into account the stability of the train operation plan. When the iterative optimization is carried out to a certain degree, the total delay satisfaction and the satisfaction of the stability of the train operation plan appear mutually exclusive.

In actual production, if one of the satisfaction with the total delay and the satisfaction with the stability of the train operation plan is emphasized, it can be achieved by adjusting the weight coefficient θ in the comprehensive optimization goal. Therefore, the train operation adjustment method designed in this section has strong flexibility.

6.6 Chapter Summary

First, this chapter analyzes the driving rules of high-speed trains under emergency conditions, including strong winds, heavy rains and floods, blizzards, earthquakes, and high temperatures.

Then, a bi-level programming model for train scheduling optimization and train operation adjustment under emergency conditions is established. The upper-level plan optimizes dispatching strategies such as train detours, train reconnections, and train outages. After the upper-layer scheduling strategy is given, the lower-layer plan adopts strategies such as changing train interval running time, train dwell time, and train crossing mode, and provides an adjusted operating schedule for the upper-layer plan to achieve overall optimization of the double-layer iteration.

And this bi-level programming model considers different objective functions and designs two algorithms to solve the train operation adjustment model. The first is to use the multi-directional plant growth simulation algorithm proposed in Chap. 6 to solve the fuzzy operation adjustment algorithm based on the tolerance method in an uncertain environment. Compared with the original algorithm, each iteration in PMGSA does not depend on the choice of the initial solution, but is only related to the current state. The improved algorithm not only increases the diversity of solution, but also speeds up the algorithm's convergence speed.

The second is the operation adjustment algorithm based on the convergent fuzzy particle swarm algorithm, which introduces fuzzy operators and convergence factors to ensure the accuracy and efficiency of the calculation results. The general test function is used to prove it, and its significant optimization performance is proved.

Finally, using the Beijing-Shanghai high-speed railway disturbance scenario, the effectiveness and feasibility of the above two algorithms for train operation adjustment problems on the railway network under emergency conditions are verified, respectively.

Based on the research work in Chaps. 4 and 5, this chapter specifically studies how to guide the organization of high-speed railway transportation in the field of practical applications. The models and algorithms designed in this chapter can provide some guidance for readers to study the adjustment of high-speed railway traffic plans under emergency conditions.

References

1. Qiang Z (1999) Study on the model and algorithm of train operation adjustment and optimization. Syst Eng 6:12–18 (in Chinese)
2. Szpigel B (1973) Optimal train scheduling on a single line railway. Oper Res Int J 7:344–351
3. D'Ariano A, Pacciarelli D, Pranzo M (2007) A branch and bound algorithm for scheduling trains in a railway network. Eur J Oper Res 183(2):643–657
4. Sauder RL, Westerman WM (1987) Computer aided train dispatching: decision support through optimization. Interfaces 13(6):24–37

5. Yang L, Li K, Gao Z (2009) Train timetable problem on a single-line railway with fuzzy passenger demand. IEEE Trans Fuzzy Syst 17(3):617–629
6. Yong Z, Xxiaojun Z, Liujie Z et al (2009) Application of analytic hierarchy process in train operation adjustment. Railway Transp Econ 3:70–73 [7] (in Chinese)
7. Jiaming C (1995) Optimization method for operation scheduling adjustment of double-track railway. J Southwest Jiaotong Univ 30(5):520–526 (in Chinese)
8. Tao W, Qi Z, Hongtao Z et al (2013) Train operation adjustment plan preparation and optimization method based on substitution diagram. China Railway Sci 34(5):126–133 (in Chinese)
9. Rodriguez J (2007) A constraint programming model for real-time train scheduling at junctions. Transp Res Part B 41(2):231–245
10. Xiaojuan L, Baoming H, Dewei L et al (2013) High_speed train operation adjustment method based on transformation maximal algebra and sequence optimization. China Railway Scie 34(6):125–130 [11] (in Chinese)
11. Qiyuan P, Songnian Z, Pei W (2001) Study on mathematical model and algorithm of network train operation diagram. J Railway 23(1):1–8 (in Chinese)
12. Yongjun C, Leishan Z (2010) Research on train operation adjustment algorithm based on sequence optimization method. J Railway 32(3):1–8 (in Chinese)
13. Cacchiani V, Caprara A, Toth P (2010) Scheduling extra freight trains on railway networks. Transp Res Part B 44(2):215–231
14. Ding Y, Liu HD, Bai Y et al (2001) A two-level otptimization model and algorithm for energy-efficient urban train operation. J Transp Syst Eng Inf Technol 11(1):96–101
15. Chang SC, Chung YC (2005) From timetabling to train regulation: a new train operation model. Inf Softw Technol 47(1):575–585
16. Ming X, Leishan X, Qi S et al (2008) Study on operation adjustment of two-line railway train based on ant colony algorithm. Logist Technol 27(6): 61–64 (in Chinese)
17. Leishan Z, Zuorui Q (1994) General algorithm of train operation planning and adjustment and its computer realization. J Railway 16(3):56–65 (in Chinese)
18. Qilong W, Hongxun Z (1996) Partial improvement of operation adjustment algorithm for layered rolling train. J Railway Sci 18:42–45 (in Chinese)
19. Wenting M, Yu D (2010) Research on railway train operation adjustment model and three-swarm cooperative particle swarm optimization. Railway Oper Technol 16(2):13–15 (in Chinese)
20. Wei S, Xiaojuan L (2009) Research on genetic algorithm based on selection in train operation adjustment. Shanxi Electr Subtechnol 1:28–29 (in Chinese)
21. Houqing D, Jinyong W, Haifeng W (2005) Tabu search algorithm for operation adjustment of two-line railway train. China Railway Sci 26(4):114–119 (in Chinese)
22. Fucai J, Siji H (2004) Research on train operation adjustment algorithm based on minimum train order entropy. Railway Acta Sinica 26(4):1–4 (in Chinese)
23. Xuesong Z, Yu Z, Siji H (1999) Train operation adjustment algorithm based on train running state derivation diagram. J Railway 21(6): 1–4 (in Chinese)
24. Cuiping Z, Cheng-hyun C, Yu-jiao T et al (2010) Graph theory model and heuristic algorithm for train operation adjustment. Sci Technol Eng 10:2560–2564, 2573 (in Chinese)
25. Cheng YH, Yang LA (2009) A fuzzy petri nets approach for railway traffic control in case of abnormality: evidence from Taiwan railway. Expert Syst Appl 36(4):8040–8048
26. Leishan Z, Siji H, Jianjun M et al (1998) Research on the method of train operation map of netted line by computer. J Railway 20(5):15–21 (in Chinese)
27. Honggang W, Zhang qi, Wang jianying, et al. High-speed railway operation adjustment model based on genetic algorithm.China railway science, 2006, 27 (3) : 96.100. (in Chinese)
28. Minghui W (2004) Research on application of rough set theory in railway operation dispatching command system. China Railway Sci 25(4):103–107 (in Chinese)
29. Chung JW, Oh SM, Choi IC (2009) A hybrid genetic algorithm for train sequencing in the Korea railway. Omega 37(3):555–565

30. Chuanjun J, Siji H, Yudong Y (2006) Research on particle swarm optimization algorithm for train operation adjustment. J Railway 28(3):6–11 (in Chinese)
31. Huihui M (2014) Research on train operation adjustment based on dual heuristic dynamic programming algorithm. Southwest Jiaotong University, Chengdu (in Chinese)
32. Lin H, Liu J, Liu WD et al (2012) Current situation and enlightenment of Japanese railway transportation organization. Railway Transp Econ 34(3):65–72 (in Chinese)
33. Liu B (2002) Theory and practice of uncertain programming. Physica Verlag, Heidelberg
34. Dajie Z, Ciguang W (2009) Discussion on the problems related to the reconnection of bullet trains on passenger dedicated line. Railway Transp Econ 31(7): 32–34 (in Chinese)
35. Zhiyan P (1999) Research on the cost and clearance of mid-speed cross-line trains in high-speed railway. Southwest Jiaotong University, Chengdu, pp 45–87 (in Chinese)
36. Jianhua C (2007) Related models and algorithms of railway passenger fare optimization. Beijing Jiaotong University, Beijing, pp 6–12 (in Chinese)
37. Tao H (2006) Research on organizational mode and cost effectiveness of high-speed railway operators. Beijing Jiaotong University, Beijing, pp 32–37 (in Chinese)
38. Johanna T, Persson AJ (2007) N-tracked railway traffic re-scheduling during disturbances. Transp Res Part B: Methodol 41(3):342–362.255
39. Goverde RM (2005) Punctuality of railway operations and timetable stability analysis. Ph.D. Dissertation. Delft University of Technology, TRAIL Research School, Delft, Netherlands
40. Goverde RMP (2007) Railway timetable stability analysis using max-plus system theory. Transp Res Part B 41(2):179–201
41. Qi-yuan P, Yu-cheng Z, Meng L (1998) Research on equalization evaluation method of train operation map. J Southwest Jiaotong University 33(4):372–377 (in Chinese)
42. Carey M, Carville S (2000) Testing schedule performance and reliability for train stations. J Oper Res Soc 51(6):666–682
43. de Kort AF, Heidergott B, Ayhan H (2003) A probabilistic approach for determing railway infrastructure capacity. Eur J Oper Res 148(3):644–661
44. Vromans MJ (2005) Reliability of railway systems. Erasmus University Rotterdam, Rotterdam, Netherlands
45. Engelhardt-Funke O, Kolonko M (2004) Analysing stability and investments in railway networks using advanced evolutionary algorithm. Int Trans Oper Res 11(4):381–394
46. Delorme X, Gandibleux X, Rodriguez J (2009) Stability evaluation of arailway timetable at station level. Eur J Oper Res 195(3):780–790
47. Hansen IA (2000) Station capacity and stability of train operations. Adv Transp Comput Railways VII 7:809–816
48. Goverde RMP (1998) Max-plus algebra approach to railway timetable design. In: Proceedings of the 6th international conference on computer aided design, manufacture and operation in the railway and other advanced mass transitsystems, Lisbon, Portugal, pp 339–350
49. Zhaoxia Y, Anzhou H, Ju L et al (1993) Study on dynamic performance of train operation diagram and its index system. J Railway 15(3): 46–56 (in Chinese)
50. Qiyuan P, Songnian Z, Haifeng Z (1998) Research on evaluation system of adjustable degree of train operation map. J Southwest Jiaotong Univ 33(4):367–371 (in Chinese)
51. D'Ariano A, Pacciarelli D, Pranzo M (2008) Assessment of flexible timetables in real-time traffic management of railway bottleneck. Transp Res Part C 16(2):232–245
52. Herrmann MT (2006) Stability of timetable and train routings through station regions. ETH, Swiss
53. Slowinski R (1986) A multicriteria fuzzy linear programming methods for water supply systems development planning. Fuzzy Sets Syst 19(3):217–237
54. Junhua C (2009) Research on the compilation and evaluation of the operation diagram of passenger dedicated line based on stability. Beijing Jiaotong Uuniversity (in Chinese)
55. Meng XL, Jia LM, Qin Y (2010) Train timetable optimizing and re-scheduling based on improved particle swarm algorithm. J Transp Res Rec 2(2197):71–79
56. Shucheng F, Dingwei W (1997) Fuzzy mathematics and fuzzy optimization. Science Press, Beijing (in Chinese)

57. Rommelfanger H (1996) Fuzzy linear programming and applications. Eur J Oper Res 92:512–527
58. Werners B (1987) An interactive fuzzy mathematical programming. Fuzzy Sets Syst 23:131–147
59. Negiota CV, Sularia M (1976) On fuzzy mathematical programming and the tolerances in planning. Econ Comput Econ Cybern Stud Res 3:83–95
60. Eberhat R, Kennedy J (1995) A new optimizer using particle swarm theory. In: Proceedings of 6th international symposium on micro machine and human science, Nagoya, pp 39–43
61. Clerc M, Kennedy J (2002) The particle swarm explosion, stability, and convergence in multidimensional complex space. IEEE Trans Evol Comput 6(1):58–73
62. Kennedy J, Eberhart CR (2001) Swarm intelligence. Morgan Kaufmann Press, San Francisco
63. Abdelbar MA, Abdelshahid S, Wunsch II CD (2005) Fuzzy PSO ageneration of particle swarm optimization. In: Proceedings of international joint conference on neural networks, Montreal, Canada, pp 1086–1091
64. Abdelbar MA, Abdelshahid S (2004) Instinct-based PSO with local search applied to satisfiability. In: Proceedings of 2004 international joint conference on neural networks, Budapest, Hungary, pp 2291–2295
65. Abdelshahid S (2004) Variations of particle swarm optimization and their experimental evaluation on maximum satisfiability. American University in Cairo, New York
66. Mendes R, Kennedy J, Neves J (2004) The fully informed particle swarm:simpler, maybe better. IEEE Trans Evol Comput 8(3):204–210
67. Yang GY (2007) A modified particle swarm optimizer algorithm. In: The eighth international conference on electronic measurement and instruments, vol 2, pp 675–679
68. Singh P (2007) Fuzzy adaptive particle swarm optimization for bidding strategy in uniform price spot market. IEEE Trans Power Syst 22(4):2152–2160
69. Saber YA,Senjyu T, Yona A et al (2007) Unit commitment computation by fuzzy adaptive Particle Swarm Optimization. IET Gener Trans Distrib 1(3):456–465
70. Esmin A (2006) Generating fuzzy rules from examples using the particle swarm optimization algorithm. In: Seventh international conference on hybrid intelligent systems, Kaiserslautern, pp 340–343
71. Esmin A, Lambert-Torres GA (2006) Fitting fuzzy membership functions using hybrid particle swarm optimization. In: Proceedings of 2006 IEEE international conference on fuzzy systems, Vancouver, BC. Canada, pp 2112–2119
72. An SF, Liu KQ, Liu B (2007) Improved weighted fuzzy reasoning algorithm based on particle swarm optimization. In: Proceedings of Sixth International Conference on Machine Learning and Cybernetics, vol 3, Hong Kong, pp 1304–1308
73. Shi Y, Eberhart CR (1998) A modified particle swarm optimizer. In: Proceedings IEEE international conference on evolutionary computation, pp 69–73
74. Clerc M, Kennedy J (2002) The particle swarm explosion, stability and convergence in multidimensional complex space. IEEE Trans Evol Comput 6(1):58–73
75. Jang J, Sun CT, Mizutani E (2002) Neuro-fuzzy and soft computing. Prentice-Hall, pp 94–99
76. Bergh F, Engelbrecht PA (2002) A new locally convergent particle swarm optimizer. In: Proceedings IEEE international conference on systems, man and cybernetics, pp 94–99
77. Trelea IC (2003) The particle swarm optimization algorithm: convergence analysis and parameter selection. Inf Process Lett 85(6):317–325
78. Xuelei M, Limin J (2009) A new class of particle swarm optimization. Control Decis 24(6):941–944, 948 (in Chinese)
79. Xuelei M (2011) Research on train operation organization theory and method under the condition of emergency. Beijing Jiaotong University, Beijing (in Chinese)
80. Siji H (2007) Theory of train operation drawing. China Railway Publishing House, pp 8–18 (in Chinese)

Chapter 7
Resource Allocation on High-Speed Railway Emergency Management

The reasonable allocation of high-speed railway emergency resources is the basis for reducing incident response time, improving emergency rescue efficiency, and avoiding waste of resources. Emergency resource allocation refers to the allocation of various types of emergence resources according to the statistical probability of the occurrence of regional accidents, the number of events, and the event level to ensure the timeliness of rescue [1]. This chapter is mainly carried out from two perspectives, static and dynamic, which are the preparation phase and respond phase of railway emergency management. The case of single emergence resource is discussed here, because multiple emergence resources can be allocated separately, if the correlation between resources is not taken into account.

Static emergency resource allocation refers to the optimization of resource allocation to ensure the resource support level of the whole railway line or the entire railway network as high as possible after the address of the emergency service facility is selected, and the structure and quantity of emergency resources are not change over time. Aiming at the characteristics of the multiple rescue in the case of multi-emergency resources, the concept of a continuous feasible solution was introduced to solve the problem based on the research of the single emergency resource problem [2]. Reference [3] proposed a new nonparametric Data Envelopment Analysis (EDA) model for optimal allocation of resources, evaluated the overall use of emergency resources in the system, and considered the decision-maker's preference to constructed a preferred DEA rescue allocation decision model. The results show that the efficiency of the emergency system can be greatly improved when the total resources are limited. The problem of reasonable allocation of resources for multiple disaster-stricken sites with a limited amount of emergency relief materials was studied in [4]. The weight of each evaluation index was calculated by using the analytic hierarchy process to analyze the questionnaires distributed to the experts, and the utility coefficient of the unit's emergency rescue materials for each affected site is obtained. Then, the total utility equation of the emergency rescue materials to the disaster site was established and solved by genetic algorithm. An example verifies

L. Jia et al., *High-Speed Railway Operation Under Emergent Conditions*,
Advances in High-speed Rail Technology,
https://doi.org/10.1007/978-3-662-63033-4_7

the validity and feasibility of the model, but the model does not take into account the complexity of actual operations and the constraints are not sufficient. Reference [5] analyzed the key role played by the efficient allocation of emergency resources before disasters in emergency rescue management. A nonlinear integer model based on rescue points and paths that takes into account factors such as travel time, transportation costs, and reliability of batch delivery was established, and solved using a non-dominated genetic algorithm and other different evolutionary algorithms of non-dominated.

Dynamic configuration is to constantly supplement resources to emergency service facilities when the structure and quantity of emergency resources required change with time, so as to maximize the level of resource guarantee. For example, in response to a sudden fire accident, when the first batch of resources only reaches 70% of the required resources, the subsequent resources are likely to be greater than 30% of the resources required in the previous stage. In reference [6], the problem of rearranging emergency vehicles in fire companied was first raised. Reference [7] calculated the performance and effectiveness of different accident sites and available resources after the earthquake, and established a dynamic optimization model of resource allocation. In reference [8], a nonlinear mathematical model, which was solve by the parallel tabu algorithm, was established for the redeployment of ambulance fleet in the real-time management of emergency medical services. Reference [9] developed a GIS-based emergency system for emergency decision making of urban geological disasters. In the process of optimizing the allocation of resources, reference [10] proposed the concept of "dynamic game network technology" to solve the key chain management, resource optimization allocation, and other major problems in the dynamic network by comprehensively considering the optimal allocation of resources for the identified project and the resource redistribution in the next stage. The resource allocation model of multiple emergency points was established by game theory, and the resource allocation scheme corresponding to each accident was solved by Nash equilibrium optimization, with the availability of resources, the severity of the accident and the number of required resources as constraints [11]. The model aimed to shorten the emergency response time and reduce the loss, but it can only be used in the case of multiple low-level and medium-level accidents at the same time, and the emergency resources should be located in the same fixed area. According to the characteristic of emergency management and the number of emergency points, the emergency resource allocation process was divided into several stages [12]. Taking the number of emergency resources dispatched at the beginning of each stage or the end of the previous stage as the stage variable, combined with the basic theory of dynamic planning, a mathematical model for optimal allocation of emergency resources was constructed. Case studies proved that the model can avoid problems such as duplication, lack of resources, and inefficient use of resources in the traditional management mode, and can effectively guide emergency resources in the emergency process. In reference [13], in order to solve the problem of emergency resource allocation when multiple accidents coexist, an improved preference order-based utility function was designed to describe the timeliness and effectiveness of rescue for each accident. The multi-accident resource allocation problem was described as a complete information non-cooperative game process, and the Gambit software was used to solve the Nash equilibrium of the game process to obtain a resource allocation scheme. In order to solve the problems of uneven distribution and resource competition in the process of emergency resource allocation in multiple disaster areas, reference [14] proposed an algorithm for emergency resource allocation based on the principle of fair priority, and established a two-level decision-making method, which makes the process of emergency resource allocation take timeliness and fairness into account. According to the principle of approaching emergency rescue nearby, a multi-disaster-multi-emergency resource allocation dynamic optimization strategy was proposed to quickly find the global optimal solution of the two-level emergency resource allocation model. In reference [15], a mathematical model and topological model of dynamic multi-stage emergency resource allocation based on the "situation information" model were established. According to the situation information and development stage of emergency, the rescue effect of each stage was quickly mastered, and the emergency resource allocation strategy was timely and accurately adjusted to achieve the dynamic rolling forward mode of emergency resource allocation.

In summary, there are fewer studies on the optimal allocation of multiple emergency resources under various emergency disaster, especially for the allocation of emergency resources on high-speed railways. Moreover, most studies only take the shortest emergency time as the optimization goal of the system, which is too single and lacks consideration of the complexity of the emergency process. In addition, some studies only put forward a macro allocation strategy without giving a specific plan, and some studies can only solve the problem of resource allocation for single emergency point rescue. In summary, there are few studies on the allocation of emergency resources to multiple disaster-affected points after disasters, and they are not in-depth enough.

7.1 Static Allocation of High-Speed Railway Emergency Resources Based on Utility Function

The static allocation of high-speed railway emergency resources is the allocation of emergency service facilities under the condition that no emergency occurs at the emergency point. The goal is to minimize the loss of resources under the conditions of economic allocation of emergency resources for high-speed railways, that is, to meet the minimum allocation of emergency resource requirements. Since the emergency resources required by emergency points are uncertain, the economic allocation of emergency resources at emergency service facilities is also uncertain. Therefore, the emergency resource utility function is introduced to analyze the amount of loss that emergency resources can reduce, and the Monte Carlo method is used to determine the expected value of the optimal static configuration of emergency resources.

7.1.1 High-Speed Railway Emergency Resource Utility Function

The high-speed railway emergency resource utility function is $f_j(x_{\mathrm{ij}}, t_{\mathrm{ij}})$, and the form and parameters of the function need to be determined according to the actual situation of the emergency. Considering the actual situation of the railway emergency and the simplicity of calculation, the emergency resources unitality function is defined as a linear function that is positively related to the number of emergency resources and negatively related to the emergency start time [16], which can be expressed as:

$$f_j(x_{\mathrm{ij}}, t_{\mathrm{ij}}) = \begin{cases} (T_j - t_{\mathrm{ij}}) \cdot \mathrm{unitSave}_j \cdot x_{\mathrm{ij}} & 0 \le t_{\mathrm{ij}} < T_j \\ -M \cdot x_{\mathrm{ij}} & t_{\mathrm{ij}} \ge T_j \end{cases} \tag{7.1}$$

where x_{ij} is the number of emergency resources allocated by emergency service facility S_i to emergency point E_j; T_j is the emergency failure time of emergency point E_j, which means that when the time for emergency resources to reach E_j is greater than or equal to T_j, the loss cannot be reduced; $\mathrm{unitSave}_j$ is the amount of loss that can be reduced per unit time when unit emergency resources are used at emergency point E_j, M is a large positive number.

7.1.2 Monte Carlo Method

The Monte Carlo method, also known as stochastic simulation method, belongs to a branch of computational mathematics. It was developed in the mid 1940s to adapt to the development of the atomic energy cause at that time. Traditional empirical methods cannot approximate the real physical process, so it is difficult to get satisfactory results. However, Monte Carlo method can simulate the real physical process and get very satisfactory results.

The basic principle and idea of the Monte Carlo method is that when the problem to be solved is the probability of the occurrence of an event, or the expected value of a random variable, it can be obtained through some "experimental" method. In general, the Monte Carlo method includes three main steps: constructing or describing the probability process, sampling from probability distributions, and establishing various estimates [17].

(1) Construct or describe the probability process

For the problems with random properties, such as particle transport, the main task is to describe and simulate the probability process. For deterministic problem, such as calculating define integrals, an artificial probabilistic process must be constructed in advance, that is, a problem that does not have random properties must be converted into a problem with random properties.

(2) Sample from probability distributions

Based on the constructed probability model, a random variable (or random vector) with a known probability distribution can be generated. The most basic and important probability distribution is the uniform distribution (rectangular distribution) on (0,1). A random number is a random variable with a uniform distribution, and a random number sequence is a sequence of independent random variables with such a distribution. Therefore, generating random numbers is a process of sampling from a uniform distribution, which can be realized by two methods. One is to produce it physically on a computer, but it is expensive and cannot be repeated. Another is to use mathematical recursion formula to generate a pseudo-random number, or a sequence of pseudo-random numbers. A variety of statistical tests have shown that pseudo-random numbers and pseudo-random number sequences have similar properties to true random numbers and random number sequences, so then can be used as true random numbers.

(3) Establish various estimates

Generally, a random variable is determined as the solution to the problem, that is, unbiased estimate. Establishing various estimators is equivalent to investigating and registering the results of simulation experiments to obtain a solution to the problem [18].

The Monte Carlo method can solve various types of problem, but in general, depending on whether it involves the behavior and results of random processes, it is divided into the following two categories:

One type is deterministic mathematical problems. Firstly, a probability model is established, whose probability distribution or mathematical expectation is the solution. Then the model is randomly sampled to generate random variables. Finally, the arithmetic average value is used as the approximate estimated value. For example, calculating multiple integrals, solving inverse matrices, solving linear algebraic equations, solving integral equations, calculating boundary value problems for partial differential equations, and calculating eigenvalues of differential operators, etc.

The second type is random mathematical problems. For example, the problem of neutron diffusion in the medium, in which neutrons are not only affected by certain determinism, but also more by randomness. For this kind of problems, the indirect simulation method of random sampling based on multiple integral or some function equations is not used in general, but the direct simulation method of sampling test based on the probability rule of actual physical situation by electronic computer. For example, nuclear physics problem, inventory problems in operations research, queuing problems in random service systems, ecological competition of animals, and the spread of infectious diseases [19].

Generation of random numbers and random variables. There are many methods to generate random numbers, the most commonly used method is the multiply–add congruence method. Choose a function g(x) to make the integer into a random number, and the equation g(x) is:

$$g(x) = (ax_0 + c) \bmod m \tag{7.2}$$

where x_0 is the initial value or seed ($x_0 > 0$), a is a multiplier ($a \geq 0$), c is a value-added ($c \geq 0$), and m is the modulus.

For binary integers with t-digits, the modulus is usually 2^t. Here, x_0, a and c are all integers and have the same value range: $x_0 < m$, $a < m$ and $c < m$. The random number sequence can be obtained from the following equation:

$$x_{m+1} = (ax_n + c) \bmod m \tag{7.3}$$

This sequence is called a linear congruence sequence. If $x_0 = a = c = 7$, the sequence is:

$$7, 6, 9, 0, 7, 6, 9, 0 \cdots \tag{7.4}$$

It can be proven that a linear congruence sequence always enters a loop set; that is, in the end, a loop of endlessly repeated number always appears. The sequence period length in Eq. (7.4) is 4. Of course, a useful sequence must be a sequence with a relatively long period.

In addition to the above-mentioned multiply–add congruence method, there are other methods, such as squaring and taking the middle method, and multiplication and congruence method [20–22].

For solving multi-dimensional or complex factors, the general calculation method is very difficult, while Monte Carlo method has great advantages, and its characteristics are as follows:

(1) Track particles directly, the physical logic is clear and easy to understand.
(2) The random sampling method is used to more realistically simulate the process of particle transport, which reflects the law of statistical fluctuations.
(3) It is not limited by the complexity of the system, such as multi-dimensional and multi-factors.
(4) The program structure of the Monte Carlo is clear and simple.
(5) It is easy to get any intermediate results and has strong application flexibility.
(6) The main weaknesses are the slow convergence speed and probabilistic nature of the error. The probability error is proportional to $\sigma/\sqrt{N}$. If the error is simply reduced by increasing the number of sampling particles N, a large amount of calculation will be added.

7.1.3 Static Allocation Model of High-Speed Railway Emergency Resources Based on Utility Function

In the allocation of railway emergency resources, for the case where multiple emergency service facilities provide a single emergency resource for multiple emergency points, it is assumed that there are m emergency points (E_1, E_2, E_3, ..., E_m) where emergency events may occur, the level of emergency events is uncertain, and there are n emergency service facilities $S_1, S_2, S_3, \ldots, S_n$. t_{ij} ($i = 1,2,3,\ldots,n$; $j = 1,2,3,\ldots,m$; $t_{ij} > 0$) is the shortest time from emergency service facility S_i to emergency point E_j, and x_{ij} is the number of emergency resources allocated by emergency service facility S_i to emergency E_j.

The optimal solution for allocating emergency resources at emergency service facilities is to minimize the total loss while meeting the emergency resource requirements.

Because the level of possible emergencies is uncertain, and the amount of emergency resources required is also uncertain. Therefore, statistical methods are used to obtain the probability of various types of emergency events and the corresponding emergency resource requirements.

$$d_j = \begin{cases} d_{j1} \sim N(D_{j1}, \sigma_{j1}) \text{ probability is } M_{j1} \\ \vdots \\ d_{jq} \sim N(D_{jq}, \sigma_{jq}) \quad \text{probability is } M_{jq} \end{cases} \tag{7.5}$$

where u is the level of the emergency and there are q kinds of emergency events, d_j is thc amount of emergency resources required by the emergency point E_j, d_{ju} is the emergency resources needed when a u-level emergency occurs at the emergency point, the probability of occurrence is M_{ju}, d_{ju} obeys the normal distribution with mean D_{ju} and variance σ_{ju} ($d_{ju} \sim N(D_{ju}, \sigma_{ju})$).

The utility function corresponding to the emergency level at E_j is $f_j(x_{ij}, t_{ij})$. According to the resource constraints and the maximization of emergency resource utility, an optimization model is established, that is,

$$\max_{x,p} \sum_{j=1}^{m}\sum_{i=1}^{n} f_j(x_{ij}, t_{ij})$$

$$s.t. \begin{cases} \sum_{i=1}^{n} x_{ij} = d_j \\ \sum_{j=1}^{m} x_{ij} = p_i \\ x_{ij} \geq 0 \\ t_{ij} \geq 0 \\ i = 1, 2, 3, \ldots, n \\ j = 1, 2, 3, \ldots, m \end{cases} \tag{7.6}$$

In Eq. (7.6), the first constraint indicates the d_j should be equal to the total resources provided to E_j, that is, the emergency resources of E_j are sufficient; the second constraint ensures that the resource availability p_i of emergency service facility S_i should be equal to the total amount of resources provided by S_i.

The Monte Carlo method is used to simulate the distribution of the number of emergency resources d_j to obtain the emergency resource availability p_i.

The steps are as follows.

(1) Let p be a zero vector.
(2) For each emergency point E_j, the emergency level u is determined by the roulette selection method. In order to select mating individuals, multiple rounds of selection are needed. Each round generates a random number uniformly distributed on [0,1], which is used as the selection pointer to determine the selected individual [23, 24].
Let sum $M = \sum_{l=1}^{q} M_{jl}$, and generate a random number *rand* between [0,1]. When $\frac{\sum_{l=1}^{u-1} M_{jl}}{\text{sum}\,M} < \text{rand} < \frac{\sum_{l=1}^{u} M_{jl}}{\text{sum}\,M}$, the emergency level is u and the corresponding utility function $f_j(x_{ij}, t_{ij})$ is determined. Then generate a value that obeys the normal distribution $N(D_{ju}, \sigma_{ju})$ as the number of emergency resources required by E_j as d_j.
(3) Solve the deterministic programming as follows, and find the optimal solution of p as p^*.

$$\max_{x,p} \sum_{j=1}^{m} \sum_{i=1}^{n} f_j(x_{ij}, t_{ij})$$

$$s.t.\begin{cases} \sum_{i=1}^{n} x_{ij} = d_j \\ \sum_{j=1}^{m} x_{ij} = p_i \\ x_{ij} \geq 0 \\ t_{ij} \geq 0 \\ i = 1, 2, 3, \ldots, n \\ j = 1, 2, 3, \ldots, m \end{cases} \tag{7.7}$$

(4) Let $p = p + p^*$.
(5) Repeat steps (2) to (4) a total of N times.
(6) Let $p = p / N$, and finally get the result p.

The calculation flowchart based on Monte Carlo method is shown in Fig. 7.1.

7.1.4 Case Study

Assuming that the number of emergency points that may occur is $m = 8$, the number of emergency service facilities is $n = 4$, and the minimum time t_{ij} from emergency service facilities point S_i to emergency point E_j is set as shown in Table 7.1:

With reference to the classification of railway emergencies in Chapter 2, five cases are classified here: no event, general event, lager event, major event, and particularly major event, with values ranging from 1 to 5. According to the statistical data, the

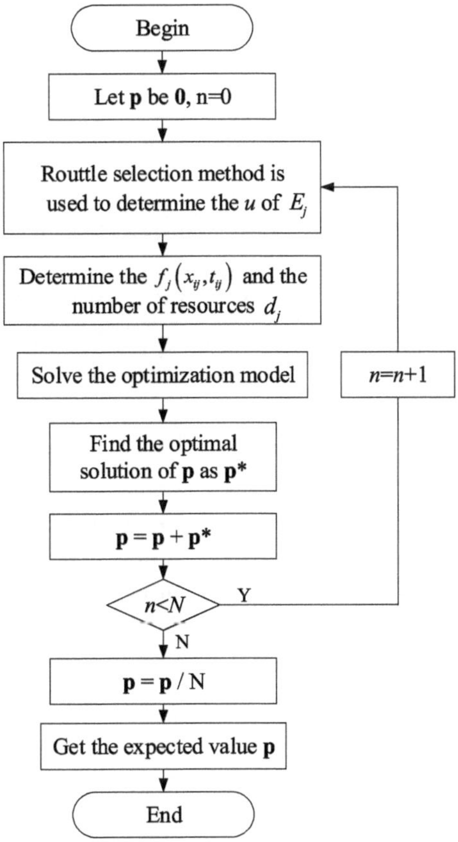

Fig. 7.1 Calculation flowchart based on Monte Carlo method

Table 7.1 Minimum time from emergency service facility point to emergency point

Emergency point / Facilities point	E_1	E_2	E_3	E_4	E_5	E_6	E_7	E_8
S_1	1.63	2.07	2.13	0.37	1.45	2.12	2.03	1.19
S_2	1.83	2.95	3.11	1.25	0.48	2.11	0.77	0.74
S_3	0.91	0.76	1.06	1.80	3.12	1.05	1.06	1.97
S_4	1.02	2.08	2.37	2.26	2.98	0.50	0.50	1.79

corresponding emergency resource demands at different levels obey the following normal distribution, and the emergency resource demands are shown in Table 7.2.

The amount of emergency resources required by the emergency point is as follows:

Table 7.2 Emergency resource requirements at different levels of emergencies

Emergency point	No event $u = 1$	General event $u = 2$	Larger event $u = 3$	Major event $u = 4$	Particular major event $u = 5$
E_1	0	N(9, 0.5)	N(10, 0.3)	N(12, 0.25)	N(15, 0.15)
E_2	0	N(8, 0.7)	N(10, 0.4)	N(13, 0.26)	N(16, 0.25)
E_3	0	N(9, 0.4)	N(11, 0.41)	N(14, 0.09)	N(17, 0.13)
E_4	0	N(10, 0.5)	N(12, 0.35)	N(15, 0.16)	N(18, 0.15)
E_5	0	N(11, 0.14)	N(13, 0.2)	N(14, 0.09)	N(17, 0.2)
E_6	0	N(12, 0.4)	N(13, 0.35)	N(15, 0.25)	N(18, 0.3)
E_7	0	N(12, 0.35)	N(14, 0.12)	N(16, 0.35)	N(17, 0.5)
E_8	0	N(11, 0.4)	N(13, 0.25)	N(16, 0.3)	N(18, 0.26)

$$d_1 = \begin{cases} d_{11} = 0 & \text{probalbility is } 0.1 \\ d_{12} \sim N(9, 0.5) & \text{probalbility is } 0.3 \\ d_{13} \sim N(10, 0.3) & \text{probalbility is } 0.3 \\ d_{14} \sim N(12, 0.25) & \text{probalbility is } 0.2 \\ d_{15} \sim N(15, 0.15) & \text{probalbility is } 0.1 \end{cases}$$

$$d_2 = \begin{cases} d_{21} = 0 & \text{probalbility is } 0.2 \\ d_{22} \sim N(8, 0.7) & \text{probalbility is } 0.3 \\ d_{23} \sim N(10, 0.4) & \text{probalbility is } 0.3 \\ d_{24} \sim N(13, 0.26) & \text{probalbility is } 0.1 \\ d_{25} \sim N(16, 0.25) & \text{probalbility is } 0.1 \end{cases}$$

$$d_3 = \begin{cases} d_{31} = 0 & \text{probalbility is } 0.6 \\ d_{32} \sim N(9, 0.4) & \text{probalbility is } 0.1 \\ d_{33} \sim N(11, 0.41) & \text{probalbility is } 0.1 \\ d_{34} \sim N(14, 0.25) & \text{probalbility is } 0.1 \\ d_{35} \sim N(17, 0.13) & \text{probalbility is } 0.1 \end{cases}$$

$$d_4 = \begin{cases} d_{41} = 0 & \text{probalbility is } 0.4 \\ d_{42} \sim N(10, 0.5) & \text{probalbility is } 0.2 \\ d_{43} \sim N(12, 0.35) & \text{probalbility is } 0.2 \\ d_{44} \sim N(15, 0.16) & \text{probalbility is } 0.1 \\ d_{45} \sim N(18, 0.15) & \text{probalbility is } 0.1 \end{cases}$$

$$d_5 = \begin{cases} d_{51} = 0 & \text{probalbility is } 0.3 \\ d_{52} \sim N(11, 0.14) & \text{probalbility is } 0.2 \\ d_{53} \sim N(13, 0.2) & \text{probalbility is } 0.2 \\ d_{54} \sim N(14, 0.09) & \text{probalbility is } 0.2 \\ d_{55} \sim N(17, 0.2) & \text{probalbility is } 0.1 \end{cases}$$

$$d_6 = \begin{cases} d_{61} = 0 & \text{probalbility is } 0.25 \\ d_{62} \sim N(12, 0.4) & \text{probalbility is } 0.25 \\ d_{63} \sim N(13, 0.35) & \text{probalbility is } 0.2 \\ d_{64} \sim N(15, 0.25) & \text{probalbility is } 0.2 \\ d_{65} \sim N(18, 0.3) & \text{probalbility is } 0.1 \end{cases}$$

$$d_7 = \begin{cases} d_{71} = 0 & \text{probalbility is } 0.35 \\ d_{72} \sim N(12, 0.35) & \text{probalbility is } 0.25 \\ d_{73} \sim N(14, 0.12) & \text{probalbility is } 0.2 \\ d_{74} \sim N(16, 0.35) & \text{probalbility is } 0.1 \\ d_{75} \sim N(17, 0.5) & \text{probalbility is } 0.1 \end{cases}$$

$$d_8 = \begin{cases} d_{81} = 0 & \text{probalbility is } 0.2 \\ d_{82} \sim N(11, 0.4) & \text{probalbility is } 0.3 \\ d_{83} \sim N(13, 0.25) & \text{probalbility is } 0.2 \\ d_{84} \sim N(16, 0.3) & \text{probalbility is } 0.2 \\ d_{85} \sim N(18, 0.26) & \text{probalbility is } 0.1 \end{cases}$$

The utility functions of emergency points in different emergency levels are as follows:

$$f_1(x_{i1}, t_{i1}) = \begin{cases} 0 & \text{u} = 1 \\ \begin{cases} (11 - t_{i1}) \cdot 15 \cdot x_{i1} & t_{i1} < 11 \\ -M \cdot x_{i1} & t_{i1} \geq 11 \end{cases} & \text{u} = 2 \\ \begin{cases} (10 - t_{i1}) \cdot 15 \cdot x_{i1} & t_{i1} < 10 \\ -M \cdot x_{i1} & t_{i1} \geq 10 \end{cases} & \text{u} = 3 \\ \begin{cases} (9 - t_{i1}) \cdot 15 \cdot x_{i1} & t_{i1} < 9 \\ -M \cdot x_{i1} & t_{i1} \geq 9 \end{cases} & \text{u} = 4 \\ \begin{cases} (8 - t_{i1}) \cdot 15 \cdot x_{i1} & t_{i1} < 8 \\ -M \cdot x_{i1} & t_{i1} \geq 8 \end{cases} & \text{u} = 5 \end{cases}$$

$$f_2(x_{i2}, t_{i2}) = \begin{cases} 0 & \text{u} = 1 \\ \begin{cases} (12 - t_{i1}) \cdot 14 \cdot x_{i2} & t_{i2} < 12 \\ -M \cdot x_{i2} & t_{i2} \geq 12 \end{cases} & \text{u} = 2 \\ \begin{cases} (11 - t_{i1}) \cdot 14 \cdot x_{i2} & t_{i2} < 11 \\ -M \cdot x_{i2} & t_{i2} \geq 11 \end{cases} & \text{u} = 3 \\ \begin{cases} (10 - t_{i1}) \cdot 14 \cdot x_{i2} & t_{i2} < 10 \\ -M \cdot x_{i2} & t_{i2} \geq 10 \end{cases} & \text{u} = 4 \\ \begin{cases} (9 - t_{i1}) \cdot 14 \cdot x_{i2} & t_{i2} < 9 \\ -M \cdot x_{i2} & t_{i2} \geq 9 \end{cases} & \text{u} = 5 \end{cases}$$

$$f_3(x_{i3}, t_{i3}) = \begin{cases} 0 & \text{u} = 1 \\ \begin{cases} (15 - t_{i1}) \cdot 20 \cdot x_{i3} & t_{i3} < 15 \\ -M \cdot x_{i3} & t_{i3} \geq 15 \end{cases} & \text{u} = 2 \\ \begin{cases} (14 - t_{i1}) \cdot 20 \cdot x_{i3} & t_{i3} < 14 \\ -M \cdot x_{i3} & t_{i3} \geq 14 \end{cases} & \text{u} = 3 \\ \begin{cases} (13 - t_{i1}) \cdot 20 \cdot x_{i3} & t_{i3} < 13 \\ -M \cdot x_{i3} & t_{i3} \geq 13 \end{cases} & \text{u} = 4 \\ \begin{cases} (12 - t_{i1}) \cdot 20 \cdot x_{i3} & t_{i3} < 12 \\ -M \cdot x_{i3} & t_{i3} \geq 12 \end{cases} & \text{u} = 5 \end{cases}$$

$$f_4(x_{i4}, t_{i4}) = \begin{cases} 0 & \text{u} = 1 \\ \begin{cases} (14 - t_{i4}) \cdot 16 \cdot x_{i4} & t_{i4} < 14 \\ -M \cdot x_{i4} & t_{i4} \geq 14 \end{cases} & \text{u} = 2 \\ \begin{cases} (13 - t_{i4}) \cdot 16 \cdot x_{i4} & t_{i4} < 13 \\ -M \cdot x_{i4} & t_{i4} \geq 13 \end{cases} & \text{u} = 3 \\ \begin{cases} (12 - t_{i4}) \cdot 16 \cdot x_{i4} & t_{i4} < 12 \\ -M \cdot x_{i4} & t_{i4} \geq 12 \end{cases} & \text{u} = 4 \\ \begin{cases} (11 - t_{i4}) \cdot 16 \cdot x_{i4} & t_{i4} < 11 \\ -M \cdot x_{i4} & t_{i4} \geq 11 \end{cases} & \text{u} = 5 \end{cases}$$

$$f_5(x_{i5}, t_{i5}) = \begin{cases} 0 & \text{u} = 1 \\ \begin{cases} (15 - t_{i5}) \cdot 16 \cdot x_{i5} & t_{i5} < 15 \\ -M \cdot x_{i5} & t_{i5} \geq 15 \end{cases} & \text{u} = 2 \\ \begin{cases} (13 - t_{i5}) \cdot 16 \cdot x_{i5} & t_{i5} < 13 \\ -M \cdot x_{i5} & t_{i5} \geq 13 \end{cases} & \text{u} = 3 \\ \begin{cases} (12 - t_{i5}) \cdot 16 \cdot x_{i5} & t_{i5} < 12 \\ -M \cdot x_{i5} & t_{i5} \geq 12 \end{cases} & \text{u} = 4 \\ \begin{cases} (11 - t_{i5}) \cdot 16 \cdot x_{i5} & t_{i4} < 11 \\ -M \cdot x_{i5} & t_{i4} \geq 11 \end{cases} & \text{u} = 5 \end{cases}$$

$$f_6(x_{i6}, t_{i6}) = \begin{cases} 0 & u=1 \\ \begin{cases} (15 - t_{i6}) \cdot 17 \cdot x_{i6} & t_{i6} < 15 \\ -M \cdot x_{i6} & t_{i6} \geq 15 \end{cases} & u=2 \\ \begin{cases} (14 - t_{i6}) \cdot 17 \cdot x_{i6} & t_{i6} < 14 \\ -M \cdot x_{i6} & t_{i6} \geq 14 \end{cases} & u=3 \\ \begin{cases} (13 - t_{i6}) \cdot 17 \cdot x_{i6} & t_{i6} < 13 \\ -M \cdot x_{i6} & t_{i6} \geq 13 \end{cases} & u=4 \\ \begin{cases} (12 - t_{i6}) \cdot 17 \cdot x_{i6} & t_{i6} < 12 \\ -M \cdot x_{i6} & t_{i6} \geq 12 \end{cases} & u=5 \end{cases}$$

$$f_7(x_{i7}, t_{i7}) = \begin{cases} 0 & u=1 \\ \begin{cases} (16 - t_{i7}) \cdot 17 \cdot x_{i7} & t_{i7} < 16 \\ -M \cdot x_{i7} & t_{i7} \geq 16 \end{cases} & u=2 \\ \begin{cases} (14 - t_{i7}) \cdot 17 \cdot x_{i7} & t_{i7} < 14 \\ -M \cdot x_{i7} & t_{i7} \geq 14 \end{cases} & u=3 \\ \begin{cases} (12 - t_{i7}) \cdot 17 \cdot x_{i7} & t_{i7} < 12 \\ -M \cdot x_{i7} & t_{i7} \geq 12 \end{cases} & u=4 \\ \begin{cases} (11 - t_{i7}) \cdot 17 \cdot x_{i7} & t_{i7} < 11 \\ -M \cdot x_{i7} & t_{i7} \geq 11 \end{cases} & u=5 \end{cases}$$

$$f_8(x_{i8}, t_{i8}) = \begin{cases} 0 & u=1 \\ \begin{cases} (15 - t_{i8}) \cdot 15 \cdot x_{i8} & t_{i8} < 15 \\ -M \cdot x_{i8} & t_{i8} \geq 15 \end{cases} & u=2 \\ \begin{cases} (13 - t_{i8}) \cdot 15 \cdot x_{i8} & t_{i8} < 13 \\ -M \cdot x_{i8} & t_{i8} \geq 13 \end{cases} & u=3 \\ \begin{cases} (12 - t_{i8}) \cdot 15 \cdot x_{i8} & t_{i8} < 12 \\ -M \cdot x_{i8} & t_{i8} \geq 12 \end{cases} & u=4 \\ \begin{cases} (10 - t_{i8}) \cdot 15 \cdot x_{i8} & t_{i8} < 10 \\ -M \cdot x_{i8} & t_{i8} \geq 10 \end{cases} & u=5 \end{cases}$$

When the number of cycles $N \geq 500$, the result is stable, and the expected value of emergency resource availability $p = [9.1467, 19.5187, 22.6364, 19.31]$ is obtained. This solution is an economic resource allocation of emergency resources that can achieve the optimization goal. The model and algorithm successfully solve the problem of static resource allocation in the case of multi-emergency points, multi-emergency service facilities, and single resources, and the maximum amount of emergency reduction loss is met on the premise of meeting emergency resource requirements.

7.2 Dynamic Allocation of Emergency Resources for High-Speed Railway Based on Nash Equilibrium

The resources required by the emergency point will change with the progress of the incident and the passage of time. Therefore, the resources of the emergency service facility need to be dynamically configured based on static configuration. This section considers the allocation of emergency resources at emergency service facilities as a non-cooperative static game and uses Nash equilibrium to dynamically configure emergency resources.

7.2.1 Introduction to Game Theory

Non-cooperative game refers to a type of game in which it is impossible for participants to reach a binding agreement, which is a mutually incompatible situation. Non-cooperative game studies how people choose to make the most of their benefits in situations where interests are mutually affected, that is, the choice of strategy [25].

Game theory solves the decision-making problem with constraints, multiple players and multiple goals, and conflicting objective functions. At present, the problem of multiple players and multiple goals decision making proposed by Von Neumann and Morgenstem in the 1940s is used as the beginning of game theory. Since then, research has focused on cooperative Game (CG). In the early 1950s, the Nash equilibrium theory proposed by Nash laid the foundation for Non-cooperative Game (NG).

From a methodological point of view, game theory is a vector optimization that includes objective function vector and strategy vector. From the perspective of game theory, it mainly studies the decision making when the behavior of decision-makers interacts under the relevant constraints, as well as the balance problem formed by decision making. According to whether there is a binding agreement, the game can be divided into CG and NG. CG emphasizes the realization of collective rationality through cooperation, and the parties to the game can obtain a reasonable Pareto solution through some cooperation and negotiation. NG emphasizes the realization of individual rationality in competition. By assuming individual behavioral rationality, individuals obtain the optimal objective function, and eventually form an equilibrium result, which is called the "Nash Equilibrium (NE)" of the game.

According to the above description and the application of game theory in various fields, the basic principles of game theory can be summarized as: conflict, optimization, and equilibrium.

(1) Conflict: Conflict is the essence of the game. In real production and life, although the conflict structure, subject, expression, and object are diversified in different environments, there are inconsistencies and differences between independent individual goals. The existence of conflicts indicates the decentralization of decision making, which makes games suitable for structurally distributed problems.

(2) Optimization: The conflict resolution is based on the decision to achieve the individual independent optimization goal. In the optimization process, CG is the problem of solving the Pareto strategy, and NG is the problem of solving the individual optimal strategy. This feature make games an effective tool for dealing with problem with complex optimization requirements.
(3) Equilibrium: The ultimate result of multi-body optimization is a "binding compromise," that is, equilibrium. In CG, equilibrium is a benefit distribution method recognized by all parties in the game; while in NG, Nash equilibrium is used to represent the final equilibrium state, as well as the behavior and relationship between the players [26, 27].

The application of game theory focuses on solving the equilibrium solution of the game. Reference [25] pointed out that in the process of obtaining a game solution, two main points of view are followed: One is the epistemic view represented by Aumann, the other is the evolutionary view represented by Binmore. Aumann et al. believed that game is "the interaction behavior of rational people," that is, under the given game mechanism, each independent decision-maker gives the optimal response according to the information he has. Binmore et al. doubted the applicability of the universal knowledge assumption about rationality. They believed that, on the one hand, due to the limitation of objective conditions, each decision-maker cannot always accurately obtain all accurate knowledge in the game environment; on the other hand, all decision-makers may not be absolutely rational subjectively. To this end, the bounded and evolutionary rationality is used to replace the absolute rationality, and a reasonable equilibrium result is formed through continuous attempts and mutual learning of the behaviors among decision-makers. Epistemology and evolution are two sides of a coin, and the views of Aumann and Binmore are complementary and can be unified in a more abstract and complex framework.

Resource allocation is a typical problem of scarce resource allocation and competition with constraints and optimization indicators. Conflicts between multiple entities are often implied in the allocation problem, which makes the allocation very similar to economic and planning issues in nature. Therefore, game theory may be a good tool for describing resources allocation problems and has been applied initially in multi-CPU scheduling, air freight scheduling, and production scheduling.

Application research of game theory follows a clear process from modeling, analyzing, and solving. Modeling is mainly based on the physical structure or distribution characteristic of optimization requirements, and describes the original problem as a multi-person-multi-objective problem. Analyzing mainly discusses the existence, uniqueness, and global performance of the equilibrium solution based on the model. The solving process mostly follows the view of knowledge theory, and the specific algorithm has strong problem dependence. CG mainly discusses the calculation and distribution mechanism of Pareto solutions; NG includes the calculation of NE solutions and the design of effective incentive mechanisms to guide individual self-interest behaviors toward global optimization. Among them, the establishment of a suitable game model is the key point, and the analysis and solving of the game equilibrium solution is the main content and also the difficulty [4].

7.2.2 Nash Equilibrium

Nash equilibrium, also known as non-cooperative game equilibrium, is an important term in game theory, named after John Nash. Assuming that there are n individuals participating in the game. Given the strategy of other people, each person chooses the best strategy of individual, so as to maximize the personal interests. An individual's optimal strategy may or may not depend on other's strategies, and all strategies constitute a Strategy Profile. NE is, in essence, a state of non-cooperative games.

(1) The existence of NE

Nash first gave the definition of game equilibrium and the theory of its existence in 1950. Since then, the existence of NE has become a major direction of game theory research, and a lot of research results have been obtained.

Existence theorem of NE I: Every finite game has at least one NE [28].

Existence theorem of NE II: In the n-person game problem, if everyone's pure strategy set S_i is a non-empty compact convex subset of Euclidean space, and each income function $a_i(s)$ is continuous with respect to s and quasi-concave with respect to s_i, there is a pure strategy NE. Glicksberg relaxes the quasi-concave condition to obtain the existence of a hybrid strategy [29–31].

Existence theorem of NE III: In the n-person game problem, if everyone's pure strategy set S_i is a non-empty compact convex subset of Euclidean space, and each income function $a_i(s)$ is continuous with respect to s, there is a mixed strategy NE. By relaxing the continuous condition of income function, Dasguta and Maskin get the existence of NE of discontinuous game [30].

Existence theorem of NE IV: In the n-person game problem, if everyone's pure strategy set S_i is a non-empty compact convex subset of a finite-dimensional Euclidean space, each income function $a_i(s)$ is quasi-concave with respect to s_i, continuous with respect to the upper semi-continuity of s, and has a continuous maximum, there is a pure strategy NE [32].

(2) The stability of NE

According to the definition of NE, the strict NE of pure strategy is stable, that is, when other participants adopt the NE strategy, any participant's strategy deviating from the equilibrium strategy has the motivation to recover to the equilibrium strategy, but the equilibrium of hybrid strategy is not stable.

(3) The multiplicity of NE

An important flaw of the concept of NE is its multiplicity, that is, many game problems have multiple NEs. Wilson studies the calculation method of equilibrium and proposes the famous odd number theorem.

Odd number theorem: Almost every finite game problem has an odd number of NEs.

Given the number of participants and strategies, a finite game problem is uniquely determined by an $m\prod_{i=1}^{n} k_i$-dimensional vector (i.e., an income matrix). The "almost" in the theorem means that it is true except for the set of zero measures in the $m\prod_{i=1}^{n} k_i$-dimensional Euclidean space [33].

(4) Refinement and selection of NE

Due to the multiplicity of NE, eliminating unreasonable equilibrium or choosing a reasonable equilibrium, has become one of the main direction for the development of game theory in recent years. Zelten proposed the perfect equilibrium of subgames in 1965 and the perfect equilibrium of trembling hands in 1975. The concepts of sequential equilibrium proposed by Krips and Wilson in 1982 further refined the concept of equilibrium. Burhan and Pierce proposed the rationalization process and rationalization strategy to eliminate unreasonable strategies in 1984. However, these concepts cannot guarantee their uniqueness and existence, and cannot solve the problem of the multiplicity of NE. Hassani and Zelten proposed the NE selection theory in 1988, but they were not confirmed by experimental results [34].

7.2.3 Dynamic Allocation Model of High-Speed Railway Emergency Resources Based on NE

In the process of railway emergency rescue, when there are multiple emergencies at the same time, and the same type of emergency resources are required, the resources of the emergency service facilities will be continuously consumed. Assume that the number of emergency resources needed for n' $(0 < n' \leq n)$ emergency service facilities $(S_1, S_2, S_3, \cdots, S_{n'})$ is $a_1, a_2, a_3, \cdots, a_{n'}$ (all greater than 0), and n represents the total number of all emergency service facilities. There are m' emergency resource collection points $C_1, C_2, C_3, \cdots, C_{m'}$, and the amount of emergency resources that can be raised is $A_1, A_2, A_3, \cdots, A_{m'}$. t'_{ij} represents the shortest path time from emergency resource point C_i to emergency service facility point S_j, and $\sum_{i=1}^{m'} A_i \geq \sum_{i=1}^{n'} a_i$ can be ensured by expanding the emergency resource mobilization area.

In this section, a non-cooperative game theory is used to build a model. The dynamic allocation of fair emergency resources is mapped to the NE in non-cooperative games [35].

x'_{ij} is the number of emergency resources dynamically allocated from C_i to S_j. The number of emergency service facilities can be considered as n' people, and the number of emergency resources dynamically allocated by S_j is a pure strategy.

Let the set of pure strategies that the jth person can adopt is $T_j = \left\{T_1^j, T_2^j, T_3^j, \cdots, T_{m_i}^j\right\}$, $j = 1, 2, 3, \cdots, n'$, m_i is the number of pure strategies of the ith person. The pure strategy space of n' individuals is $T = T_1 \times T_2 \times T_3 \times \cdots \times T_{n'}$, while that for n' individuals except the jth individual is $T^{-j} = T_1 \times \cdots \times T_{j-1} \times T_{j+1} \times \cdots \times T_{n'} = \left\{T_1^{-j}, T_2^{-j}, \cdots, T_{g_i}^{-j}\right\}$, where $g_j = m_1 \times \cdots \times m_{j-1} \times m_{j+1} \times \cdots \times m_{n'}$. Any pure strategy combination of strategy space is $t = \left\{T_{j_1}^1, T_{j_2}^2, T_{j_3}^3, \cdots, T_{j_{n'}}^{n'}\right\}$, $T_{j_i}^i \in T_i, i = 1, 2, 3, \cdots, n'$.

The generated policy set satisfies the following constraints.

$$\begin{cases} 0 \le x'_{ij} \le \min\{a_j, A_i\} \\ 0 \le \sum_{i=1}^{m'} x'_{ij} \le a_j, \\ x'_{\mathrm{ij}} \text{ is an integer} \end{cases} \tag{7.8}$$

The return function of person j in the game is defined as follows:

$$\mathrm{payoff}_j(t) = \sum_{i=1}^{m'} \lambda_i \frac{1}{t'_{\mathrm{ij}}} x'_{\mathrm{ij}} \tag{7.9}$$

where $\lambda_i = \begin{cases} -1 & \sum_{j=1}^{n'} x'_{ij} > A_i \\ 1 & \text{otherwise} \end{cases}$.

The return matrix of person j in the game is defined as follows:

$$U_j = \begin{bmatrix} \mathrm{payoff}_j\left(T_1^j, T_1^{-j}\right) & \mathrm{payoff}_j\left(T_1^j, T_2^{-j}\right) & \cdots & \mathrm{payoff}_j\left(T_1^j, T_{g_j}^{-j}\right) \\ \mathrm{payoff}_j\left(T_2^j, T_1^{-j}\right) & \mathrm{payoff}_j\left(T_2^j, T_2^{-j}\right) & \cdots & \mathrm{payoff}_j\left(T_2^j, T_{g_j}^{-j}\right) \\ \vdots & \vdots & \ddots & \vdots \\ \mathrm{payoff}_j\left(T_{m_j}^j, T_1^{-j}\right) & \mathrm{payoff}_j\left(T_{m_j}^j, T_2^{-j}\right) & \cdots & \mathrm{payoff}_j\left(T_{m_j}^j, T_{g_j}^{-j}\right) \end{bmatrix} \tag{7.10}$$

Solving the fair emergency resource allocation scheme is to find the NE of the n person non-cooperative game. Combining the background of the model and the literature [28], it can be seen that the NE solution of the limited strategy game of the model exists.

The model is actually a NE that solves pure strategies for n'-person static games.

$$\underline{T}^{i-1} = \begin{cases} T_1 \times T_2 \times \cdots \times T_{i-1} & 1 < i \le n \\ \emptyset & i = 1 \end{cases} \tag{7.11}$$

$$\overline{T}^{i+1} = \begin{cases} T_{i+1} \times T_{i+2} \times \cdots \times T_{n'} & 1 \le i < n \\ \emptyset & i = n \end{cases} \tag{7.12}$$

$$T = \begin{cases} T_1 \times \overline{T}^{i+1} & i = 1 \\ \underline{T}^{i-1} \times T_{n'} & i = n \\ \underline{T}^{i-1} \times T_i \times \overline{T}^{i+1} & 1 < i < n \end{cases} \tag{7.13}$$

Without loss of generality, when $1 < i < n : \forall t^{i-1} \in \underline{T}^{i-1}$, $t^{i-1} = \left(T_{j_1}^1, T_{j_2}^2, \cdots, T_{j_{i-1}}^{i-1}\right)$, $T_{j_k}^k \in T_k$, $k = 1, 2, \cdots, i-1$; $\forall t^{i+1} \in \overline{T}^{i+1}$, $t^{i+1} = \left(T_{j_{i+1}}^{i+1}, T_{j_{i+2}}^{i+2}, \cdots, T_{j_{n'}}^{n'}\right)$, $T_{j_k}^k \in T_k, k = i+1, i+2, \cdots, n'$.

For the ith $(i = 1, 2, \cdots, n')$ person, if $t^* = \left(T_{j_1}^{1^*}, T_{j_2}^{2^*}, \cdots, T_{j_{n'}}^{n'^*}\right) = \left(t^{i-1^*}, T_{j_i}^{i^*}, t^{i+1^*}\right)$ exists such that $\text{payoff}_i(t^*) = \text{payoff}_i\left(t^{i-1^*}, T_{j_i}^{i^*}, t^{i+1^*}\right) \geq \text{payoff}_i\left(t^{i-1^*}, T_{j_i}^{i}, t^{i+1^*}\right), \forall T_{j_i}^i \in T_i\left(i = 1, 2, \cdots, n'\right)$, then $t^* = \left(T_{j_1}^{1^*}, T_{j_2}^{2^*}, \cdots, T_{j_{n'}}^{n'^*}\right)$ is called the NE point.

Note $\max T_i\left(t^{i-1}, t^{i+1}\right) = \left\{\left(t^{i-1}, T_{j_i}^i, t^{i+1}\right) \middle| \text{payoff}_i\left(t^{i-1}, T_{j_i}^i, t^{i+1}\right) = \max_{T_k^i \in T_i} \text{payoff}_i\left(t^{i-1}, T_k^i, t^{i+1}\right)\right\}, \max T_i = \bigcup_{\left(t^{i-1}, t^{i+1}\right) \in \underline{T}^{i-1} \times \overline{T}^{i+1}} \max T_i\left(t^{i-1}, t^{i+1}\right)$, then the pure strategy combination in the set $\bigcap_{i=1}^{n'} \max T_i$ is equivalent to the pure strategy NE of the finite strategy static game.

Proof is as follows [36].

(1) Adequacy

If $t^* = \left(T_{j_1}^{1^*}, T_{j_2}^{2^*}, \cdots, T_{j_{n'}}^{n'^*}\right) \in \bigcap_{i=1}^{n'} \max T_i$, then for $i = 1, 2, \cdots, n'$, there is $t^* = \left(T_{j_1}^{1^*}, T_{j_2}^{2^*}, \cdots, T_{j_{n'}}^{n'^*}\right) \in \max T_i\left(t^{i-1^*}, t^{i+1^*}\right)$. By definition $\text{payoff}_i(t^*) = \text{payoff}_i\left(t^{i-1^*}, T_{j_i}^{i^*}, t^{i+1^*}\right) = \max_{T_k^i \in T_i} \text{payoff}_i\left(t^{i-1^*}, T_k^i, t^{i+1^*}\right) \geq \text{payoff}_i\left(t^{i-1^*}, T_k^i, t^{i+1^*}\right)\left(\forall T_k^l \in T_i\right)$. According to the definition of NE, $t^* = \left(T_{j_1}^{1^*}, T_{j_2}^{2^*}, \cdots, T_{j_{n'}}^{n'^*}\right)$ is the NE point.

(2) Necessity

Let $t^* = \left(T_{j_1}^{1^*}, T_{j_2}^{2^*}, \cdots, T_{j_{n'}}^{n'^*}\right)$ be the NE point. From the definition of NE, it can be known that for $i = 1, 2, \cdots, n'$, there is $\text{payoff}_i(t^*) = \text{payoff}_i\left(t^{i-1^*}, T_{j_i}^{i^*}, t^{i+1^*}\right) = \max_{T_k^i \in T_i} \text{payoff}_i\left(t^{i-1^*}, T_k^i, t^{i+1^*}\right) \geq \text{payoff}_i\left(t^{i-1^*}, T_k^i, t^{i+1^*}\right)\left(\forall T_k^i \in T_i\right)$. That is, $t^* = \left(T_{j_1}^{1^*}, T_{j_2}^{2^*}, \cdots, T_{j_{n'}}^{n'^*}\right) \in \max T_i\left(t^{i-1^*}, t^{i+1^*}\right)$, so $t^* = \left(T_{j_1}^{1^*}, T_{j_2}^{2^*}, \cdots, T_{j_{n'}}^{n'^*}\right) \in \bigcap_{i=1}^{n'} \max T_i$.

In summary, by solving $\bigcap_{i=1}^{n'} \max T_i$, the pure strategy NE for n' human static games can be obtained. when multiple NE points are obtained, the NE point with the largest sum of returns is used as the model solution.

7.2.4 Case Study

Assume that the number of emergency resource collection points $m' = 3$, the number of emergency service facilities $n' = 4$, and the minimum time t_{ij}' and the number of resources from the emergency resource collection point C_i to the emergency service facility S_j are shown in Table 7.3.

4 emergency service facilities correspond to 4 persons in the game respectively, and the kth pure strategy $T_k^j = \left(x_{1j}', x_{2j}', x_{3j}'\right)$ of the jth person indicates that the

Table 7.3 Minimum time t_{ij}' and the number of resources from C_i to S_j

	S_1	S_2	S_3	S_4	Available resources A_j
C_1	4	10	5	8	2
C_2	2	6	3	10	4
C_3	5	7	12	7	1
Required resources a_i	2	1	2	2	

Table 7.4 Results of dynamic allocation of emergency resources

	S_1	S_2	S_3	S_4	Available emergency resources
C_1	0	0	0	2	2
C_2	2	0	2	0	4
C_3	0	1	0	0	1
Required emergency resources	2	1	2	2	

number of emergency resources allocated dynamically form 3 emergency resources mobilization points to emergency service point S_j is $x_{1j}', x_{2j}', x_{3j}'$ respectively.

The pure strategy set of person 1 in the game is $T_1 = \{T_1^1, T_2^1, T_3^1, T_4^1, T_5^1\}$, where $T_1^1 = (2, 0, 0)$, $T_2^1 = (0, 2, 0)$, $T_3^1 = (1, 1, 0)$, $T_4^1 = (1, 0, 1)$, $T_5^1 = (0, 1, 1)$.

The pure strategy set of person 2 in the game is $T_2 = \{T_1^2, T_2^2, T_3^2\}$, where $T_1^2 = (1, 0, 0)$, $T_2^2 = (0, 1, 0)$, $T_3^2 = (0, 0, 1)$.

The pure strategy set of person 3 in the game is $T_3 = \{T_1^3, T_2^3, T_3^3, T_4^3, T_5^3\}$, where $T_1^3 = (2, 0, 0)$, $T_2^3 = (0, 2, 0)$, $T_3^3 = (1, 1, 0)$, $T_4^3 = (1, 0, 1)$, $T_5^3 = (0, 1, 1)$.

The pure strategy set of person 4 in the game is $T_4 = \{T_1^4, T_2^4, T_3^4, T_4^4, T_5^4\}$, where $T_1^4 = (2, 0, 0)$, $T_2^4 = (0, 2, 0)$, $T_3^4 = (1, 1, 0)$, $T_4^4 = (1, 0, 1)$, $T_5^4 = (0, 1, 1)$.

Finally, the NE point with the largest sum of individual income is $t^* = \{T_2^1, T_3^2, T_2^3, T_1^4\}$, and the corresponding income is $(1, 0.6667, 0.25, 0.1429)$, respectively. The results of the dynamic allocation of emergency resources are shown in Table 7.4.

It can be seen from Table 7.4 that the non-cooperative game theory of NE can effectively solve the problem of dynamic allocation of emergency resources and achieve the maximum total return under the condition of fair distribution of emergency resources [37].

7.3 Chapter Summary

This chapter mainly discusses the allocation of emergency resources for high-speed railway from both static and dynamic aspects.

In the static allocation of emergency resources, the resource allocation of single resources and multiple emergency points based on the utility function is studied. The utility function considers the amount of emergency resources and emergency start time, and satisfies the goal of minimizing the loss of emergency resources under the condition of economic allocation of emergency resources of high-speed railway. In view of the uncertainty of the type emergency and the uncertainty of the number of emergency resources required, Monte Carlo method is used to simulate the distribution of random variables and solve the optimal solution of the model. The model considers the demand and supply of emergency resources and the amount of resources available and the total demand at the emergency point, so that the resource allocation of each emergency service facility can not only meet the emergency demand but also maximize the loss of resources, which is more practical than the previous mathematical model.

The dynamic allocation of high-speed railway emergency resources is to dynamically configure the resources of emergency service facilities based on the development of the incident when an emergency occurs. This is a non-cooperative game process, so the non-cooperative game theory is used to establish the dynamic allocation model of emergency resources to meet the requirements of fair distribution of emergency resources, and the NE of the non-cooperative game model is used to solve the problem.

Finally, the validity of these two models is verified by examples, which provides a reliable reference and effective guidance for the allocation of emergency resources for high-speed railways.

References

1. Gan C, Qian Z, Qi H et al (2010) Allocation of emergency rescue resources for freeway traffic [J]. China Safety Sci J 20(1):165–170 (in Chinese)
2. Gengxin D, Qingli Da (2000) The study of combinatorial scheduling problem in emergency systems [J]. Syst Eng Theory Pract 20(9):52–55 (in Chinese)
3. Fang L (2008) Resource allocation of emergency system based on the DEA model with preference information [J]. Syst Eng Theory Pract 28(5):98–104 (in Chinese)
4. Mingang Z, Zengshou C, Shuang Li (2010) A novel model for resource assignment among multiple disaster places in emergency management [J]. Indus Eng J 13(1):85–89 (in Chinese)
5. Wang HJ, Du LJ, Ma SH (2014) Multi-objective open location-routing model with split delivery for optimized relief distribution in post-earthquake [J]. Transport Res Part E, 160–179
6. Kolesar P, Walker EW (1974) An algorithm for the dynamic relocation of fire companies [J]. Oper Res 22(2):249–274
7. Fiedrich F, Gehbauer F, Rickers U (2000) Optimized resource allocation for emergency response after earthquake disasters [J]. Saf Sci 35(1):41–57
8. Gendreau M, Laporte G, Semet F (2001) A dynamic model and parallel tabu search heuristic for real-time ambulance relocation [J]. Parallel Comput 27(12):1641–1653
9. Chen, Chen L, Zhang J et al (2002) Development of evacuation route and shelter system for debris flow disaster [C]. Study on environmental protection of mountain disasters on both sides of the Taiwan Straits, (in Chinese)
10. Hong C, Lei Ji, Aiqun C (2003) Discussion on 'dynamic game network technology' caused by emergencies [J]. Project Manage Technol 1:11–14 (in Chinese)
11. Shetty SR (2004) An event single game solution for resource allocation in a multi-crisis environment [D]. University of South Florida, Tampa Bay
12. Xiaomeng Z, Lizhen J, Yunlong Z (2007) An innovated quantitative model of the optimal resource allocation in case of emergency [J]. J Saf Environ 7(6):113–115 (in Chinese)
13. Jing Z, Shifei S, Rui Y (2007) Preference-order-based game modeling of multiple emergency resource allocation [J]. J Tsinghua Univ (Sci Technol) 47(12):2172–2175 (in Chinese)

14. Suisheng W, Yan W (2008) Emergency resources allocation among multiple disaster places under fair priority principle [J]. Oper Res Manage Sci 17(3):16–21 (in Chinese)
15. Xia J, Wang W, Xia J (2012) On the dynamic and rolling emergency resources allocation based on context-response [J]. J Hubei Univ Technol. (in Chinese)
16. Vladimir M, Charles R (1996) The queuing maximal availability location problem: a model for the sitting of emergency vehicles [J]. Eur J Oper Res 93:110–120
17. Morin LR (1988) Monte carlo simulation in the radiological sciences [M]. CRC Press, Boca Raton
18. Jenkins T, Nelson W, Rindi A (1998) Monte carlo transport of electrons and photos [D]. Plenum Press, New York
19. Zhang K, Gu C (1987) Computational physics [M]. Shanghai: Fudan University Press, (in Chinese)
20. Zou X, Jin Z (1993) Computataional solid state physics[M]. Wuhan University Press, (in Chinese)
21. Chen Yu (2007) The monte carlo simulation of surface segregation of the pd based alloys [D]. Hunan Normal University, Changsha (in Chinese)
22. Brotcorne L, Laporte G, Semet F (2003) Ambulance location and relocation models[J]. Eur J Oper Res 147:451–463
23. Quanxin W, Xianfeng L, Qiang H et al (2008) The comparison of different selection methods in genetic algorithms [J]. J Commun Comput 5(8):61–65 (in Chinese)
24. Weiying Z (2004) Game theory and information economics [M]. Shanghai Renmin Chubanshe, Shanghai (in Chinese)
25. Shuguang Z (2000) China's economy (1998) [M]. Shanghai Renmin Chubanshe, Shanghai (in Chinese)
26. Changjun W (2006) Study on model and algorithm of production scheduling based on noncooperative game [D]. Shanghai Jiao Tong University, Shanghai (in Chinese)
27. Nash J (1951) Non-cooperative games [J]. Ann Math 54(2):286–295
28. Debreu D (1952) A social equilibrium existence theorem [C]. Proc Natl Acad Sci 38:886–893
29. Gilicksberg IL (1952) A further generalization of the Kakutani fixed point theorem with application to nash equilibrium points [C]. Proc Natl Acad Sci 38:170–174
30. Fan K (1952) Fixed point and minmax theorems in locally convex topological linear spaces [C]. Proc Natl Acad Sci 38:121–126
31. Dasgupta P, Maskin E (1986) The existence of equilibrium in discontinuous economic games [J]. Rev Econ Stud 53(1):1–26
32. Wilson R (1971) Computing equilibrium of n-person games [J]. SIAM J Appl Math 21(1):80–87
33. Baoming Li, Jiazhuang L (2000) Utility function and nash equilibrium [J]. Math Econ 17(4):21–28 (in Chinese)
34. Zhenglong Li (2004) An existence distinguishing method for pure strategy nash equilibrium existence in n-person static games [J]. Oper Res Manage Sci 13(1):33–37 (in Chinese)
35. Jöfberg J (2004) Yalmip: a toolbox for modeling and optimization in MATLAB [C]. Proceedings of the CACSD conference, Taipei, Taiwan
36. Zhou HJ, Jia LM, Qin Y (2009) Metadata specification of railway video information and its application in video monitoring system for Qinghai-Tibet railway [C]. Proceedings of international symposium on computer network and multimedia technology, 1018–1021
37. Huijuan Z (2011) Plan management and resource allocation optimization on railway emergency management [D]. Beijing Jiaotong University, Beijing (in Chinese)

Chapter 8
Emergency Resource Scheduling Optimization for High-Speed Railways

After a railway emergency occurs, the emergency command center needs to make emergency resource scheduling decisions immediately. Emergency decision makers need to determine the scheduling of subsequent resources based on the demand for emergency resources at this stage, and schedule emergency resources in multiple stages as events develop and change. Therefore, emergency resource scheduling is a dynamic process. Given the layout and configuration of emergency resources, it is an issue of concern for emergency managers to formulate an optimized scheduling plan to ensure the timeliness of emergency resource scheduling. Therefore, it is necessary to formulate emergency resource scheduling scheme in advance for specific scenarios of emergencies, so as to start the emergency resource scheduling scheme in the first time and ensure the timeliness of emergency rescue actions.

In the process of railway emergency rescue, the allocation of emergency resources can effectively ensure the timely implementation of emergency work, the effective use of emergency resources, and the safety and smoothness of railway transportation. Therefore, this chapter will study how to establish and solve optimal scheduling model of railway emergency resources. According to different situations, the corresponding emergency resource scheduling model is established and solved, including the situation of a single emergency point with a single resource, the situation of a single emergency point with multiple resources, and the situation with multiple emergency points and multiple resources.

At present, the research on emergency resource scheduling is mainly focused on the following aspects.

(1) The situation of a single emergency point with a single resource. In the early days, a linear programming model was established to study the distribution of bulk food in emergency systems, with the goal of minimizing the transportation cost or maximizing the distribution of food [1]. The emergency problem was considered in the radioactive dangerous goods transportation optimization model, but once an accident occurs, the model always made the nearest rescue

L. Jia et al., *High-Speed Railway Operation Under Emergent Conditions*,
Advances in High-speed Rail Technology,
https://doi.org/10.1007/978-3-662-63033-4_8

point to participate in the emergency [2]. By decomposing the road network into several sub-networks and taking the shortest total path of the sub-network as the optimization goal, the genetic algorithm was used to solve the problem of selecting parallel paths [3]. Literature [4] analyzed the optimization of maintenance resource scheduling under limited conditions, especially the problem of path selection, but focused on processing deterministic information, which has great limitations in applications. In reference [5], the post-disaster emergency problem was transformed into the solution of the shortest network, which improved the rescue efficiency and reduced the loss.

(2) The situation of a single emergency point with multiple resources. In view of the speed and continuity of emergency resource consumption, reference [6] discussed the problem of one-time consumption and continuous consumption, established a multiple rescue point selection model with "the earliest emergency start time" as the goal, and further discussed the two-level goal planning problem of "the shortest emergency time and the least number of emergency rescue points". By defining the number of contacts and its order relationship in the set-pair analysis, the earliest model and algorithm for the emergency time of a multi-resource continuous consumption emergency system was established, and the scheduling problem of the multi-resource continuous consumption emergency system with uncertain time was solved [7]. Literature [8] discussed the two-stage random planning framework in the preparation of disaster relief transportation plans. In the first stage, based on the information obtained initially, a model for transporting different resources to the disaster area through different modes of transportation was established, with the goal of minimizing transportation costs and the amount of goods converted between different modes of transportation. In the second stage, a second linear programming model was established based on the possibility of different outcomes, with the goal of minimizing losses due to total expenses, transportation mode conversion costs, and holdings and shortages in the second stage. In the literature [9], a fuzzy set theory was proposed to solve the emergency resource scheduling problem. However, since the time from the rescue point to the emergency location still appears at a certain value, the emergency scheduling scheme cannot fully reflect the reality. According to the continuous consumption emergency, considering the initial amount of emergency resources at the rescue point, aiming at the cost of rescue and the loss caused by the delay of rescue, a mathematical model of emergency resource scheduling based on particle swarm optimization was established, and the influence of different parameter combinations on the solution efficiency was discussed [10]. Literature [11] believed that the material dispatching should adopt the corresponding dispatching method according to the situation of material consumption, so a dynamic material dispatching model was established, which improved the emergency rescue efficiency. Under the condition of transportation constraints, a model with the shortest time and the fewest rescue points for multiple resource points was established [12]. Literature [13] used Markov decision-making methods to study the dynamic optimization process of emergency resources scheduling schemes under public emergencies and obtained optimal emergency resource

scheduling schemes in specific scenarios to ensure the timeliness and support dynamic dispatch of emergency resources.

(3) The situation with multiple emergency points and multiple resources. The concept of "opportunity cost of rescue" was proposed, and a rescue vehicle configuration model was established based on this concept to achieve "minimum opportunity cost" [14]. In literature [6], the emergency resource allocation problem in emergency management was systematically studied from the aspects of multiple rescue points and route selection, a generalized "resource scheduling model" was established, and the solution method of the optimal scheme under the condition of define and uncertain network information was given. A multi-objective scheduling model based on "minimum number of rescue points" and "earliest emergency start time" was established in an emergency system with multiple resources and multiple rescue points, and the ideal point method was used to solve the model [15]. Literature [16] proposed a game model and algorithm for the emergency resource allocation problem with multiple accident points and multiple rescue points and designed a preference-based algorithm. An algorithm to adjust the resource competition of multiple disaster sites by adjusting the plan of each disaster site was proposed in literature [17], but this algorithm has certain limitations for solving the multi-objective problem. In order to solve the problem of emergency resource scheduling when multiple accidents exist simultaneously, an improved utility function based on preference order was designed to characterize the timeliness and effectiveness of rescue for each accident [18]. In reference [19], a multi-objective genetic algorithm based on greedy search algorithm was proposed, which can automatically adjust the allocation of available resources, generate a series of feasible emergency logistics solutions, and solve the problems of demand shortage, minimized transportation time, minimized transportation cost, etc.

There are also some preliminary studies on railway emergency resource dispatching. Considering the different priorities of emergency points, the resource scheduling model of railway emergency multi-emergency points was established on the basis of the earliest single emergency point activation time model. By converting to a linear model, the model can be easily solved using LINGO software. For the railway emergency resource scheduling problem, a multi-objective planning model based on the shortest time and the lowest cost was established, and the optimal scheduling scheme was obtained by using the ideal point method, which provided the possibility for the application of this model in railway rescue work.

In summary, the research on resource scheduling has gradually deepened from the initial deterministic to the uncertain type, which includes the uncertainty of resource demand and resource supply. The methods used mainly include 0–1 programming, integer programming, mixed-integer programming, heuristic methods, and queuing theory. The main factors considered include time, cost, uncertainty, etc. The main industries involved include police, fire protection, medical rescue, natural disasters, production accidents, etc. In general, although quantitative models have been improved in this field, dynamic models need to be further studied, especially for railway emergency resources dispatching.

8.1 Emergency Resource Scheduling with Single Emergency Point and Single Resource

In the railway emergency rescue process, the actual emergency resource consumption rate is not constant but changes accordingly with the development of the emergency. Therefore, the dispatch of emergency resources should take corresponding measures based on the dynamic consumption of resources. Based on the work of Liu Chunlin and others, this section uses the railway emergency resources as the background to establish a two-level optimization mathematical model with the earliest emergency start time and the minimum number of emergency service facilities under the condition of non-constant rate consumption. The rationality of the model and the effectiveness of the algorithm are verified using the given resource consumption curve.

8.1.1 Emergency Resource Scheduling Model with Single Emergency Point and Single Resource

Assume that $S_1, S_2, S_3, \cdots, S_n$ are n emergency service facilities, E is the emergency point, d is the emergency resource demand, and the emergency resource reserve of S_i is p_i ($p_i > 0$), $i = 1,2,\ldots,n$. The consumption rate of emergency resources can be expressed by a continuous function or a piecewise function, which is defined as $v(t)$. The time required from S_i to E is t_i ($t_i > 0$), $t_1 \le t_2 \le \cdots \le t_n$.

Either solution φ can be expressed as a set:

$$\varphi = \{(S_{i_1}, x_{i_1}), (S_{i_2}, x_{i_2}), \cdots, (S_{i_m}, x_{i_m})\} \tag{8.1}$$

where $0 < x_{i_k} \le p_{i_k}, \sum_{k=1}^{m} x_{i_k} = d, \{i_1, i_2, \cdots, i_m\} \subset \{1, 2, \cdots, n\}$. To facilitate discussion, let $i_0 = 0$, $t_0 = 0$, $p_0 = 0$.

Due to the non-constant rate consumption of emergency resources, it is necessary to ensure that the amount of resources reached at any time meets the requirements. Therefore, defining scheme φ is continuously feasible with regard to the emergency start time.

Definition 8. Solution φ is continuously feasible with regard to the emergency start time, if for any $t \in [s, s + \Delta t]$, the following formula is satisfied.

$$\sum_{k \in \left\{ j \middle| t_{i_j} \le t,\ j=1,2,\cdots,m \right\}} x_{i_k} \ge \int_s^t v(t)\mathrm{d}t \tag{8.2}$$

The left end of Eq. (8.1) represents the amount of resources reached at time t, and the right end represents the amount of resources consumed at time t. Definition 8.1 describes the solution φ regarding the emergency start time s continuously feasible as the amount of resources reached at any time is not less than the continuous consumption of resources from the emergency start time s to t.

Let the set of emergency start time be χ_s, and the earliest emergency start time is s^*, that is

$$s^* = \min_{\chi_s \neq \varnothing} s \tag{8.3}$$

Theorem 8.1 *If* $t_1 \leq t_2 \leq \cdots \leq t_n$, *the earliest emergency start time of solution* φ*is*

$$s^* = \max_{k \in (1,2,\cdots,p)} \left\{ t_k - \sum_{i \in \{i | t_i < t_k, i=0,1,2,\cdots,p\}} \Delta t_i \right\} \tag{8.4}$$

where t_k is the time it takes for emergency resources to reach the emergency point from the kth emergency service facility, and Δt_i is the time that the emergency resource reserve p_i of the ith emergency service facility runs out at the rate $v(t)$, which can be calculated by $p_i = \int_0^{\Delta t_i} v(t)dt$. A feasible solution should satisfy $\sum_{k=0}^{p-1} p_{i_k} < d \leq \sum_{k=0}^{p} p_{i_k}$, and then, the corresponding optimal solution is

$$\varphi^* = \left\{ \left(S_{i_1}, p_{i_1}\right), \left(S_{i_2}, p_{i_2}\right), \cdots, \left(S_{i_p}, d - \sum_{k=0}^{p-1} p_{i_k}\right) \right\} \tag{8.5}$$

In this scheme, S_{i_1}, the nearest emergency service facility point to emergency point E, is selected to participate in the emergency. If p_{i_1} is less than d, S_{i_2}, the nearest emergency service facility point except S_{i_1}, is selected to participate in the emergency, and so on, until the demand of emergency resources is met. For Theorem 8.1, literature [20] has given strict mathematics proof in the case of constant rate consumption, and the proof in the case of non-constant rate consumption is similar.

However, the above plan only considers the earliest emergency start time, not the least cost. In the process of emergency rescue, the best plan is to minimize the number of emergency service facilities involved in the emergency without delaying the start time of the emergency. Because the number of emergency service facilities involved in the emergency is directly related to the stability and reliability of the system, and from the perspective of cost, it is also directly related to the total cost of the system. Let $N(\varphi)$ be the number of emergency service facilities. For the above scheme φ^*, obviously $N(\varphi^*) = p$ holds.

Since $s^* = \min_{\chi_s \neq \varnothing} s$, the scheme φ^* is optimized to obtain the following scheme:

$$\varphi^{**} : \min_{\varphi \in \chi_{s^*}} N(\varphi) \tag{8.6}$$

Only the case of $\chi_{s^*} \neq \varnothing$ is discussed, because the scheme φ^{**} is meaningful under this condition.

Under the premise of ensuring that the emergency resources meet the needs, the number of emergency service facilities is regarded as the upper-level problem, and the emergency start time is regarded as the lower-level problem. A two-layer optimization mathematical model is established, which has the earliest emergency start time and the least number of emergency service facilities involved in the emergency.

$$
\begin{aligned}
&\min_{\varphi \in \chi_{s^*}} N(\varphi) \\
&\text{s.t.} \begin{cases}
s^* = \min\limits_{\chi_s \neq \varnothing} s \\
\text{s.t.} \begin{cases}
p_i \geq 0, t_i > 0, i = 1, 2, 3, \cdots, n \\
\sum\limits_{i=1}^{n} p_i \geq d \\
\sum\limits_{k=0}^{p-1} p_{i_k} < d \leq \sum\limits_{k=0}^{p} p_{i_k} \\
0 < x_{i_k} \leq p_{i_k} \\
\sum\limits_{k=1}^{p} x_{i_k} = d \\
\sum\limits_{k \in \left\{ j \middle| t_{i_j} \leq t,\ j=1,2,\cdots,p \right\}} x_{i_k} \geq \int_s^t v(t)dt
\end{cases}
\end{cases}
\end{aligned}
\tag{8.7}
$$

The goal of this model is to determine the number of emergency service facilities participating in emergency rescue and the number of emergency resources provided under the condition of ensuring emergency resource requirements, so that the emergency start time is the earliest and the number of emergency service facilities is the smallest.

Constraint 1 indicates that there are n emergency service facilities equipped with emergency resources. Constraint 2 indicates that the total number of resources available for the n emergency service facilities is sufficient to meet emergency needs. Constraint 3 refers to the principle of proximity of emergency service facility points, that is, the emergency service facility points close to the emergency point are required to participate in the emergency first and then increase in order from near to far, until the demand is met. Constraint 4 indicates that the number of emergency resources provided by each emergency service facility point of the scheme determined according to the proximity principle is less than or equal to the amount of available resources. Condition 5 constrains the total amount of emergency resources provided by all emergency service facilities participating in the emergency equal to the demand d. Constraint 6 means that the amount of emergency resources arriving at time t must be greater than or equal to the amount of resource consumption at time t, that is, the situation where emergency activities stop due to undelivered resources cannot occur.

This model is a two-layer optimization mathematical model with the earliest emergency start time and the least number of emergency service facilities under the condition of non-constant rate consumption. The ergodic search method can be used to solve the above model. The steps are as follows.

(1) According to the principle of proximity, ensure $\sum_{k=0}^{p-1} p_{i_k} < d \leq \sum_{k=0}^{p} p_{i_k}$, and get the scheme φ^*.

(2) According to $s^* = \max\limits_{k\in(1,2,\cdots,p)} \left\{ t_k - \sum_{i\in\{i|t_i<t_k, i=0,1,2,\cdots,p\}} \Delta t_i \right\}$, the earliest emergency start time, $\min\limits_{\chi_s\neq\varnothing} s$, is solved.

(3) Solve $\min\limits_{\varphi\in\chi_{s^*}} N(\varphi)$, and the algorithm is as follows (this algorithm is called **Algorithm 1**).

① $u = s^*, k = 1, D = 0, I = \varnothing$.
② Solve i_k to make it meet the condition $\max\limits_{i\in\{j|t_j\leq u, j\notin I\}} p_i = p_{i_k}, I = I + \{i_k\}$, turn to step ③.
③ If $D + p_{i_k} < d$, then $D = D + p_{i_k}, u = u + \Delta t_{i_k}, k = k + 1$, turn to step ②; otherwise, $x_{i_k} = d - D$, turn to step ④.
④ $\varphi^{**} = \left\{ (S_{i_1}, p_{i_1}), (S_{i_2}, p_{i_2}), \cdots, (S_{i_{k-1}}, p_{i_{k-1}}), (S_{i_k}, x_{i_k}) \right\}$.

The **Algorithm 1** flowchart is shown in Fig. 8.1.

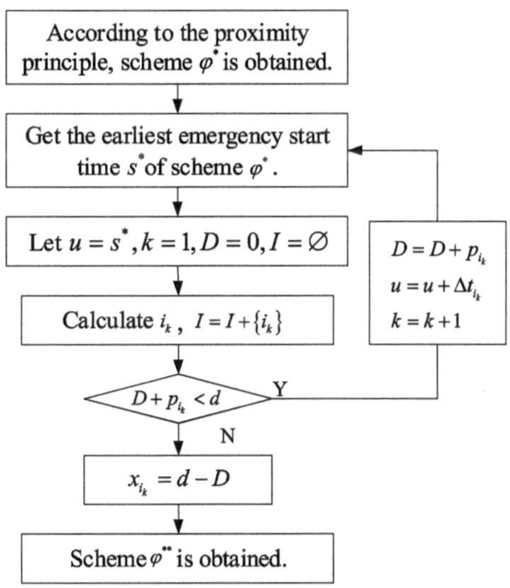

Fig. 8.1 Flowchart of **Algorithm 1**

8.1.2 Mathematical Proof

Chunlin Liu and others have given a strict mathematical proof in the case of constant rate consumption [6]. Based on this, this section will give a mathematical proof of the optimality of **Algorithm 1** for the case of non-constant rate consumption. In order to facilitate the discussion of the problem, the step ② of **Algorithm 1** is slightly modified to get a more general **Algorithm 2**. The specific steps are as follows.

① $u = s^*, h = 1, D = 0, J = \varnothing$.
② If any $j_h \in \{j | t_j \le u, j \notin I\}$, $J = J + \{j_h\}$, turn to step ③.
③ If $D + p_{j_h} < d$, then $D = D + p_{j_h}$, $u = u + \Delta t_{j_h}$, $h = h + 1$, turn to step ②; otherwise, $x_{j_h} = d - D$, turn to step ④.
④ $\varphi^{**} = \{(S_{j_1}, p_{j_1}), (S_{j_2}, p_{j_2}), \cdots, (S_{j_{h-1}}, p_{j_{h-1}}), (S_{j_h}, x_{j_h})\}$.

Let the set of all possible schemes generated by this algorithm be χ_s'.

Lemma 8.1 *When $\chi_s \ne \varnothing$, any scheme obtained by* **Algorithm 2***is continuously feasible.*

It is proved that the scheme obtained by **Algorithm 2** can ensure that the amount of resources arrived at any time is not less than the continuous consumption of resources at the emergency start time s to t. From Definition 8.1, it can be seen that any scheme obtained by **Algorithm 2** is continuous and feasible. Because **Algorithm 1** is a special case of **Algorithm 2**, the scheme obtained by **Algorithm 1** is also continuous and feasible.

Lemma 8.2 *When $\chi_s \ne \varnothing$, there exists a scheme $\varphi \in \chi_s'$, which makes* Eq. (8.6) *optimal.*

Let scheme $\varphi' = \{(S_{i_1}, x_{i_1}), (S_{i_2}, x_{i_2}), \cdots, (S_{i_m}, x_{i_m})\}$ be the optimal scheme, where $0 < x_{i_k} \le p_{i_k}$, $\sum_{k=1}^{m} x_{i_k} = d$. Without loss of generality, let $i_1 < i_2 < \cdots < i_m$, then $t_{i_1} \le t_{i_2} \le \cdots \le t_{i_m}$. It can be seen that $\sum_{k=1}^{m-1} p_{i_k} < \sum_{k=1}^{m} x_{i_k} = d \le \sum_{k=1}^{m} p_{i_k}$, scheme $\varphi'' = \left\{(S_{i_1}, p_{i_1}), (S_{i_2}, p_{i_2}), \cdots, (S_{i_{m-1}}, p_{i_{m-1}}), \left(S_{i_m}, d - \sum_{k=1}^{m-1} p_{i_k}\right)\right\}$ is feasible. Because the amount of resources that have arrived at time t in scheme φ'' must not be less than the amount of resources that have arrived at time t in scheme φ'. Obviously, $\varphi'' \in \chi_s'$, so φ'' is the optimal solution of Eq. (8.6).

Theorem 8.2 When $\chi_s \ne \varnothing$, the solution obtained by **Algorithm 1** is the optimal solution of Eq. (8.6).

When $\chi_s \ne \varnothing$, let $\{(S_{i_1}, p_{i_1}), (S_{i_2}, p_{i_2}), \cdots, (S_{i_{k-1}}, p_{i_{k-1}}), (S_{i_k}, x_{i_k})\}$ be the solution obtained by **Algorithm 1**, and let $\{(S_{j_1}, p_{j_1}), (S_{j_2}, p_{j_2}), \cdots, (S_{j_{h-1}}, p_{j_{h-1}}), (S_{j_h}, x_{j_h})\}$ be any solution obtained by **Algorithm 2**. To prove **Theorem 2**, just prove that $k \le h$, and the proving process is as follows.

When $h = 1$, it is obvious that $k = 1, k \le h$.

When $h \geq 2$, using the method of counter evidence, assuming $k > h$, then $\sum_{m=1}^{h} p_{i_m} < d = \sum_{m=1}^{h-1} p_{j_m} + x_{j_h} \leq \sum_{m=1}^{h} p_{j_m}$, and we get $\sum_{m=1}^{h} p_{i_m} < \sum_{m=1}^{h} p_{j_m}$.

The result of derivation is in contradiction with $\sum_{m=1}^{h} p_{i_m} < \sum_{m=1}^{h} p_{j_m}$, which proves that $k \leq h$. The derivation process is as follows:

$$p_{i_1}, p_{i_2}, \cdots, p_{i_h} \tag{8.8}$$

$$p_{j_1}, p_{j_2}, \cdots, p_{j_h} \tag{8.9}$$

From the two-level optimization mathematical model [see Eq. (8.7)], it can be known that at least one element in (8.8) is smaller than the corresponding position in (8.9), and it is assumed that it appears at g for the first time, that is,

$$\begin{aligned} p_{i_1} &\geq p_{j_1} \\ p_{i_2} &\geq p_{j_2} \\ &\vdots \\ p_{i_{g-1}} &\geq p_{j_{g-1}} \\ p_{i_g} &< p_{j_g} \end{aligned} \tag{8.10}$$

Obviously, $p_{i_1} \geq p_{j_1}$ holds, so $2 \leq g \leq h$.

$$p_{i_1}, p_{i_2}, \cdots, p_{i_g} \tag{8.11}$$

$$p_{i_1}, p_{i_1}, \cdots, p_{i_h} \tag{8.12}$$

Adjust the sequence (8.12) so that the elements in (8.11) are greater than or equal to the elements in the corresponding position in (12). Suppose that after q rearrangements, only two element positions are swapped each time to obtain the sequence $p_{j_1}^{(q)}, p_{j_2}^{(q)}, \cdots, p_{j_g}^{(q)}$, such that $p_{i_m} \geq p_{j_m}^{(q)}$, $m = 1, 2, \cdots, g$, where $j_1^{(q)}, j_2^{(q)}, \cdots, j_g^{(q)}$ is a permutation of $j_1, j_2, \cdots, j_g$. Proof is as follows:

Calculate p_{i_g}, according to **Algorithm 1**:

$$p_{i_g} = \max_{i \in \{j | t_j \leq u', j \notin I_{g-1}\}} p_i, I_{g-1} = \{i_1, i_2, \cdots, i_{g-1}\}, u' = s + \sum_{m=1}^{g-1} \Delta t_{i_m}$$

Calculate j_g, according to **Algorithm 2**:

$$j_g \in \{j | t_j \leq u'', j \notin J_{g-1}\}, J_{g-1} = \{j_1, j_2, \cdots, j_{g-1}\}, u'' = s + \sum_{m=1}^{g-1} \Delta t_{j_m}$$

It is proved in detail by the following three steps:

(1) Since $p_{i_m} = \int_0^{\Delta t_{im}} v(t)dt$, $p_{j_m} = \int_0^{\Delta t_{jm}} v(t)dt$, it can be known from Eq. (8.10) that, $p_{i_m} \geq p_{j_m}$, then $\Delta t_{i_m} \geq \Delta t_{j_m}$, and $u' \geq u''$, so $j_g \notin I_{g-1}$. Because $j_g \in I_{g-1}$, let $j_g = i_c$, $1 \leq c < g$, and then, change the order of $j_1, \cdots, j_c, \cdots, j_g$ to get $j_1^{(1)}, \cdots, j_c^{(1)}, \cdots, j_g^{(1)}$, where $j_1^{(1)} = j_g$, $j_g^{(1)} = j_c$, $j_m^{(1)} = j_m$, $(m \neq c, g)$. After reversing the order, $p_{j_1}^{(q)}, p_{j_2}^{(q)}, \cdots, p_{j_g}^{(q)}$ is obtained from $p_{j_1}, p_{j_2}, \cdots, p_{j_g}$. Obviously, $p_{i_m} \geq p_{j_m}^{(1)}$, $m = 1, 2, \cdots, g-1$ (when $m = c$, the equal sign holds).

(2) If $p_{i_g} < p_{j_g}^{(1)}$, then $j_g^{(1)} \in I_{g-1}$ (If $j_g^{(1)} \notin I_{g-1}$, let $u = s + \sum_{m=1}^{c-1} \Delta t_{j_m}$, it can be known that $u' \geq u'' \geq u$, and then, $j_g^{(1)} \in \{j | t_j \leq u, j \notin J_{c-1} \cup I_{g-1}\} \subset \{j | t_j \leq u', j \notin I_{g-1}\}$. $p_{j_g}^{(1)} \leq p_{i_g}$, which contradicts the known condition). Because $j_g \in I_{g-1}$, let $j_g^{(1)} = i_e$, $1 \leq e < g$, and then, change the order of $j_1^{(1)}, \cdots, j_e^{(1)}, \cdots, j_g^{(1)}$ to get $j_1^{(2)}, \cdots, j_e^{(2)}, \cdots, j_g^{(2)}$, where $j_e^{(2)} = j_g^{(1)}$, $j_g^{(2)} = j_e^{(1)}$, $j_m^{(2)} = j_m^{(1)}$, $(m \neq e, g)$. After reversing the order, $p_{j_1}^{(2)}, p_{j_2}^{(2)}, \cdots, p_{j_g}^{(2)}$ is obtained from $p_{j_1}^{(1)}, p_{j_2}^{(1)}, \cdots, p_{j_g}^{(1)}$. Obviously, $p_{i_m} \geq p_{j_m}^{(2)}$, $m = 1, 2, \cdots, g-1$.

(3) If $p_{i_g} \geq p_{j_g}^{(2)}$, repeat the above process, ensuring that the "order" relationship of the elements in the first (*g*-1) positions is not changed at each time. Therefore, at most (*g*-1) times of adjustment can make $p_{i_g} \geq p_{j_g}^{(g-1)}$. Because after (*g*-1) adjustments, set $\left\{j_1^{(g-1)}, j_2^{(g-1)}, \cdots, j_{g-1}^{(g-1)}\right\} = J_{g-1} = \{i_1, i_2, \cdots, i_{g-1}\} = I_{g-1}$, so $j_g^{(g-1)} \neq j_1^{(g-1)}, \cdots, j_g^{(g-1)} \neq j_{g-1}^{(g-1)}$, that is $j_g^{(g-1)} \in \{j | t_j \leq u', j \notin J_{g-1} = I_{g-1}\}$, we get $p_{i_g} \geq p_{j_g}^{(g-1)}$.

In summary, the $p_{j_1}^{(q)}, p_{j_2}^{(q)}, \cdots, p_{j_g}^{(q)}$ obtained after q ($q < g$) adjustments makes $p_{i_m} \geq p_{j_m}^{(q)}$, $m = 1, 2, \cdots, g$.

According to this analogy, it can be proved that sequence (8.9) can adjust its order continuously so that its elements are not greater than those in corresponding position in (8.8), that is to say, it contradicts with $\sum_{m=1}^{h} p_{i_m} < \sum_{m=1}^{h} p_{j_m}$, so it can be seen that $k > h$ is not tenable. It can be proved that $k \leq h$, that is, theorem 8–2 is proved.

8.1.3 Case Study

Assuming a major traffic event occurs on a railway line, the railway network is shown in Fig. 8.2, and the ☆ marked point is the emergency points *E*. Some simulation data is shown in Table 8.1. The amount of emergency service demand $d = 50$. During the entire emergency rescue process, the resource consumption rate first increases and then decreases with time, so it is assumed that the resource consumption rate is $v(t) = -0.6t^2 + 30t$.

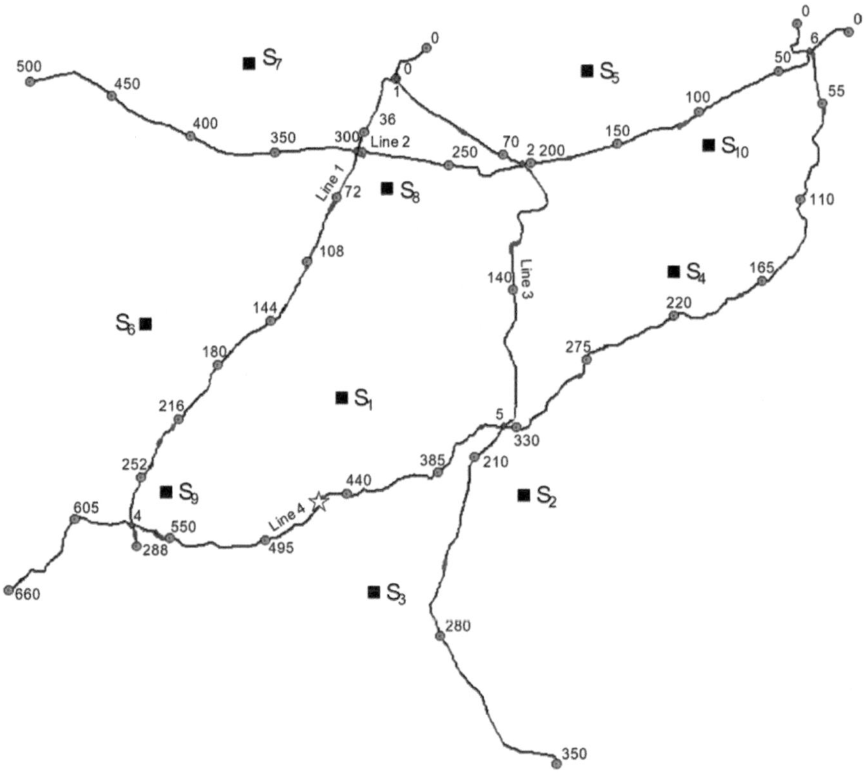

Fig. 8.2 Railway network

Table 8.1 Simulation data

	S_1	S_2	S_3	S_4	S_5	S_6	S_7	S_8	S_9	S_{10}
t_i/h	0.18	0.9	2.8	2.9	3.0	3.2	3.5	3.7	4.0	4.0
p_i	18	9	5	4	4	21	10	9	3	8

The solution process is as follows.

(1) According to the principle of proximity, ensure $\sum_{k=0}^{p-1} p_{i_k} < d \le \sum_{k=0}^{p} p_{i_k}$, and get the scheme φ^*.

$$\varphi^* = \{(S_1, 18), (S_2, 9), (S_3, 5), (S_4, 4), (S_5, 4), (S_6, 10)\}$$

(2) According to Theorem 8.1, the earliest emergency starting time $s^* = \max\limits_{k\in(1,2,\cdots,6)} \left\{t_k - \sum_{i\in\{i|t_i<t_k, i=0,1,2,\cdots,6\}} \Delta t_i\right\}$ is obtained, where Δt_i is obtained by $p_i = \int_0^{\Delta t_i} v(t)dt$. Take Δt_1 as an example to illustrate the specific process.

$$\int_0^{\Delta t_1} (-0.6t^2 + 30t)\mathrm{d}t = 18$$

$$-0.2\Delta t_1^3 + 15\Delta t_1^2 = 18$$

$$\Delta t_1 = 1.10 \text{ or } \Delta t_1 = 74.98 \text{ or } \Delta t_1 = -1.09$$

The last two solutions are not in line with the actual situation and are discarded, so $\Delta t_1 = 1.10$. In the same way, $\Delta t_2 = 0.77$, $\Delta t_3 = 0.58$, $\Delta t_4 = 0.52$, $\Delta t_5 = 0.52$, $\Delta t_6 = 1.20$.

$$\begin{aligned} s^* &= \max_{k \in (1,2,\cdots,6)} \left\{ t_k - \sum_{i \in \{i | t_i < t_k, i=0,1,2,\cdots,6\}} \Delta t_i \right\} \\ &= \max\{0.18, -0.2, 0.93, 0.45, 0.03, -0.29\} \\ &= 0.93 \end{aligned}$$

(3) Solve the $\min_{\varphi \in X_{s^*}} N(\varphi)$ on the premise that the earliest emergency start time is obtained.

① $u = s^* = 0.93, k = 1, D = 0, I = \varnothing$.
② There is only one case where $t_1 < u$, so $I = \{1\}$. Because $D + p_1 = 18 < d$, then $D = D + p_1 = 18, u = u + \Delta t_1 = 0.93 + 1.1 = 2.03, k = k + 1 = 2$. So, $I = \{1, 2, 3, 6\}$, and then, $D + p_6 = 53 > d, x_6 = d - D = 18$.
③ The scheme $\varphi^{**} = \{(S_1, 18), (S_2, 9), (S_3, 5), (S_6, 18)\}$ is obtained.

Under the condition that the minimum emergency start time is 0.93, compared with scheme φ^*, which is only based on the principle of proximity, scheme φ^{**} reduces two emergency service facility points S_4 and S_5, reduces the emergency cost, and improves the stability and reliability of the system.

8.2 Emergency Resource Scheduling with Single Emergency Point and Multi-resources

In the railway emergency rescue activities, sometimes multiple emergency service facilities are needed to provide multiple resources to participate in the emergency. In order to minimize the losses caused by emergencies, emergency rescue activities should be started early, that is, the earliest emergency start time is required. To reduce transportation costs as much as possible, reasonable selection of emergency service facilities should be made to minimize the number of emergency service facilities participating in rescue, and priority should be given to emergency service facilities close to the emergency point. In this section, a multi-objective optimization model based on the earliest emergency start time and the minimum number of emergency service facilities participating in an emergency is established for the issue of railway emergency resource dispatch with multiple resources being consumed at one time.

8.2.1 *Emergency Resource Scheduling Model with Single Emergency Point and Multi-resources*

1. **Model building**

Unified consumption resources refers to resources that can be used for emergency rescue activities after all the resources required for emergency have arrived at the emergency point, such as rescue trains and their auxiliary equipment and rescue team personnel. Assume that $S_1, S_2, S_3, \cdots, S_n$ are n emergency service facilities, E is the emergency point. Emergency point E requires a total of m $(m > 1)$ resources, respectively $X_1, X_2, X_3, \cdots, X_m$, and the quantity demand is $x_1, x_2, x_3, \cdots, x_m$.

Let the availability of the jth $(1 \le j \le m)$ resource at the ith $(1 \le i \le n)$ emergency service facility point be p_{ij}, and the quantity of the jth resource available at the ith emergency service facility point be x_{ij}, where $\sum_{i=1}^{n} p_{\text{ij}} \ge x_j$, $p_{\text{ij}} \ge x_{\text{ij}}$, $j = 1,2,\ldots,m$. For any S_i, $p_{\text{ij}} \ge 0$, and not all 0. The transport time from S_i to E is $t_i (t_i > 0)$, so let us say that $t_1 \le t_2 \le \cdots \le t_n$.

A multi-objective optimization model based on the earliest emergency start time and the minimum number of emergency service facilities is established. Under the premise of ensuring emergency resource requirements, the model is used to solve the optimal scheduling scheme, so that the two objectives are synthetically optimal.

The emergency resource scheduling scheme $\boldsymbol{\varphi}$ can be expressed in matrix form as

$$\varphi = \begin{bmatrix} x_{11} & x_{21} & \cdots & x_{n1} \\ x_{12} & x_{22} & \cdots & x_{n2} \\ \vdots & \vdots & \ddots & \vdots \\ x_{1m} & x_{2m} & \cdots & x_{nm} \end{bmatrix} \tag{8.13}$$

where $0 \le x_{ij} \le p_{ij}$, $\sum_{i=1}^{n} x_{\text{ij}} = x_j$. Column i represents the number of various resources involved in emergency response at S_i, and row j represents the number of jth resources provided by each emergency service facility point. Select k emergency service facilities to participate in the emergency, set as $S_{d_1}, S_{d_2}, \cdots, S_{d_k}$. The actual quantity of the jth resource provided by them is $x_{d_1 j}, x_{d_2 j}, \cdots, x_{d_k j}$, where $\sum_{i=1}^{k} x_{d_i j} = x_j$, and the set of all feasible scheduling schemes is Ω.

For a certain dispatching scheme $\boldsymbol{\varphi}$, let the number of emergency service facilities participating in the emergency be $N(\boldsymbol{\varphi})$, and the emergency start time be $T(\boldsymbol{\varphi})$. According to the definition of the unified resource consumption emergency system, the emergency start time refers to the time when all resources required for the emergency arrive at the emergency point. Therefore, $T(\boldsymbol{\varphi}) = \max\{t_i\}, i = d_1, d_2, \cdots, d_k$, the single emergency point—multi-resource emergency resource scheduling model of the unified resource consuming emergency system is expressed as

$$\begin{cases} \min\{T(\boldsymbol{\varphi})\} \\ \min\{N(\boldsymbol{\varphi})\} \end{cases}$$

$$\text{s.t.}\begin{cases} \sum_{i=1}^{k} x_{d_i j} = x_j \\ x_{d_i j} \le p_{d_i j} \end{cases} \tag{8.14}$$

This model is a typical multi-objective programming problem. The ideal point method can be used to find the positive and negative ideal points of the formula (8.14), that is, the optimal and worst solutions of the following two objective functions are obtained.

$$\min\{T(\boldsymbol{\varphi})\}$$

$$\text{s.t.}\begin{cases} \sum_{i=1}^{k} x_{d_i j} = x_j \\ x_{d_i j} \le p_{d_i j} \end{cases} \tag{8.15}$$

$$\min\{N(\boldsymbol{\varphi})\}$$

$$\text{s.t.}\begin{cases} \sum_{i=1}^{k} x_{d_i j} = x_j \\ x_{d_i j} \le p_{d_i j} \end{cases} \tag{8.16}$$

Let ψ_T and ψ_T' be the optimal solution and the worst solution of Eq. (8.15), and ψ_N and ψ_N' be the optimal solution and worst solution of Eq. (8.16). The closeness of the non-inferior scheme ψ to the positive ideal point is

$$R = \omega_1 \frac{\psi_N}{N(\varphi)} + \omega_2 \frac{\psi_T}{T(\varphi)} \tag{8.17}$$

The closeness of the non-inferior scheme ψ to the negative ideal point is

$$r = \omega_1 \frac{N(\varphi)}{\psi_N'} + \omega_2 \frac{T(\varphi)}{\psi_T'} \tag{8.18}$$

Among them, ω_1 and ω_2 are the weights of the number of emergency service facilities participating in the emergency and the emergency start time, respectively, and satisfy the normalization condition $\omega_1 + \omega_2 = 1$. Due to the urgency of railway emergency, emergency activities should be carried out as soon as possible in order to troubleshoot the fault as soon as possible and restore the line. Therefore, $\omega_1 < \omega_2$

is taken. The relative closeness of each non-inferior scheme to the ideal point is

$$\varepsilon = \frac{R}{r + R} \qquad 0 \le \varepsilon \le 1 \tag{8.19}$$

The scheme with the largest value is the optimal scheduling scheme.

2. **Optimal solution and worst solution of emergency start time**

In the emergency system of unified consumption resources, the maximum emergency start time is the required ψ_T , and the specific steps are as follows:

The emergency service facilities $S_1, S_2, S_3, \cdots, S_n$ are arranged in ascending order of the transportation time t_i, then for the jth resource, set

$$\sum_{z=1}^{q_j-1} x_{zj} < x_j \le \sum_{z=1}^{q_j} x_{zj} \tag{8.20}$$

Among them, $1 \le q_j \le n$, and the optimal scheduling scheme for the jth resource with the earliest emergency start time as the optimization goal is

$$\varphi_j = \left\{ (S_1, x_{1j}), (S_2, x_{2j}), \cdots, \left(S_{q_j}, x_{qj} - \sum_{z=1}^{q_j-1} x_{zj} \right) \right\} \tag{8.21}$$

Then, the earliest emergency start time of the resource is t_{q_j}, which is the largest of the transportation time involved in the emergency. Similarly, the earliest emergency start time of other resources can be obtained, and the optimal solution of emergency start time is $\psi_T = \max\{t_{q_1}, t_{q_2}, \cdots, t_{q_m}\}$.

According to the definition of the unified consumption resource emergency system, the maximum value of the emergency start time is not greater than the maximum transportation time t_n of each emergency service facility, and then, the worst solution of the emergency start time is $\psi_T' = t_n$. For any scheduling scheme, the earliest emergency start time must be between ψ_T and ψ_T'.

3. **The best solution and the worst solution to the number of emergency service facilities involved in the emergency**

In emergency rescue activities, the number of emergency service facilities participating in an emergency is not greater than the total number of emergency service facilities, and its worst solution is $\psi_N' = n$.

For the jth resource, any scheduling scheme $\varphi_j = \{(S_{k_1}, x_{k_1 j}), (S_{k_2}, x_{k_2 j}), \cdots, (S_{k_q}, x_{k_q j})\}$, where $\{k_1, k_2, \cdots, k_{q_j}\} \subset \{1, 2, \cdots, n\}$, if $w_j(1 \le w_j \le q \le n)$ exists in the scheme and $\sum_{i=1}^{w_j-1} x_{k_i j} < x_j < \sum_{i=1}^{w_j} x_{k_i j}$ is satisfied, then w_j is the minimum number of emergency service facilities required by the resource. Therefore, by calculating the minimum number of emergency service facilities required for each resource, the maximum value V of the minimum number of emergency service facilities required for all resources can be obtained, that is, $V = \max\{w_1, w_2, \cdots, w_m\}$. In this way, any scheme $\boldsymbol{\varphi}$ that can meet the resource emergency needs can meet the requirement of $n \geq N(\boldsymbol{\varphi}) \geq V$.

Let the number of emergency service facilities initially selected be V', and then, the algorithm for solving ψ_N is as follows.

(1) $V' = V$;
(2) Find all dispatching plans for V' emergency service facilities;
(3) If there are feasible scheduling schemes for V' emergency service facilities, the algorithm will be terminated;
(4) If there is no feasible scheduling scheme, $V' = V' + 1$, turn to step (2).

The V' obtained at the end of the algorithm is the optimal solution ψ_N for the number of emergency service facilities involved in the emergency.

4. **Solving the optimal scheduling scheme**

When solving the optimal scheduling plan, the number of emergency service facilities participating in the emergency and the earliest emergency start time must be considered. If the number of emergency service facilities in scheme $\boldsymbol{\varphi}_1$ is less than that in scheme $\boldsymbol{\varphi}_2$, and the earliest emergency start time in scheme $\boldsymbol{\varphi}_1$ is longer than that in scheme $\boldsymbol{\varphi}_2$, then it is impossible to judge which scheme is better intuitively. In this case, ideal point method can be used to calculate their relative proximity to ideal point, and the scheme with the largest ε is the optimal scheduling scheme. The optimal scheduling algorithm is as follows.

(1) Calculate ψ_T, ψ_T', ψ_T, and ψ_T';
(2) Let the set of emergency service facilities be $M_1 = \{S_1, S_2, S_3, \cdots, S_n\}$, $V' = \psi_N$;
(3) In the set M_1, the scheduling scheme of V' emergency service facility points meeting emergency needs is found, and the scheme with the earliest emergency start time is selected. If the scheme is not unique, the scheme which is close to the emergency service facility points from the perspective of reducing transportation costs is the better scheme;
(4) The earliest emergency start time t_1' of the better scheme is solved;
(5) If the transportation time of the selected scheme is greater than or equal to the emergency service facility point of t′, the emergency start time must be greater than t', so the scheme containing $(V' + 1)$ emergency service facilities must not be a better scheme. Therefore, the emergency service facility set M_1 is modified to remove the emergency service facility with a transportation time greater than or equal to t_1', and a new set M_2 is obtained;
(6) Let $V' = V + 1$;
(7) Find the scheduling scheme of V' emergency service facilities in M_2 to meet the emergency needs. If there is no scheduling scheme that meets the needs, go to step (6);
(8) If there are scheduling schemes to meet the requirements, the minimum emergency start time t_2' of these schemes is calculated;

(9) Modify the set of emergency service facilities M_2, remove the emergency service facilities with a transportation time greater than or equal to t_2', and obtain a new set M_3, so that $V' = V' + 1$;

(10) Find the scheduling scheme of V' emergency service facilities in M_3 to meet the emergency needs. If there is no scheduling scheme that meets the needs, go to step (6);

(11) When the set of emergency service facilities $M_n < V'$, the algorithm terminates;

(12) Calculate the ε of all feasible schemes, and the scheme with the largest ε is the optimal scheduling scheme.

8.2.2 *Case Study*

Assume that an emergency occurs on a railway line, as shown in Fig. 8.2, ☆ marked point is emergency points E. X_1, X_2, and X_3 are the three kinds of emergency resources needed, with the demand of 30, 32, and 27, respectively. $S_1, S_2, S_3, \cdots, S_{10}$ are optional emergency service facility points. The transportation time from each emergency service facility point to the emergency point is shown in Table 8.2, and the emergency resource availability of each emergency service facility point is shown in Table 8.3.

1. **Calculate ψ_T, ψ_T', ψ_N, and ψ_N'**

Since the maximum transportation time from the emergency service facility to the emergency site is 4.00 and the total number of emergency service facilities is 10, $\psi_T' = 4.00$ and $\psi_N' = 10$. Since all resources meet $\sum_{p=1}^{4} x_{pj} < x_j \le \sum_{p=1}^{5} x_{pj}$, the transportation time of emergency service facility point S_5 is ψ_T, so, $\psi_T = 3.00$. The minimum number of emergency service facilities required by resource point $p_1 = p_2 = p_3 = 2$, i.e., $V = 2$. When $V' = 2$, the emergency service facility points S_3 and S_{10} are selected, and the scheduling scheme meets the emergency resource demand. So $\psi_N = 2$.

Table 8.2 Transportation time from emergency service facility point to the emergency facility point

	S_1	S_2	S_3	S_4	S_5	S_6	S_7	S_8	S_9	S_{10}
t_i	0.18	0.9	2.8	2.9	3.0	3.2	3.5	3.7	4.0	4.0

Table 8.3 Emergency resource availability of each emergency service facility point

	S_1	S_2	S_3	S_4	S_5	S_6	S_7	S_8	S_9	S_{10}
S_1	3	8	11	5	4	12	14	12	2	20
S_2	1	9	17	5	6	7	10	8	5	15
S_3	4	7	10	1	5	8	13	7	6	17

2. **Calculate the ε of all feasible schemes**

The weights of the number of emergency service facilities and the emergency start time are taken as 0.4 and 0.6, respectively. When $V' = 2$, the available emergency service facilities are $S_1, S_2, S_3, \cdots, S_{10}$, and the feasible scheduling scheme is

$$\varphi_1 = \begin{bmatrix} 0\ 0\ 11\ 0\ 0\ 0\ 0\ 0\ 0\ 19 \\ 0\ 0\ 17\ 0\ 0\ 0\ 0\ 0\ 0\ 15 \\ 0\ 0\ 10\ 0\ 0\ 0\ 0\ 0\ 0\ 17 \end{bmatrix}$$

That is $R_1 = 0.85,\ r_1 = 0.68,\ \varepsilon_1 = 0.5556$.

When $V' = 3$, the available emergency service facilities are $S_1, S_2, S_3, \cdots, S_9$, and the feasible scheduling scheme is

$$\varphi_2 = \begin{bmatrix} 0\ 8\ 11\ 0\ 0\ 0\ 11\ 0\ 0 \\ 0\ 9\ 17\ 0\ 0\ 0\ 6\ 0\ 0 \\ 0\ 7\ 10\ 0\ 0\ 0\ 10\ 0\ 0 \end{bmatrix} \quad \varphi_3 = \begin{bmatrix} 0\ 0\ 11\ 0\ 0\ 12\ 7\ 0\ 0 \\ 0\ 0\ 17\ 0\ 0\ 7\ 8\ 0\ 0 \\ 0\ 0\ 10\ 0\ 0\ 8\ 9\ 0\ 0 \end{bmatrix}$$

$$\varphi_4 = \begin{bmatrix} 0\ 0\ 11\ 0\ 0\ 0\ 14\ 5\ 0 \\ 0\ 0\ 17\ 0\ 0\ 0\ 10\ 5\ 0 \\ 0\ 0\ 10\ 0\ 0\ 0\ 13\ 4\ 0 \end{bmatrix}$$

According to the principle that the number of emergency service facilities participating in the emergency is the same, the earliest emergency start time is selected, and scheme $\boldsymbol{\varphi}_4$ is eliminated. Scheme $\boldsymbol{\varphi}_2$ and scheme $\boldsymbol{\varphi}_3$ have the same closeness, but from the perspective of transportation costs, the closer scheme $\boldsymbol{\varphi}_2$ is better. So, $R_1 = 0.781,\ r_1 = 0.645,\ \varepsilon_1 = 0.5477$.

When $V' = 4$, the available emergency service facilities are $S_1, S_2, S_3, \cdots, S_6$, and the feasible scheduling scheme is

$$\varphi_5 = \begin{bmatrix} 3\ 8\ 11\ 0\ 0\ 8 \\ 1\ 9\ 17\ 0\ 0\ 5 \\ 4\ 7\ 10\ 0\ 0\ 6 \end{bmatrix} \quad \varphi_6 = \begin{bmatrix} 0\ 8\ 11\ 0\ 4\ 7 \\ 0\ 9\ 17\ 0\ 6\ 0 \\ 0\ 7\ 10\ 0\ 5\ 5 \end{bmatrix}$$

The closeness of scheme $\boldsymbol{\varphi}_5$ and scheme $\boldsymbol{\varphi}_6$ is the same, so scheme $\boldsymbol{\varphi}_5$ with closer distance is selected, i.e., $R_5 = 0.7625,\ r_5 = 0.64,\ \varepsilon_5 = 0.5437$.

When $V' = 5$, the available emergency service facilities are S_1, S_2, S_3, S_4, S_5, and the feasible scheduling scheme is

$$\varphi_7 = \begin{bmatrix} 3\ 8\ 11\ 5\ 3 \\ 1\ 9\ 17\ 5\ 0 \\ 4\ 7\ 10\ 1\ 5 \end{bmatrix}$$

That is $R_7 = 0.76,\ r_7 = 0.65,\ \varepsilon_7 = 0.539$.

When $V' = 6$, the available emergency service facilities are S_1, S_2, S_3, S_4. Since $M_n < V'$, the algorithm terminates.

3. Choose the best solution

Choose the scheduling scheme with the largest ε, and then, the optimal scheduling scheme is

$$\varphi_1 = \begin{bmatrix} 0\ 0\ 11\ 0\ 0\ 0\ 0\ 0\ 0\ 0\ 19 \\ 0\ 0\ 17\ 0\ 0\ 0\ 0\ 0\ 0\ 0\ 15 \\ 0\ 0\ 10\ 0\ 0\ 0\ 0\ 0\ 0\ 0\ 17 \end{bmatrix}$$

The case results show that the optimal scheduling solution can be solved by using a multi-objective optimization model based on the earliest emergency start time and the smallest number of emergency service facilities, which can provide effective decision support for emergency dispatch.

8.3 Emergency Resource Scheduling with Multiple Emergency Points and Multiple Resources

In the case of multiple emergency points, the emergency rescue situation of all emergency points should be considered comprehensively. Therefore, the utility function of railway emergency resources (see Eq. 7.1) is used to establish an optimization model with the maximum amount of total loss that can be reduced at all emergency points as the objective function.

8.3.1 *Emergency Resource Scheduling Model with Multiple Emergency Points and Multi-Resources*

Suppose each emergency service facility point has w kinds of emergency resources. For the kth ($k = 1, 2, \cdots, w$) kind of resources, the resources available at emergency service facility point S_i are p_i^k, the demand for emergency point E_j is d_j^k, t_{ij}^k, ($i = 1, 2, 3, \cdots, n$; $j = 1, 2, 3, \cdots, m$) is the shortest transport time from S_i to E_j, and v_{ij} is the upper limit of the total resources from S_i to E_j.

According to the problem description, in order to establish the emergency resource scheduling model based on the utility function, the following variables are defined [21].

x_{ij}^k is the number of the kth emergency resources from S_i to E_j.

T_j^k is the emergency failure time of emergency resource R_k required by emergency point E_j, that is, when the time of R_k reaching E_j is greater than or equal to T_j^k, the loss cannot be reduced.

unitSave_j^k is the amount of loss per unit time that can be reduced when unit emergency resource R_k is used for emergency point E_j.

When $t_{ij}^k \geq T_j^k$, the scheme of dispatching unit emergency resources R_k from S_i to E_j is unreasonable in actual emergency management, so the amount of loss that can be reduced is set to –M, where M is a large positive number.

$f_j^k(x_{ij}^k, t_{ij}^k)$ indicates the amount of loss that can be reduced by S_i dispatching x_{ij}^k emergency resources R_k to emergency point E_j, whose form and parameters need to be determined according to the actual situation of the emergency. The emergency resources utility function is defined as a linear function that is positively related to the number of emergency resources and negatively related to the emergency start time, which can be expressed as

$$f_j^k(x_{ij}^k, t_{ij}^k) = \begin{cases} (T_j^k - t_{ij}^k) \cdot \text{unitSave}_j^k \cdot x_{ij}^k & 0 \leq t_{ij}^k < T_j^k \\ -\text{M} \cdot x_{ij}^k & t_{ij}^k \geq T_j^k \end{cases} \tag{8.22}$$

The emergency resource scheduling model with multiple emergency points and multiple resources is

$$\max_x \sum_{k=1}^{w} \sum_{j=1}^{m} \sum_{i=1}^{n} f_j^k(x_{ij}^k, t_{ij}^k)$$

$$\text{s.t.} \begin{cases} \sum\limits_{i=1}^{n} x_{ij}^k = d_j^k \\ \sum\limits_{j=1}^{m} x_{ij}^k \leq p_i^k \\ \sum\limits_{k=1}^{w} x_{ij}^k \leq v_{ij} \\ \sum\limits_{i=1}^{n} p_i^k \geq \sum\limits_{j=1}^{m} d_j^k \\ x_{ij}^k \geq 0, t_{ij}^k \geq 0 \\ i = 1, 2, 3, \cdots, n;\ j = 1, 2, 3, \cdots, m;\ k = 1, 2, 3, \cdots, w \end{cases} \tag{8.23}$$

The objective function of the model is to maximize the amount of loss that can be reduced.

The meaning of the constraints is as follows.

(1) Constraint $\sum_{i=1}^{n} x_{ij}^k = d_j^k$ indicates that the total amount of the kth resource dispatched to emergency point E_j by all emergency service facilities is equal to the demand for the kth resource at the emergency point;
(2) Constraint $\sum_{j=1}^{m} x_{ij}^k \leq p_i^k$ indicates that the total amount of the kth resource dispatched by the emergency service facility S_i to all emergency points is not greater than the available amount of the kth resource in S_i;

(3) Constraint $\sum_{k=1}^{w} x_{ij}^{k} \leq v_{ij}$ indicates that the total amount of various emergency resources dispatched to emergency point E_j must meet the upper limit requirement;
(4) $\sum_{i=1}^{n} p_i^k \geq \sum_{j=1}^{m} d_j^k$ constrains the availability of the kth resource of all emergency service facilities to meet the demand of the kth resource of all emergency services.

When the emergency resource utility function is defined as a linear function that is positively related to the amount of emergency resources and negatively related to the emergency start time, the model can be solved using the simplex method. When the emergency resource utility function is defined as a nonlinear function, it can be solved using the Newton method, DFP method, BFGS method, and variable-scale method.

This section uses the powerful YALMIP toolbox that runs in the MATLAB environment to solve optimization problems [22]. It can solve convex optimization and even non-convex optimization problems, including linear, quadratic, second-order cone, semi-positive definite, integer, and multi-parameter.

In general, $t_{ij}^k \geq T_j^k$ corresponds to the amount of scheduling resources $x_{ij}^k = 0$. If $x_{ij}^k \neq 0$, the model has no solution.

8.3.2 Case Study

Assume that the number of emergency points where an emergency may occur is $m = 4$, the number of emergency service facilities $n = 3$, and the number of emergency resources $w = 2$.

For the first type of emergency resources, the emergency resource requirements of each emergency point, emergency failure time, the amount of loss that can be reduced by unit emergency resources per unit time, and the availability of emergency resources at each emergency service facility point are shown in Table 8.4.

For the second type of emergency resources, the emergency resource requirements of each emergency point, emergency failure time, the amount of loss that can be reduced by unit emergency resources per unit time, and the availability of emergency

Table 8.4 Data of type 1 emergency resources

	E_1	E_2	E_3	E_4	p_i^1
S_1	3	11	3	10	7
S_2	1	8	2	8	4
S_3	7	4	12	5	10
d_j^1	3	6	5	6	
T_j^1	8	9	12	11	
unitSave_j^1	15	14	20	16	

resources at each emergency service facility point are shown in Tables 8.5. Due to the limitation of traffic volume, the upper limit of the total amount of emergency resources from emergency service facilities to emergency potins is shown in Table 8.6.

After calculating f_{ij}^1, the results are shown in Table 8.7.

After calculating f_{ij}^2, the results are shown in Table 8.8.

Table 8.5 Data of type 2 emergency resources

	E_1	E_2	E_3	E_4	p_i^2
S_1	3	11	3	10	3
S_2	1	8	2	8	6
S_3	7	4	12	5	11
d_j^2	4	7	4	5	
T_j^2	8	10	12	10	
unitSave_j^2	20	15	16	12	

Table 8.6 Upper limit of total resources from emergency service facility point to emergency point

	E_1	E_2	E_3	E_4
S_1	3	2	7	2
S_2	5	4	2	3
S_3	2	10	2	12

Table 8.7 Results after calculating f_{ij}^1

	E_1	E_2	E_3	E_4	p_i^1
S_1	75	-M	180	16	7
S_2	105	14	200	48	4
S_3	15	70	-M	96	10
d_j^1	3	6	5	6	

Table 8.8 Results after calculating f_{ij}^2

	E_1	E_2	E_3	E_4	p_i^2
S_1	100	−M	144	−M	3
S_2	140	30	160	24	6
S_3	20	90	−M	60	11
d_j^2	4	7	4	5	

Table 8.9 Dispatching of the first type of emergency resources

	E_1	E_2	E_3	E_4	p_i^1
S_1	2	0	4	0	7
S_2	1	2	1	0	4
S_3	0	4	0	6	10
d_j^1	3	6	5	6	

Table 8.10 Dispatching of the second type of emergency resources

	E_1	E_2	E_3	E_4	p_i^2
S_1	0	0	3	0	3
S_2	4	1	1	0	6
S_3	0	6	0	5	11
d_j^2	4	7	4	5	

The emergency resource scheduling model with multiple emergency points and multiple resources was established and solved by using YALIMP toolbox, as shown in Tables 8.9 and 8.10.

The optimal target value is 4 081, that is, the maximum loss that can be reduced through the optimal scheduling of emergency resources is 4 081, which shows that emergency resource scheduling based on utility function can effectively solve the problem of emergency resource scheduling with multiple emergency points and multiple resources [23].

8.4 Chapter Summary

This chapter mainly discusses the problem of emergency resource scheduling for high-speed railways. Firstly, the problem of emergency resource scheduling with non-constant rate consumption of single resource and single emergency point is studied, and the two-stage problem of the minimum number of emergency service facilities under the condition of continuity and the earliest emergency time is solved. Then, the emergency resource scheduling problem with single emergency point and multi-resource consumption at one time is discussed, and a multi-objective optimization model based on the minimum emergency start time and the minimum number of emergency facility service points is established, which is solved by ideal point method. Finally, the emergency resource scheduling problem of multiple emergency points and multiple resources based on the utility function is studied, and a mathematical model is established with the maximum possible loss that can be reduced as the objective function, and the optimal scheduling scheme is obtained.

References

1. Knott R (1998) The logistics of bulk relief supplies [J]. Disasters 11(4):113–115
2. George F (1998) Routing and emergency response team sitting for high level radioactive waste shipment [J]. IEEE Trans Eng Manage 45(2):141–152
3. Sannomiya N, Tatemura K (2003) Application of genetic algorithm to a parallel path selection problem [J]. Int J Syst Sci 27(2):269–274
4. Sittithumwata A, Soudib F, Tomsovic K (2004) Optimal allocation of distribution maintenance resources with limited information [J]. Electr Power Syst Res 68(3):208–220
5. Altay N, Green GW (2006) OR/MS research in disaster operations management [J]. Eur J Oper Res 175(1):475–493
6. Chunlin L (2000) Research on models and methods of emergency material dispatch in emergency management [D]. Southeast University, Nanjing (in Chinese)
7. Shuping G, Sanyang L (2003) Scheduling problem in multi-resource emergency systems based on the connection number [J]. Syst Eng Theory Pract 23(6):113–116 (in Chinese)
8. Barbarosoglu G, Arda Y (2004) A two-stage stochastic programming framework for transportation planning in disaster response [J]. J Oper Res Soc 55(1):43–53
9. Belmont E, Fried MB, Gonen SJ et al (2006) Emergency preparedness, response and recovery checklist: beyond the emergency management plan [J]. Hosp Prep Bioterror 37(4):295–331
10. Pan Yu, Jia Yu, Qingli Da (2007) Emergency resources scheduling on continuous consumption system based on particle swarm optimization [J]. J Syst Eng 22(5):556–560 (in Chinese)
11. Rajagopalan KH, Saydam C, Xiao J (2008) A multiperiod set covering location model for dynamic redeployment of ambulances [J]. Comput Oper Res 35(3):814–826
12. Nagumey A (2009) A system-optimization perspective for supply chain network integration: The horizontal merger case[J]. Transp Res Part E 45(1):1–15
13. Wei W, Mao L, Li W (2010) The dyynamic optimal method of emergency resources deployment planning based on markov decision processes [J]. J Nankai Univ (Natural Science Edition). 43(3):18–23 (in Chinese)
14. Sherali HD, Subramanian S (1999) Opportunity cost-based models for traffic incident response problem [J]. J Transp Eng 125(3):176–185
15. Wang Yu, Jianmin He (2002) Research on multi-resource dispatch in emergency system [J]. J Southeast Uni (Natural Science Edition). 32(3):1–4 (in Chinese)
16. Shetty SR (2004) An event driven single game solution for resource allocation in a multi-crisis environment [D]. Tampa: Ulliv South Florida
17. Polychronakis M, Anagnostakis K, Markatos E (2006) Network-level polymorphic shellcode detection using emulation [C]. Proceedings of the conference on deteetiorl of intrusions and malware & vulnerability assessment
18. Zhang J, Shen S, Yang R (2007) Preference-order-based game modeling of multiple emergency resource allocation [J]. J Tsinghua Univ (Sci Technol) 47(12):2171–2175 (in Chinese)
19. Chang FS, Wu JS, Lee CN et al (2014) Greedy-search-based multi-objective genetic algorithm for emergency logistics scheduling [J]. Expert Syst Appl 41(6):2947–2956
20. Hongrui N, Ping Li, Fuzhang W (2009) Research on modeling and simulation of multi-emergency-areas model for emergency resource dispatch in railway emergency events [J]. Railway Comput Appl 18(12):20–22 (in Chinese)
21. Xingshen X (2010) Study on the scheduling problem of railway emergency materials [J]. Transp Technol Econ 12(2):52–55 (in Chinese)
22. He J, Liu C, Cao J et al (2005) Emergency management and emergency systems: site selection, scheduling, and algorithm [M]. (in Chinese)
23. Huijuan Z (2011) Plan management and resource allocation optimization on railway emergency management [D]. Beijing Jiaotong University, Beijing (in Chinese)

Chapter 9
Decision Support System and Demonstration Application of High-Speed Railway Train Operation Under Emergency Conditions

At present, the way that China's railway operation management departments respond to railway emergencies still remains at the level of paper-based document inquiry and internal telephone communication, showing problems such as low efficiency of emergency response and lack of effective communication between railways. Therefore, this chapter constructs the acquisition and deduction of special operating conditions based on closed-loop control, the generation of high-speed train digital plan, the matching of passenger demand and transportation resources, and the integrated intelligent optimization platform of line plan and train operation plan adjustment, and designs and develops a high-speed railway train operation decision support system under emergency conditions.

The system covers three subsystems that are independent and interconnected: high-speed railway special operating conditions acquisition and generation subsystem, high-speed railway transportation organization emergency plan preparation and decision support subsystem, and high-speed line planning subsystem under special operating conditions. The system realizes comprehensive data analysis, decision support, emergency command, and collaborative decision-making functions and meets the diversified needs of the railway dispatching management department.

The high-speed railway train operation decision support system under emergencies has taken an important step in the development of China's railway emergency management, laying a solid foundation for the comprehensive realization of electronic and information-based railway intelligent transportation, effectively improving China's large organizational efficiency of high-speed rail transportation under scale-forming conditions. In addition, two application demonstration projects were promoted based on this system. Among them, the railway transportation organization emergency response decision support system (high-speed railway) has been the core part of the railway transportation information integration platform of China Railway Corporation, providing key technology and platform support for railway transportation organization emergency response decision;

L. Jia et al., *High-Speed Railway Operation Under Emergent Conditions*,
Advances in High-speed Rail Technology,
https://doi.org/10.1007/978-3-662-63033-4_9

the national railway emergency plan management system has developed 106 functional modules for the whole road, and realized digital management of 24 ministerial-level plans, 116 bureau-level plans and 625 station-level plans, and the application effect is remarkable. This chapter will introduce the architecture and database design of the high-speed railway train operation decision support system, and introduce the system requirements, system structure, system interface of three subsystems, and two application demonstration projects, so as to show the system comprehensively and stereoscopically.

9.1 System Architecture

With the continuous expansion of the system scale and increasing complexity, the impact of system architecture on system performance is also increasing. Therefore, it is important to choose and design a reasonable architectural pattern. This section will introduce the overall architecture, subsystem design, and network environment structure of the high-speed railway train operation decision support system under emergency conditions.

9.1.1 System Architecture

The overall architecture of the high-speed railway train operation decision support system under emergency conditions is shown in Fig. 9.1, which shows the interaction relationship between the subsystems, and the data flow and control information transmission relationship. The system includes presentation layer, application layer and data layer. The system takes the structural characteristics of "high cohesion, low coupling" as the design goal, and considers the needs of a "one-stop" workflow. The various subsystems interact through data interfaces and standardized data design to build a cooperative operation mode. [1].

1. **Presentation layer**

The presentation layer uses the unified GIS and RIA front-end technologies to establish interaction with users.

2. **Application layer**

The application layer establishes three subsystems that can work independently or interactively. When a railway emergency occurs, a special operation scenario is established by the special operating conditions acquisition and generation subsystem of the high-speed railway. According to the emergency plan prepared by the high-speed railway transportation organization and the plan matched with the decision support subsystem, analysis and calculation are performed in the high-speed line

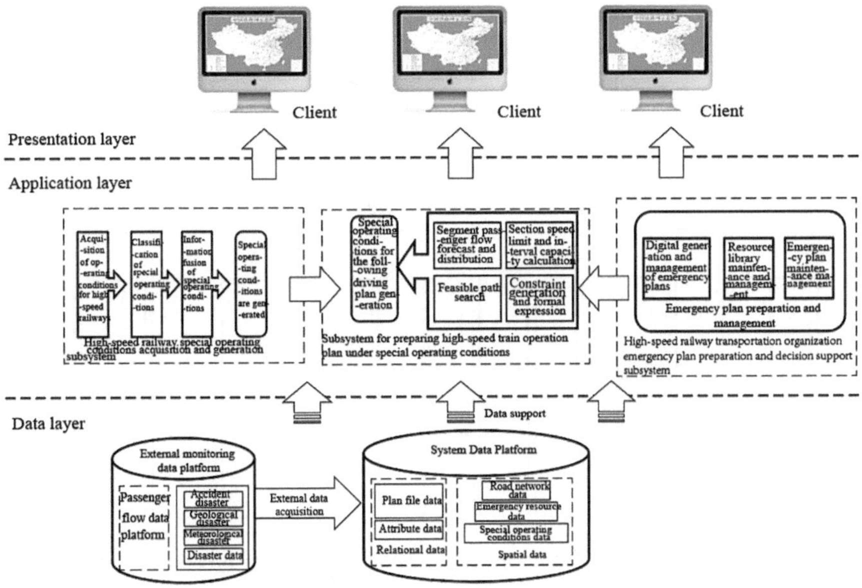

Fig. 9.1 Overall architecture of high-speed railway train operation decision support system under emergency conditions

planning preparation subsystem under special operating conditions to generate a set of intelligent line plans to form the integrated generation method of emergency response and decision making.

3. **Data layer**

The system data platform built by the data layer provides data support for the application layer and obtains the required data information by establishing a data interface with the external monitoring data platform.

9.1.2 System Subsystem Design

The functional structure of the high-speed railway train operation decision support system under special operating conditions is shown in Fig. 9.2. It is mainly divided into three subsystems, including the high-speed railway special operating conditions acquisition and generation subsystem, the high-speed railway transportation organization emergency plan preparation and decision making support subsystem, and high-speed line planning subsystem under special operating conditions.

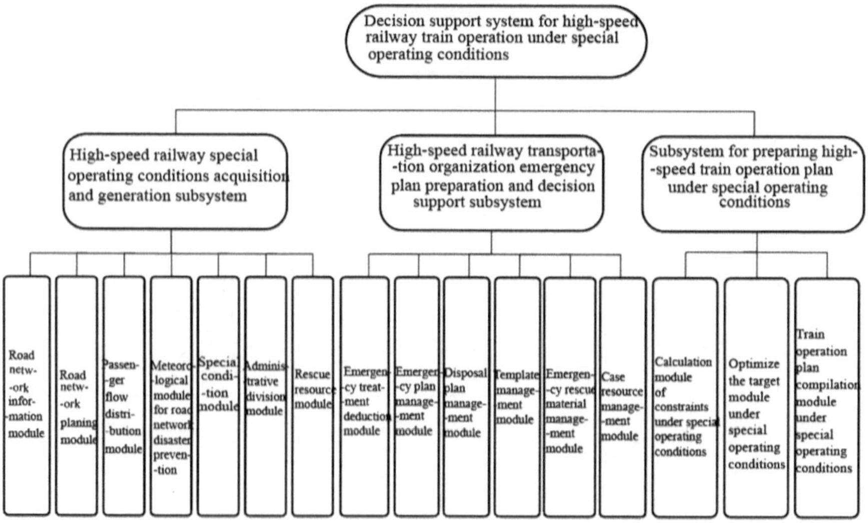

Fig. 9.2 Functional structure of high-speed railway train operation decision support system under special operating conditions

1. High-speed railway special operating conditions acquisition and generation subsystem

The high-speed railway special operating conditions acquisition and generation subsystem is mainly divided into seven modules, including road network information module, road network planning module, passenger flow distribution module, road network disaster prevention meteorological module, special condition module, administrative division module, rescue resource module. As an important support of other subsystems, the subsystem obtains the basic data of special operating conditions from the external monitoring system by establishing a standardized data interface, and cleans, filters, transforms and corrects it, and finally obtains standardized special operating conditions information. On this basis, special operation scenarios and scene libraries are established, and multimedia data technology and GIS spatial data technology are integrated to realize diversified generation and display of special operation scenarios. Therefore, the high-speed railway special operating conditions acquisition and generation subsystem plays a cornerstone role in comprehensive display and information acquisition in the whole system.

2. High-speed railway transportation organization emergency plan preparation and decision support subsystem

The high-speed railway transportation organization emergency plan preparation and decision support subsystem are divided into six modules, including emergency treatment deduction module, emergency plan management module, disposal plan management module, template management module, emergency rescue material management module, and case resource management module. The subsystem analyzes and decomposes existing railway emergency plans, and compiles and manages various emergency plans required for special operating conditions of high-speed railways. At the same time, the subsystem adopts the "scenario-response" type emergency plan matching technology, dynamically matches the corresponding emergency plan for different special operating conditions, and dynamically matches the corresponding disposal according to the "sample set-based disposal plan fusion extraction" method. The program provides decision support for the entire system.

3. High-speed line planning subsystem under special operating conditions

The high-speed line planning subsystem under special operating conditions includes three parts: the constraint condition calculation module under special operating conditions, the optimization target generation module under special operating conditions, and the line planning module under special operating conditions. According to the "based on series–parallel system operating reliability analysis and calculation method of road network," "high-speed rail network operation reliability analysis method based on the demand change," and "based on Markov chain railway interval through capacity calculation method under the condition of special operations", the subsystem calculates the constraint conditions, such as the road network operation reliability, the section carrying capacity, etc. Similarly, the "K-shortest path algorithm for mileage limitation and the improved K-shortest path algorithm based on sufficient operating conditions" are also used for train allocation in the line plan, and combined with "line plan optimization objective generation method based on fuzzy membership" are used to generate the optimization target for line plan. In addition, the subsystem designed and implemented the constraint condition generation module for the line plan based on the "train synthesis method for train running route constraints based on fuzzy comprehensive optimization," which laid the foundation for the further optimization and evaluation of line plan. As the core link of the "one-stop" workflow involving the most independent innovation algorithms, the largest amount of data calculations, and the highest degree of intelligent decision making, the high-speed line planning subsystem plays an indispensable role in the entire system under special operating conditions.

9.1.3 System Network Environment Structure

The Browser/Server (B/S) architecture is adopted in the decision support system of high-speed railway trains and its subsystems under emergency conditions, as shown in Fig. 9.3. At the presentation layer, interaction with users is achieved through mainstream web browsers (such as Internet Explorer, Firefox, and Chrome). In addition, because the system uses Flex to implement the client application interface, the Adobe Flash Player plug-in must be installed on the browser. When the user needs to perform data interaction with the system, the user request is transmitted from the Web browser of the presentation layer to the Web server of the application layer through the Internet (Internet, also known as the Extranet) or Intranet (Intranet, also

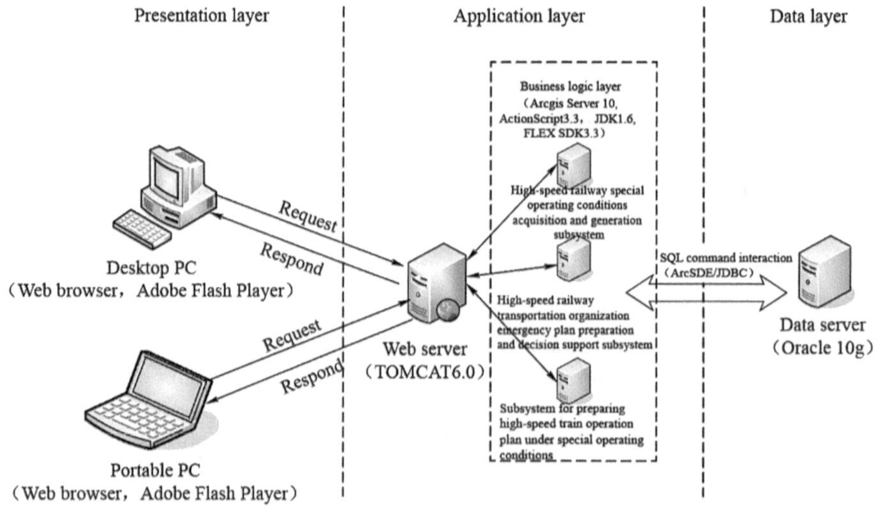

Fig. 9.3 Network environment architecture of high-speed railway train operation decision support system under emergency conditions

known as the Intranet). The web server used in the application layer of this system is preliminarily loaded into the TOMCAT 6.0. There are 3 subsystems under the TOMCAT application directory, supported by Arcgis Server10, ActionScript 3.3, JDK 1.6, and FLEX SDK 3.3. The system realizes GIS-based spatial technology application logic, J2EE-based business logic and data interaction, and Flex-based RIA front-end response and feedback, providing an important platform for effective interconnection and collaborative operation of various subsystems. At the same time, as the core component of the application layer, the Web server needs to perform data interaction with the data layer in a timely manner. In the data layer, with Oracle 10g as the central data platform, and through the JDBC standard application interface, a data interaction channel from the application layer to the data layer is established; relying on ArcSDE, it supports GIS spatial data interaction and provides powerful data support for the efficient and stable operation of the entire system.

9.1.4 Non-functional Requirements

The so-called non-functional requirements refer to the characteristics that software products must have in addition to the functional requirements in order to meet the user's business needs. Non-functional requirements for software products include system performance, reliability, availability, maintainability, security, extensibility, and adaptability to technology and business. This section will briefly analyze some of the non-functional requirements of the high-speed railway train operation decision support system.

1. **Performance**
   Performance mainly includes the following requirements:

   (1) For general business processing operations, the system response time should be within 10 s;
   (2) For query and map rendering processing, the system response time should be within 15 s.

2. **Reliability**
   Reliability mainly includes the following requirements:

   (1) The system architecture should ensure that there is no single point of failure, and the key processing points should consider load balancing;
   (2) The abnormal operation of the entire system due to a single point of failure or system upgrade must be avoided;
   (3) The system should be available 24 h a day, 7 days a week;
   (4) The extension time should be less than 4%;
   (5) The average time between failures should exceed 3 months;
   (6) The correctness of business logic under normal and extreme conditions should be guaranteed.

3. **Availability**
   Availability mainly includes the following requirements:

   (1) Convenient keyboard operation, data input should have a memory function to minimize repeated input;
   (2) Various coding inputs need to support multiple methods, such as direct input coding, input mnemonic code, simple spelling, and list selection;
   (3) The interface design should be beautiful, and the operation style should be consistent. The application interface should meet the requirements of different screen resolutions such as 800 × 600, 1 024 × 768, 1280 × 1024.

4. **Maintainability**
   Maintainability mainly includes the following requirements:

   (1) The system can be easily modified and upgraded, and is readable, modifiable, testable, etc.;
   (2) System components have automatic deployment and automatic upgrade functions;
   (3) The system has the function of remote analysis and elimination of errors function.

5. **Security**
   Security mainly includes the following requirements:

   (1) The necessary technology should be used to ensure the high availability of the database;

(2) The important data records in the database should adopt the checksum and other methods to ensure that the data is not illegally modified;
(3) Using database backup products, periodically backing up data on the database server for accident and disaster recovery;
(4) Perform data file backup, periodically import the data into the file, and store it on the CD.

9.2 System Database Design

Data is an important part of the high-speed railway train operation decision support system under emergencies. Therefore, a well-organized data resource structure and data resource relationship are of great significance for the efficient operation of the system.

9.2.1 Concept Design

Conceptual design is the key to the entire database design. The main features of the conceptual design include:

(1) Being able to truly and fully reflect the real world (including the connection between things and things) is a real model of the real world.
(2) It is easy to understand so you can exchange ideas with people who are not familiar with the computer.
(3) Easy to change, it is easy to modify and augment the conceptual model when the application environment and application requirements change.
(4) It is easy to convert into various data models such as relationships, meshes, and hierarchies.

The conceptual design is mainly to obtain the conceptual model of the database. This section uses the physical contact method proposed by P. P. S. Chen in 1976; that is, using the E-R diagram to describe the real-world conceptual model, a detailed conceptual design description is given by taking the subsystem for preparing a high-speed line plan under special operating conditions as an example.

Due to the large number of physical attribute maps of the high-speed line planning subsystem under special operating conditions, this section only draws the physical attribute maps of "segment," "emergency scene," and "line plan," as shown in Fig. 9.4. The field marked with an underscore is the primary key.

The conceptual structure design adopts the bottom-up approach. Firstly, the local conceptual structure is designed by abstract data and then integrated to obtain the global conceptual structure. The E-R diagram of the database is shown in Figs. 9.5, 9.6, and 9.7.

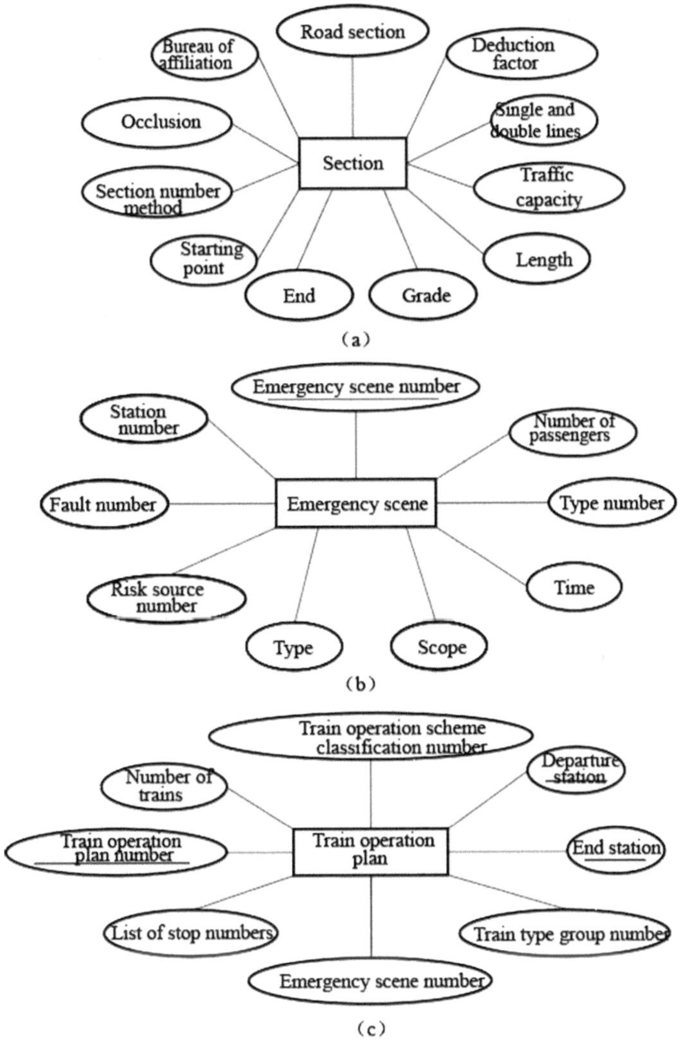

Fig. 9.4 Database entity attribute map

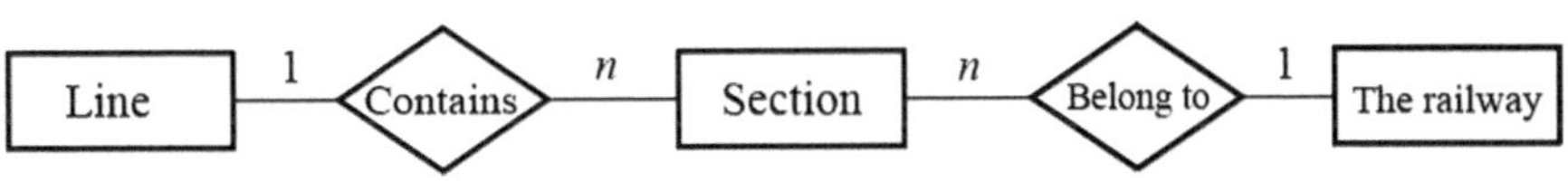

Fig. 9.5 Part of the E-R diagram in database (1)

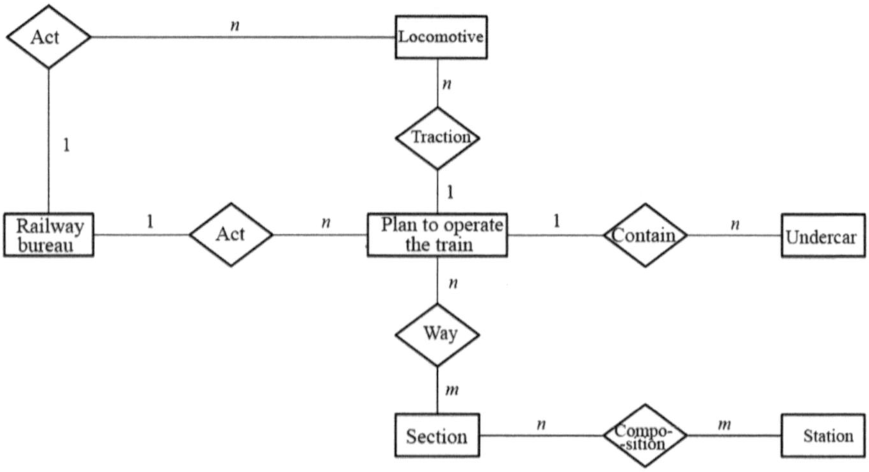

Fig. 9.6 Part of the E-R diagram in database (2)

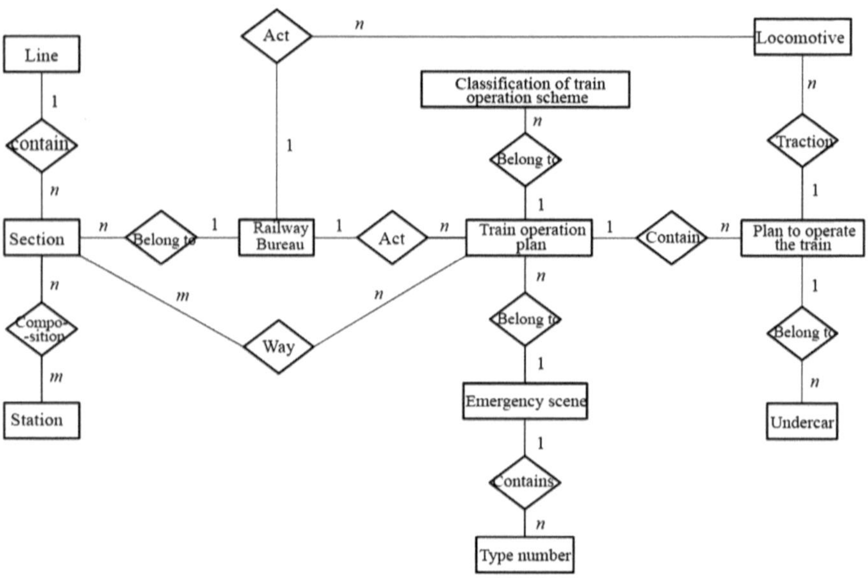

Fig. 9.7 Part of the E-R diagram in database (3)

9.2.2 *Logic Design*

The logical structure of the high-speed railway train operation decision support system under emergencies is shown in Fig. 9.8.

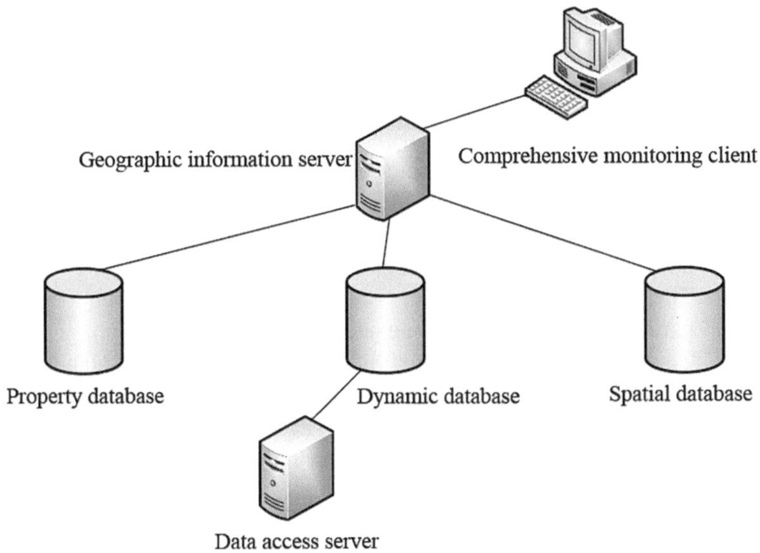

Fig. 9.8 Logical structure of high-speed railway train operation decision support system under emergency conditions

The high-speed railway train operation decision support system under the emergency conditions is mainly divided into attribute database, spatial database and dynamic database according to the logical structure [2, 3].

1. Attribute database

   The attribute database provides attribute information of various business systems. The attribute database does not exist as an entity in the GIS platform of the high-speed railway train operation decision support system under emergency conditions, but exists in a reference manner, and data fusion and sharing are achieved by referring to various other business systems. A large amount of attribute data is stored and shared in a database. The physical existence of the attribute database depends on the physical existence of each business database. Therefore, different user names/passwords are used to control access to different business databases. The attribute database includes information such as emergency plan information, line plan information, and so on [4].

   (1) Emergency plan information

   The emergency plan information includes the plan basic information table, the plan template information table, the plan template node table, the plan evaluation information table, the disposal plan basic table, the case basic information table, the expert basic information table, the expert resume information table, the rescue organization information table, and the rescue organization personnel table, the distribution table of resources and materials, the information sheet of the responsible person for relief materials, the list of relief materials, the list of rescue equipment, and the basic table of the regulations. The specific contents are as follows.

(1) The basic information table of the plan consists of the plan number, the plan name, the plan category number, the abstract, the preparer, the preparation time, the contingency plan content XML, the plan content, and the plan type.
(2) The plan template information table consists of the model number, model name, plan model type, compiler, and compilation time.
(3) The plan template node table consists of the model number, node number, parent node number, node name, and description.
(4) The plan evaluation information form is composed of a plan number, an expert number, an evaluation content, and a score.
(5) The basis of the disposal plan consists of the disposal plan number, the disposal plan name, the disposal plan category number, the disposal plan level number, the summary, the compiler, the preparation time, the plan text, and the plan storage path.
(6) The case basic information table consists of case number, case name, case type number, case level number, case implementation organization, case content, and case structured storage.
(7) The expert basic information table consists of expert number, expert name, expert type, gender, date of birth, ethnicity, work unit, professional title, special expertise in emergency response, professional, mobile phone, office phone, fax, E-mail, home phone, Home address, working hours, ID number, highest education, graduate school, department in charge, major, correspondence address, confirmation time, and remarks.
(8) The expert resume information form consists of the plan number, expert number, photograph, work resume, and remarks.
(9) The rescue organization information form consists of the organization number, organization name, organization type number, organization level number, location, and administrator.
(10) The personnel list of the rescue organization consists of the personnel number, organization number, job category, name, mobile phone, and office phone.
(11) The distribution table of resources and materials reserves consists of the stock number, name, type, competent department, level, location, person in charge, contact number, stock item number, storage capacity, storage unit, watch phone, and profile.
(12) The information sheet of the person in charge of relief materials consists of the personnel number, name, mobile phone, office phone number, and organization number.
(13) The relief materials list consists of relief material number, reserve number, material name, unit of measurement, and quantity.

(14) The rescue equipment schedule consists of equipment number, equipment name, equipment quantity, equipment type, and competent department.

(15) The basic table of the regulatory library consists of the regulatory number, the name of the regulation, the type of the regulation, the number of the regulation level, the preparation unit, the preparation time, and the details of the regulations.

(2) Line plan information

The line plan information includes line plan and operation condition association table, train formation information table, line plan basic information table, network topology structure information table, model algorithm type table, model algorithm basic information table, index system table, index weight table, unit target basic information table, unit target weight information table, comprehensive target information table, constraint basic information table, line plan evaluation result information table, line plan indicator expert evaluation form, train operation plan basic table, train distribution information table, train operation plan evaluation form, the passenger flow integration table, the passenger flow fluctuation coefficient table, and the path information table. The specific contents are as follows.

(1) The line plan and operation condition association table consists of the line plan classification and classification number, operation condition category, operation condition level, and remarks.

(2) The train formation information table consists of train type grouping number, group number, vehicle type, number of personnel, and remarks.

(3) The basic information table of the line plan consists of the line plan number, the line plan classification number, the originating station, the final arrival station, the stop number list, the number of trains, and the remarks.

(4) The network topology information table consists of the arc segment number, starting point, end point, length (km), nature, belonging line, line level, road network passing capacity, connectivity, traction type, and remarks.

(5) The model algorithm type table consists of the model algorithm type number, type name, and description.

(6) The model algorithm basic information table is composed of a model algorithm number, a method name, a model algorithm type number, a corresponding method name, and a description.

(7) The indicator system table consists of the indicator system number, indicator number, parent node number, indicator name, and description.

(8) The indicator weight table consists of the indicator number, the operating environment category, the operating environment level, the indicator name, the indicator weight, and the description.

(9) The unit target basic information table is composed of a unit target number, a corresponding stored procedure name, a name, and a description.
(10) The unit target weight information table is composed of a unit target number, an operation condition category, an operation condition level, a weight, and a description.
(11) The comprehensive target information table is composed of a comprehensive target number, a corresponding stored procedure name, a name, and a description.
(12) The constraint basic information table consists of a number, a name, a stored procedure, a category, and a description.
(13) The line plan evaluation result information table is composed of the line plan number, the evaluation method number, the quantitative evaluation score, and the qualitative evaluation result.
(14) The line plan indicator expert evaluation form consists of the line plan number, indicator number, expert number and score.
(15) The basic schedule of the train operation plan consists of the job number, start time, end time, start mileage, end mileage, occupied track, and occupied switch.
(16) The train distribution information table consists of the station number, type, train type group number, number of idle trains, and time.
(17) The train operation plan evaluation form consists of the train operation plan number, the train operation plan evaluation plan number, the quantitative evaluation score, and the qualitative evaluation result.
(18) The passenger flow integration table consists of the originating station number, the arrival station number, the number of passenger flows, the nature of the passenger flow, and the time.
(19) The passenger flow fluctuation coefficient table consists of the number, the operating condition category, the operating condition level, the passenger flow fluctuation coefficient value, and the description.
(20) The path information table consists of the path number, the path level, the starting station number, the final station number, the path including the arc number, and the path including the station number and the path attribute.

2. Spatial database

Spatial data is stored using an Oracle database and uses SDE's spatial data engine. In order to control the rights of the geographic information database, different user names/passwords are designed for different types of users. The spatial database stores various 2D geographic information data, including basic vector libraries and infrastructure information.

(1) Basic Vector Gallery

The basic vector gallery covers the national road network line table, power line information table, telephone pole information table, national station table, road bureau management boundary table, provincial area, provincial administrative center, regional area, regional administrative center, county area, county administrative center, township administrative centers, villages, contour lines, elevation points, vegetation, geology, water systems, and roads. The specific contents are as follows.

(1) The national road network line table consists of the line code, the name of the line, the starting point of the line, the ending point of the line, the starting point of the line, the ending city of the line, and the remarks.
(2) The power line information table consists of code, voltage, and remarks.
(3) The telephone pole information table consists of code, location, and remarks.
(4) The national station list consists of station code, station name, center mileage, and remarks.
(5) The road bureau management boundary table consists of the administrative code of the road bureau, the name of the road bureau, the administrative location, and the remarks.
(6) The provincial area consists of administrative code, name, level code, and remarks.
(7) The provincial administrative center consists of administrative code, name, level code, and remarks.
(8) The regional area is composed of administrative code, name, level code, and remarks.
(9) The regional administrative center consists of administrative code, name, level code, and remarks.
(10) The county are is composed of administrative codes, names, level codes, and remarks.
(11) The county administrative center consists of administrative code, name, level code, and remarks.
(12) The township administrative center consists of administrative code, name, level code, and remarks.
(13) The village consists of an administrative code, a name, a level code, and a remark.
(14) The contour line consists of elevation values and remarks.
(15) The elevation point consists of elevation point coding and elevation values.
(16) Vegetation consists of vegetation codes, forests, grasslands, urban green spaces, others, vegetation names, descriptive information, and remarks.
(17) Geology consists of geological code, geological type code, geological level code, geological name, description information, and remarks.

(18) The water system consists of water code, water type code, water grade code, water system name, description information, and remarks.
(19) The road consists of road code, road type code, road grade code, road name, description information, and remarks.

(2) Infrastructure information

The infrastructure information includes line data table, curve data table, slope information table, section information table, switch information table, track information table, bridge information table, tunnel information table, signal information table, station information table, hub information table, contact line table, platform information table, station building information table, EMU depot information table, video equipment information table, EMU information table, risk source basic table, risk source monitoring basic table, substation information table, and station and line association table. The specific content is as follows.

(1) The line data table consists of line number, line name, associated road station number, starting mileage, line level, direction, ending mileage, line maximum speed limit, line type, signal system, and remarks.
(2) The curve data table consists of the curve number, its own line number, curve radius, starting point mileage, end point mileage, and remarks.
(3) The slope information table consists of the slope number, the line number, the slope, the location, and the remarks.
(4) The section information table is composed of a section number, a section name, an associated line number, a starting mileage, an ending mileage, and a remark.
(5) The switch information table consists of the number, the number of the station, the line number, the type number, the center mileage, the description information, and the remarks.
(6) The track information table consists of the route number, the line number, the station or segment number, the type of use, the starting mileage, the ending mileage, and the remarks.
(7) The bridge information table consists of the bridge number, name, line number, bridge type number, center mileage, bridge structure information, and remarks.
(8) The tunnel information table consists of the tunnel number, name, line number, type number, center mileage, description information, and remarks.
(9) The signal information table consists of the signal number, the signal name, the line number, the type number, the center mileage, the description information, and the remarks.
(10) The station information table consists of the station number, name, type, road station number, affiliated hub, connected train segment, pick-up capability, departure capacity, receiving passenger capacity, sending passenger capacity, center mileage, station level, and remarks.

(11) The hub information table consists of a hub number, a hub name, and a remark.
(12) The contact line list consists of the contact line number, the associated hub, the starting station, the final arrival station, and the remarks.
(13) The station information table consists of the station number, station name, station number, location, and remarks.
(14) The station building information table in the station consists of the building number, building name, station number, location, and remarks.
(15) The EMU depot information table consists of the name of EMU depot/station, the number of EMU depot, the number of connection points, the section number of the section, the operation mode of the EMU depot, the operation time, the use of each yard of the EMU depot, the number of tracks, the effective length, the number of cars in the EMU depot, and remarks.
(16) The video device information table consists of the device number, the station number, the line number, the device type, the web access URL, the web access port, the location, and the remarks.
(17) The EMU information table consists of vehicle number, vehicle type, capacity, self-weight, whether it is sleeping car, number of groups, maximum speed limit, average speed, and remarks.
(18) The risk source base table consists of risk source code, risk source name, type code, level code, longitude, latitude, range, description information, and remarks.
(19) The risk source monitoring basic table consists of risk source coding, type coding, warning level code, alarm level code, description information, and remarks.
(20) The substation information table consists of code, location, and remarks.
(21) The station and line association table consists of line id, line name, station sequence number, station id, station and section name, and remarks.

3. Dynamic database

Dynamic data resources are stored in the Oracle database, using the username/password access control method. The dynamic database includes real-time access to special operating conditions information, climate and meteorological information, train dynamic information, and passenger flow dynamic information. Special operating condition information, train dynamic information, and passenger flow dynamic information are obtained through the railway operation management data platform, and climate and meteorological information is mainly obtained by importing external meteorological data.

(1) Special operating conditions information
The special operating conditions information covers the fault real-time table, the geological real-time table, and the risk source real-time table. The specific contents are as follows.

(1) The fault real-time table consists of the fault number, fault type number, fault level number, fault name, level fault description, time, longitude, latitude, and remarks.
(2) The geological real-time table consists of type number, level number, degree of influence, range of influence, status, description, longitude, latitude, time, and remarks.
(3) The risk source real-time table consists of the risk source number, monitoring type number, level number, risk source warning alarm status, influence range, description, longitude, latitude, time, and remarks.

(2) Climate and meteorological information

The climate and meteorological information covers the meteorological real-time table. The meteorological real-time table consists of type number, level number, influence degree, influence range, status, description, longitude, latitude, time, and remarks.

(3) Train dynamic information

The train dynamic information covers real-time basic information tables of large-scale events, real-time tables of external systems, real-time tables, and real-time tables of stations. The specific contents are as follows.

(1) The real-time basic information table for large-scale events consists of station number, activity number, activity name, start time, end time, and remarks.
(2) The external system real-time table consists of the external system number, external system name, system status, time, and remarks.
(3) The real-time table consists of line number, fault type number, fault level number, fault name, level fault description, time, longitude, latitude, and remarks.
(4) The station real-time table consists of the fault number, fault type number, fault level number, fault name, level fault description, time, longitude, latitude, and remarks.

(4) Passenger flow dynamic information

The passenger flow dynamic information covers the real-time table of passenger flow information. The real-time table of passenger flow information consists of the originating station number, the arrival station number, the number of passengers, the nature of the passenger flow, the time, and the remarks.

9.3 High-Speed Railway Special Operating Conditions Acquisition and Generation Subsystem

9.3.1 System Requirements Analysis

Under special operating conditions, high-speed railway dispatch decision-makers need to improve the event information as the data basis for driving plans in special operating scenarios [4]. The main goal of the special operating conditions acquisition and generation subsystem of high-speed railway is to integrate various information, such as emergency information and road network information, and display the space in three dimensions through the GIS platform.

In summary, the requirements for the acquisition and generation of special operating conditions for high-speed railways mainly include GIS map operation and display, road network information query and display, special operational scenario generation, and rescue resource information query and display, as shown in Fig. 9.9.

1. GIS map operation and display requirements

Based on the map of China and the important railway route map based on the Beijing-Shanghai high-speed rail line, a variety of map tools are used to comprehensively display the map information. It provides basic operations such as zooming in, zooming out, and eagle eye, and displaying the entire road network information, administrative division information, station information, and operating environment information in a vector diagram.

2. Road network information query and display requirements

It can be explained from the following three aspects.

(1) In the aspect of line information inquiry and display, the user can query the basic information of the line, the schematic diagram of the line, the information of the station, the distribution information of the railway department, the information of the road station, the information of the city, the information of the city rescue resources, the information of the section, the information of the bridge section and shows it on the GIS map.
(2) In the information inquiry and display of the station, the user can obtain the relevant information of the queried station. For example, the basic information of the station, the schematic diagram of the station, the information of the rescued resources, the information of the railway department, the information of the arrival and departure train, and the information of the station passing line, train station information, bus station information, public transportation information, which can be displayed on the GIS map.
(3) In the information inquiry and display of the railway bureau, the user can query the basic information of the railway bureau, the information of the station and section in the railway bureau, the information of the line, the information of the signal device in the railway bureau, and display it on the GIS map.

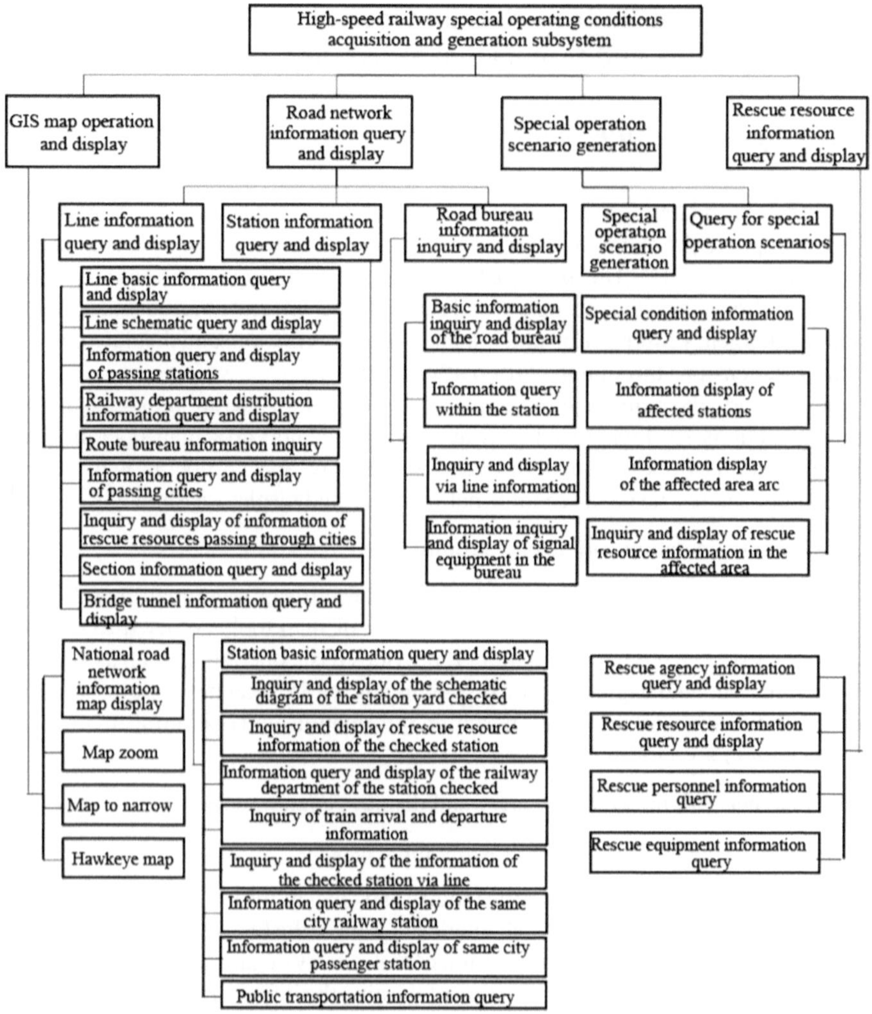

Fig. 9.9 High-speed railway special operating conditions acquisition and generation subsystem requirements map

3. Special operational scenario generation requirements

By providing special condition information, for the station, line and section information in the affected area, the manual input function is provided according to the classified special operation scenarios, so as to realize the query of special condition information, the station information, arc segment information and rescue resource information in the affected area, and display it on the GIS map.

4. Rescue resource information query and display requirements

Realize the inquiry of rescue resource information including rescue agency information, rescue resource information, rescue personnel information, rescue equipment information, and display the information of the rescue organization on the GIS map.

9.3.2 System Structure Design

The functional structure of the special operating conditions acquisition and generation subsystem of high-speed railway is shown in Fig. 9.10. It is mainly divided into seven modules, including railway network information module, railway network video module, passenger flow distribution module, railway network disaster prevention meteorological module, special condition module, administrative division module, rescue resource module. The railway network information module, the railway network video module, the passenger flow distribution module, and the railway network disaster prevention meteorological module provide the function of querying and displaying railway network information (railway network basic information, railway network video information, passenger flow information, weather information) based on the GIS map; the special condition module and the administrative division module support the acquisition of the special condition information and the data conversion input, by drawing the affected area on the map and combining the event, weather, passenger flow and other information to generate a special operation scene, realizing the query and display function under special condition based on the GIS map (the query methods include type-level query, time-based query, and administrative division query [5–8]); rescue resource module provides information

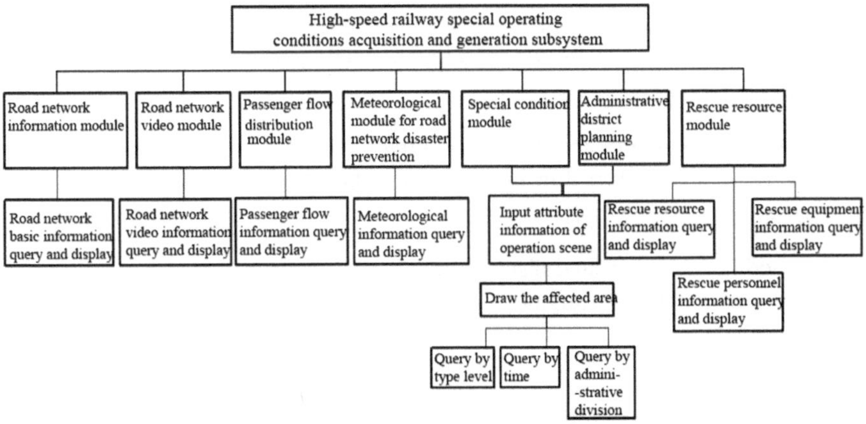

Fig. 9.10 High-speed railway special operating conditions acquisition and generation subsystem functional structure

retrieval and display of rescue resources (rescue resources, rescue personnel, rescue equipment) based on GIS map.

1. Railway network information module

The module includes line information, station information, and railway bureau information. The line information is mainly used to query the line level, the maximum speed limit, the signal system, and other information through the query methods of the railway bureau and the station; the station information is mainly used to query the center mileage and railway bureau information through the query methods of the railway bureau and the city; the railway bureau information mainly refers to query the telephone and address information by the name of the railway bureau.

2. Railway network video module

The module can realize real-time video screen monitoring of each railway bureau, line, station section, and station for inquiry.

3. Passenger flow distribution module

The module can realize the query and display of the railway network passenger flow, comprehensively analyze the railway network passenger flow according to category and type, and display it in the form of a list and a chart.

4. Railway network disaster prevention meteorological module

The module can realize the query and display of the meteorological information of the whole railway network and provides the basis for the operational decision analysis personnel according to the meteorological conditions.

5. Special condition module

The module can automatically determine the special condition level according to the category and type of the special conditions currently occurring, generate map information, display it in the form of GIS and can be related to the railway bureau, line, section, station, and other information affected by it.

6. Administrative division module

The module can query and display information such as line information, station information, and passenger flow information in the city.

7. Rescue resource module

The module can query and display the information of the entire network rescue resources, personnel and equipment.

9.3.3 System Interface

The high-speed railway special operating conditions acquisition and generation subsystem is based on the map of China and the Beijing-Shanghai high-speed rail and other important railway routes. The rescue resource information, railway network information, special condition information, etc., are displayed in the form of vector diagrams and schematic diagrams, as shown in Figure 9.11. At the same time, it supports the basic operations of the map, including zooming in, zooming out, eagle eye, previous view, back view, etc. [9].

The interface of the system is mainly a comprehensive display of line information, station information, signal information, passenger flow information, operation information (meteorological information, geological information), which can realize real-time operational information (meteorological information, geological information) and online status information of external systems display; display station, line, and equipment information in the form of GIS; display station information, line information, equipment information along the line, real-time information of each monitoring indicator, real-time information of operating environment, and online status information of each subsystem in the form of graphs and lists.

1. Railway network information module
(1) Line information

The line information is divided into two ways: search by railway bureau and search by station. According to the railway bureau search, you need to select the nature of the line and the railway bureau to determine the line number and display it in the table below. According to the station search, you need to select the province and city information and display the line in the table below. Double-click on the line to view the details, as shown in Fig. 9.12. Line details include 12 function buttons, and the information window can be minimized though the "Minimize" button 

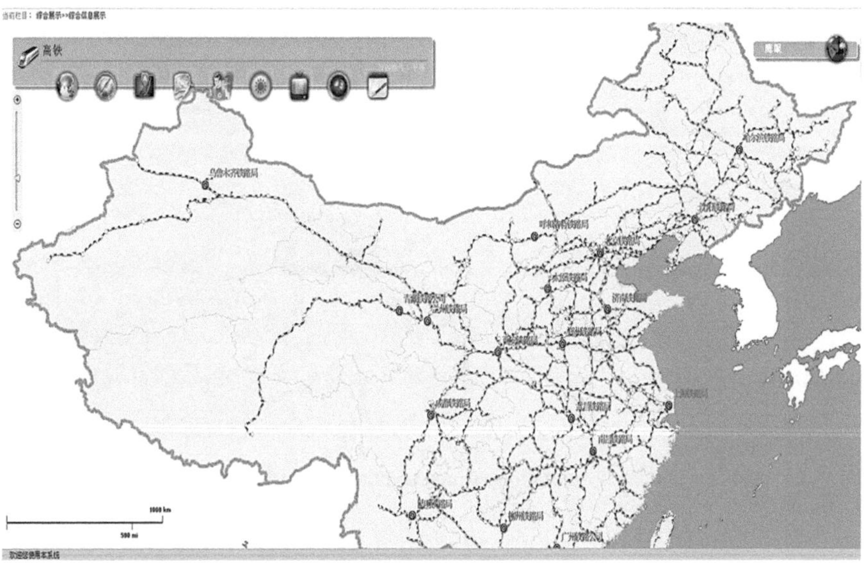

Fig. 9.11 High-speed railway special operating conditions acquisition and generation subsystem interface

Fig. 9.12 High-speed railway special operating conditions acquisition and generation subsystem line information interface

(2) Station information

The station information can be displayed in the following list by the way of the railway bureau or by city query, as shown in Fig. 9.13.

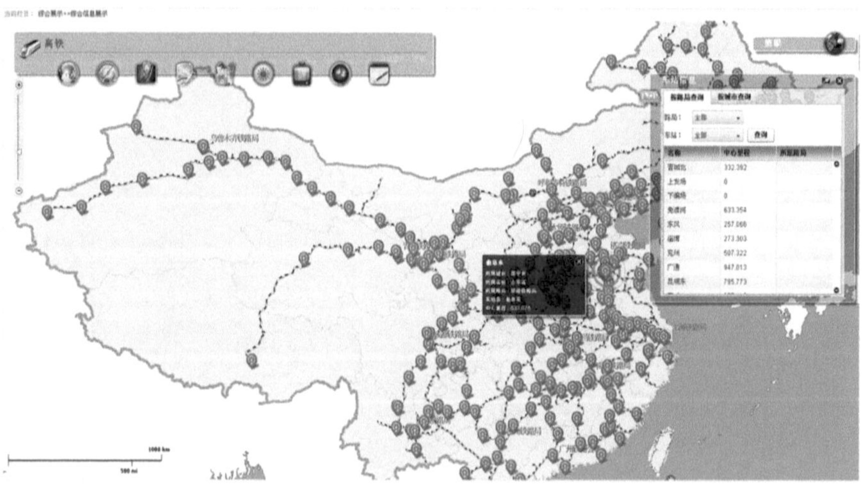

Fig. 9.13 High-speed railway special operating conditions acquisition and generation subsystem station information interface

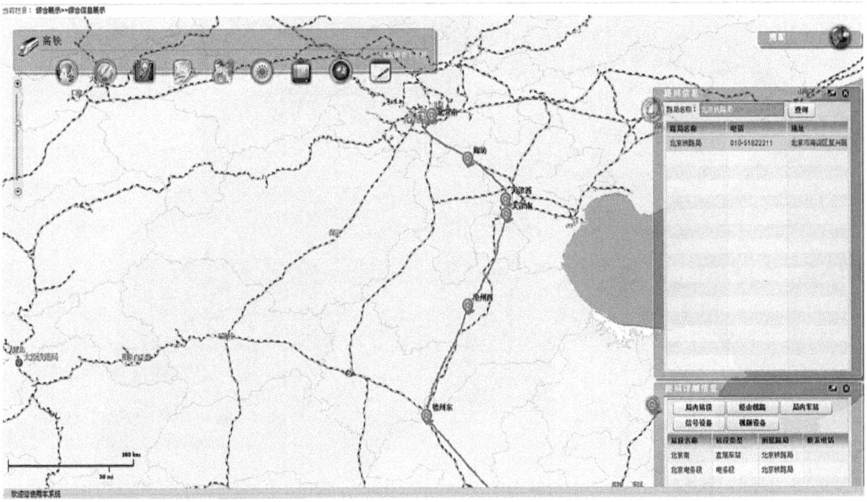

Fig. 9.14 High-speed railway special operating conditions acquisition and generation subsystem railway bureau information interface

(3) Railway bureau information

The railway bureau information can be used to search for railway bureau information by searching for keywords and is listed in a list form. Double-click on a row to display the Railway Office Details window, as shown in Fig. 9.14. The station details include five function buttons.

2. Railway network video module

The railway network video module monitors the real-time video images of each bureau, line, station and depot, and station for future reference.

The railway network video can be queried according to the line and station, as well as the station and depot, to display the real-time video pictures along the line, as shown in Fig. 9.15.

Enter the line name and station name, and click the "query" button to display all videos Double-click a video to display video recording, as shown in Fig. 9.16. You can also enter the name of the railway station and the name of the station to query the video.

3. Passenger flow distribution module

The module is a query and display of the railway network passenger flow, and comprehensively analyzes the railway network passenger flow according to the category and type, and displays it in the form of a list and a chart.

In the "Passenger Flow Distribution" dialog box, you can choose to query by the arrival station or by administrative division, and display the queried passenger flow information in the form of a list or a histogram. Double-click a row in the list to view the details, as shown in Fig. 9.17. It includes detailed information, comparison

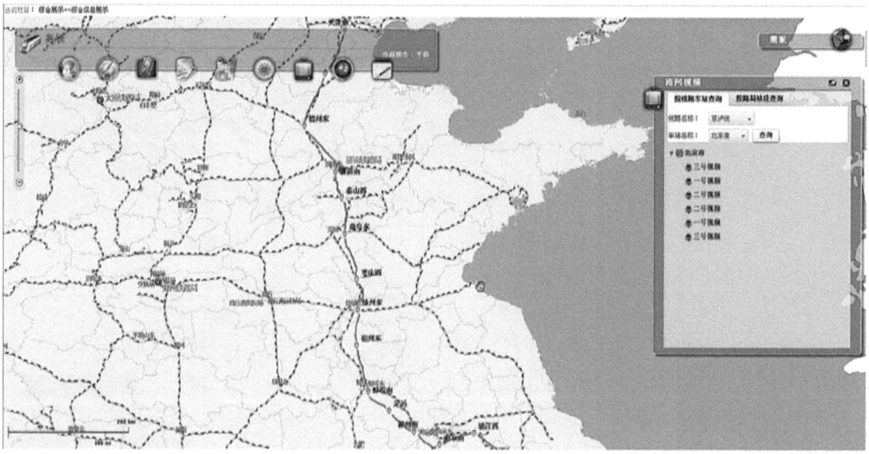

Fig. 9.15 High-speed railway special operating conditions acquisition and generation subsystem railway network video interface

Fig. 9.16 High-speed railway special operation condition acquisition and generation subsystem video query interface

of arrival and departure flow, comparison of received passenger flow, comparison of sent passenger flow, comparison of arrival passenger flow, and comparison of departure passenger flow.

4. Railway network disaster prevention meteorological module

The module queries and displays the meteorological information of the whole railway network and can display various meteorological information visually, as shown in Fig. 9.18, and provides basis for the operational decision analysis personnel according to meteorological conditions.

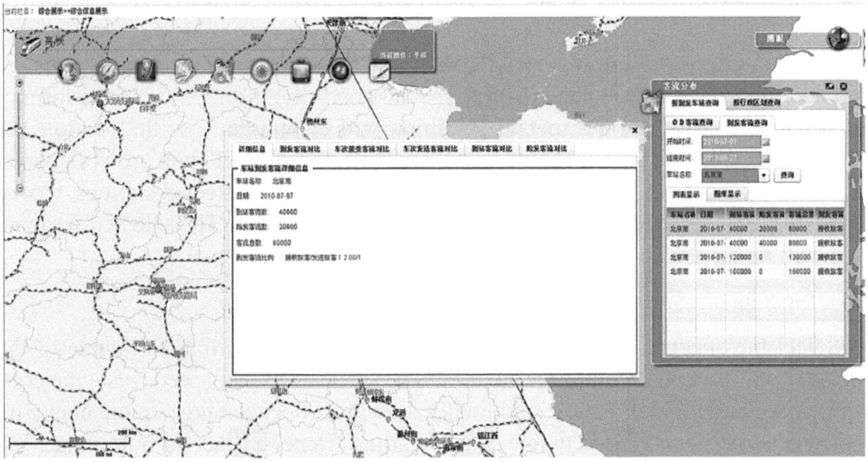

Fig. 9.17 High-speed railway special operating conditions acquisition and generation subsystem passenger flow distribution interface

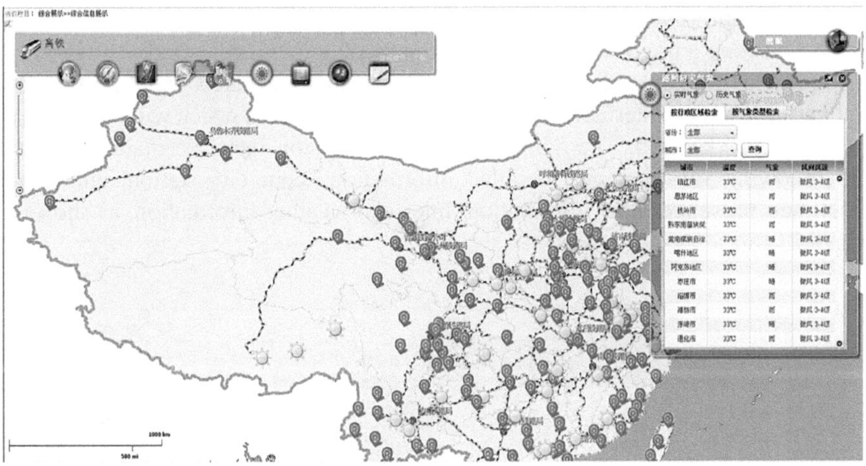

Fig. 9.18 High-speed railway special operation conditions acquisition and generation subsystem road network disaster prevention meteorological interface

5. Special condition module

According to the category and type of special conditions currently occurring, the system automatically determines the special condition level, generates map information, displays it in the form of GIS, and can be associated with the information of the railway bureau, line, depot, station, and so on. It mainly includes special condition modules and special condition generation modules.

(1) Special condition module

You can choose to query special conditions by time of occurrence, type level, and administrative division and display special condition information in the form of a list. Double-clicking on a special condition will display its specific location and information, and the "Special Condition Details" dialog box will pop up, as shown in Fig. 9.19. The details attribute types include affected areas, affected stations, rescue resources, and affected arcs, and support multiple selections. The selected attribute type will be displayed on the map.

(2) Special condition generation module

Select the special condition category and type, enter the time and name of the special condition, and determine its duration, sustained wind speed, instantaneous wind speed and wind direction, and finally draw its influence range on the map with a map drawing tool (Rectangle, Polygonal frame, Round box), and click the "Generate" button to generate it, as shown in Fig. 9.20.

6. Administrative division module

The administrative division module comprehensively displays the route information, station information, passenger flow information, etc., within the city.

By selecting a province and city, you can display the administrative divisions in maps and lists. Double-click on the selected administrative area, and the "Administrative sub-division interface" dialog box will pop up below, which will be displayed in blocks on the map. The administrative sub-division interface mainly includes the following sub-functions: detailed information, same-city station, emergency resources, passenger flow information, lines and weather information, as shown in Fig. 9.21.

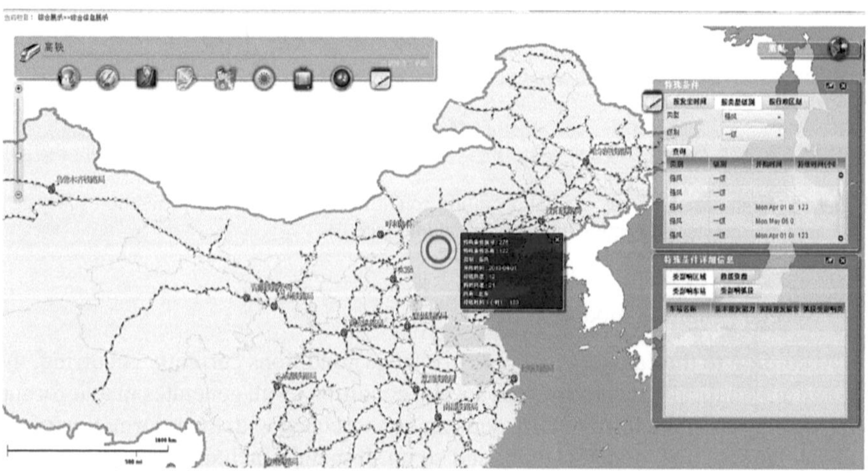

Fig. 9.19 Special condition interface for acquisition and generation subsystem of special operating conditions for high-speed railway

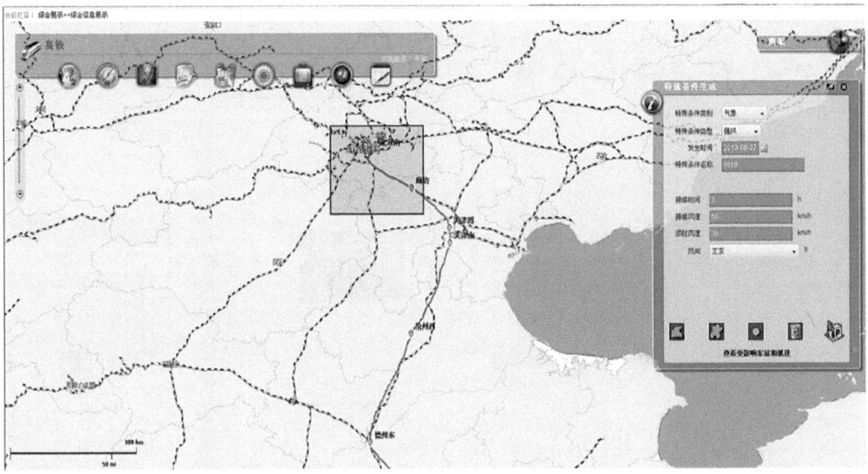

Fig. 9.20 High-speed railway special operating conditions acquisition and generation subsystem special condition generation interface

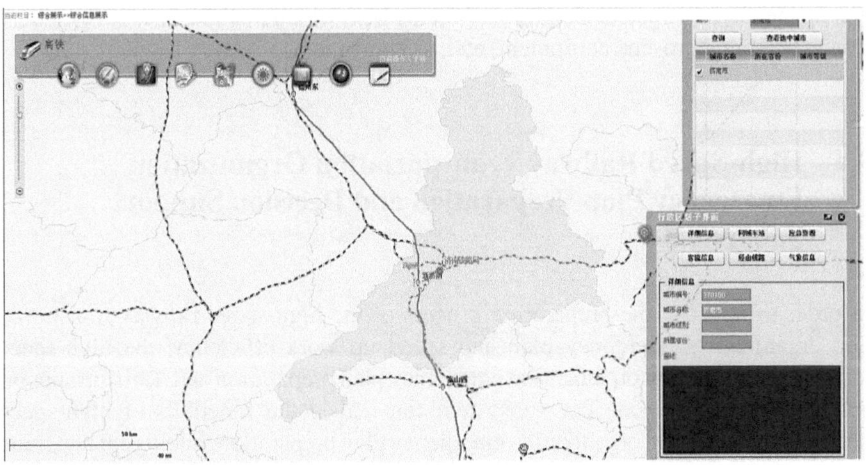

Fig. 9.21 High-speed railway special operating conditions acquisition and generation subsystem administrative division interface

7. Rescue resource module

The rescue resource module is a query and display of the rescue resources, personnel, and equipment information of the whole railway network.

You can choose to view the relief materials of the administrative division or the railway station and depot, and display them in the form of a list. Double-click on a row of records to display the specific information and location of the relief supplies, and

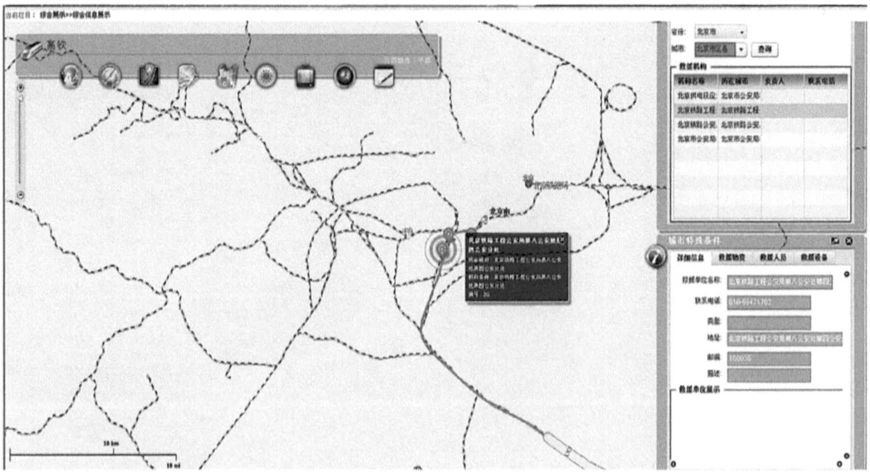

Fig. 9.22 High-speed railway special operating conditions acquisition and generation subsystem rescue resource interface

pop up the "Urban Special Conditions" dialog box, which includes rescue resources, rescue personnel, rescue equipment, etc., as shown in Fig. 9.22.

9.4 High-Speed Railway Transportation Organization Emergency Plan Preparation and Decision Support Subsystem

In order to improve the preparation process of the high-speed railway transportation organization emergency plan and speed up work efficiency, the high-speed railway transportation organization emergency plan preparation and decision support subsystem is developed. The subsystem can realize the functions of high-speed railway transportation organization emergency plan preparation, intelligent matching and generation emergency plan, and emergency plan decision support.

9.4.1 System Requirements Analysis

The contingency plan, also known as the emergency plan, is a pre-planned plan or scheme for ensuring prompt, orderly and effective emergency and rescue operations and reducing accident losses for possible accidents or disasters. The emergency plan is an action plan for the emergency rescue command organization, resource management, and disaster relief operations based on the identification of relevant hazards and the related impacts of the disaster. The railway emergency plan refers to a set of practical, rapid, effective, and orderly action plan to all kinds of emergency situations for the prevention and control of various emergencies in order to ensure the safety and smooth transportation of railways.

China's research on railway emergency plan management is still in the initial stage and most of the emergency plans are still stored in paper or electronic files in a decentralized manner. The content of the plan itself is professional and complicated, and there is no unified writing system, so the query and application are very inconvenient, resulting in responsiveness and emergency management far beyond the requirements of practical applications. At present, the main problems in the management of railway emergency plans in China are as follows.

(1) It is difficult to update and modify the plan. With the implementation of the plan, many problems exposed in the drills and trainings need to be adjusted and updated. The current plan storage and management mode will make the railway plan update work with a large manpower and material cost.
(2) The retrieval and inquiry of the plan are not easy. The plans for decentralized storage in various bureaus and departments are difficult to be uniformly consulted, and the storage in the form of printed texts is very difficult to retrieve other plans, which objectively isolates the generated plans.
(3) The sharing of the plan is poor. Based on the division of labor and unified linkage of the railway system, each plan will inevitably lead to functional cross-cutting, and the decentralized management of the plan is obviously not suitable for the unified configuration management of emergency site resources.
(4) Not conducive to the timely handling of emergencies. Due to the lack of good information sharing, the emergency command organization cannot quickly make judgments on which plans are most suitable for on-site rescue and related after-care measures.
(5) Difficulties in the study of the plan and the training of the plan. As a guide for the on-site staff, the plan requires the relevant personnel to be proficient. The format of the plan files is not uniform, the semantics are not clear, and the content conflicts, which makes the learning and training of the plan very difficult.

Therefore, the demand analysis for the preparation and management of high-speed railway emergency plans should focus on the digital generation and management of high-speed railway emergency plans, the unified data management in resource pools, the intelligent matching of high-speed railway emergency plans, and the generation of emergency response plan based on GIS-based spatial data technology.

According to the above four functional requirements, the subsystem should be able to realize the establishment of the emergency plan template, to the preparation of the emergency plan; from the query and retrieval of the emergency plan to the intelligent matching of the emergency plan; the dynamic establishment of the emergency disposal plan through the GIS spatial data technology and format the output process. The schematic diagram of the process is shown in Fig. 9.23.

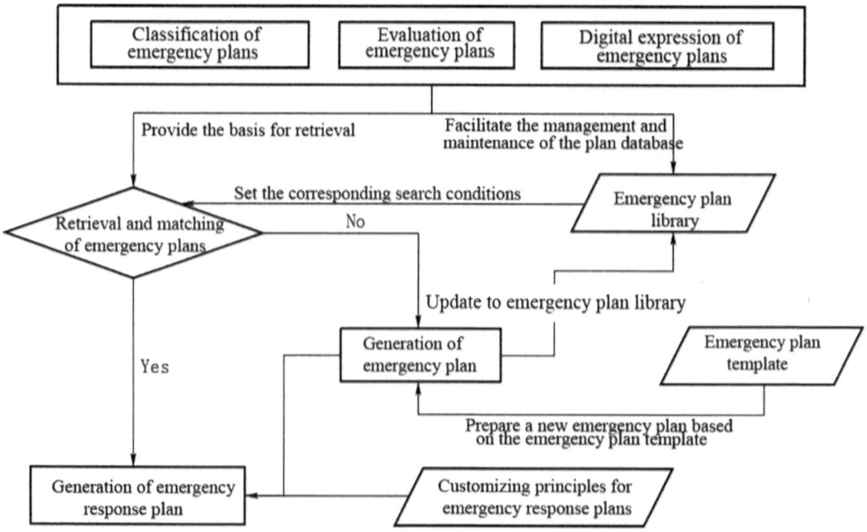

Fig. 9.23 Schematic diagram of the process of preparing emergency plan and decision support subsystem for high-speed railway transportation organization

9.4.2 System Structure Design

According to the demand analysis, the subsystem is designed with six modules, including emergency treatment deduction module, emergency plan management module, disposal plan management module, template management module, emergency rescue material management module, and case resource management module, as shown in Fig. 9.24. Through the collaborative interaction between the six modules, it is possible to quickly locate emergency plans and disposal plans that match the emergency.

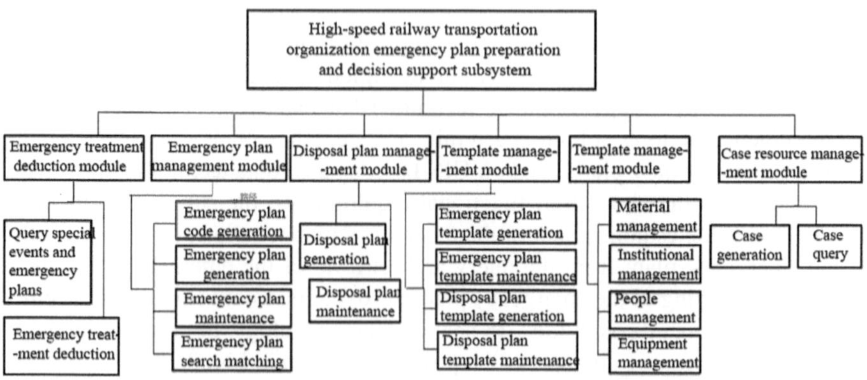

Fig. 9.24 High-speed railway transportation organization emergency plan preparation and decision support subsystem functional structure diagram

1. Emergency treatment deduction module

The module can complete the dynamic deduction of emergency response, realize special events and emergency plan inquiry, and emergency treatment deduction. By inputting the category, level and time of occurrence of special conditions, the corresponding emergency plan can be obtained. Further, the scope of the accident location and the location of the rescue organization can be displayed on the map, and the supply and demand data of the relief materials, equipment and personnel are displayed in a graphical form.

2. Emergency plan management module

The module can complete the digital display, generation, management, and retrieval matching of the emergency plan, realize the emergency plan code generation, emergency plan generation, emergency plan maintenance, and emergency plan search matching.

3. Disposal plan management module

The disposal plan management module can complete the generation and management of the disposal plan and realize the disposal plan generation and disposal plan maintenance.

4. Template management module

The module adds emergency plan and disposal plan while developing emergency plan template and disposal plan template and maintains the emergency plan and disposal plan. The module is divided into four sub-modules, namely emergency plan template generation, emergency plan template maintenance, disposal plan template generation, and disposal plan template maintenance.

5. Emergency rescue material management module

The module can complete the inquiry and management of emergency relief materials. The module is divided into four sub-modules: material management, organization management, personnel management, and equipment management.

6. Case Resource Management Module

The module can complete the generation of new cases and query and maintenance of the case. The module is divided into two sub-modules: case generation and case inquiry.

9.4.3 System Interface

The high-speed rail transport organization emergency plan preparation and decision support subsystem is a standard web application, also known as the B/S system. Its function is to realize the management of emergency plans and cases, and to realize the deduction of emergency response. The interfaces of the six modules are as follows.

1. Emergency treatment deduction module
(1) Query special events and emergency plans

The query special event interface is shown in Fig. 9.25.
The query emergency response interface is shown in Fig. 9.26.

(2) Emergency treatment deduction

The emergency treatment deduction interface is shown in Fig. 9.27.
The contact information of the emergency rescue organization is displayed below the emergency treatment deduction interface. When you click on the name of the

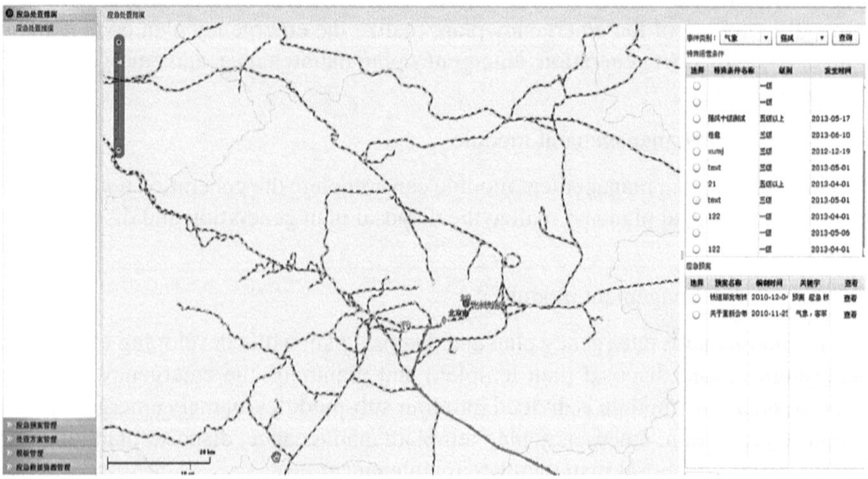

Fig. 9.25 Query special event interface

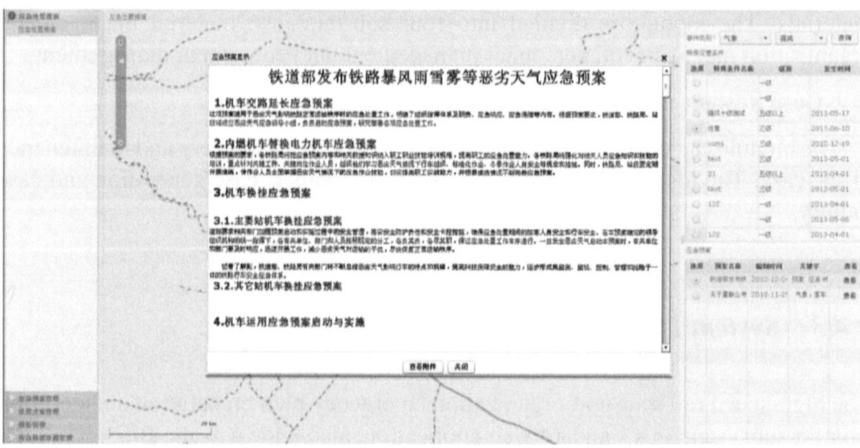

Fig. 9.26 Query emergency plan interface

Fig. 9.27 Emergency disposal deduction interface

emergency rescue organization, the geographical location and material possession information of the organization will be displayed on the map, as shown in Fig. 9.28.

2. Emergency plan management module

(1) Emergency plan code generation

The emergency plan code generation interface is shown in Fig. 9.29.

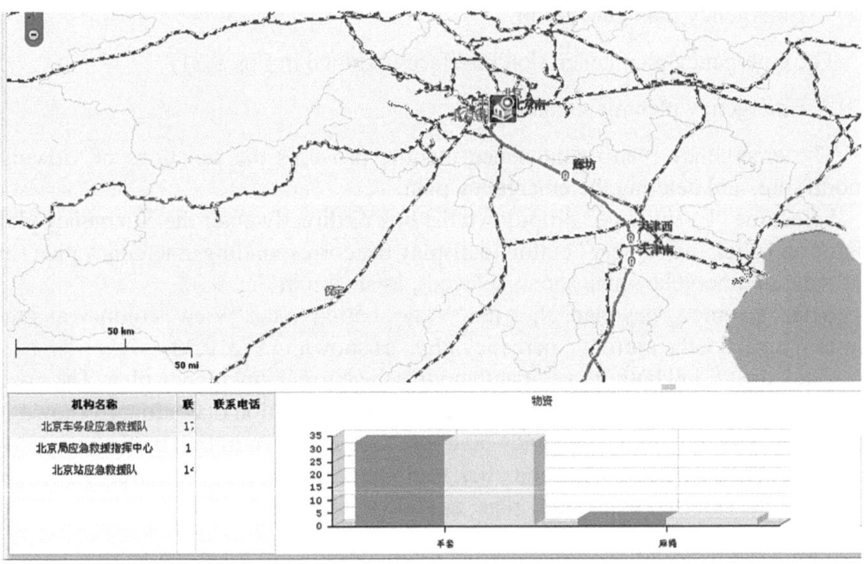

Fig. 9.28 Rescue agency location and material ownership interface

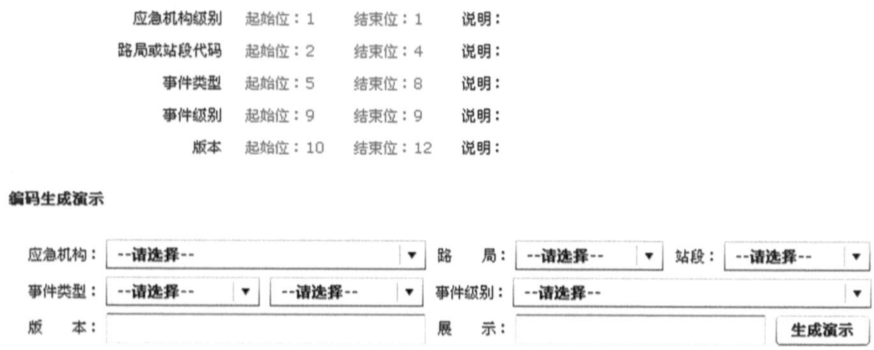

Fig. 9.29 Emergency plan code generation interface

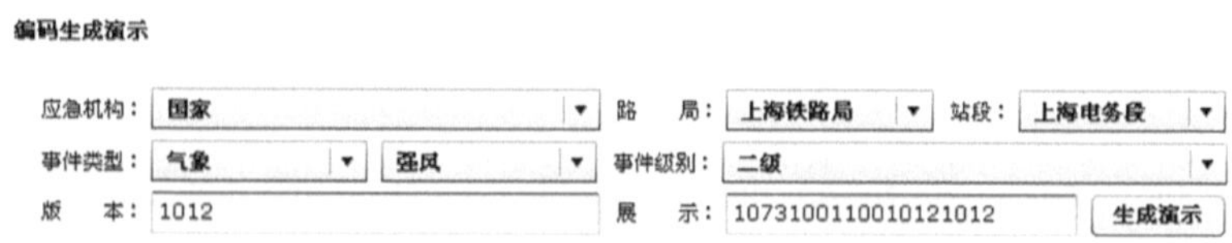

Fig. 9.30 Coding interface of the specific version of the emergency plan

Click the "Generate Demo" button to generate the code for the specific version of the emergency plan, as shown in Fig. 9.30.

(2) Emergency plan generation

The emergency plan generation interface is shown in Fig. 9.31.

(3) Emergency plan maintenance

The emergency plan maintenance module provides the functions of viewing, modifying, and deleting the emergency plan.

Select the "Event Type" drop-down list box or directly enter the emergency plan name and click the "Query" button to display the corresponding emergency plan (by default, all emergency plans are displayed), as shown in Fig. 9.32.

Select an emergency plan, click the "View" button in the "View" column, and the system displays the current emergency plan, as shown in Fig. 9.33.

Click the "Edit" button to edit and modify the current emergency plan. The operation is basically the same as the emergency plan generation operation, as shown in Fig. 9.34. The modified emergency plan interface is shown in Fig. 9.35.

Select the emergency plan and click the "Delete" button in the lower left corner to delete the selected emergency plan, as shown in Fig. 9.36.

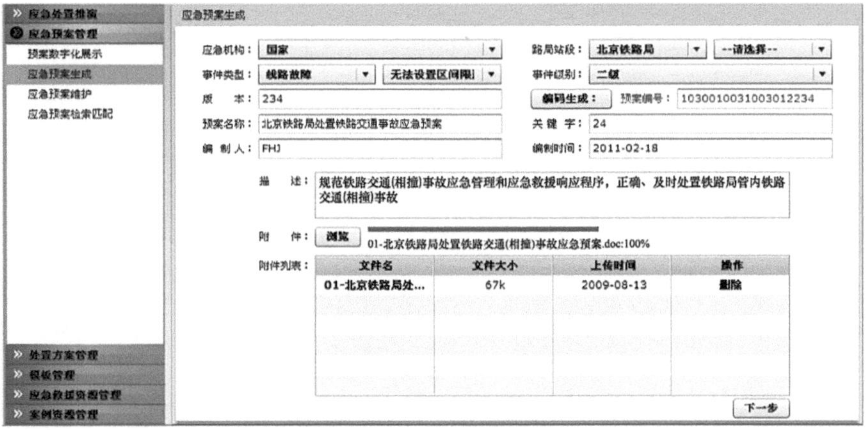

Fig. 9.31 Emergency plan generation interface

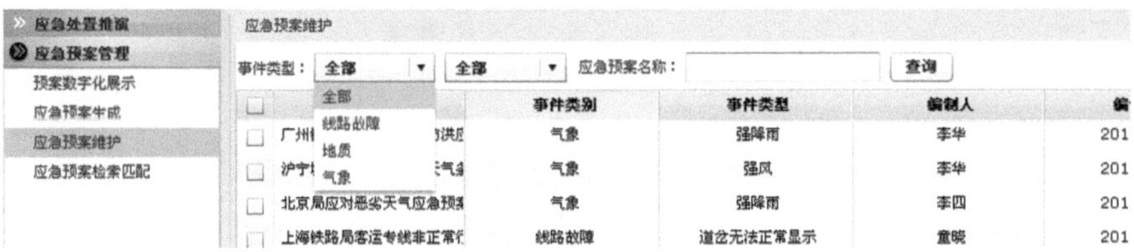

Fig. 9.32 Emergency plan query interface

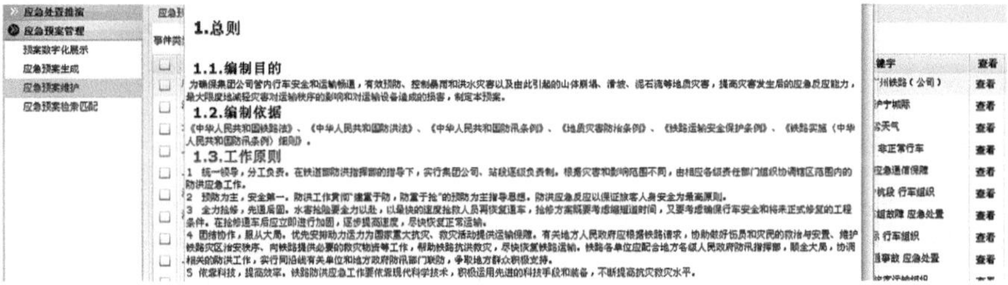

Fig. 9.33 Current contingency plan interface viewed

(4) Emergency plan search matching

The emergency plan search matching interface is shown in Fig. 9.37.

Select the target plan in the retrieved list and double-click to open the view, as shown in Fig. 9.38.

If necessary, you can also click the "View Attachments" button to find the source file of the emergency plan, as shown in Fig. 9.39.

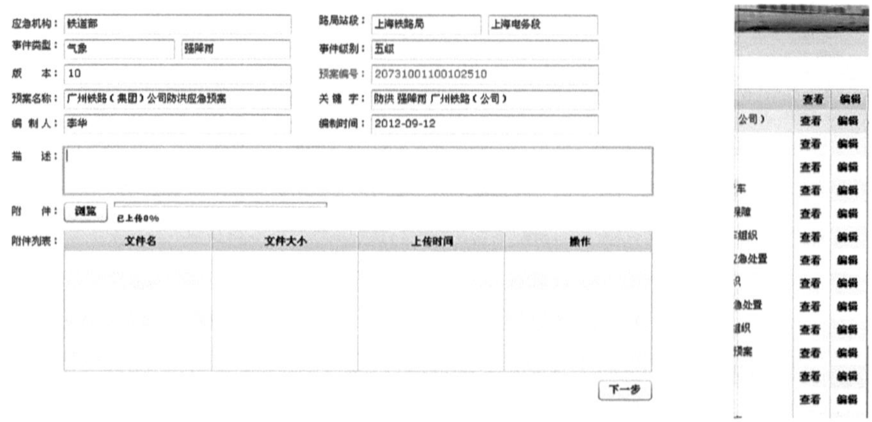

Fig. 9.34 Edit and modify the emergency plan interface

Fig. 9.35 Modified emergency plan interface

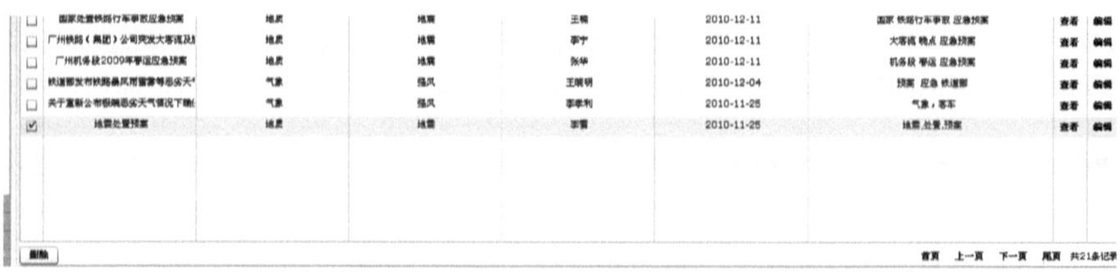

Fig. 9.36 Delete emergency plan interface

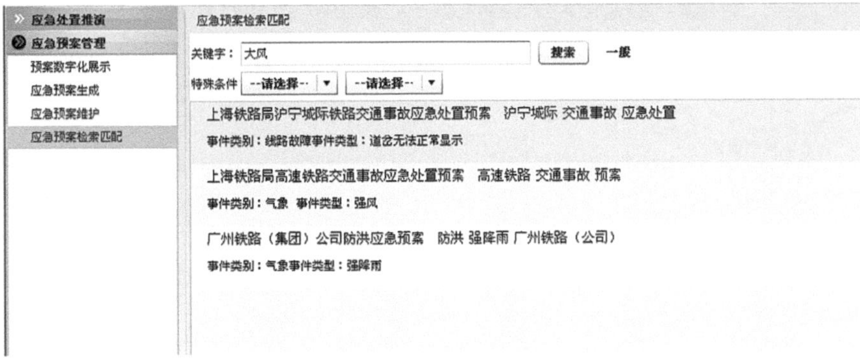

Fig. 9.37 Emergency plan search matching interface

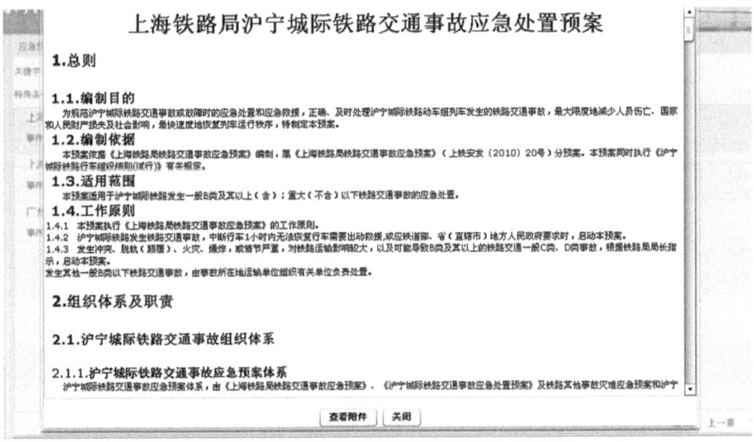

Fig. 9.38 View emergency plan interface

3. Disposal plan management module

The main function of the disposal plan management module is to generate a disposal plan and maintain and manage the generated disposal plan.

(1) Disposal plan generation

The disposal plan generation interface is shown in Fig. 9.40.

(2) Disposal plan maintenance

Select the "Event Type" drop-down list box or directly enter the name of the disposal plan, click the "Query" button to display the corresponding disposal plan (by default, all disposal plans are displayed), as shown in Fig. 9.41.

Select a disposal plan, click the "View" button in the "View" column, the system displays the current disposal plan, and view the disposal plan flowchart, as shown in Fig. 9.42.

Fig. 9.39 Source file interface of the emergency plan

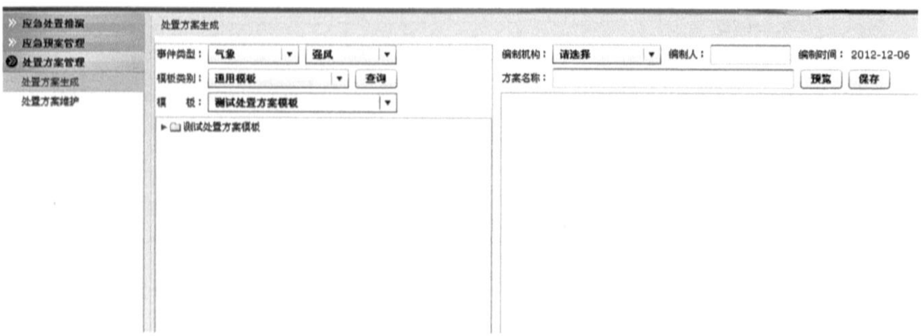

Fig. 9.40 Disposition plan generation interface

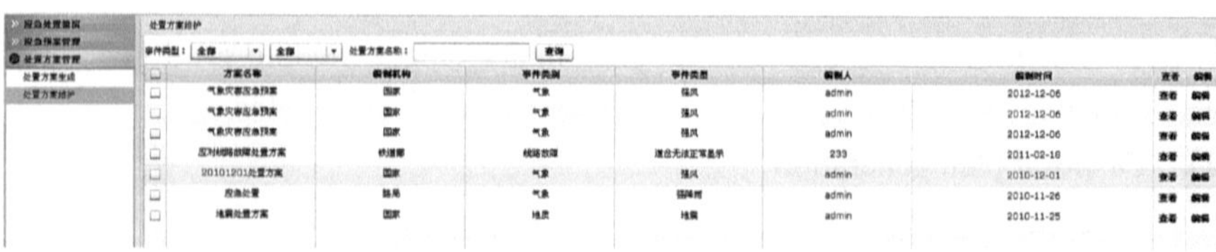

Fig. 9.41 Disposition plan query interface

Click the "Edit" button to edit and modify the current disposal plan. After that, the operation is basically the same as the disposal plan generation operation, as shown in Fig. 9.43.

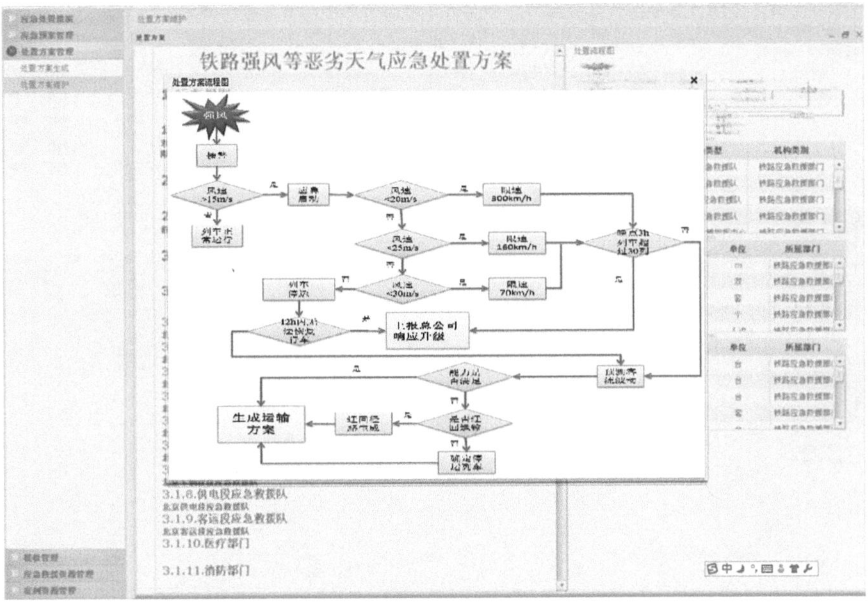

Fig. 9.42 Disposal plan view information interface

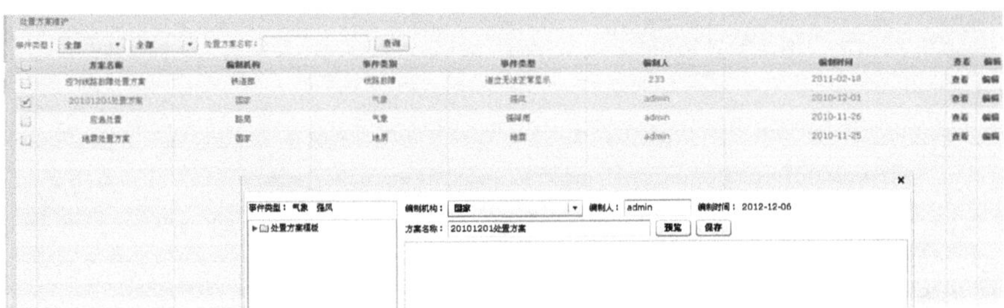

Fig. 9.43 Disposition plan editing interface

Select the emergency plan and click the "Delete" button in the lower left corner to delete the selected emergency plan, as shown in Fig. 9.44.

4. Template management module
(1) Emergency plan template generation

The emergency plan template generation interface is shown in Fig. 9.45.

Click the "Add Template" button, select "Add Subtitle," and edit the name, as shown in Fig. 9.46; click the "Rename" button to modify the name of the emergency plan template, as shown in Fig. 9.47.

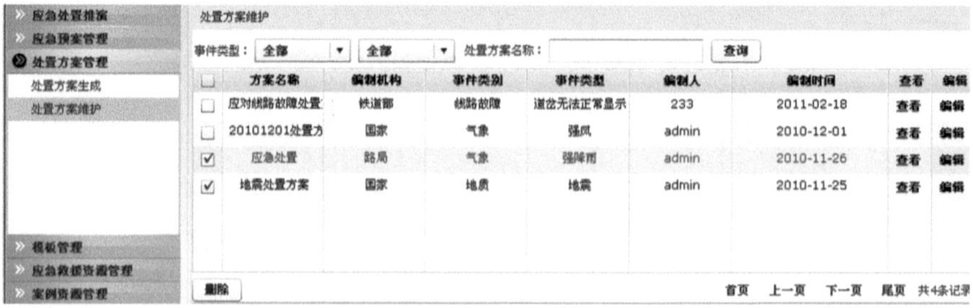

Fig. 9.44 Disposal plan deletion interface

Fig. 9.45 Emergency plan template generation interface

Fig. 9.46 Emergency plan template to add subtitle interface

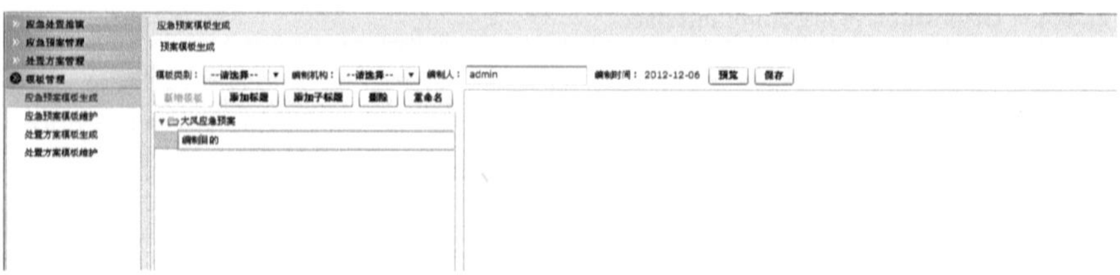

Fig. 9.47 Emergency plan template rename interface

Fig. 9.48 Emergency plan template maintenance interface

(2) Emergency plan template maintenance

Select the "Preview Template Category" and "Compilation Organization" drop-down list boxes, or select the "Time Range" prepared by the plan template, and click the "Query" button to display the corresponding emergency plan template, as shown in Fig. 9.48.

The emergency plan template can be maintained by clicking the "View" "Edit" or "Delete" button.

(3) Disposal plan template generation

Select the "Event Type," "Template Category," "Compilation" drop-down list box, enter the compiler and template name directly, select the "Disposal Plan Template Generation" entry on the left, and click the "Preview" button to generate the disposal plan template, as shown in Fig. 9.49.

(4) Disposal plan template maintenance

Select the "Disposition Scheme Template Category" and "Compilation Organization" drop-down list boxes, or select the "Time Range" for the planning template. Click the "Query" button to display the corresponding disposal plan template, as shown in Fig. 9.50.

You can maintain the disposition plan template by clicking the "View", "Edit" or "Delete" button.

Fig. 9.49 Disposition plan template preview generation interface

Fig. 9.50 Disposition plan template maintenance interface

5. Emergency rescue material management module
(1) Material management

Select the "Institution Type" and "Material Type" drop-down list boxes, or directly enter the material name and click the "Query" button to display the corresponding material reserve information, as shown in Fig. 9.51.

Click the "Detail" button to view the material reserve details, as shown in Fig. 9.52.

In the Device Details dialog, you can modify the material information and save it.

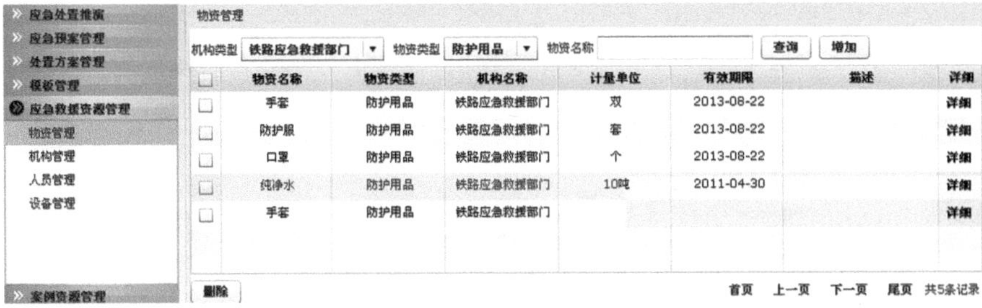

Fig. 9.51 Material management query interface

Fig. 9.52 Material reserve details interface

(2) Institutional management

Select the "Organization Type" drop-down list box, or directly enter the organization name and click the "Query" button to display the corresponding organization information, as shown in Fig. 9.53.

Click the "Detail" button to view the basic information and contact information of the rescue organization, and you can click the "Modify" button to maintain the information, as shown in Fig. 9.54.

Click the "Device" or "Materials" button to view the equipment or materials information of the organization, as shown in Fig. 9.55, and click the "Modify" button to update the information of the equipment or materials, as shown in Fig. 9.56.

(3) Personnel management

Enter the name of the organization or the name of the person directly, and click the "Query" button to search for the organization personnel, as shown in Fig. 9.57.

Click the "Detail" button to view the personnel details and contact information, as shown in Fig. 9.58.

In the "Personnel Information" dialog box, select the "Modify" button to modify the personnel information. After the editing is complete, click the "Save" button.

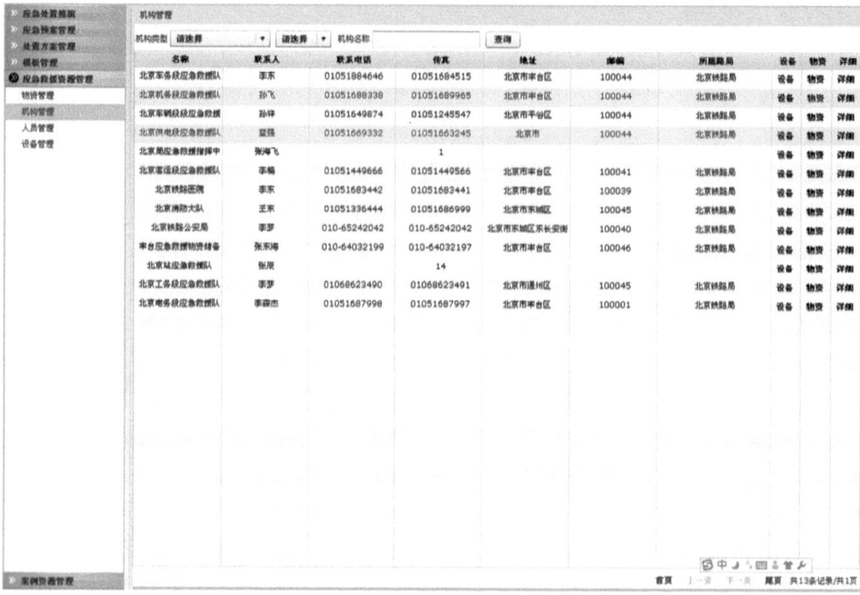

Fig. 9.53 Organization management query interface

Fig. 9.54 Organization basic information maintenance interface

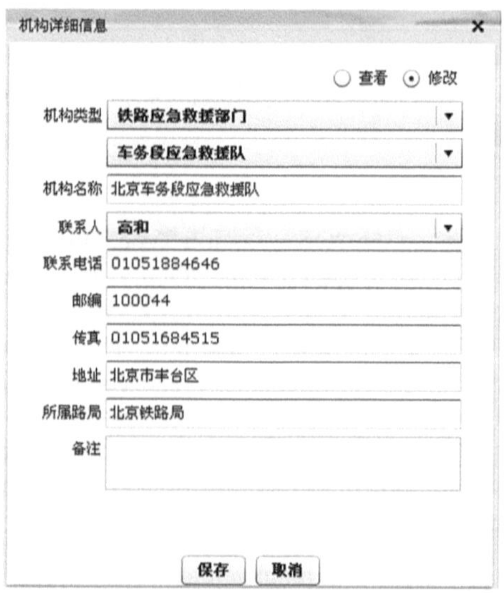

Click the “Add” button to add the personnel information. When the addition is complete, click the “Save” button, as shown in Fig. 9.59.

The person can be updated by clicking the “Delete” button, as shown in Fig. 9.60.

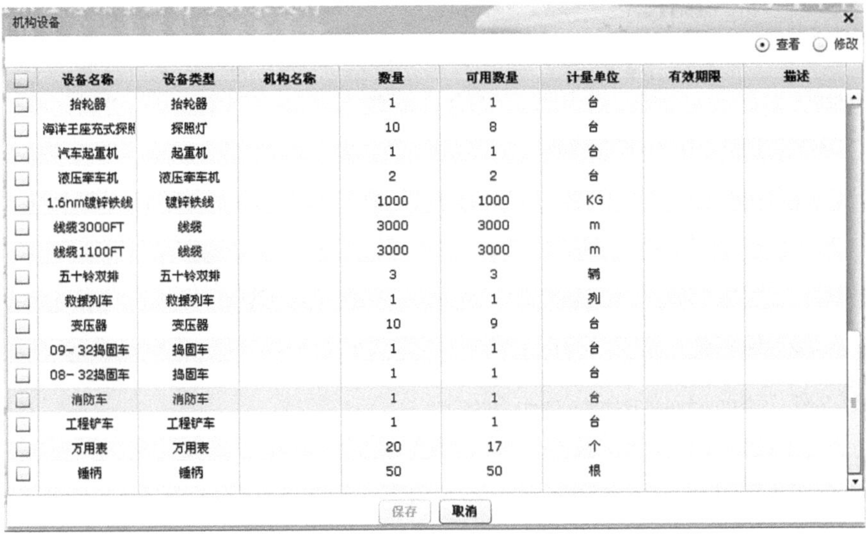

Fig. 9.55 Organization "equipment" "materials" query interface

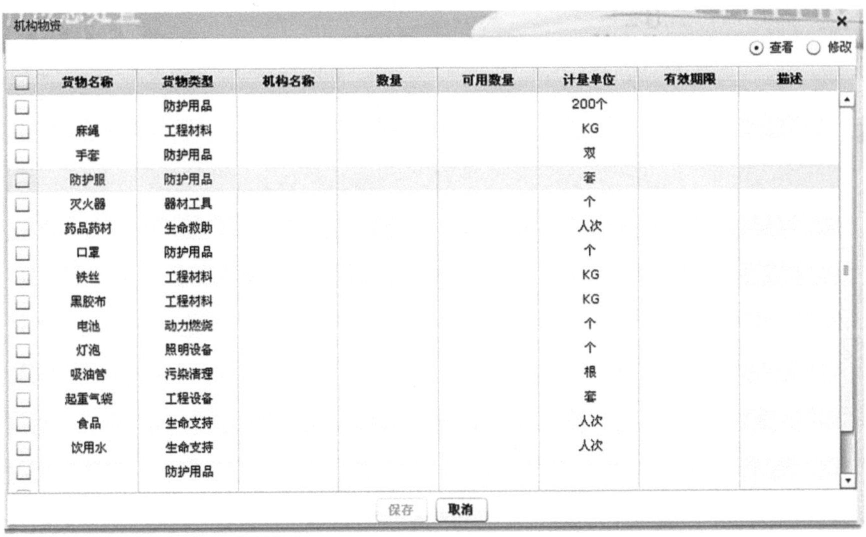

Fig. 9.56 Rescue organization "equipment" "materials" modification interface

(4) Equipment management

Select the "Institution Type," "Device Category," "Device Type" drop-down list box, or directly enter the device name and click the "Query" button to display the corresponding device information, as shown in Fig. 9.61.

Click the "Detail" button to view the device details, as shown in Fig. 9.62.

Fig. 9.57 Personnel management query interface

Fig. 9.58 Personnel information interface

Fig. 9.59 Add staff interface

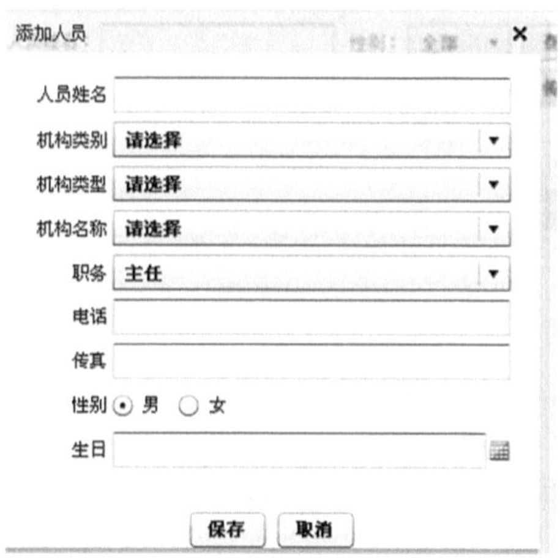

Fig. 9.60 Update staff interface

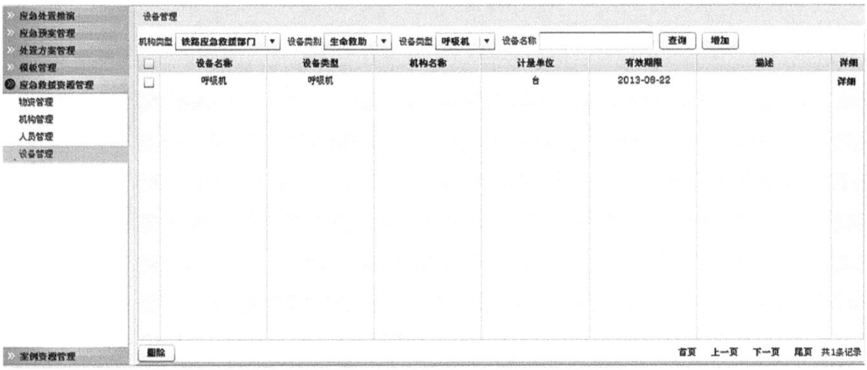

Fig. 9.61 Device management interface

In the "Device Details" dialog box, select the "Modify" button to modify the device information. After editing, click the "Save" button.

Click the "Add" button to add device information. When the addition is complete, click the "Save" button, as shown in Fig. 9.63.

You can cull the device by clicking the "Delete" button.

6. Case Resource Management Module

(1) Case generation

Select the "Special Condition Category," "Special Condition Type," and "Special Condition Level" drop-down list boxes, and click the "Query" button to display the corresponding case, as shown in Fig. 9.64 and Fig. 9.65.

Click the "Generate Case" button to view information about the relevant plan and case name, compiler, etc.

Select the plan and click the "View" button to view the plan, as shown in Fig. 9.66, then enter the case name, compiler, and memo information on the right side, click the "Save" button to close, as shown in Fig. 9.67.

Fig. 9.62 Device detailed interface

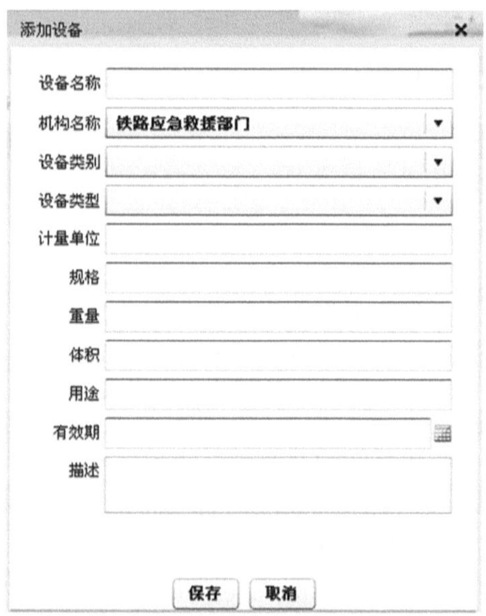

Fig. 9.63 Add device interface

You can view the source files of the plan by clicking the "View Attachments" button at the bottom of the "Emergency Plan Display" dialog box.

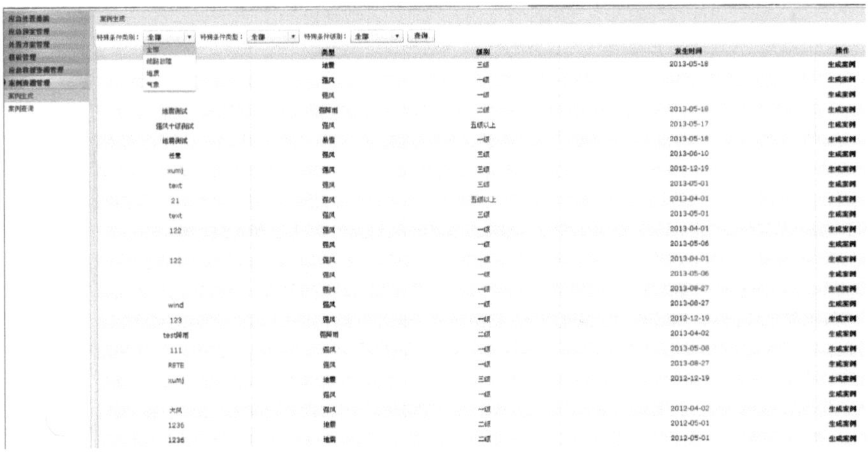

Fig. 9.64 Case query Interface 1

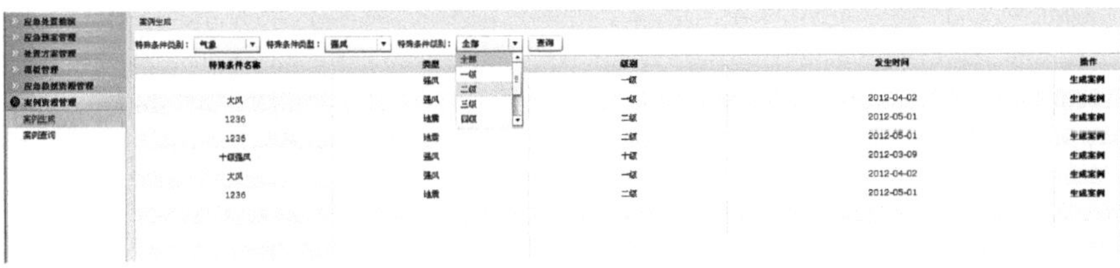

Fig. 9.65 Case query Interface 2

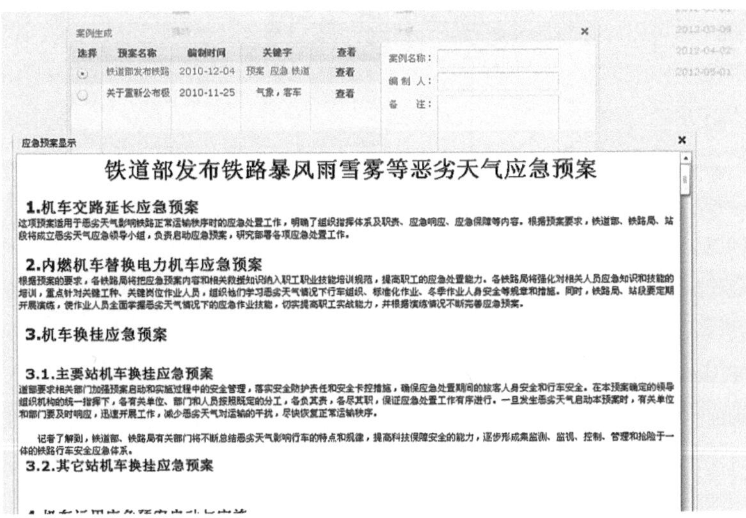

Fig. 9.66 Pre-view interface

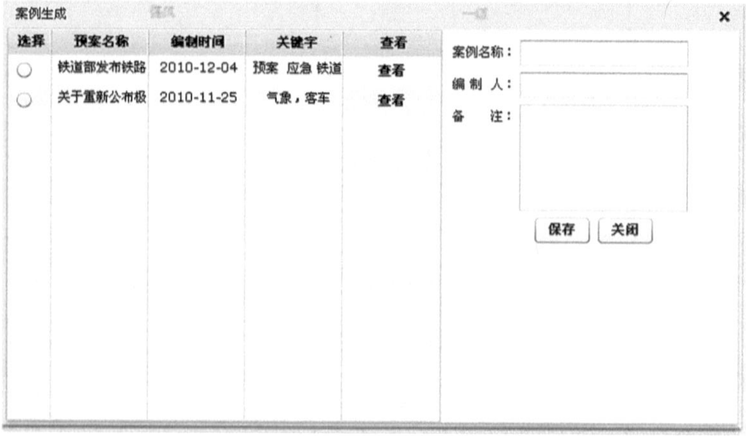

Fig. 9.67 Generate case-related information interface

(2) Case inquiry

Select the "Special Condition Category", "Special Condition Type", and "Special Condition Level" drop-down list boxes, and click the "Query" button to display the corresponding case, as shown in Fig. 9.68.

Click the "View Plan" button to view the details of the plan, as shown in Fig. 9.69.

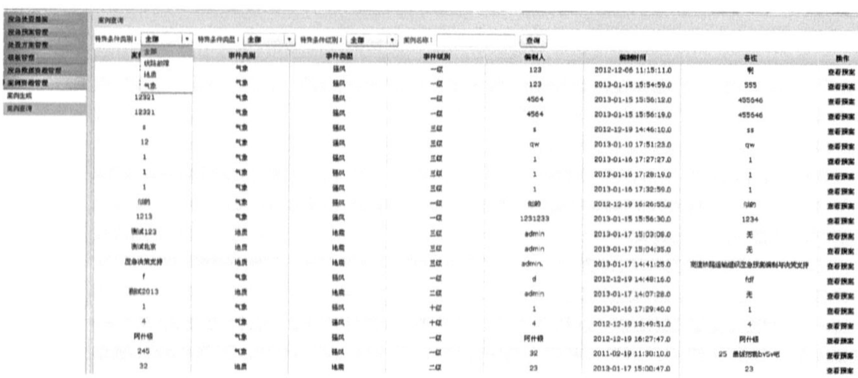

Fig. 9.68 Case query interface

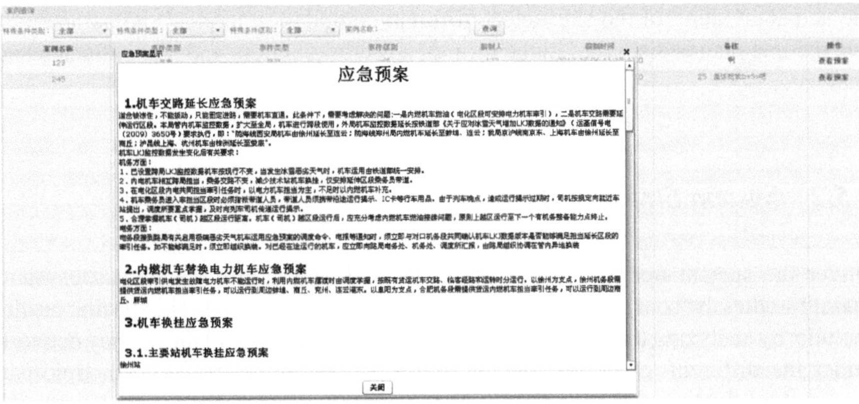

Fig. 9.69 Plan details interface

9.5 High-Speed Line Planning Subsystem Under Special Operating Conditions

The functions of the high-speed line planning subsystem under special operating conditions include the constraint condition calculation module under special operating conditions, the optimization target generation module under special operating conditions, and the line plan optimization and evaluation module under special operating conditions.

9.5.1 System Requirements Analysis

The line plan refers to the plan for determining the passenger train running section, train type, route, number of operation pairs and the rolling stock. It is an important part of the train operation organization.

The demand analysis of high-speed train operation planning subsystem under special operating conditions mainly includes: 1 Constraint calculation requirements under special operating conditions, by analyzing the influencing factors affecting the reliability of stations and sections, define and calculate the connectivity reliability of stations, sections and railway networks, and obtain the section carrying capacity under special operating conditions; 2 optimization goals generation requirements under special operating conditions, analyze the relationship between the emergency conditions and the line plan, and combine the expert scoring method to generate comprehensive objectives for different types and levels of the line plan generation and optimization under special operating conditions; 3 optimization and evaluation of line plan requirements, according to the characteristics of the emergencies and the actual situation of the site, select different compilation models and parameters, and the line plan is displayed in the form of a vector diagram, and it is evaluated and analyzed.

9.5.2 *System Structure Design*

Under the special operating conditions, the high-speed line planning subsystem mainly studies the constraints and optimization target calculation. The specific results include: by analyzing the connectivity and operation reliability of the railway network under the different levels of emergencies, the constraint calculation module can realize the calculation of the passing capacity of the railway section, the generation of the feasible path set of the train operation and the passenger flow forecasting, etc., and provide standardized constraints for line plan settings under special operating conditions; under the special operating conditions, the optimization target generation module can realize the setting of single target and multi-target weight; the line plan optimization and evaluation module can realize the parameter setting, formal expression, and evaluation of the line planning model.

Its functional architecture is shown in Fig. 9.70.

1. Constraint calculation module under special operating conditions

The module can define and calculate the connectivity reliability of stations, sections, and railway networks by analyzing the influencing factors affecting the reliability of stations and sections. On this basis, define and calculate the travel time reliability of railway network and the operation reliability of high-speed railway network.

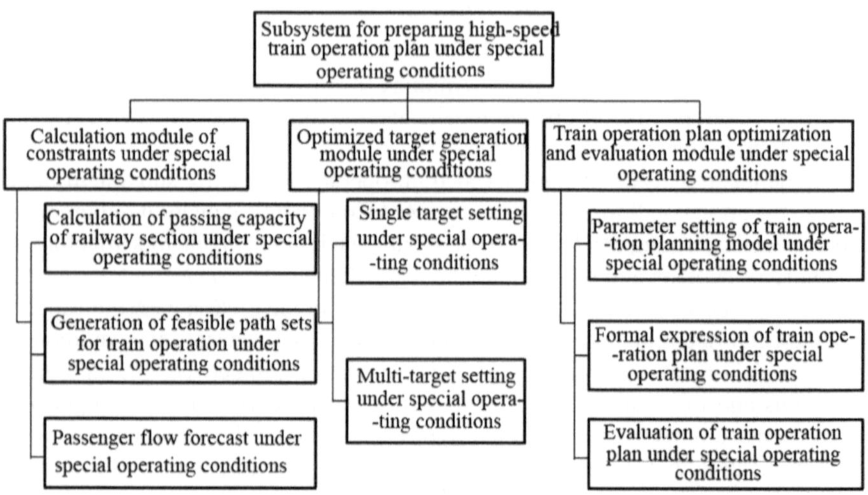

Fig. 9.70 Functional architecture of high-speed line planning subsystem under special operating conditions

Based on the analysis of the impact of special operating conditions on the railway network, it is concluded that the railway section carrying capacity under special operating conditions is a Markov process. According to this principle, the system realizes the calculation of section carrying capacity based on pessimistic strategy and fuzzy strategy. Finally, the feasible route set of train operation is generated and displayed, and the passenger flow on the feasible path is predicted.

2. Optimization target generation module under special operating conditions

The module can realize single target setting and multi-target weight setting under special operating conditions, so as to optimize the high-speed line plan under special operating conditions.

3. Line plan optimization and evaluation module under special operating conditions

The module can display the special operation condition scenarios built by 3D technology, match different line plan models and parameters according to different scenarios, realize the formal expression of the line plan, and establish an evaluation index system for line plan under the emergency conditions to facilitate decision-makers' judgment and selection.

9.5.3 System Interface

The following is a hypothesis based on the line plan of the Dezhou East-Jinan West and Jinan West-Dezhou East High-speed trains under special operating conditions (see Table 9.1) and further introduces the high-speed train operation programming subsystem under special operating conditions.

First enter the main interface of the system operation, Fig. 9.71 shows the main process of the generation and optimization of the line plan.

Table 9.1 Special operating conditions scenario assumptions

Starting time	2012-03-09	Wind direction	Zhengxi
Special condition number	244	Special condition name	Ten strong winds
Duration/h	11	Continuous rainfall	no
Affected section	Starting mileage: 327.980 Ending mileage: 419.445 Section name: Dezhou East—Jinan West Starting mileage: 419.445 Ending mileage: 327.980 Section name: Jinan West—Dezhou East		
Affected station	Dezhou East		

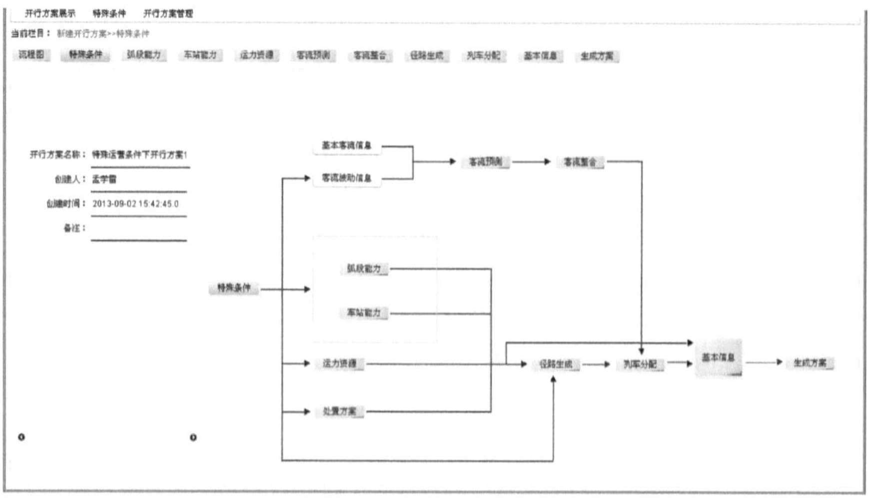

Fig. 9.71 Special condition line plan generation and optimization interface

1. Constraint calculation module under special operating conditions

The module can realize: analyze the connectivity and reliability of the railway network under different operating conditions of different levels; calculate the passing capacity of the railway section under special operating conditions; generate a set of feasible routes under special operating conditions; predict the passenger flow.

(1) Calculation of railway section passing capacity under special operating conditions
According to the type and level of the emergency, query the driving rules to obtain the speed limit value under the current event condition, and calculate the passing capacity of the arc segment. The specific process is as follows.

 (1) Click the "Arc Capability" button to enter the arc segment capability interface, as shown in Fig. 9.72.
 (2) Click the "Calculate" button to get the arc segment capability. The calculation list is shown in Fig. 9.73.

According to the type and level of emergencies, the throat throughput ability of the station under the current event conditions is queried, and the station passing ability calculation method is adopted under special operating conditions to obtain the passing ability of the station. The specific process is as follows.

 (1) Click the "Station Capacity" button to enter the station capability interface, as shown in Fig. 9.74.
 (2) The station capacity calculation list is shown in Fig. 9.75.

Click the "Capacity Resource" button to get the capacity resources stored in the system, as shown in Fig. 9.76.

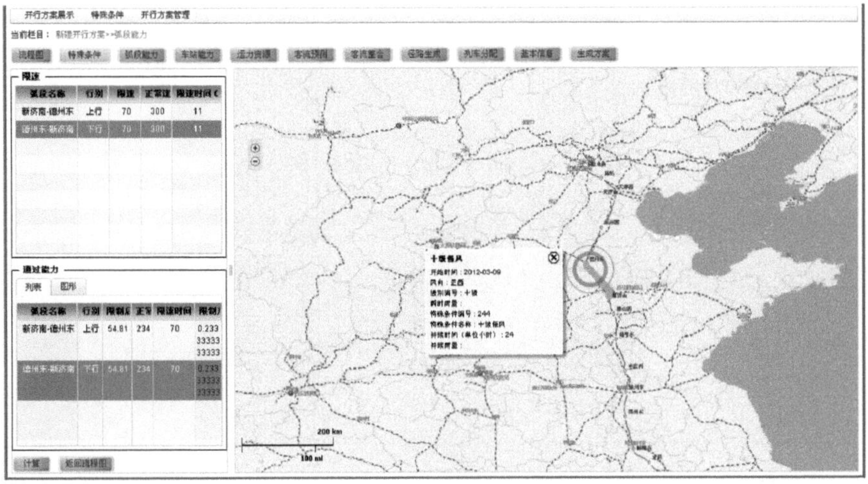

Fig. 9.72 Arc segment capability interface

Fig. 9.73 Arc segment capability calculation list

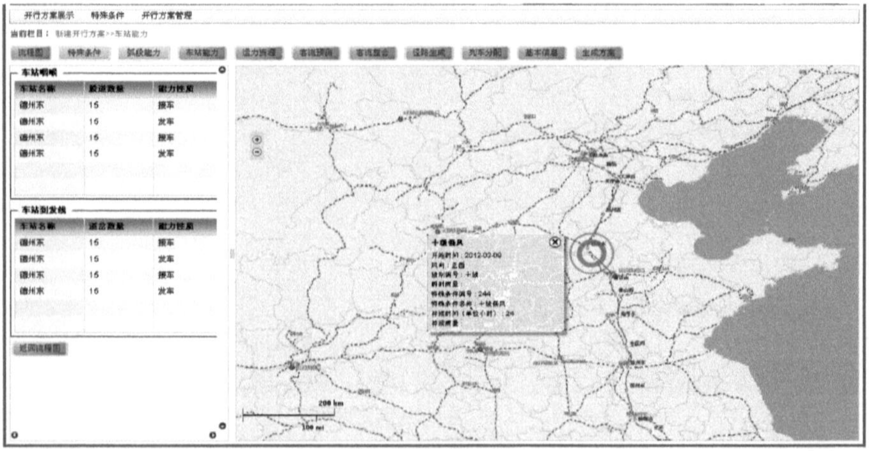

Fig. 9.74 Station capability interface

Fig. 9.75 Station capacity calculation list

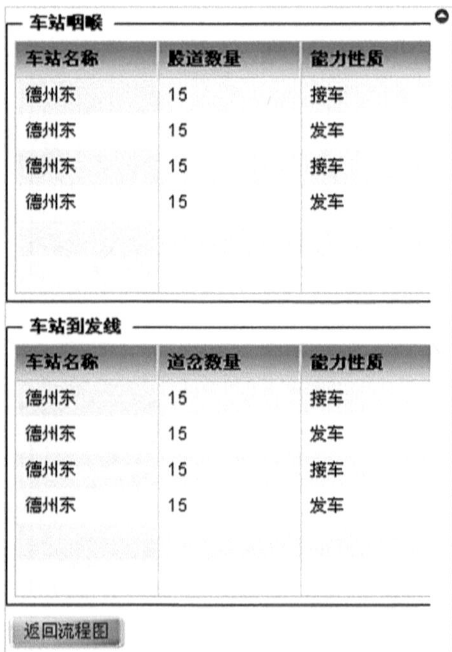

车站咽喉

车站名称	股道数量	能力性质
德州东	15	接车
德州东	15	发车
德州东	15	接车
德州东	15	发车

车站到发线

车站名称	道岔数量	能力性质
德州东	15	接车
德州东	15	发车
德州东	15	接车
德州东	15	发车

返回流程图

(2) Feasible route set generation under special operating conditions

According to the characteristics and scope of the emergency, the train running route under the emergency conditions is calculated, the feasible route set is

Fig. 9.76 List of storage capacity resources in the system

车站名	列车类型	可用	列车数	可	运行	
徐州东	短编组动	17	17	2	2	删除
德州东	短编组动	26	26	3	3	删除
南京南	长编组动	12	12	3	3	删除
虹桥:长	长编组动	23	23	4	4	删除
徐州东	长编组动	30	30	4	4	删除
天津西	短编组动	20	20	4	4	删除
天津西	动卧	11	11	5	5	删除
北京南	动卧	12	12	6	6	删除
德州东	动卧	23	23	6	6	删除
虹桥:动	动卧	25	25	7	7	删除
德州东	长编组动	27	27	7	7	删除
天津西	长编组动	14	14	7	7	删除
北京南	长编组动	15	15	7	7	删除
北京南	短编组动	21	21	7	7	删除
虹桥:短	短编组动	28	28	8	8	删除
北京南	动卧	10	10	8	8	删除
南京南	短编组动	13	13	9	9	删除
徐州东	动卧	12	12	10	10	删除

提交

generated, and then the distribution of the transport resources is carried out. The specific process is as follows.

(1) Click the "Route Generation" button to enter the generation route set interface. As shown in Fig. 9.77, enter the route length and generate a new route, as shown in Fig. 9.78.
(2) Click the "Train Assignment" button to assign the train running route, as shown in Fig. 9.79. The distribution route list is shown in Fig. 9.80.
(3) Passenger flow forecast under special operating conditions

Timely obtain the passenger flow aggregation in the current time period, automatically generate the passenger flow growth coefficient according to the type, level and influence range of the emergent event, and realize the passenger flow forecast in a certain period in the future. The specific process is as follows.

(1) Obtain current passenger flow data, as shown in Fig. 9.81.
(2) Click the "Customer Flow Prediction" button to obtain the passenger flow prediction result, as shown in Fig. 9.82.

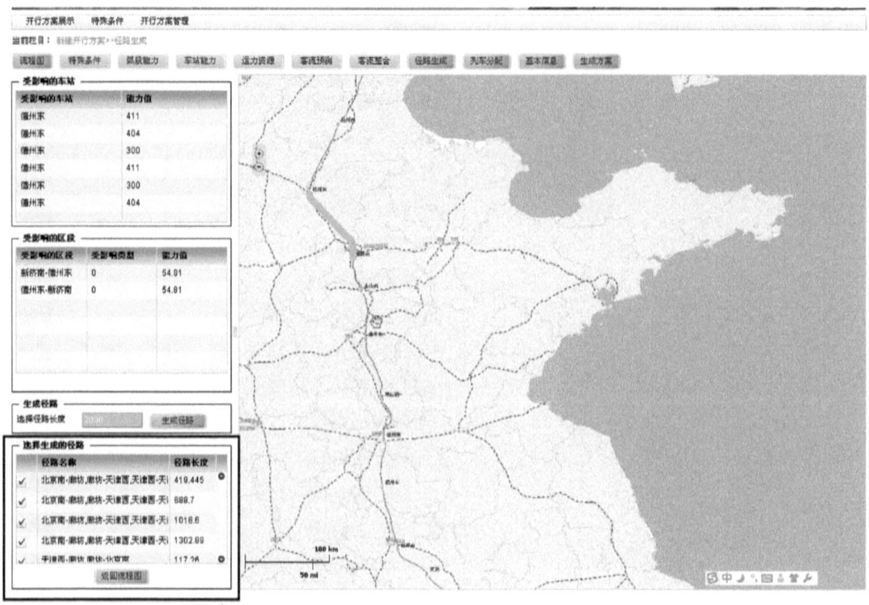

Fig. 9.77 Generate route set interface

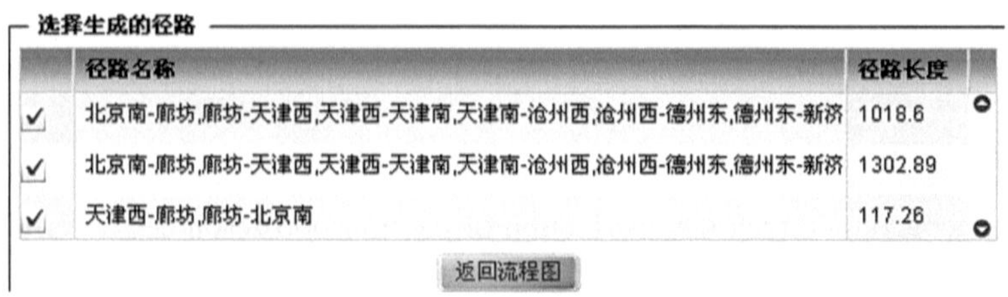

Fig. 9.78 route set list interface

(3) Click the "Customer Flow Integration" button to integrate the passenger flow, as shown in Fig. 9.83.

2. Optimization target generation module under special operating conditions

(1) Single target setting under special operating conditions

According to the type and characteristics of the emergency, select a single optimization target, as shown in Fig. 9.84.

(2) Multi-target weight setting under special operating conditions

According to the type, level and scope of the incident, multiple targets can be selected and the weight parameters can be set, as shown in Fig. 9.85.

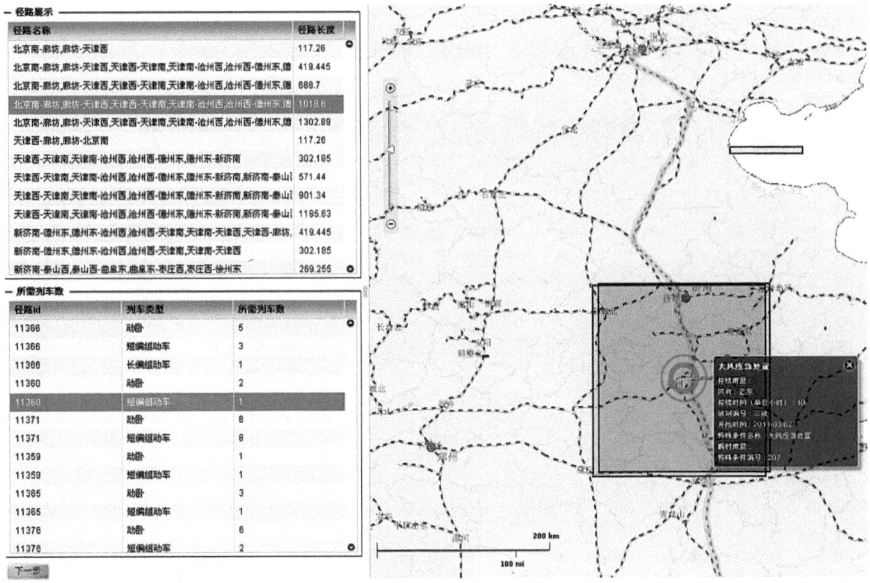

Fig. 9.79 Train running route assignment interface

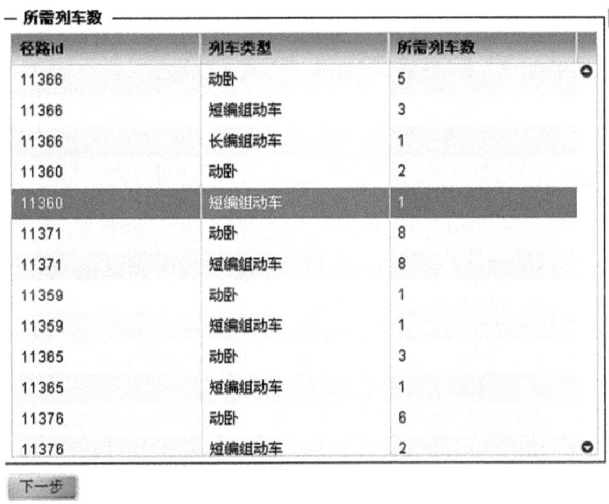

– 所需列车数

径路id	列车类型	所需列车数
11366	动卧	5
11366	短编组动车	3
11366	长编组动车	1
11360	动卧	2
11360	短编组动车	1
11371	动卧	8
11371	短编组动车	8
11359	动卧	1
11359	短编组动车	1
11365	动卧	3
11365	短编组动车	1
11376	动卧	6
11376	短编组动车	2

下一步

Fig. 9.80 Train distribution plan list on the trail

3. Line plan optimization and evaluation module under special operating conditions

(1) Parameter setting of line plan programming model under special operating conditions

According to the characteristics of emergencies and the real-time requirements of event processing, the corresponding programming models and parameters

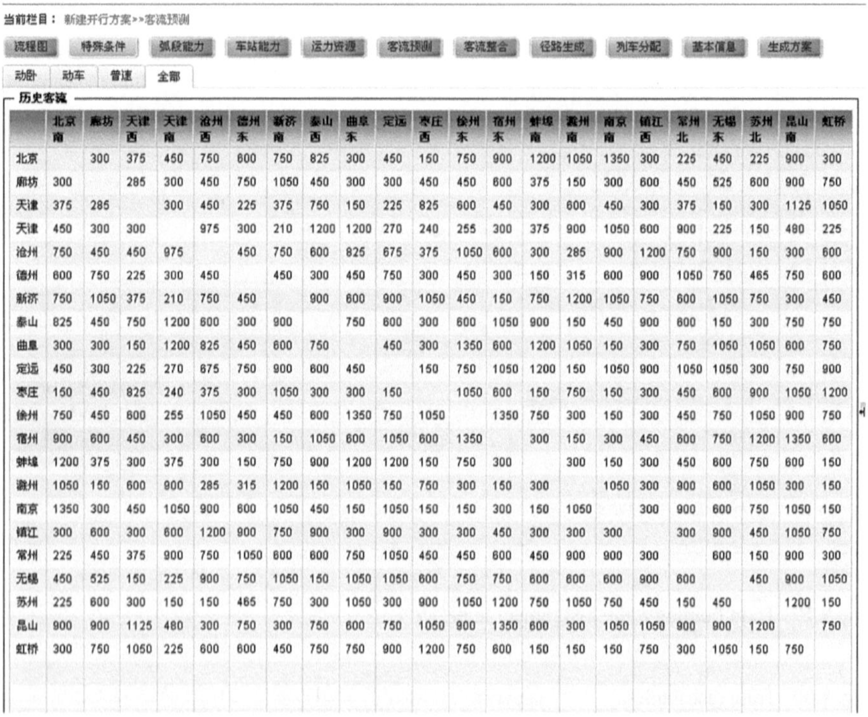

	北京南	廊坊	天津西	天津南	沧州西	德州东	新济南	泰山西	曲阜东	定远	枣庄西	徐州东	宿州东	蚌埠南	滁州南	南京南	镇江西	常州北	无锡东	苏州北	昆山南	虹桥
北京		300	375	450	750	600	750	825	300	450	150	750	900	1200	1050	1350	300	225	450	225	900	300
廊坊	300		285	300	450	750	1050	450	300	300	450	450	600	375	150	300	600	450	525	600	900	750
天津	375	285		300	450	225	375	750	150	225	825	600	450	300	600	450	300	375	150	300	1125	1050
天津	450	300	300		975	300	210	1200	1200	270	240	255	300	375	900	1050	600	900	225	150	480	225
沧州	750	450	450	975		450	750	600	825	675	375	1050	600	300	285	900	1200	750	900	150	300	600
德州	600	750	225	300	450		450	300	450	750	300	450	300	150	315	600	900	1050	750	465	750	600
新济	750	1050	375	210	750	450		900	600	900	1050	450	150	750	1200	1050	750	600	1050	750	300	450
泰山	825	450	750	1200	600	300	900		750	600	300	600	1050	900	150	450	900	600	150	300	750	750
曲阜	300	300	150	1200	825	450	600	750		450	300	1350	600	1200	1050	150	300	750	1050	1050	600	750
定远	450	300	225	270	675	750	900	600	450		150	750	1050	1200	150	1050	900	1050	1050	300	750	900
枣庄	150	450	825	240	375	300	1050	300	300	150		1050	600	150	750	150	300	450	600	900	1050	1200
徐州	750	450	600	255	1050	450	450	600	1350	750	1050		1350	750	300	150	300	450	750	1050	900	750
宿州	900	600	450	300	600	300	150	1050	600	1050	600	1350		300	150	300	450	600	750	1200	1350	600
蚌埠	1200	375	300	375	300	150	750	900	1200	1200	150	750	300		300	150	300	450	600	750	600	150
滁州	1050	150	600	900	285	315	1200	150	1050	150	750	300	150	300		1050	300	900	600	1050	300	150
南京	1350	300	450	1050	900	600	1050	450	150	1050	150	150	300	150	1050		300	900	600	750	1050	150
镇江	300	600	300	600	1200	900	750	900	300	900	300	300	450	300	300	300		300	900	450	1050	750
常州	225	450	375	900	750	1050	600	600	750	1050	450	450	600	450	900	900	300		600	150	900	300
无锡	450	525	150	225	900	750	1050	150	1050	1050	600	750	750	600	600	600	900	600		450	900	1050
苏州	225	600	300	150	150	465	750	300	1050	300	900	1050	1200	750	1050	750	450	150	450		1200	150
昆山	900	900	1125	480	300	750	300	750	600	750	1050	900	1350	600	300	1050	1050	900	900	1200		750
虹桥	300	750	1050	225	600	600	450	750	750	900	1200	750	600	150	150	150	750	300	1050	150	750	

Fig. 9.81 Current passenger flow data interface

can be selected, such as traditional mathematical planning methods or artificial intelligence computing methods. The model parameters required for each computing method are different, and the user can set according to the actual situation of the site (see Fig. 9.86), and the optimization result of the line plan is calculated (see Fig. 9.87).

(2) Formal expression of the line plan under special operating conditions

The line plan can be displayed in the form of a schematic diagram or a vector diagram. In the function module of the line plan, click the "Line Plan Diagram" button, and the line plan will be displayed in a schematic manner, as shown in Fig. 9.88.
Click the "Start Line Plan Vector" button, and the line plan will be displayed as a vector diagram, as shown in Fig. 9.89.

(3) Evaluation of the operation plan under special operating conditions

For one or more operation plans in the list of line plans, a single plan evaluation or a comparison evaluation of multiple plans can be performed respectively. Figures 9.90 and 9.91 are the evaluation results of a single line plan and multiple line plans, respectively.

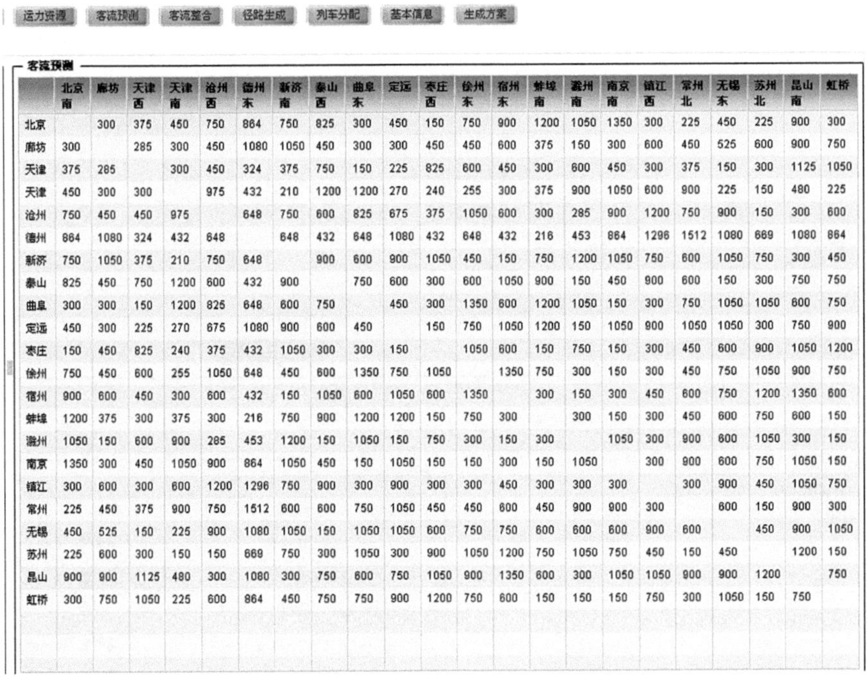

	北京南	廊坊	天津西	天津南	沧州西	德州东	新济南	泰山西	曲阜东	定远	枣庄西	徐州东	宿州东	蚌埠南	滁州南	南京南	镇江西	常州北	无锡东	苏州北	昆山南	虹桥
北京		300	375	450	750	864	750	825	300	450	150	750	900	1200	1050	1350	300	225	450	225	900	300
廊坊	300		285	300	450	1080	1050	450	300	300	450	450	600	375	150	300	600	450	525	600	900	750
天津	375	285		300	450	324	375	750	150	225	825	600	450	300	600	450	300	375	150	300	1125	1050
天津	450	300	300		975	432	210	1200	1200	270	240	255	300	375	900	1050	600	900	225	150	480	225
沧州	750	450	450	975		648	750	600	825	675	375	1050	600	300	285	900	1200	750	900	150	300	600
德州	864	1080	324	432	648		648	432	648	1080	432	648	432	216	453	864	1296	1512	1080	669	1080	864
新济	750	1050	375	210	750	648		900	600	900	1050	450	150	750	1200	1050	750	600	1050	750	300	450
泰山	825	450	750	1200	600	432	900		750	600	300	600	1050	900	150	450	900	600	150	300	750	750
曲阜	300	300	150	1200	825	648	600	750		450	300	1350	600	1200	1050	150	300	750	1050	1050	600	750
定远	450	300	225	270	675	1080	900	600	450		150	750	1050	1200	150	1050	900	1050	1050	300	750	900
枣庄	150	450	825	240	375	432	1050	300	300	150		1050	600	150	750	150	300	450	600	900	1050	1200
徐州	750	450	600	255	1050	648	450	600	1350	750	1050		1350	750	300	150	300	450	750	1050	900	750
宿州	900	600	450	300	600	432	150	1050	600	1050	600	1350		300	150	300	450	600	750	1200	1350	600
蚌埠	1200	375	300	375	300	216	750	900	1200	1200	150	750	300		300	150	300	450	600	750	600	150
滁州	1050	150	600	900	285	453	1200	150	1050	150	750	300	150	300		1050	300	900	600	1050	300	150
南京	1350	300	450	1050	900	864	1050	450	150	1050	150	150	300	150	1050		300	900	600	750	1050	150
镇江	300	600	300	600	1200	1296	750	900	300	900	300	300	450	300	300	300		300	900	450	1050	750
常州	225	450	375	900	750	1512	600	600	750	1050	450	450	600	450	900	900	300		600	150	900	300
无锡	450	525	150	225	900	1080	1050	150	1050	1050	600	750	750	600	600	600	900	600		450	900	1050
苏州	225	600	300	150	150	669	750	300	1050	300	900	1050	1200	750	1050	750	450	150	450		1200	150
昆山	900	900	1125	480	300	1080	300	750	600	750	1050	900	1350	600	300	1050	1050	900	900	1200		750
虹桥	300	750	1050	225	600	864	450	750	750	900	1200	750	600	150	150	150	750	300	1050	150	750	

Fig. 9.82 Passenger flow prediction result interface

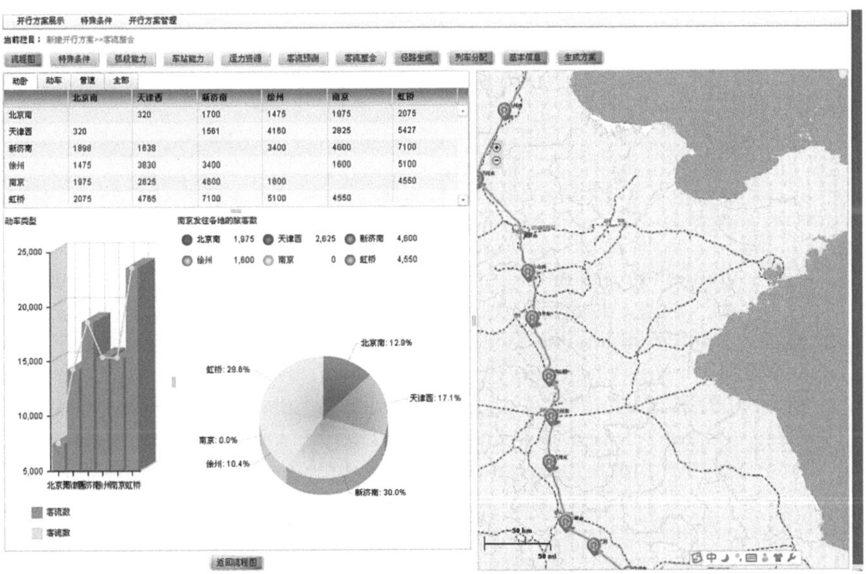

	北京南	天津西	新济南	徐州	南京	虹桥
北京南		320	1700	1475	1975	2075
天津西	320		1561	4160	2825	5427
新济南	1898	1638		3400	4600	7100
徐州	1475	3830	3400		1600	5100
南京	1975	2625	4600	1600		4550
虹桥	2075	4765	7100	5100	4550	

Fig. 9.83 Integrated data interface for passenger flow under emergencies

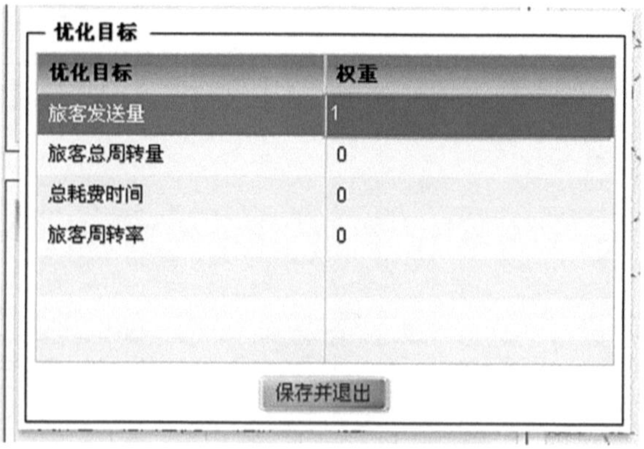

Fig. 9.84 Line plan single target setting interface

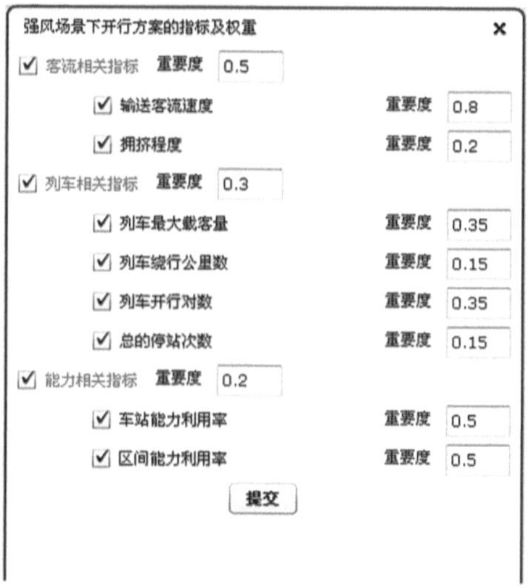

Fig. 9.85 Index system construction and weight setting

The evaluation results of the single line plan shown in Fig. 9.90 give the names of the previously determined index systems and their reasonableness in the upper right part of the interface, and the lower part gives the explanation of the evaluation results of the line plan.

In Fig. 9.91, the multi-program evaluation results are compared and evaluated for the selected two line plans. The table in the upper right corner compares the dimensionless values of the three-level indicator system of the two schemes and gives the calculation results of the secondary indicators. The bar chart below shows the visual comparison of the three evaluations of the two options (train-related indicators, passenger-related indicators, and capacity-related indicators). When the mouse moves over the column of the histogram, the corresponding values will also be displayed.

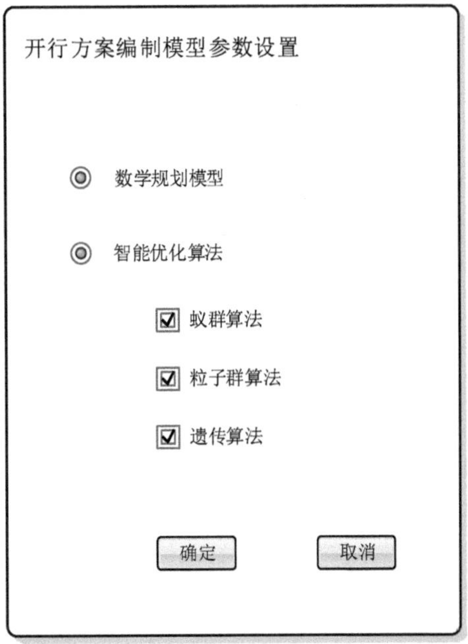

Fig. 9.86 Line planning model parameter setting interface

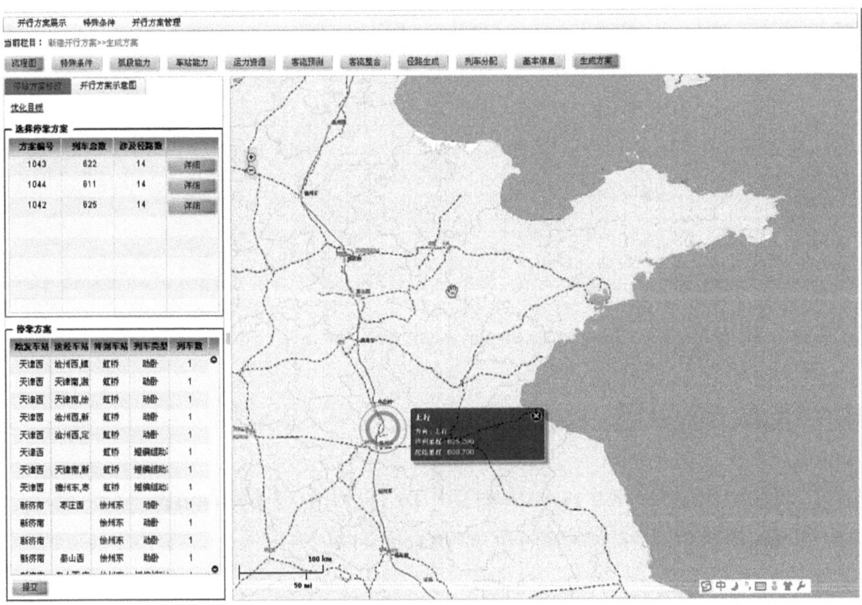

Fig. 9.87 Line plan optimization result interface

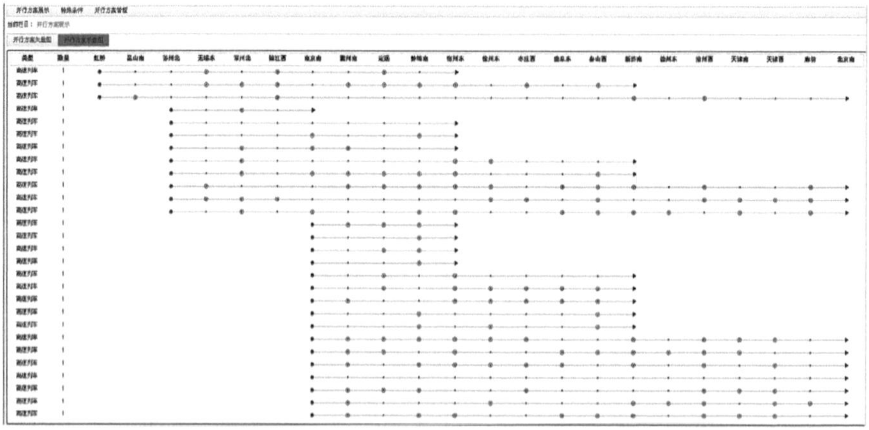

Fig. 9.88 Schematic diagram of the line plan

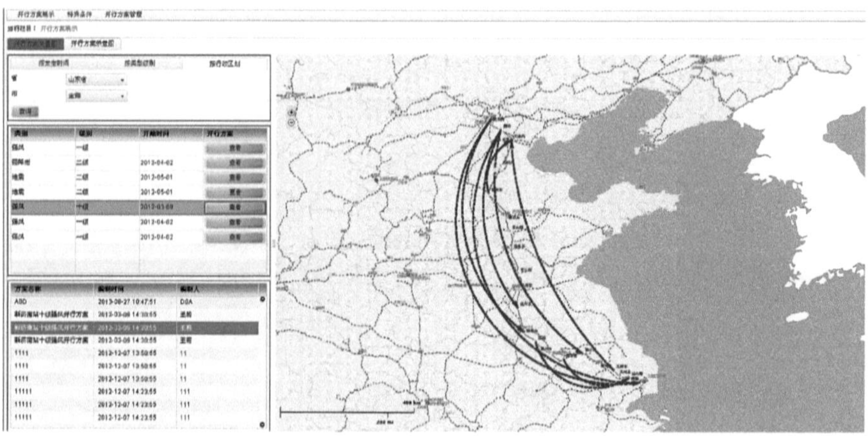

Fig. 9.89 Line plan vector illustration display interface

9.6 Demonstration Application

9.6.1 Demonstration Application of National Railway Emergency Plan Management System

In order to better improve the railway emergency plan management work and strengthen the railway emergency plan system construction, China Railway Corporation and Beijing Jiaotong University jointly carried out the research work of the "Railway Emergency Plan Management Research" project, aiming to analyze the

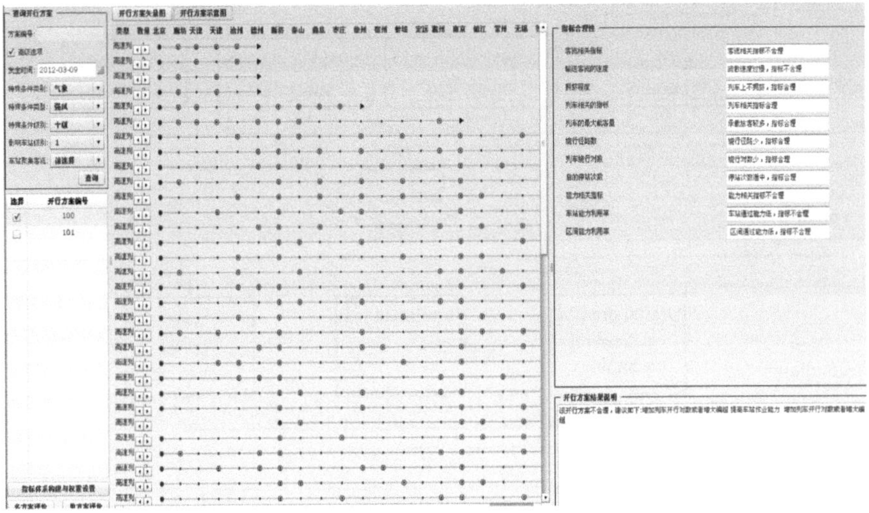

Fig. 9.90 Evaluation results of the line plan

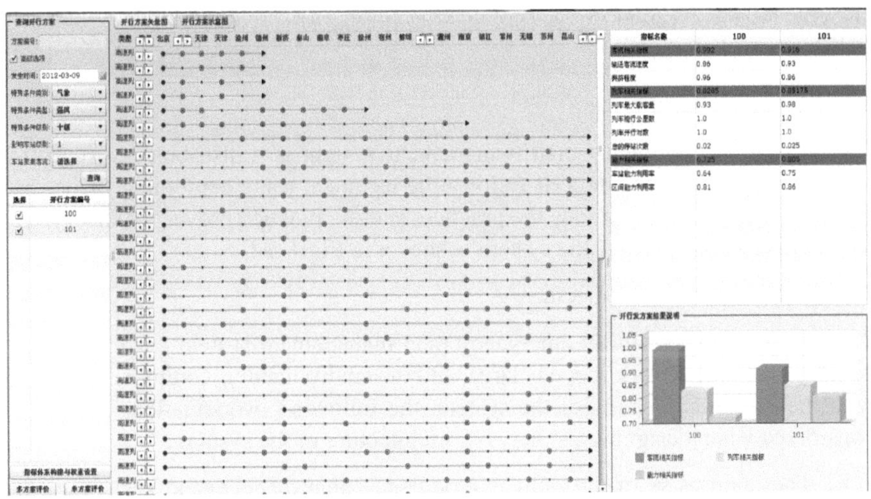

Fig. 9.91 Multi-program evaluation results

positioning, scope and content of the railway emergency plan; studying the supporting theoretical methods for the preparation of the railway emergency plan, the comprehensive evaluation of the plan and the daily management of the plan; establishing a relatively complete theory, method and framework for railway emergency and plan management; developing the railway emergency plan management system. To further strengthen the scientific, systematic, standardized, and standardized management of railway emergency plans.

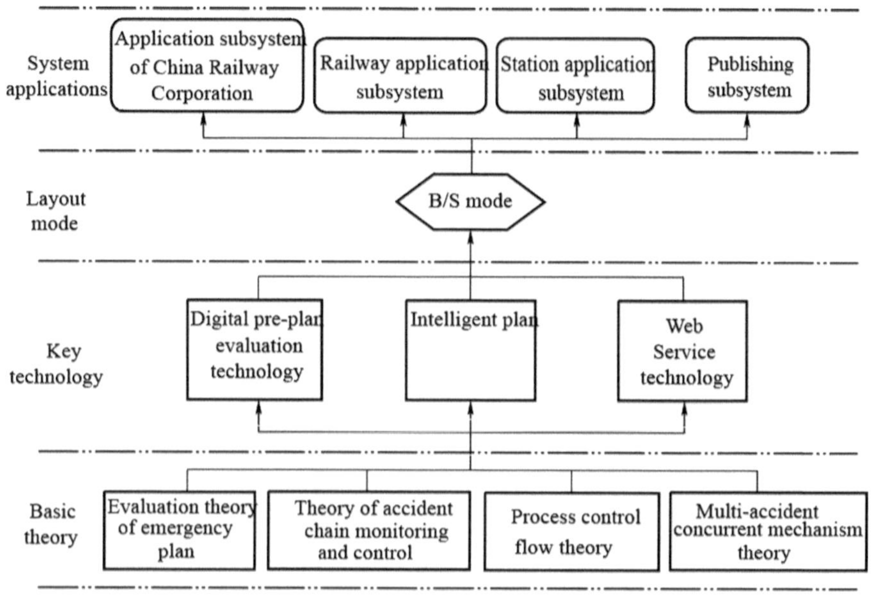

Fig. 9.92 Technical route of railway emergency plan management system

1. System technical route

The railway emergency plan management system is a distributed system that integrates functions and data. It follows the technical route combining theory and practice, and the theory guides the practice, and the practice proves the theory. The technical research route of the system is shown in Fig. 9.92.

2. System network architecture

The users of the railway emergency plan management system cover the whole country, and the network scope is large, and the environment is complex. Combined with the actual application of the system, the following two situations need to be considered when designing the network architecture of the system.

(1) The main users of the railway emergency plan management system are the China Railway Corporation and the business departments of the Railway Administration and stations under its jurisdiction. The network environment is the railway network.
(2) Within the China Railway Corporation, the Railway Administration or the station, the network environment is local area network.

In view of the above situation, the system design adopts the b/s mode, which is mainly responsible for the creation, reporting, approval, release, evaluation, inquiry, statistics, user creation and configuration authority, railway emergency plan work dynamics, laws and regulations and the release of related news, etc. The system is mainly used for the daily work of users.

In summary, the network topology of the system is shown in Fig. 9.93.

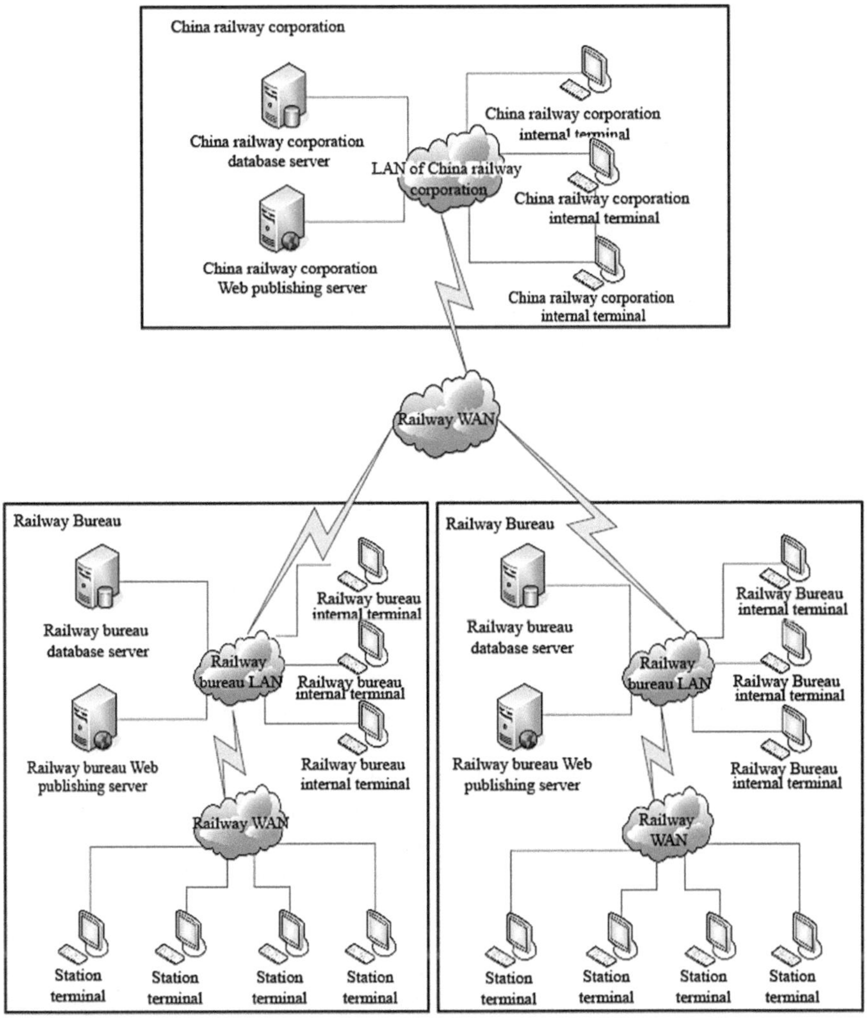

Fig. 9.93 Railway contingency plan management system network topology

3. System architecture

According to the relevant national emergency plan-related policies and regulations, combined with the specific organization and network system of China Railway Corporation, the logical structure of the system is shown in Fig. 9.94.

The system can be roughly divided into three levels: data layer, system call layer, and application layer.

(1) The data layer provides data required for system operation, including plan information, evaluation method information, accident information, risk source information, and work dynamic information, etc.

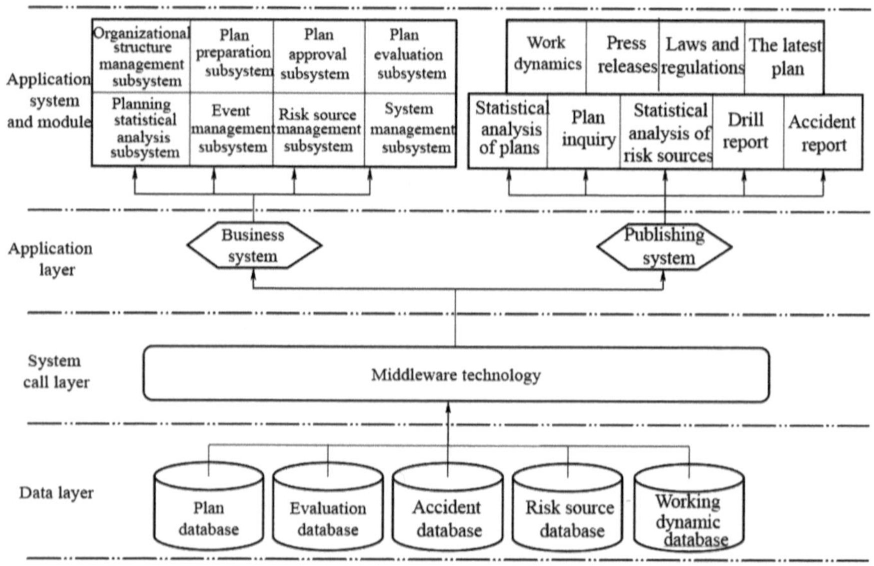

Fig. 9.94 Logical structure of railway emergency plan management system

(2) The system call layer uses the Web Service technology to publish the functions of the application layer in the form of text through the encapsulation of the system call layer, so that the system can exchange information and cooperate with the existing system of the China Railway Corporation.

(3) The application layer system can be roughly divided into two categories, namely the business system and the release system. The business system is mainly for the emergency plan business personnel, including 8 subsystems. They are organizational structure management, plan preparation, plan approval, plan evaluation, plan statistical analysis, event management, risk source management, and system management. The publishing system is aimed at all users in the railway network, including 9 parts, which are work dynamics, news release, laws and regulations, latest plans, planned statistical analysis, plan inquiry, risk source statistical analysis, drill report, and accident report.

For different users, the system grants different permissions to implement user hierarchical management and data hierarchical access control to ensure the security of system data.

4. System function design

According to the needs analysis, the system should be divided into two categories: business system and release system. Its module structure is shown in Fig. 9.95.

The above functions mainly realize the daily management, decision support, statistical analysis, and information release of the railway emergency plan management system.

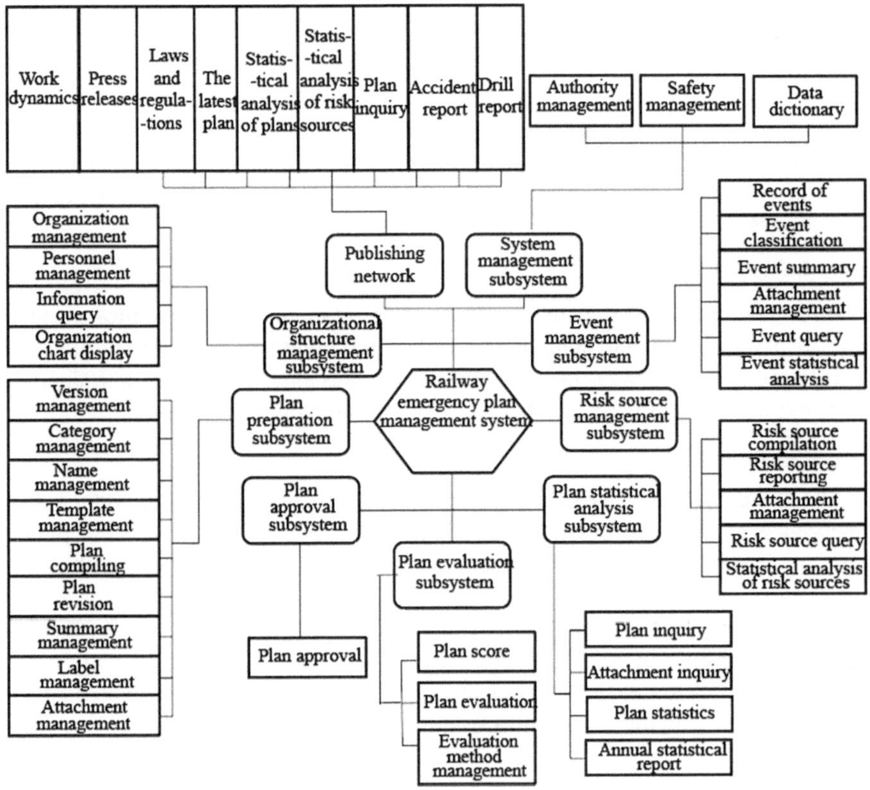

Fig. 9.95 National railway emergency plan management system module structure

(1) Daily management

Daily management refers to the daily management of the work related to the emergency plan, including various daily tasks such as institutional management, plan preparation, plan approval, plan evaluation, plan statistical analysis, event classification, event statistical analysis, risk source reporting, legal and regulatory issues, work dynamic reports, and the production of annual statistical reports. The above functions make the railway emergency plan management work normative and improve the efficiency of the plan management work.

(2) Decision support

Decision support is mainly used when the accident occurs, and the system can provide comprehensive plan information, simple past accident handling information, risk source profile information and so on, so as to provide comprehensive information and operational guidance for leaders to report accidents, handle accidents, avoid dangers, and other decisions. It can speed up the handling of accidents and improve the effectiveness of accident rescue.

(3) Statistical analysis

The statistical analysis mainly includes two aspects, namely the statistical analysis of the plan and the statistical analysis of the risk source. The system has the functions of statistical analysis of plans and risk sources from various aspects and angles through structured, streamlined, digital management plans and risk source reports. In addition, it supports digital and graphical display of statistical results and provides brief analysis.

(4) Information release

Information release refers to the release of pre-planning work dynamics, accident news, drill news, laws and regulations, and the latest plans through the release system, so as to promote exchanges between railway bureaus and stations, and introduce the relevant management work of emergency plans to the outside world.

5. User interface

The home page interface of the system consists of system login, important notice, emergency plan-related documents, railway news, emergency big vision, picture news, plan statistics, weather conditions and common link modules, as shown in Fig. 9.96.

The emergency plan management interface, as shown in Fig. 9.97, consists of five parts: view, new, score, review, and revision.

Figure 9.98 shows the actual application of the railway emergency plan management system in Qinghai-Tibet Railway Company.

Fig. 9.96 Railway emergency plan management system home page interface

Fig. 9.97 Emergency plan management interface

Fig. 9.98 Actual application

9.7 Railway Transportation Organization Emergency Response Decision Support System (High-Speed Railway) Demonstration Application

On the basis of the theoretical research and system realization of the project, the emergency response theory system is further refined. Combined with the actual needs of the dispatching business site, part of the results of the project realization system is transformed into the current railway transportation organization emergency response decision support system (High-speed railway). The railway transportation organization emergency response decision support system (high-speed railway) is part of the railway transportation information integration platform jointly supervised by the China Railway Corporation Transportation Bureau and the Information Technology Center, and is trial run at the China Railway Corporation.

With the approach of the "Overall Implementation of Railway Transport Dispatching Command Information by 2020" proposed by the "Overall Plan of Railway Informatization," the pace of railway informatization construction will also accelerate, especially in the realization of special operation scenarios. Under this background, the railway transportation organization emergency response decision support system (high-speed railway) is also gradually improving, and gradually realizes the goal of information, automation, and intelligence of railway transportation organization emergency response decision making. As a part of the railway transportation information integration platform, the railway transportation organization emergency response decision support system (high-speed railway) is of great significance in the trial operation of the China Railway Corporation. With the continuous improvement and gradual improvement of the system, it will be extended to various bureaus and even the whole railway network. At the same time, it will play its role in the emergency response decision support of existing lines, passenger dedicated lines, and even urban rail transit.

The user interface of the railway transportation organization emergency response decision support system (high-speed railway) is mainly divided into the emergency event input query interface, the event comprehensive display interface and the back-off event processing modification interface. Figures 9.99 and 9.100 are the input query interfaces for two types of different attribute emergencies. After the incident is selected by the attribute structure on the left side, the details can be filled in the upper right part of the field to complete the incident. After the information is reported, the

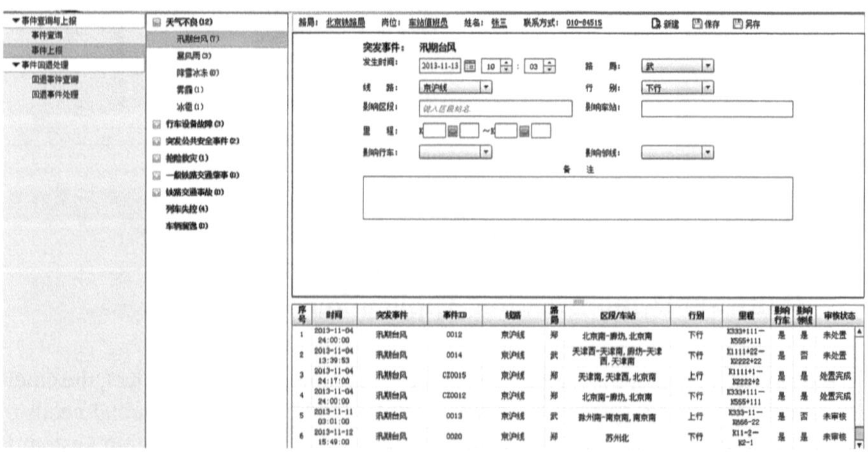

Fig. 9.99 Incident entry query interface (for bad weather, line faults, etc.)

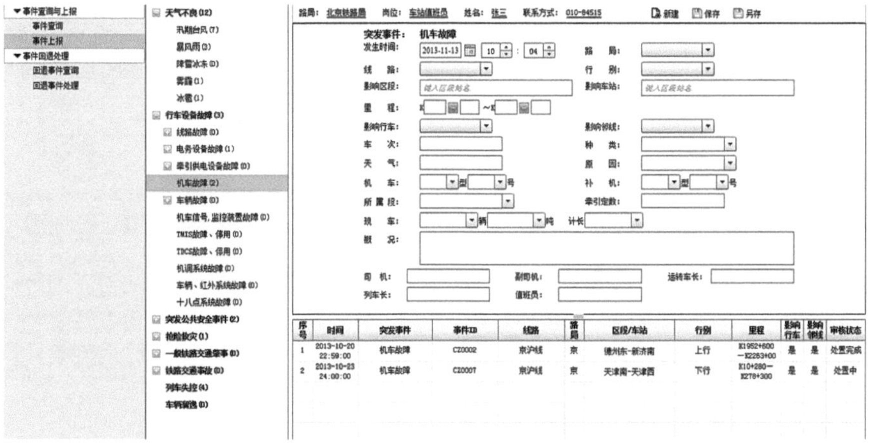

Fig. 9.100 Incident entry query interface (for railway traffic accidents, rolling stock failures, etc.)

saved data will appear in the comprehensive event display interface at the bottom of Fig. 9.100. At the same time, the interface can also query the previously reported events, as shown in Fig. 9.101.

The interface shown in Fig. 9.102 clearly shows all the reported or collected incident information and displays it in the map. After the emergency information is reviewed and confirmed by the dispatching personnel, according to the audit result, it will be returned to the input personnel or enter the "Event Processing List" to start the disposal process. And each piece of data in the "Event Processing List" corresponds to its disposal status, which is convenient for the leader to check the progress of the event processing and urge the relevant dispatching station to dispose of it in time.

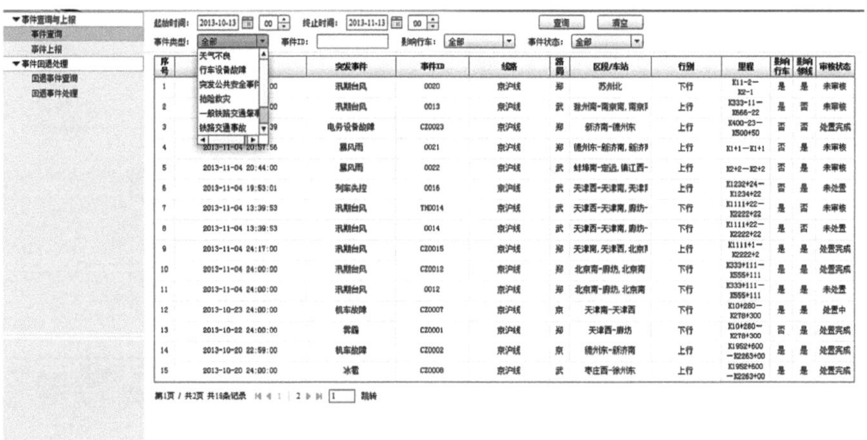

Fig. 9.101 Incident escalation query

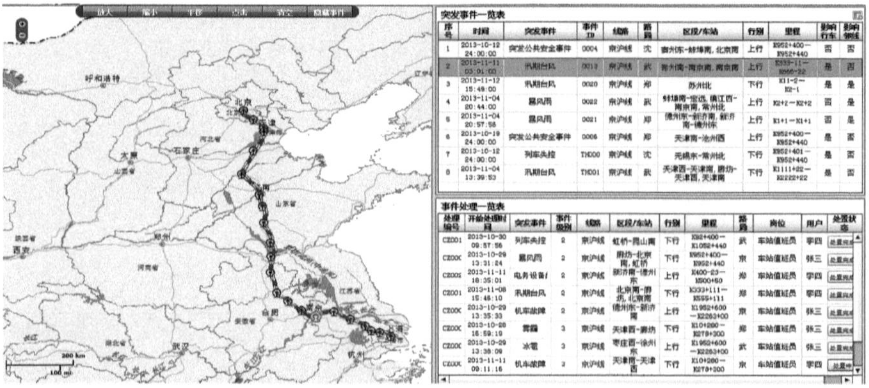

Fig. 9.102 Event comprehensive display interface

The interfaces shown in Figs. 9.103 and 9.104 are for the event entry personnel to query and modify the event that has been retreats after the auditor audits. After processing, the event will be repushed to the emergency event comprehensive display interface.

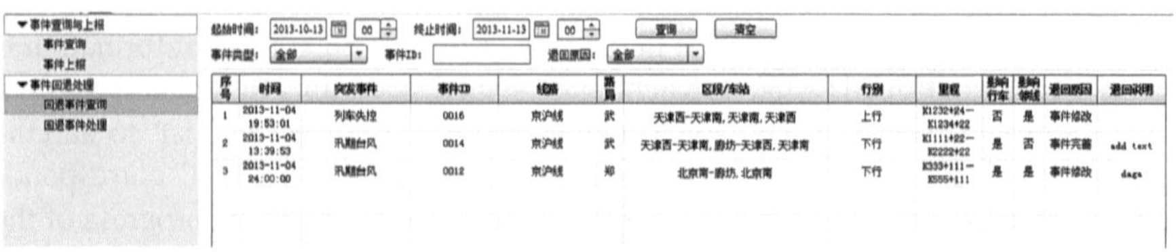

Fig. 9.103 Rollback event query interface

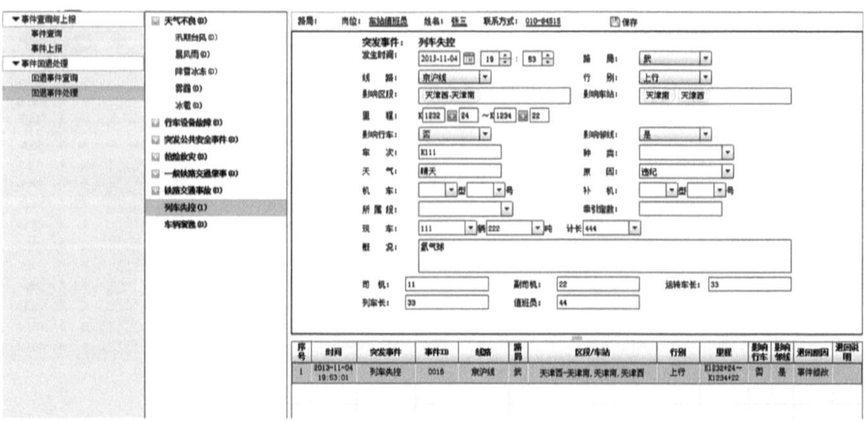

Fig. 9.104 Rollback event modification interface

9.8 Implementation Effect

At present, the research results of these systems have been promoted on a large scale. Among them, the railway transportation organization emergency response decision support system (high-speed railway) has been used as the core part of the railway transportation information integration platform of China Railway Corporation, and it is trial run at China Railway Corporation. The system utilizes core technologies such as railway emergency scene generation and information fusion and emergency coordinated disposal to systematically manage the generation, management, display, and collaborative emergency response workflow of railway emergencies, and significantly improve the railway transportation organization. The efficiency and ability of emergency management facilitate and improve the standardization and convenience of the operation of the business personnel incident reporting process. At present, the system functions mainly include the reporting and query of emergency information, the query and modification of the fallback event, the point-line diagram of the emergency and the emergency prompt on GIS map, the sub-item display of the unhandled events and the events under disposal, and the state transition of the events under disposal. It realizes the workflow of railway accident entry, display, inspection, audit (return, re-audit), processing, and disposal, strongly supports the emergency response of railway dispatchers and related leaders, standardizes the reporting and handling of emergencies, prevents adverse phenomena such as delays in handling false alarm events and untimely handling of emergencies, and provides key technical and platform support for railway transportation organization emergency response decision making.

The national railway emergency plan management system has been put into operation on the spot. After the system is put into operation, the work is stable and reliable, the function meets the requirements, the operation interface is visualized, and the information organization is scientific. The National Railway Emergency Plan Management System is a distributed system that integrates functions and integrates data. Based on the background of user coverage, wide network range and complex environment, a star network structure with the China Railway Corporation as the center and the railway bureau as the node is constructed. The system architecture is clear and reasonable, and the professional model is rich. It realizes the intelligent management of emergency plan, the dynamic management of emergency resources, the daily management of hazard source and case base management, and improves the efficiency of emergency plan management. In addition, the successful deployment and application of the system in China Railway Corporation and the typical railway bureau have effectively improved the management level of the railway emergency plan, and provided technical means and system platform support for the formation and continuous application of railway emergency plan management mechanism, mode and process with Chinese characteristics, and the realization of the integration, standardization and information of the national railway emergency management.

Up to now, the national railway emergency plan management system has established 6 emergency plan databases for the whole railway, deployed 6 data exchange platforms, and developed 106 functional modules; it has realized the digital management of 24 ministerial-level plans, 116 bureau-level plans and 625 station segment-level plans. The system covers China Railway Corporation Emergency Office, 4 railway bureaus, 104 stations, 168 departments, and about 1200 users.

9.9 Chapter Summary

This chapter mainly introduces the decision support system for high-speed railway train operation under emergencies, as well as some demonstration applications. The implementation and promotion of these systems have yielded a rich set of results. In the theoretical research, the method of obtaining and deriving special operating conditions based on closed-loop control, the generation of high-speed train digital plan, the matching of passenger demand and transportation resources, and the programming of the line plan are constructed. At the same time, the decision support system for high-speed railway train operation under emergency conditions is designed and developed, covering three independent and interrelated subsystems: high-speed railway special operating conditions acquisition and generation subsystem, high-speed railway transportation organization emergency plan preparation and the decision support subsystem, and the high-speed line planning subsystem under special operating conditions. The system can realize comprehensive data analysis, decision support, emergency command and collaborative decision-making functions, meet the diversified needs of the railway dispatching management department, lay a solid foundation for the full realization of electronic and information-based railway intelligent transportation, and effectively improved the organization efficiency of high-speed railway transportation under the conditions of large-scale network construction in China.

References

1. Dennis A, Wixom HB, Roth MR (2008) Systems analysis & design, 4th edn. Wiley, New York
2. Kennedy M, Goodchild FM (2009) Introducing geographic information systems with ArcGIS: a workbook approach to learning GIS, 2nd edn. Wiley, New York
3. Price J (2005) Oracle database 10g SQL. McGraw-Hill Companies Inc., New York
4. Qin Y, Jia L, Wang L et al (2013) Key technologies and systems for the design of high-speed railway trains under special operating conditions: technical Report [r]. State Key Laboratory of Rail Traffic Control and Safety, Beijing Jiaotong University, Beijing (in Chinese)
5. Qin Y, Jia L, Wang L et al (2013) Key technologies and systems for the design of high-speed railway trains under special operating conditions: acceptance report [r]. State Key Laboratory of Rail Traffic Control and Safety, Beijing Jiaotong University, Beijing (in Chinese)
6. Qin Y, Jia L, Wang L et al (2013) Key technologies and systems for the design of high-speed railway trains under special operating conditions: work report [r]. State Key Laboratory of Rail Traffic Control and Safety, Beijing Jiaotong University, Beijing (in Chinese)

7. Qin Y, Jia L, Wang L et al (2008). Design decision support system for high-speed railway trains under special operating conditions: demand analysis [r]. State Key Laboratory of Rail Traffic Control and Safety, Beijing Jiaotong University, Beijing (in Chinese)
8. Qin Y, Jia L, Wang L et al (2009) Design decision support system for high-speed railway train operation plan under special operating conditions: design specification [r]. State Key Laboratory of Rail Traffic Control and Safety, Beijing Jiaotong University, Beijing (in Chinese)
9. Qin Y, Jia L, Wang L et al (2013) Key technologies and systems for the design of high-speed railway trains under special operating conditions: user Manual (in Chinese)

Chapter 10
Open Question

Railways are not only a popular means of transportation, but also the main artery of the national economy. They play an irreplaceable role in economic and social development and improvement of people's livelihood. Since its inception of the high-speed railway, it has been favored by various countries with its characteristics of low energy consumption, high speed, high transport capacity, and high comfort. However, with the rapid development of high-speed railways, it also highlights the characteristics of complex technology, strong time–space coupling of the network, and sensitivity to emergencies. Transportation organizations under the condition of emergency will play an increasingly important role in the future management of high-speed railway network operations.

This chapter mainly introduces the open issues that need to be solved urgently by high-speed railway transportation organizations under the condition of emergency. These problems can be divided into three types of problems. The first type of problems mainly involves the interactions between the various components of the high-speed railway system and the mechanisms and laws related to system behavior. The second type of problems mainly includes the high-speed railway-related resources allocation, layout and scheduling under normal or abnormal circumstances, that is, transportation organization mode and strategy-related issues. The third type of problem focuses on high-speed railways through information technology to improve service and emergency-related capabilities. These three types of issues cover the connotation and extension of the entire high-speed railway system, especially the four aspects of prevention, preparation, response, and recovery of railway emergency management. Some of these problems are the core content of transportation organizations under the condition of emergency, while others are important supplements and conditions. Therefore, these issues are organized into a chapter for scholars to study together.

L. Jia et al., *High-Speed Railway Operation Under Emergent Conditions*,
Advances in High-speed Rail Technology,
https://doi.org/10.1007/978-3-662-63033-4_10

10.1 The First Category: Mechanism and Regularity Issues

Firstly, from the interaction between various components of the high-speed railway system and the mechanisms and laws related to system behavior, it includes the mechanism of spatio-temporal propagation affected by emergencies, the deterioration of the overall capacity of the railway network under emergencies and the train operation rules of the high-speed railway under the conditions of the event, which will lead to structural changes in the high-speed railway system, and ultimately lead to changes in system behavior, the need to develop a new transport organization strategy.

1. The mechanism of temporal and spatial propagation affected by emergencies

As a subset of generalized emergencies, the sudden events of the railway system have the basic characteristics of general emergencies, but they have their own particularities because of their different objects and influence objects, such as resonance among the events and expansion of impact scale introduced in Chap. 2. Under the conditions of high-speed railway network operation, the impact of emergencies will spread along the two dimensions of time space and cyberspace. Especially in future, in the case of coordinated transportation with existing railways and urban rail transit, the spread of the impact of emergencies will be more complicated. The research on this issue can be started from the media affected by emergencies, especially for the typical and severe emergencies, focusing on the propagation path and activity law of the impact, and fully depicting the temporal and spatial diffusion process affected by the emergency. It provides the basis for operational management and emergency management under emergency conditions.

2. The deterioration law of the overall capacity of the road network under emergency conditions

Transport capacity is the foundation for the development and implementation of all transport organization strategies and emergency management strategies. Accurate estimation of transport capacity under emergency conditions is the first step in carrying out emergency rescue work. The research on this problem, first of all, we need to analyze the coupling relationship between key nodes, lines, and networks under the condition of network structure. Then we can accurately identify the system capacity bottleneck on the basis of predicting the impact process of sudden events. The prediction technology of key hub distribution capacity and the road passage the capacity and the effective carrying capacity of the road network, are studied to explore the this multi-scale capability change and the interaction mechanism between the "point-line-network", so as to realize the scientific description of the deterioration process of the effective carrying capacity of the entire road network.

3. High-speed railway driving rules under emergency conditions

The train operation rules are a driving specification for the components of the high-speed railway system (drivers, attendants, bullet trains, etc.). The development of operation rules needs to consider various complicated conditions such as the line environment and equipment status. Under the conditions of emergencies, the development of high-speed railway operation rules is more complicated with the spread of incidents and the loss of road network capacity. The solution to this problem can be considered from two aspects: guiding principles and architecture.

In terms of guiding principles for the development of operation rules, the traditional rules focus on the driver's code of conduct and the principles of use and maintenance of various types of equipment and facilities under general conditions. Although some emergency measures are involved in emergency situations, they are all formulated separately by various railway management departments and are not systematic. The high-speed railway operation rules under emergencies pay more attention to how to quickly restore normal driving order in the event of an emergency and the failure of traditional driving rules. The fundamental principle of rulemaking is to reduce the overall loss and reflect the "people-oriented" concept. This fundamental change in guiding principles and concepts makes the architecture and implementation of high-speed railway driving rules under emergencies very different from traditional driving rules.

In terms of the architecture of the driving rules, the traditional driving rules are generally formulated and organized according to the type of operation, which can be divided into five parts: basic requirements, grouping trains, shunting work, train blocking method, and train operation. The high-speed railway operation rules under the emergency conditions should establish a hierarchical rule system due to the complex types of emergencies and various disposal measures. The first layer is the general guiding principle, which mainly explains the main purpose, basic requirements and overall process of the operation work in emergency conditions. The second layer is the rule of characterization scope, which mainly describes the transportation organization boundary involved in the emergency, including the space boundary (such as the scope of the event, the range of the train, the scope of the equipment and facilities) and the object boundary (such as management department, emergency rescue department, affected passengers in the emergency conditions). Only when the scope of application of the operation rules is determined, the next step can be carried out to implement specific disposal rules, which is the content of the third layer. The rules of the third layer should cover all possible emergency scenarios as much as possible and give specific train running routes, driving speeds, and emergency rescue procedures.

The different types and levels of emergencies, geographical and climatic differences, and the interaction of multiple events make it difficult to formulate operation rules under emergencies. At present, no country or organization can provide a complete set of high-speed railway operating rules under emergency conditions. However, it is an indispensable research content for future railway development to formulate perfect and systematic operating rules of high-speed railway under the condition of emergency.

10.2 The Second Category: Transportation Organization Mode and Strategy

The second type of problem aims to form a network-wide transport organization model and emergency response strategy that adapts to the affected area through the rational allocation of various resources. This is the core content of high-speed railway operation management and emergency management under emergency conditions. Specifically, it includes the prediction and distribution of high-speed railway passenger flow under emergency conditions, the allocation of high-speed railway transportation resources under emergency conditions, the customization, evaluation, and optimization of high-speed railway operation schemes under emergency conditions, as well as the formulation, evaluation and optimization of the emergency disposal plan for high-speed railways.

1. High-speed railway passenger flow forecast under emergency conditions

The occurrence of railway emergencies will generally lead to sudden changes in the demand for a large number of passengers. Under the conditions of emergencies, traditional passenger flow prediction methods are often ineffective. In addition, the lack of historical data to refer to, the uncertainty of the impact of emergencies and other factors, greatly increasing the complexity of passenger flow forecasting. To solve this problem, we can focus on the analysis of passenger flow characteristics, explore the temporal and spatial evolution of passenger flow distribution under the influence of unexpected events, and conduct sensitivity analysis on the key factors of sensitivity caused by emergencies, so as to accurately predict the competition relationship of multiple transportation modes. The passenger travel mode selection under the action of the passengers is constructed to construct the passenger flow forecasting model under the multi-scale of "point-line-net," which provides the basis for the passenger flow distribution of high-speed railway under the emergency conditions.

2. High-speed railway passenger flow allocation under emergency conditions

The distribution of passenger flow in high-speed railways under emergencies is quite different from the distribution of passengers under general conditions. First of all, in the principle of passenger flow allocation, under the emergency conditions, the safe and rapid evacuation of passengers should be the primary premise. The factors such as passenger consumption level, riding preference, and comfort are not considered as key contents in the general conditions. They might be ignored directly under the malignant emergency condition. This difference in guiding philosophy will inevitably lead to huge differences in distribution strategies. In addition, under the emergency conditions, the passenger flow changes of the high-speed railway are nonlinear fluctuations, which often cause a large amount of passenger flow. In the case of limited capacity, it is very difficult to achieve reasonable configuration of the whole network. To solve this problem, we can firstly define the reasonable passenger flow distribution principle, analyze the transportation demand transfer law under the influence of emergency events, and establish a high-speed railway passenger flow distribution model under the emergency conditions.

3. High-speed railway capacity allocation under emergency conditions

The optimal allocation of railway transportation capacity resources is to optimize the combination of various fixed and mobile transportation resources so that the transportation capacity is compatible with the transportation needs. The allocation of transportation resources is the basis for the formulation of the operation plan and is related to the smooth implementation of various transportation plans. Under the emergency conditions, the allocation of transport resources is not only restricted by the distribution structure of existing high-speed railway fixed facilities, but also affected by emergencies and passenger flow allocation. In addition, under the conditions of bad emergencies, a large number of scheduling problems of off-road emergency resources will also be involved. In a complex dynamic environment, the rational configuration and scheduling of multiple types of resources is a very complicated problem. The allocation of capacity resources should be combined with transportation needs. Under the conditions of emergencies, we first clarify the allocation principle of transportation resources, explore the cooperative allocation strategy of mobile equipment, fixed facilities and emergency resources, and study the measurement index and evaluation methods of "point-line-network" multi-scale capacity matching problem, and finally the feedback optimization of transportation demand and transportation capacity resource allocation is realized.

4. Customization, evaluation, and optimization of high-speed railway operation schemes under emergency conditions

The principles and strategies of high-speed railway operation organization under emergencies are quite different from those under general conditions. The core is "people-oriented," with emphasis on ensuring global security. This change in guiding principles has led to fundamental changes in the optimization objectives and constraints of train operation program development. To solve this problem, we should first establish the mapping relationship between the impact degree of different types of emergencies and events and the optimization objectives and constraints of the plan, and sort out the inheritance and feedback relationship between the train running route and the train operation plan; it is necessary to establish a set of high-speed railway operation plan evaluation indicators suitable for emergency conditions. Through feedback optimization of "assessment-optimization-assessment", a operation plan that can be truly applied to the actual operating environment on site under emergencies is finally formed.

5. Formulation, evaluation, and optimization of high-speed railway emergency disposal plan

Under the condition of vicious emergency, the high-speed railway emergency disposal plan is an important extension of the driving plan, and also an important basis for emergency disposal and emergency management. First of all, it should clarify the relevant emergency agencies, team composition, and their responsibilities. At the same time, it is necessary to integrate the resources inside and outside the railway, explore the coupling relationship between emergency disposal points and driving organization strategies, and study the evaluation, sequencing and splicing techniques of disposal points. The operation plan should be coordinated to minimize the number of casualties and property losses caused by emergencies.

10.3 The Third Category: System Intelligent Services

The third type of problem is how to use information and intelligent means to provide efficient management solutions for railway management departments and to provide the public with comprehensive and intelligent information services. It mainly includes the cooperative transportation of various transportation modes under the conditions of emergency and high-speed railway comprehensive information services under emergencies.

1. Collaborative transportation of multiple transportation modes under emergency conditions

Under the conditions of emergencies, the loss of high-speed railway transportation capacity is serious. This can be used to integrate resources of various modes of transportation such as railways, suburban railways, and urban rail transit by means of informationization, and use big data and cloud computing technologies to properly match each. The transportation capacity and capacity of the transportation mode can realize the balanced distribution based on transportation demand under the emergency conditions. In addition, the competitive relationship of various modes of transportation under the conditions of emergencies is analyzed, and the collaborative organization, optimization, and evaluation model of transportation organization with scalable targets and parameters are constructed to realize the coordinated transportation of multiple modes of transportation under emergency conditions.

2. High-speed railway comprehensive information service under emergency conditions

Once an emergency occurs, it will often attract widespread public attention. However, at present, both existing railways and high-speed railways have problems such as information independence and lack of services. The timely feedback of the impact of emergencies changes in transportation demand, and recovery of transportation capacity directly affects the efficiency of emergency response and the implementation of driving plans. Under the conditions of emergencies, information technologies such as mobile Internet and cloud computing can be used to dynamically track the status of high-speed railway transportation systems, explore the coordinated processing and release mechanism of vehicle information for emergencies, and study the comprehensive information push technology for the public, to build a high-speed railway accident information reporting service platform under emergencies and to communicate emergency response information to the public in a timely and transparent manner based on the new generation communication technology to provide passengers with safe, efficient and intelligent transportation services.

10.4 Chapter Summary

This chapter mainly introduces the open problems that need to be solved urgently under the conditions of emergency. The high-speed railway system is a complex giant system, and the above problems are only used to start a discussion. Readers can establish their own research paths based on their own research directions and focus on relevant research work.

Appendix A
The Rescheduled Timetables

See Tables A.1, A.2, A.3, A.4, A.5, A.6, A.7, A.8, A.9, A.10, A.11, A.12, A.13, A.14, A.15, A.16 and A.17.

Table A.1 Rescheduled timetable by no-relaxation method in case of single train delay at single section

Train No.	K115		L15		K101		G105	
Station	Arrival	Departure	Arrival	Departure	Arrival	Departure	Arrival	Departure
Nanjing Nan	11.2000	11.2000	11.2632	11.2632	11.3006	11.3206	11.4152	11.4352
Zhenjing Xi	11.3500	11.3500	11.3822	11.3822	11.4628	11.4628	11.5814	11.5814
Changzhou Bei	12.1143	12.1741	12.1400	12.1400	12.1700	12.2000	12.2000	12.2259
Wuxi Dong	12.2714	12.3645	12.2415	12.2515	12.3015	12.3115	12.3314	12.3414
Suzhou Bei	12.4127	12.4127	12.3007	12.3007	12.3607	12.3607	12.3907	12.3907
Kunshan Nan	12.4658	12.4658	12.3549	12.3549	12.4119	12.4119	12.4449	12.4449
Hongqiao	12.5505	−1.0000	12.4344	−1.0000	12.4944	−1.0000	12.5244	−1.0000

L. Jia et al., *High-Speed Railway Operation Under Emergent Conditions*, Advances in High-speed Rail Technology,
https://doi.org/10.1007/978-3-662-63033-4

Table A.2 Rescheduled timetable by relaxation method in case of single train delay at one section

Train No.	K115		L15		K101		G105	
Station	Arrival	Departure	Arrival	Departure	Arrival	Departure	Arrival	Departure
Nanjing Nan	11.2000	11.2000	11.2632	11.2632	11.3006	11.3206	11.4152	11.4352
Zhenjing Xi	11.3500	12.0044	11.3754	11.3754	11.4628	12.0314	11.5814	11.5814
Changzhou Bei	12.1330	12.1330	12.0830	12.0830	12.1600	12.1700	12.1100	12.1100
Wuxi Dong	12.2310	12.2410	12.1810	12.1810	12.2640	12.2640	12.2040	12.2040
Suzhou Bei	12.2845	12.2845	12.2344	12.2344	12.3115	12.3115	12.2615	12.2615
Kunshan Nan	12.3407	12.3407	12.2906	12.2906	12.3637	12.3637	12.3137	12.3137
Hongqiao	12.4157	−1.0000	12.3634	−1.0000	12.4427	−1.0000	12.3927	−1.0000

Table A.3 Rescheduled timetable in case of single train stochastic delay at one section

Train No.	K115		L15		K101		G105	
Station	Arrival	Departure	Arrival	Departure	Arrival	Departure	Arrival	Departure
Nanjing Nan	11.2000	11.2000	11.2632	11.2632	11.3006	11.3206	11.4152	11.4352
Zhenjing Xi	11.3700	11.3700	11.3939	11.3939	11.4728	11.4728	11.5814	11.5814
Changzhou Bei	12.0243	12.0725	12.0455	12.0455	12.0708	12.1308	12.0952	12.0952
Wuxi Dong	12.1858	12.1858	12.1520	12.1520	12.2253	12.2353	12.1946	12.2046
Suzhou Bei	12.2336	12.2336	12.2030	12.2030	12.2835	12.2835	12.2547	12.2547
Kunshan Nan	12.2856	12.2856	12.2515	12.2515	12.3309	12.3309	12.3050	12.3050
Hongqiao	12.3605	−1.0000	12.3250	−1.0000	12.4358	−1.0000	12.4027	−1.0000

Table A.4 Rescheduled timetable by no-relaxation method in case of two trains delay at multi-sections

Train No.	K115		L15		K101		G105	
Station	Arrival	Departure	Arrival	Departure	Arrival	Departure	Arrival	Departure
Nanjing Nan	11.2000	11.2000	11.2632	11.2632	11.3006	11.3206	11.4152	11.4352
Zhenjing Xi	11.3500	11.3500	11.3822	11.3822	11.4628	11.4628	11.5814	11.5814
Changzhou Bei	12.0700	12.2530	12.1400	12.1400	12.1700	12.2000	12.2000	12.2259
Wuxi Dong	12.3555	12.3555	12.2415	12.2515	12.3015	12.3115	12.3314	12.3414
Suzhou Bei	12.4147	12.4147	12.3007	12.3007	12.3607	12.3607	12.3907	12.3907
Kunshan Nan	12.4709	12.4709	12.3549	12.3549	12.4119	12.4119	12.4449	12.4449
Hongqiao	12.5515	−1.0000	12.4344	−1.0000	12.4944	−1.0000	12.5244	−1.0000

Table A.5 Rescheduled timetable by relaxation method in case of two trains delay at multi-sections

Train No.	K115		L15		K101		G105	
Station	Arrival	Departure	Arrival	Departure	Arrival	Departure	Arrival	Departure
Nanjing Nan	11.2000	11.2000	11.2632	11.2632	11.3006	11.3206	11.4152	11.4352
Zhenjing Xi	11.3500	11.4928	11.3822	11.3822	11.4628	11.4628	11.5814	11.5814
Changzhou Bei	12.0700	12.1300	12.0100	12.0100	12.0400	12.0400	12.1030	12.1030
Wuxi Dong	12.2315	12.2315	12.1125	12.1225	12.1415	12.1515	12.2045	12.2045
Suzhou Bei	12.2807	12.2807	12.1706	12.1706	12.2007	12.2007	12.2536	12.2536
Kunshan Nan	12.3349	12.3349	12.2248	12.2248	12.2549	12.2549	12.3118	12.3118
Hongqiao	12.4227	−1.0000	12.3043	−1.0000	12.3441	−1.0000	12.3927	−1.0000

Table A.6 Rescheduled timetable in case of two trains stochastic delay at multi-sections

Train No.	K115		L15		K101		G105	
Station	Arrival	Departure	Arrival	Departure	Arrival	Departure	Arrival	Departure
Nanjing Nan	11.2000	11.2000	11.2632	11.2632	11.3006	11.3206	11.4152	11.4352
Zhenjing Xi	11.3500	11.5010	11.3754	11.3754	11.4728	11.4728	11.5914	11.5914
Changzhou Bei	12.1143	12.1648	12.0326	12.0326	12.0608	12.0608	12.1310	12.1310
Wuxi Dong	12.2742	12.2742	12.1338	12.1318	12.1553	12.1653	12.2345	12.2345
Suzhou Bei	12.3345	12.3345	12.1913	12.1913	12.2128	12.2128	12.2815	12.2815
Kunshan Nan	12.3924	12.3924	12.2438	12.2438	12.2709	12.2709	12.3420	12.3420
Hongqiao	12.4716	−1.0000	12.3210	−1.0000	12.3441	−1.0000	12.4310	−1.0000

Table A.7 Rescheduled timetable by no-relaxation method in case of single trains delay and single section speed limitation

Train No.	K115		L15		K101		G105	
Station	Arrival	Departure	Arrival	Departure	Arrival	Departure	Arrival	Departure
Nanjing Nan	11.2000	11.2000	11.2632	11.2632	11.3006	11.3206	11.4152	11.4352
Zhenjing Xi	12.3807	12.3807	12.4439	12.4439	12.5013	12.5013	13.0159	13.0159
Changzhou Bei	13.1600	13.3015	13.1900	13.1900	13.2200	13.2500	13.2500	13.2800
Wuxi Dong	13.4030	13.4030	13.2915	13.2915	13.3515	13.3615	13.3815	13.3915
Suzhou Bei	13.4507	13.4507	13.3507	13.3507	13.4107	13.4107	13.4407	13.4407
Kunshan Nan	13.5149	13.5149	13.4049	13.4049	13.4649	13.4649	13.4949	13.4949
Hongqiao	13.5944	−1.0000	13.4844	−1.0000	13.5444	−1.0000	13.5744	−1.0000

Table A.8 Rescheduled timetable obtained by relaxation method in case of single trains delay and single section speed limitation

Train No.	K115		L15		K101		G105	
Station	Arrival	Departure	Arrival	Departure	Arrival	Departure	Arrival	Departure
Nanjing Nan	11.2000	11.2000	11.2632	11.2632	11.3006	11.3206	11.4152	11.4352
Zhenjing Xi	12.2506	12.2506	12.3138	12.3138	12.3712	12.3712	12.4858	12.4858
Changzhou Bei	13.0530	13.1530	13.0800	13.0800	13.1030	13.1030	13.1300	13.1300
Wuxi Dong	13.2545	13.2545	13.1724	13.1824	13.2045	13.2815	13.2315	13.2315
Suzhou Bei	13.3037	13.3037	13.2252	13.2252	13.3306	13.3306	13.2807	13.2807
Kunshan Nan	13.3619	13.3619	13.2805	13.2805	13.3849	13.3849	13.3349	13.3349
Hongqiao	13.4414	−1.0000	13.3520	−1.0000	13.4644	−1.0000	13.4144	−1.0000

Table A.9 Rescheduled timetable obtained in case of single trains delay and single section stochastic speed limitation

Train No.	K115		L15		K101		G105	
Station	Arrival	Departure	Arrival	Departure	Arrival	Departure	Arrival	Departure
Nanjing Nan	11.2000	11.2000	11.2632	11.2632	11.3006	11.3206	11.4152	11.4352
Zhenjing Xi	12.3406	12.3406	12.4038	12.4038	12.4612	12.4612	12.5758	12.5758
Changzhou Bei	13.1212	13.2355	13.1541	13.1541	13.1823	13.1823	13.2104	13.2104
Wuxi Dong	13.3410	13.3410	13.2525	13.2636	13.2806	13.3700	13.3145	13.3145
Suzhou Bei	13.4005	13.4005	13.3059	13.3059	13.4140	13.4240	13.3622	13.3622
Kunshan Nan	13.4600	13.4600	13.3623	13.3623	13.4805	13.4805	13.4146	13.4146
Hongqiao	13.5335	−1.0000	13.4354	−1.0000	13.5634	−1.0000	13.4917	−1.0000

Table A.10 Rescheduled timetable obtained by no-relaxation method in case of multi-trains delay and all sections stochastic speed limitation

Train No.	K115		L15		K101		G105	
Station	Arrival	Departure	Arrival	Departure	Arrival	Departure	Arrival	Departure
Nanjing Nan	11.2000	11.2000	11.2632	11.2632	11.3006	11.3206	11.4152	11.4352
Zhenjing Xi	11.4602	11.4602	11.5234	11.5234	11.5808	11.5808	12.0954	12.0954
Changzhou Bei	12.3100	12.4530	12.3400	12.3400	12.3700	12.4000	12.4000	12.4300
Wuxi Dong	13.0733	13.0833	12.5633	12.5733	13.0233	13.0233	13.0533	13.0633
Suzhou Bei	13.1916	13.1916	13.0816	13.0816	13.1416	13.1416	13.1716	13.1716
Kunshan Nan	13.3148	13.3148	13.2048	13.2048	13.2648	13.2648	13.2948	13.2948
Hongqiao	13.4913	−1.0000	13.3813	−1.0000	13.4413	−1.0000	13.4713	−1.0000

Table A.11 Rescheduled timetable obtained by relaxation method in case of multi-trains delay and all sections stochastic speed limitation

Train No.	K115		L15		K101		G105	
Station	Arrival	Departure	Arrival	Departure	Arrival	Departure	Arrival	Departure
Nanjing Nan	11.2000	11.2000	11.2632	11.2632	11.3006	11.3206	11.4152	11.4352
Zhenjing Xi	11.4258	11.4258	11.4930	11.4930	11.5504	11.5504	12.0650	12.0650
Changzhou Bei	12.2530	12.3530	12.2800	12.2800	12.3030	12.3030	12.3300	12.3300
Wuxi Dong	12.5524	12.5524	12.4754	12.4754	12.5024	12.5824	12.5323	12.5323
Suzhou Bei	13.0751	13.0751	12.5821	12.5821	13.0951	13.0951	13.0521	13.0521
Kunshan Nan	13.1843	13.1843	13.0924	13.0924	13.2054	13.2054	13.1624	13.1624
Hongqiao	13.3645	−1.0000	13.2446	−1.0000	13.3816	−1.0000	13.3246	−1.0000

Table A.12 Rescheduled timetable obtained in case of multi-trains delay and all sections stochastic speed limitation

Train No.	K115		L15		K101		G105	
Station	Arrival	Departure	Arrival	Departure	Arrival	Departure	Arrival	Departure
Nanjing Nan	11.2000	11.2000	11.2632	11.2632	11.3006	11.3206	11.4152	11.4352
Zhenjing Xi	11.4428	11.5919	11.5100	11.5100	11.5634	11.5634	12.0820	12.0820
Changzhou Bei	12.3330	12.4139	12.2615	12.2615	12.3045	12.3045	12.3900	12.3900
Wuxi Dong	13.0311	13.0311	12.4726	13.4726	12.5156	12.5259	13.0056	13.0056
Suzhou Bei	13.1315	13.1315	12.5845	12.5845	13.0257	13.0257	13.1100	13.1100
Kunshan Nan	13.2502	13.2502	13.0932	13.0932	13.1444	13.1444	13.2247	13.2247
Hongqiao	13.4124	−1.0000	13.2654	−1.0000	13.3106	−1.0000	13.3909	−1.0000

Table A.13 Rescheduled timetable obtained by non-relaxation method in case of comprehensive conditions

Train No.	G325		G311		G313	
Station	Arrival	Departure	Arrival	Departure	Arrival	Departure
Zhengjiang Xi	8.1322	8.1322	8.3422	8.3422	9.0422	9.0422
Changzhou Bei	8.4132	8.4132	9.0232	9.0332	9.3232	9.3332
Wuxi Dong	9.0733	9.0733	9.2933	9.2933	9.5933	9.5933
Train No.	G301		G303		G305	
Station	Arrival	Departure	Arrival	Departure	Arrival	Departure
Zhengjiang Xi	9.3922	9.3922	10.1322	10.1322	10.3622	10.3622
Changzhou Bei	10.0732	10.0732	10.4132	10.4132	11.0432	11.0432
Wuxi Dong	10.3333	10.3333	11.0733	11.0733	11.3033	11.3033

Table A.14 Rescheduled timetable obtained by relaxation method in case of comprehensive conditions

Train No.	G325		G311		G313	
Station	Arrival	Departure	Arrival	Departure	Arrival	Departure
Zhengjiang Xi	8.1322	8.1322	8.3422	8.3422	9.0422	9.0422
Changzhou Bei	8.3659	8.3659	8.5759	8.5859	9.2759	9.2859
Wuxi Dong	8.5848	8.5848	9.2048	9.2048	9.5048	9.5048
Train No.	G301		G303		G305	
Station	Arrival	Departure	Arrival	Departure	Arrival	Departure
Zhengjiang Xi	9.3922	9.3922	10.1322	10.1322	10.3622	10.3622
Changzhou Bei	10.0259	10.0259	10.3659	10.3659	10.5959	10.5959
Wuxi Dong	10.2448	10.2448	10.5848	10.5848	11.2148	11.2148

Table A.15 Rescheduled timetable obtained in case of uncertain conditions

Train No.	G325		G311		G313	
Station	Arrival	Departure	Arrival	Departure	Arrival	Departure
Zhengjiang Xi	8.1322	8.1322	8.3422	8.3422	9.0422	9.0422
Changzhou Bei	8.4016	8.4016	9.0116	9.0216	9.3116	9.3216
Wuxi Dong	9.0617	9.0617	9.2817	9.2817	9.5817	9.5817
Train No.	G301		G303		G305	
Station	Arrival	Departure	Arrival	Departure	Arrival	Departure
Zhengjiang Xi	9.3922	9.3922	10.1322	10.1322	10.3622	10.3622
Changzhou Bei	10.0616	10.0616	10.4016	10.4016	11.0316	11.0316
Wuxi Dong	10.3217	10.3217	11.0617	11.0617	11.2917	11.2917

Table A.16 Scheduled timetable of Shanghai-Nanjing Intercity Railway in case of comprehensive experiments ($\boldsymbol{\alpha} = \mathbf{0.3}$)

Train No.	G325		G311		G313	
Station	Arrival	Departure	Arrival	Departure	Arrival	Departure
Changzhou	8.3817	8.4400	8.5917	9.0017	9.2917	9.3017
Wuxi	9.0118	9.0218	9.1133	9.1133	9.4133	9.4133
Suzhou	9.1042	9.1042	9.1957	9.2057	9.4957	9.5057
Kunshan Nan	9.1746	9.1746	9.2810	9.2810	10.0300	10.0300
Train No.	G301		G323		G7005	
Station	arrival	departure	arrival	departure	arrival	departure
Changzhou	10.0417	10.0417	8.0800	8.0900	9.3600	9.3600
Wuxi	10.1659	10.1759	8.2150	8.2150	9.4600	9.4733
Suzhou	10.2624	10.2624	8.3014	8.3114	9.5600	9.5600
Kunshan Nan	10.3328	10.3328	8.3928	8.3928	10.0600	10.0600

Table A.17 Scheduled timetable of Shanghai-Nanjing Intercity Railway in case of comprehensive experiments ($\boldsymbol{\alpha} = \mathbf{0.7}$)

Train No.	G325		G311		G313	
Station	Arrival	Departure	Arrival	Departure	Arrival	Departure
Changzhou	8.3817	8.4400	8.5917	9.0017	9.2917	9.3017
Wuxi	9.0118	9.0218	9.1133	9.1133	9.4325	9.4325
Suzhou	9.1042	9.1042	9.1957	9.2057	9.5900	10.000
Kunshan Nan	9.1746	9.1746	9.2810	9.2810	10.1000	10.1000
Train No.	G301		G323			
Station	Arrival	Departure	Arrival	Departure		
Changzhou	10.0417	10.0417	8.0800	8.0900		
Wuxi	10.1659	10.1759	8.2150	8.2150		
Suzhou	10.2624	10.2624	8.3014	8.3114		
Kunshan Nan	10.3328	10.3328	8.3928	8.3928		

Note XX.YYZZ denotes XX hour YY minute ZZ second

Appendix B
The Scheduled Timetables

See Tables B.1, B.2, B.3, and B.4.

L. Jia et al., *High-Speed Railway Operation Under Emergent Conditions*,
Advances in High-speed Rail Technology,
https://doi.org/10.1007/978-3-662-63033-4

Table B.1 Scheduled timetable of high-speed railway

Train No.	G301		G303		G305		G101	
Station	Arrival	Departure	Arrival	Departure	Arrival	Departure	Arrival	Departure
Xuzhou Dong	8.0900	8.1100	8.4500	8.4600	9.1400	9.1600	9.3300	9.3500
Suzhou Dong	8.2549	8.2549	9.0049	9.0049	9.3049	9.3049	9.4949	9.4949
Bengbu Nan	8.4500	8.4600	9.2000	9.2100	9.4610	9.4610	10.0900	10.1000
Ding yuan	8.5814	8.5814	9.3314	9.3314	9.5524	9.5524	10.2214	10.2214
Chuzhou Nan	9.0903	9.0903	9.4403	9.4403	10.0613	10.0613	10.3303	10.3303
Nanjing Nan	9.2300	9.2500	9.5800	9.5900	10.2000	10.2200	10.4700	10.4900
Train No.	G103		G105		G107		G109	
Station	Arrival	Departure	Arrival	Departure	Arrival	Departure	Arrival	Departure
Xuzhou Dong	9.5200	9.5300	10.1800	10.1900	10.5300	10.5500	11.1300	11.1500
Suzhou Dong	10.0749	10.0749	10.3423	10.3423	11.0949	11.0949	11.2949	11.2949
Bengbu Nan	10.2700	10.3400	10.5300	10.5400	11.2900	11.3000	11.4900	11.5000
Ding yuan	10.4614	10.4614	11.0558	11.0558	11.4214	11.4214	12.0214	12.0214
Chuzhou Nan	10.5703	10.5703	11.1647	11.1647	11.5303	11.5303	12.1303	12.1303
Nanjing Nan	11.1100	11.1300	11.3006	11.3206	12.0700	12.0900	12.2700	12.2900
Train No.	G111		G113					
Station	Arrival	Departure	Arrival	Departure				
Xuzhou Dong	11.3300	11.3500	11.5300	11.5500				
Suzhou Dong	11.4949	11.4949	12.0949	12.0949				
Bengbu Nan	12.0900	12.1000	12.2900	12.3000				
Train No.	G111		G113					
Station	Arrival	Departure	Arrival	Departure				

(continued)

Table B.1 (continued)

Train No.	G111		G113					
Station	Arrival	Departure	Arrival	Departure				
Ding yuan	12.2214	12.2214	12.4214	12.4214				
Chuzhou Nan	12.3303	12.3303	12.5303	12.5303				
Nanjing Nan	12.4700	12.4900	13.0700	13.0900				

Note (1) The data of Table B.1 is from the research results of "The high-speed train operation plans optimization key technologies and system" project. (2) XX.YYZZ denotes XX hour YY minute ZZ second

Table B.2 Scheduled timetable of normal speed existed railway

Train No.	K58/5		K518/5		K101/4/1		D88/5	
Station	Arrival	Departure	Arrival	Departure	Arrival	Departure	Arrival	Departure
Xuzhou	9.1600	9.3500	9.3000	9.4100	9.4400	9.4700	10.0200	10.0300
Gaojiaying	9.4300	9.4300	9.4900	9.4900	9.5500	9.5500	10.0929	10.0929
Sanpu	9.4700	9.4700	9.5300	9.5300	9.5900	9.5900	10.1202	10.1202
Caocun	9.5445	9.5445	10.0045	10.0045	10.0800	10.2100	10.1624	10.1624
Jiagou	10.0430	10.0430	10.1200	10.2400	10.3245	10.3245	10.2049	10.2049
Fuliji	10.1700	10.1700	10.3830	10.3830	10.4515	10.4515	10.2700	10.2700
Suzhou	10.2445	10.2445	10.4615	10.4615	10.5300	10.5300	10.3230	10.3230
Suzhou Nan	10.2800	10.4000	10.4815	10.4815	10.5500	10.5500	10.3353	10.3353
Xisipo	10.4845	10.4845	10.5500	10.5500	11.0145	11.0145	10.3757	10.3757
Luling	10.5245	10.5245	10.5900	10.5900	11.0545	11.0545	10.4018	10.4018
Tangnanji	11.0130	11.0130	11.0745	11.0745	11.1430	11.1430	10.4528	10.4528
Guzhen	11.0430	11.0430	11.1045	11.1045	11.1730	11.1730	10.4732	10.4732
Liancheng	11.0845	11.0845	11.1500	11.1500	11.2145	11.2145	10.5033	10.5033
Xinmaqiao	11.1245	11.1245	11.1900	11.1900	11.2545	11.2545	10.5327	10.5327
Caolaoji	11.2315	11.2315	11.2930	11.2930	11.3615	11.3615	11.0117	11.0117
Bengbu	11.3700	11.4200	11.4300	11.4800	11.5000	11.5800	11.0830	11.0930
Bengbu Dong	11.4900	11.4900	11.5500	11.5500	12.0500	12.0500	11.1500	11.1500
Fengyang	11.5800	11.5800	12.0400	12.0400	12.1400	12.1400	11.2152	11.2152
Train No.	K58/5		K518/5		K101/4/1		D88/5	
Station	Arrival	Departure	Arrival	Departure	Arrival	Departure	Arrival	Departure

(continued)

Table B.2 (continued)

Train No.	K58/5		K518/5		K101/4/1		D88/5	
Station	Arrival	Departure	Arrival	Departure	Arrival	Departure	Arrival	Departure
Bangqiao	12.0330	12.0330	12.0930	12.0930	12.1930	12.1930	11.2558	11.2558
Xiaoxihe	12.0800	12.0800	12.1400	12.1400	12.2400	12.2400	11.2922	11.2922
Minggang	12.1900	12.1900	12.2500	12.2500	12.3500	12.3500	11.3740	11.3740
Bianzhuang	12.2300	12.2300	12.2900	12.2900	12.3900	12.3900	11.4030	11.4030
Guandian	12.2900	12.2900	12.3500	12.3500	12.4500	12.4500	11.4434	11.4434
Sanjie	12.3500	12.3500	12.4100	12.4100	12.5100	12.5100	11.4902	11.4902
Zhangbaling	12.4330	12.4330	12.4930	12.4930	12.5930	12.5930	11.5416	11.5416
Shaheji	12.4930	12.4930	12.5530	12.5530	13.0530	13.0530	11.5813	11.5813
Chuzhou	12.5830	12.5830	13.0430	13.0430	13.1430	13.1430	12.0413	12.0413
Danzhi	13.0530	13.0530	13.1130	13.1130	13.2130	13.2130	12.0841	12.0841
Dongge	13.1500	13.1500	13.2100	13.2100	13.3100	13.3100	12.1530	12.1530
Yongnizhen	13.2230	13.2230	13.3000	13.3300	13.3930	13.4800	12.1846	12.1846
Linchang	13.2800	13.2800	13.4030	13.4030	13.5700	14.0600	12.2237	12.2237
Nanjing	13.3800	13.5100	13.5100	14.1300	14.1800	14.2300	12.3000	12.3200
Train No.	1230/27		10135		10625		11301	
Station	Arrival	Departure	Arrival	Departure	Arrival	Departure	Arrival	Departure
Xuzhou	9.5600	10.0900	10.1800	10.1800	10.2800	10.2800	10.3800	10.3800
Gaojiaying	10.1700	10.1700	10.2600	10.2600	10.3600	10.3600	10.4600	10.4600
Sanpu	10.2200	10.2500	10.3200	10.3200	10.4200	10.4200	10.5200	10.5200

(continued)

Table B.2 (continued)

Train No.	1230/27		10135		10625		11301	
Station	Arrival	Departure	Arrival	Departure	Arrival	Departure	Arrival	Departure
Caocun	10.3500	10.3500	10.4500	10.4500	10.5500	10.5500	11.0500	11.0500
Jiagou	10.4445	10.4445	10.5800	10.5800	11.0800	11.0800	11.2000	11.4100
Fuliji	10.5715	10.5715	11.1700	11.2200	11.2700	11.5100	12.0100	12.0100
Suzhou	11.0600	11.0800	11.3700	11.3700	12.0600	12.0600	12.1300	12.1300
suzhouNan	11.1200	11.1200	11.4200	11.5900	12.0900	12.0900	12.1600	12.1600
Xisipo	11.1845	11.1845	12.1300	12.1300	12.2000	12.2000	12.2700	12.2700
Luling	11.2245	11.2245	12.2000	12.2000	12.2700	12.2700	12.3330	12.3330
Tangnanji	11.3130	11.3130	12.3300	12.3300	12.4000	12.4000	12.4630	12.4630
Train No.	1230/27		10135		10625		11301	
Station	Arrival	Departure	Arrival	Departure	Arrival	Departure	Arrival	Departure
Guzhen	11.3430	11.3430	12.3730	12.3730	12.4700	12.4900	12.5500	12.5600
Liancheng	11.3845	11.3845	12.4330	12.4330	12.5800	12.5800	13.0500	13.0500
Xinmaqiao	11.4245	11.4245	12.5000	12.5000	13.0430	13.0430	13.1130	13.1130
Caolaoji	11.5315	11.5315	13.0400	13.0400	13.1830	13.1830	13.2530	13.2530
Bengbu	12.0700	12.1500	13.1700	13.1700	13.3100	13.3100	13.3930	13.3930
Bengbu Dong	12.2200	12.2200	13.2700	13.3900			13.4900	14.0100
Fengyang	12.3100	12.3100	13.5700	13.5700			14.2100	14.3100
Bangqiao	12.3630	12.3630	14.0630	14.0630			14.4330	14.4330
Xiaoxihe	12.4200	12.5400	14.1330	14.1330			14.5300	15.2900

(continued)

Table B.2 (continued)

Train No.	1230/27		10135		10625		11301	
Station	Arrival	Departure	Arrival	Departure	Arrival	Departure	Arrival	Departure
Minggang	13.0700	13.0700	14.3400	14.4800			15.5000	15.5000
Bianzhuang	13.1100	13.1100	14.5900	15.4300			15.5800	15.5800
Guandian	13.1700	13.1700	15.5500	15.5500			16.0900	16.2600
Sanjie	13.2300	13.2300	16.0800	16.2900			16.3930	16.3930
Zhangbaling	13.3130	13.3130	16.4500	16.4500			16.5230	16.5230
Shaheji	13.3730	13.3730	16.5300	16.5300			17.0030	17.0030
Chuzhou	13.4800	13.5100	17.0530	17.0530			17.1300	17.1300
Danzhi	14.0000	14.0000	17.1600	17.1800			17.2400	17.2500
Dongge	14.1000	14.1000	17.3600	17.3600			17.4300	17.4300
Yongnizhen	14.1730	14.1730	17.4500	17.4500			17.5200	17.5200
Linchang	14.2400	14.2900	18.0200	20.0700			18.5100	20.4100
Nanjing	14.4100	14.5300	20.2300	20.2300			20.5700	20.5700
Train No.	23005		11305		T114/1		K190/87	
Station	Arrival	Departure	Arrival	Departure	Arrival	Departure	Arrival	Departure
Xuzhou	10.5800	10.5800	11.3200	11.3200				
Gaojiaying	11.0600	11.0600	11.4000	11.4000			8.0000	8.0000
Sanpu	11.1400	11.3100	11.4530	11.4530			8.0400	8.0400
Caocun	11.4700	11.4700	11.5700	11.5700			8.1145	8.1145
Jiagou	12.0000	12.0000	12.1100	12.3500	8.1130	8.1130	8.2130	8.2130

(continued)

Table B.2 (continued)

Train No.	23005		11305		T114/1		K190/87	
Station	Arrival	Departure	Arrival	Departure	Arrival	Departure	Arrival	Departure
Fuliji	12.1900	12.2900	12.5330	12.5330	8.2247	8.2247	8.3400	8.3400
Suzhou	12.4400	12.4400	13.0430	13.0430	8.2907	8.2907	8.4300	8.4300
suzhouNan	12.4800	12.4800	13.0730	13.0730	8.3042	8.3042	8.4500	8.4500
Xisipo	12.5900	12.5900	13.1730	13.1730	8.3633	8.3633	8.5145	8.5145
Luling	13.0600	13.0600	13.2330	13.2330	8.4000	8.4000	8.5545	8.5545
Tangnanji	13.1900	13.1900	13.3630	13.3630	8.4712	8.4712	9.0430	9.0430
Guzhen	13.2330	13.2330	13.4100	13.4100	8.4937	8.4937	9.0730	9.0730
Liancheng	13.2930	13.2930	13.4900	13.5900	8.5307	8.5307	9.1145	9.1145
Xinmaqiao	13.3600	13.3600	14.0800	14.0800	8.5632	8.5632	9.1545	9.1545
Caolaoji	13.5000	13.5000	14.2100	14.2100	9.0547	9.0547	9.2615	9.2615
Bengbu	14.0400	14.0400	14.3500	14.3500	9.1700	9.2100	9.4000	9.4800
Bengbu Dong	14.3700	15.4300	15.1500	15.5300	9.2645	9.2645	9.5530	9.5530
Fengyang	16.0100	16.0100	16.1100	16.1100	9.3430	9.3430	10.0630	10.0630
Bangqiao	16.1030	16.1030	16.2030	16.2030	9.3920	9.3920	10.1300	10.1300
Xiaoxihe	16.1730	16.1730	16.2730	16.2730	9.4300	9.4300	10.1800	10.1800
Minggang	16.3530	16.3530	16.4530	16.4530	9.5150	9.5150	10.3200	10.3500
Bianzhuang	16.4130	16.4130	16.5130	16.5130	9.5450	9.5450	10.4100	10.4100
Guandian	16.5030	16.5030	17.0030	17.0030	9.5930	9.5930	10.4900	10.5200
Sanjie	17.0100	17.0100	17.1100	17.1100	10.0415	10.0415	11.0030	11.0030

(continued)

Table B.2 (continued)

Train No.	23005		11305		T114/1		K190/87	
Station	Arrival	Departure	Arrival	Departure	Arrival	Departure	Arrival	Departure
Zhangbaling	17.1400	17.1400	17.2400	17.2400	10.1015	10.1015	11.1000	11.1100
Shaheji	17.2200	17.2200	17.3200	17.3200	10.1445	10.1445	11.1900	11.1900
Chuzhou	17.3700	17.3900	17.4700	18.2300	10.2145	10.2145	11.2800	11.2800
Danzhi	17.5300	18.2700	18.3700	19.3700	10.2815	10.2815	11.3500	11.3500
Dongge	18.4700	19.5100	19.5700	20.0000	10.3615	10.3615	11.4430	11.4430
Yongnizhen	20.0300	20.0300	20.1400	20.2700	10.4300	10.4300	11.5300	12.0400
Linchang	20.1300	22.4700	20.4000	22.5400	10.4800	10.4800	12.1300	12.2600
Nanjing	23.0230	23.0230	23.1000	23.1000	10.5800	11.0600	12.3800	13.0000

Train No.	T54		K8471		K374/1		T118/5	
Station	Arrival	Departure	Arrival	Departure	Arrival	Departure	Arrival	Departure
Fuliji	8.0547	8.0547						
Suzhou	8.1207	8.1207						
suzhouNan	8.1342	8.1342	8.0300	8.0300				
Xisipo	8.1933	8.1933	8.0945	8.0945				
Luling	8.2300	8.2300	8.1345	8.1345				
Tangnanji	8.3012	8.3012	8.2400	8.3500	8.0500	8.0700		
Guzhen	8.3237	8.3237	8.4000	8.4000	8.1200	8.1200		
Liancheng	8.3607	8.3607	8.4415	8.4415	8.1800	8.2000		
Xinmaqiao	8.3932	8.3932	8.4815	8.4815	8.2600	8.2600		

(continued)

Table B.2 (continued)

Train No.	T54		K8471		K374/1		T118/5	
Station	Arrival	Departure	Arrival	Departure	Arrival	Departure	Arrival	Departure
Caolaoji	8.4847	8.4847	9.0000	9.1000	8.3630	8.3630		
Bengbu	9.0000	9.0200	9.2500	9.2900	8.5000	9.0800		
Bengbu Dong	9.0745	9.0745	9.3600	9.3600	9.1500	9.1500	8.0345	8.0345
Fengyang	9.1530	9.1530	9.4500	9.4500	9.2400	9.2400	8.1130	8.1130
Bangqiao	9.2020	9.2020	9.5030	9.5030	9.2930	9.2930	8.1620	8.1620
Xiaoxihe	9.2400	9.2400	9.5500	9.5500	9.3400	9.3400	8.2000	8.2000
Minggang	9.3250	9.3250	10.0600	10.0600	9.4600	9.5600	8.2850	8.2850
Bianzhuang	9.3550	9.3550	10.1000	10.1000	10.0200	10.0200	8.3150	8.3150
Guandian	9.4030	9.4030	10.1600	10.1600	10.0800	10.0800	8.3630	8.3630
Sanjie	9.4515	9.4515	10.2200	10.2200	10.1400	10.1400	8.4115	8.4115
Zhangbaling	9.5115	9.5115	10.3030	10.3030	10.2230	10.2230	8.4715	8.4715
Shaheji	9.5545	9.5545	10.3630	10.3630	10.2830	10.2830	8.5145	8.5145
Chuzhou	10.0245	10.0245	10.4700	10.5000	10.3730	10.3730	8.5845	8.5845
Danzhi	10.0915	10.0915	10.5900	10.5900	10.4430	10.4430	9.0515	9.0515
Dongge	10.1715	10.1715	11.0830	11.0830	10.5400	10.5400	9.1315	9.1315
Yongnizhen	10.2400	10.2400	11.1600	11.1600	11.0300	11.0400	9.2000	9.2000
Linchang	10.2830	10.2830	11.2130	11.2130	11.1130	11.1130	9.2430	9.2430
Nanjing	10.3700	10.4300	11.3200	−1.0000	11.2200	11.4100	9.3300	9.4600

(continued)

Table B.2 (continued)

Train No.	T284/1		T65		T131/4/1		K559/8	
Station	Arrival	Departure	Arrival	Departure	Arrival	Departure	Arrival	Departure
Fengyang	8.0330	8.0330						
Bangqiao	8.0820	8.0820	8.0020	8.0020				
Xiaoxihe	8.1200	8.1200	8.0400	8.0400				
Minggang	8.2050	8.2050	8.1250	8.1250	8.0650	8.0650		
Bianzhuang	8.2350	8.2350	8.1550	8.1550	8.0950	8.0950		
Guandian	8.2830	8.2830	8.2030	8.2030	8.1430	8.1430		
Sanjie	8.3315	8.3315	8.2515	8.2515	8.1915	8.1915		
Zhangbaling	8.3915	8.3915	8.3115	8.3115	8.2515	8.2515	8.0430	8.0430
Shaheji	8.4345	8.4345	8.3545	8.3545	8.2945	8.2945	8.1200	8.1500
Chuzhou	8.5045	8.5045	8.4245	8.4245	8.3645	8.3645	8.2800	8.2800
Danzhi	8.5715	8.5715	8.4915	8.4915	8.4315	8.4315	8.3600	9.2200
Dongge	9.0515	9.0515	8.5715	8.5715	8.5115	8.5115	9.3330	9.3330
Yongnizhen	9.1200	9.1200	9.0400	9.0400	8.5800	8.5800	9.4100	9.4100
Linchang	9.1630	9.1630	9.0830	9.0830	9.0230	9.0230	9.4800	9.4800
Nanjing	9.2500	9.3300	9.1700	−1.0000	9.1100	9.1400	9.5800	10.0500
Train No.	T166/3		T140/37					
Station	Arrival	Departure	Arrival	Departure				
Danzhi	8.0430	8.0430						

(continued)

Table B.2 (continued)

Train No.	T166/3		T140/37					
Station	Arrival	Departure	Arrival	Departure				
Dongge	8.1130	8.1130						
Yongnizhen	8.1800	8.1800						
Linchang	8.2205	8.2205						
Nanjing	8.3000	8.3300	8.0800	8.1100				

Note (1) The data of Table B.2 is from the timetable of railway network on 1/8/2010. (2) XX.YYZZ denotes XX: YY: ZZ

Table B.3 Rescheduled timetable of route 1

Train No.	G301		G303		G305		G101	
Station	Arrival	Departure	Arrival	Departure	Arrival	Departure	Arrival	Departure
Xuzhou Dong	8.0900	8.1100	8.4500	8.4600	9.1400	9.1600	9.3300	9.3500
Suzhou Dong	8.4208	8.4208	9.1708	9.1708	9.4708	9.4708	10.0538	10.0538
Bengbu Nan	9.1200	9.1300	9.4630	9.4730	10.1600	10.1700	10.3500	10.3600
Bengbu	9.2100	9.2100	9.5600	9.5600	10.2530	10.2530	10.4330	10.4330
Bengbu Dong	9.2630	9.2630	10.0130	10.0130	10.3130	10.3130	10.4930	10.4930
Fengyang	9.3330	9.3330	10.0830	10.0830	10.3800	10.3800	10.5530	10.5530
Banqiao	9.3730	9.3730	10.1300	10.1300	10.4200	10.4200	10.5900	10.5900
Xiaoxihe	9.4130	9.4130	10.1630	10.1630	10.4630	10.4630	11.0230	11.0230
Mingguang	9.4930	9.4930	10.2430	10.2430	10.5400	10.5400	11.1030	11.1030
Bianzhuang	9.5300	9.5300	10.2800	10.2800	10.5700	10.5700	11.1330	11.1330
Guandian	9.5730	9.5730	10.3200	10.3200	11.0030	11.0030	11.1700	11.1700
Sanjie	10.0130	10.0130	10.3600	10.3600	11.0400	11.0400	11.2100	11.2100
Zhangbaling	10.0630	10.0630	10.4130	10.4130	11.0900	11.0900	11.2530	11.2530
Shaheji	10.1000	10.1000	10.4600	10.4600	11.1230	11.1230	11.3000	11.3000
Chuzhou	10.1530	10.1530	10.5130	10.5130	11.1830	11.1830	11.3630	11.3630
Danzhi	10.2030	10.2030	10.5530	10.5530	11.2230	11.2230	11.4100	11.4100
Dongge	10.2630	10.2630	11.0200	11.0200	11.2900	11.2900	11.4800	11.4800
Yongningzhen	10.3030	10.3030	11.0630	11.0630	11.3300	11.3300	11.5130	11.5130
Linchang	10.3430	10.3430	11.1100	11.1100	11.3700	11.3700	11.5530	11.5530
Nanjing	10.4130	10.4130	11.1730	11.1730	11.4330	11.4330	**12.0230**	**12.0230**

(continued)

Table B.3 (continued)

Train No.	G301		G303		G305		G101	
Station	Arrival	Departure	Arrival	Departure	Arrival	Departure	Arrival	Departure
Nanjing Nan	10.4830	10.5030	11.2430	11.2530	11.5030	11.5230	**12.0930**	**12.1130**
Train No.	G103		G105		G107		G109	
Station	Arrival	Departure	Arrival	Departure	Arrival	Departure	Arrival	Departure
Xuzhou Dong	9.5200	9.5300	10.1800	10.1900	10.5300	10.5500	11.1300	11.1500
Suzhou Dong	10.2408	10.2408	10.4942	10.4942	11.2608	11.2608	11.4538	11.4538
Bengbu Nan	10.5300	10.5400	11.1940	11.2040	11.5530	11.5630	**12.1430**	**12.1530**
Bengbu	11.0130	11.0130	11.2904	11.2904	**12.0430**	**12.0430**	**12.2330**	**12.2330**
Bengbu Dong	11.0700	11.0700	11.3434	11.3434	**12.0930**	**12.0930**	**12.2830**	**12.2830**
Fengyang	11.1400	11.1400	11.4134	11.4134	**12.1600**	**12.1600**	**12.3530**	**12.3530**
Banqiao	11.1800	11.1800	11.4504	11.4504	**12.2000**	**12.2000**	**12.4000**	**12.4000**
Xiaoxihe	11.2200	11.2200	11.4904	11.4904	**12.2330**	**12.2330**	**12.4430**	**12.4430**
Mingguang	11.2930	11.2930	11.5734	11.5734	**12.3130**	**12.3130**	**12.5230**	**12.5230**
Bianzhuang	11.3330	11.3330	**12.0034**	**12.0034**	**12.3530**	**12.3530**	**12.5530**	**12.5530**
Guandian	11.3800	11.3800	**12.0504**	**12.0504**	**12.4000**	**12.4000**	**13.0000**	**13.0000**
Sanjie	11.4130	11.4130	**12.0834**	**12.0834**	**12.4430**	**12.4430**	**13.0330**	**13.0330**
Zhangbaling	11.4700	11.4700	**12.1334**	**12.1334**	**12.4900**	**12.4900**	**13.0900**	**13.0900**
Shaheji	11.5100	11.5100	**12.1734**	**12.1734**	**12.5300**	**12.5300**	**13.1230**	**13.1230**
Chuzhou	11.5630	11.5630	**12.2304**	**12.2304**	**12.5830**	**12.5830**	**13.1830**	**13.1830**
Danzhi	**12.0100**	**12.0100**	**12.2704**	**12.2704**	**13.0300**	**13.0300**	**13.2330**	**13.2330**

(continued)

Table B.3 (continued)

Train No.	G103		G105		G107		G109	
Station	Arrival	Departure	Arrival	Departure	Arrival	Departure	Arrival	Departure
Dongge	**12.0730**	**12.0730**	**12.3404**	**12.3404**	**13.1000**	**13.1000**	**13.3000**	**13.3000**
Yongningzhen	**12.1200**	**12.1200**	**12.3834**	**12.3834**	**13.1330**	**13.1330**	**13.3430**	**13.3430**
Linchang	**12.1700**	**12.1700**	**12.4334**	**12.4334**	**13.1830**	**13.1830**	**13.3930**	**13.3930**
Nanjing	**12.2400**	**12.2400**	**12.4934**	**12.4934**	**13.2430**	**13.2430**	**13.4630**	**13.4630**
Nanjing Nan	**12.3100**	**12.3300**	**12.5630**	**12.5830**	**13.3130**	**13.3300**	**13.5330**	**13.5530**

Train No.	G111		G113					
Station	Arrival	Departure	Arrival	Departure				
Xuzhou Dong	11.3300	11.3500	11.5300	11.5500				
Suzhou Dong	**12.0538**	**12.0538**	**12.2508**	**12.2508**				
Bengbu Nan	**12.3430**	**12.3530**	**12.5500**	**12.5600**				
Bengbu	**12.4330**	**12.4330**	**13.0430**	**13.0430**				
Bengbu Dong	**12.4930**	**12.4930**	**13.1030**	**13.1030**				
Fengyang	**12.5630**	**12.5630**	**13.1730**	**13.1730**				
Banqiao	**13.0100**	**13.0100**	**13.2100**	**13.2100**				
Xiaoxihe	**13.0530**	**13.0530**	**13.2500**	**13.2500**				
Mingguang	**13.1300**	**13.1300**	**13.3300**	**13.3300**				
Bianzhuang	**13.1630**	**13.1630**	**13.3600**	**13.3600**				
Guandian	**13.2100**	**13.2100**	**13.4000**	**13.4000**				
Sanjie	**13.2500**	**13.2500**	**13.4330**	**13.4330**				

(continued)

Table B.3 (continued)

Train No.	G111		G113					
Station	Arrival	Departure	Arrival	Departure				
Zhangbaling	**13.2930**	**13.2930**	**13.4830**	**13.4830**				
Shaheji	**13.3300**	**13.3300**	**13.5200**	**13.5200**				
Chuzhou	**13.3900**	**13.3900**	**13.5730**	**13.5730**				
Danzhi	**13.4330**	**13.4330**	**14.0230**	**14.0230**				
Dongge	**13.4930**	**13.4930**	**14.0830**	**14.0830**				
Yongningzhen	**13.5300**	**13.5300**	**14.1230**	**14.1230**				
Linchang	**13.5700**	**13.5700**	**14.1700**	**14.1700**				
Nanjing	**14.0330**	**14.0330**	**14.2400**	**14.2400**				
Nanjing Nan	**14.1030**	**14.1230**	**14.3100**	**14.3300**				

Table B.4 Rescheduled timetable of route 2

Train No.	K58/5		K518/5		K101/4/1		D88/5	
Station	Arrival	Departure	Arrival	Departure	Arrival	Departure	Arrival	Departure
Xuzhou	9.1600	9.3500	9.3000	9.4100	9.4400	9.4700	10.0200	10.0300
Gaojiaying	9.4300	9.4300	9.4900	9.4900	9.5500	9.5500	10.0929	10.0929
Sanpu	9.4700	9.4700	9.5300	9.5300	9.5900	9.5900	10.1202	10.1202
Caocun	9.5445	9.5445	10.0045	10.0045	10.0800	10.2100	10.1624	10.1624
Jiagou	10.0430	10.0430	10.1200	10.2400	10.3245	10.3245	10.2049	10.2049
Fuliji	10.1700	10.1700	10.3830	10.3830	10.4515	10.4515	10.2700	10.2700
Suzhou	10.2445	10.2445	10.4615	10.4615	10.5300	10.5300	10.3230	10.3230
SuzhouNan	10.2800	10.4000	10.4815	10.4815	10.5500	10.5500	10.3353	10.3353
Xishipo	10.4845	10.4845	10.5500	10.5500	11.0145	11.0145	10.3757	10.3757
Luling	10.5245	10.5245	10.5900	10.5900	11.0545	11.0545	10.4018	10.4018
Tangnaji	11.0130	11.0130	11.0745	11.0745	11.1430	11.1430	10.4528	10.4528
Guzhen	11.0430	11.0430	11.1045	11.1045	11.1730	11.1730	10.4732	10.4732
Liancheng	11.0845	11.0845	11.1500	11.1500	11.2145	11.2145	10.5033	10.5033
Xinmaqiao	11.1245	11.1245	11.1900	11.1900	11.2545	11.2545	10.5327	10.5327
Caolaoqiao	11.2315	11.2315	11.2930	11.2930	11.3615	11.3615	11.0117	11.0117
Bengbu	11.3700	11.4200	11.4300	11.4800	11.5000	11.5800	11.0830	11.0930
Bengbu Dong	11.4800	11.4800	11.5400	11.5400	**12.0300**	**12.1530**	11.1500	11.1500
Fengyang	11.5700	11.5700	**12.0300**	**12.0300**	**12.2430**	**12.2430**	11.2152	11.2152
Banqiao	**12.0300**	**12.0300**	**12.0900**	**12.0900**	**12.3030**	**12.3030**	11.2558	11.2558
Xiaoxihe	**12.0730**	**12.0730**	**12.1330**	**12.1330**	**12.3500**	**12.3500**	11.2922	11.2922

(continued)

Table B.4 (continued)

Train No.	K58/5		K518/5		K101/4/1		D88/5	
Station	Arrival	Departure	Arrival	Departure	Arrival	Departure	Arrival	Departure
Mingguang	**12.1630**	**12.1630**	**12.2230**	**12.2230**	**12.4400**	**12.4400**	11.3740	11.3740
Bianzhuang	**12.2030**	**12.2030**	**12.2630**	**12.2630**	**12.4800**	**12.4800**	11.4030	11.4030
Guandian	**12.2500**	**12.2500**	**12.3100**	**12.3100**	**12.5400**	**13.0600**	11.4434	11.4434
Sanjie	**12.3030**	**12.3030**	**12.3730**	**12.5030**	**13.1230**	**13.1230**	11.4902	11.4902
Zhangbaling	**12.3730**	**12.3730**	**12.5800**	**12.5800**	**13.1930**	**13.1930**	11.5416	11.5416
Shaheji	**12.4230**	**12.4230**	**13.0230**	**13.0230**	**13.2430**	**13.2430**	11.5813	11.5813
Chuzhou	**12.5100**	**13.0430**	**13.1000**	**13.1100**	**13.3100**	**13.3100**	**12.0413**	**12.0413**
Danzhi	**13.1000**	**13.1000**	**13.1730**	**13.3000**	**13.3630**	**13.4930**	**12.0841**	**12.0841**
Dongge	**13.2000**	**13.2000**	**13.4000**	**13.4000**	**13.5930**	**14.0330**	**12.1530**	**12.1530**
Yongningzhen	**13.2400**	**13.2400**	**13.4400**	**13.4400**	**14.0830**	**14.1030**	**12.1846**	**12.1846**
Linchang	**13.2900**	**13.2900**	**13.4900**	**14.0300**	**14.1930**	**14.2300**	**12.2237**	**12.2237**
Nanjing	**13.3800**	**13.5300**	**14.1200**	**14.1500**	**14.3200**	**14.3700**	**12.3000**	**12.3200**
Train No.	1230/27		10135		10625		11301	
Station	Arrival	Departure	Arrival	Departure	Arrival	Departure	Arrival	Departure
Xuzhou	9.5600	10.0900	10.1800	10.1800	10.2800	10.2800	10.3800	10.3800
Gaojiaying	10.1700	10.1700	10.2600	10.2600	10.3600	10.3600	10.4600	10.4600
Sanpu	10.2200	10.2500	10.3200	10.3200	10.4200	10.4200	10.5200	10.5200
Caocun	10.3500	10.3500	10.4500	10.4500	10.5500	10.5500	11.0500	11.0500
Jiagou	10.4445	10.4445	10.5800	10.5800	11.0800	11.0800	11.2000	11.4100

(continued)

Table B.4 (continued)

Train No.	1230/27		10135		10625		11301	
Station	Arrival	Departure	Arrival	Departure	Arrival	Departure	Arrival	Departure
Fuliji	10.5715	10.5715	11.1700	11.2200	11.2700	11.5100	**12.0100**	**12.0100**
Suzhou	11.0600	11.0800	11.3700	11.3700	**12.0600**	**12.0600**	**12.1300**	**12.1300**
suzhouNan	11.1200	11.1200	11.4200	**11.5900**	**12.0900**	**12.0900**	**12.1600**	**12.1600**
Xisipo	11.1845	11.1845	**12.1300**	**12.1300**	**12.2000**	**12.2000**	**12.2700**	**12.2700**
Luling	11.2245	11.2245	**12.2000**	**12.2000**	**12.2700**	**12.2700**	**12.3330**	**12.3330**
Tangnanji	11.3130	11.3130	**12.3300**	**12.3300**	**12.4000**	**12.4000**	**12.4630**	**12.4630**
Guzhen	11.3430	11.3430	**12.3730**	**12.3730**	**12.4700**	**12.4900**	**12.5500**	**12.5600**
Liancheng	11.3845	11.3845	**12.4330**	**12.4330**	**12.5800**	**12.5800**	**13.0500**	**13.0500**
Xinmaqiao	11.4245	11.4245	**12.5000**	**12.5000**	**13.0430**	**13.0430**	**13.1130**	**13.1130**
Caolaoji	11.5315	11.5315	**13.0400**	**13.0400**	**13.1830**	**13.1830**	**13.2530**	**13.2530**
Bengbu	**12.1030**	**12.1030**	**13.1700**	**13.1700**	**13.3100**	**13.3100**	**13.3930**	**13.3930**
Bengbu Dong	**12.1700**	**12.3500**	**13.2700**	**13.3900**			**13.4900**	**14.0100**
Fengyang	**12.4500**	**12.4500**	**13.5700**	**13.5700**			**14.2100**	**14.3100**
Banqiao	**12.5200**	**12.5200**	**14.0630**	**14.0630**			**14.4330**	**14.4330**
Xiaoxihe	**12.5700**	**13.1130**	**14.1330**	**14.1330**			**14.5300**	**15.2900**
Mingguang	**13.2230**	**13.2230**	**14.3400**	**14.4800**			**15.5000**	**15.5000**
Bianzhuang	**13.2730**	**13.2730**	**14.5900**	**15.4300**			**15.5800**	**15.5800**
Guandian	**13.3330**	**13.4600**	**15.5500**	**15.5500**			**16.0900**	**16.2600**
Sanjie	**13.5300**	**13.5300**	**16.0800**	**16.2900**			**16.3930**	**16.3930**

(continued)

Table B.4 (continued)

Train No.	1230/27		10135		10625		11301	
Station	Arrival	Departure	Arrival	Departure	Arrival	Departure	Arrival	Departure
Zhangbaling	**14.0100**	**14.0100**	**16.4500**	**16.4500**			**16.5230**	**16.5230**
Shaheji	**14.0700**	**14.0700**	**16.5300**	**16.5300**			**17.0030**	**17.0030**
Chuzhou	**14.1500**	**14.1800**	**17.0530**	**17.0530**			**17.1300**	**17.1300**
Danzhi	**14.2500**	**14.2500**	**17.1600**	**17.1800**			**17.2400**	**17.2500**
Dongge	**14.3600**	**14.3600**	**17.3600**	**17.3600**			**17.4300**	**17.4300**
Yongnizhen	**14.4100**	**14.4100**	**17.4500**	**17.4500**			**17.5200**	**17.5200**
Linchang	**14.4800**	**14.5300**	**18.0200**	**20.0700**			**18.5100**	**20.4100**
Nanjing	**15.0300**	**15.1500**	**20.2300**	**20.2300**			**20.5700**	**20.5700**
Train No.	23005		11305		T114/1		K190/87	
Station	Arrival	Departure	Arrival	Departure	Arrival	Departure	Arrival	Departure
Xuzhou	10.5800	10.5800	11.3200	11.3200				
Gaojiaying	11.0600	11.0600	11.4000	11.4000			8.0000	8.0000
Sanpu	11.1400	11.3100	11.4530	11.4530			8.0400	8.0400
Caocun	11.4700	11.4700	11.5700	11.5700			8.1145	8.1145
Jiagou	**12.0000**	**12.0000**	**12.1100**	**12.3500**	8.1130	8.1130	8.2130	8.2130
Fuliji	**12.1900**	**12.2900**	**12.5330**	**12.5330**	8.2247	8.2247	8.3400	8.3400
Suzhou	**12.4400**	**12.4400**	**13.0430**	**13.0430**	8.2907	8.2907	8.4300	8.4300
SuzhouNan	**12.4800**	**12.4800**	**13.0730**	**13.0730**	8.3042	8.3042	8.4500	8.4500
Xisipo	**12.5900**	**12.5900**	**13.1730**	**13.1730**	8.3633	8.3633	8.5145	8.5145

(continued)

Table B.4 (continued)

Train No.	23005		11305		T114/1		K190/87	
Station	Arrival	Departure	Arrival	Departure	Arrival	Departure	Arrival	Departure
Luling	**13.0600**	**13.0600**	**13.2330**	**13.2330**	8.4000	8.4000	8.5545	8.5545
Tangnanji	**13.1900**	**13.1900**	**13.3630**	**13.3630**	8.4712	8.4712	9.0430	9.0430
Guzhen	**13.2330**	**13.2330**	**13.4100**	**13.4100**	8.4937	8.4937	9.0730	9.0730
Liancheng	**13.2930**	**13.2930**	**13.4900**	**13.5900**	8.5307	8.5307	9.1145	9.1145
Xinmaqiao	**13.3600**	**13.3600**	**14.0800**	**14.0800**	8.5632	8.5632	9.1545	9.1545
Caolaoji	**13.5000**	**13.5000**	**14.2100**	**14.2100**	9.0547	9.0547	9.2615	9.2615
Bengbu	**14.0400**	**14.0400**	**14.3500**	**14.3500**	9.2700	9.2900	9.4000	9.4800
Bengbu Dong	**14.3700**	**15.4300**	**15.1500**	**15.5300**	9.3430	9.3430	9.5530	9.5530
Fengyang	**16.0100**	**16.0100**	**16.1100**	**16.1100**	9.4130	9.4130	10.0230	10.1430
Bangqiao	**16.1030**	**16.1030**	**16.2030**	**16.2030**	9.4600	9.4600	10.1900	10.1900
Xiaoxihe	**16.1730**	**16.1730**	**16.2730**	**16.2730**	9.5030	9.5030	10.2230	10.2230
Minggang	**16.3530**	**16.3530**	**16.4530**	**16.4530**	9.5900	9.5900	10.3200	10.3500
Bianzhuang	**16.4130**	**16.4130**	**16.5130**	**16.5130**	10.0215	10.0215	10.4100	10.4100
Guandian	**16.5030**	**16.5030**	**17.0030**	**17.0030**	10.0630	10.0630	10.4900	10.5200
Sanjie	**17.0100**	**17.0100**	**17.1100**	**17.1100**	10.1130	10.1130	11.5800	11.1000
Zhangbaling	**17.1400**	**17.1400**	**17.2400**	**17.2400**	10.1700	10.1700	11.1700	11.1700
Shaheji	**17.2200**	**17.2200**	**17.3200**	**17.3200**	10.2045	10.2045	11.2400	11.3600
Chuzhou	**17.3700**	**17.3900**	**17.4700**	**18.2300**	10.2530	10.2530	11.4630	11.4630
Danzhi	**17.5300**	**18.2700**	**18.3700**	**19.3700**	10.3030	10.3030	11.5430	12.1500

(continued)

Table B.4 (continued)

Train No.	23005		11305		T114/1		K190/87	
Station	Arrival	Departure	Arrival	Departure	Arrival	Departure	Arrival	Departure
Dongge	**18.4700**	**19.5100**	**19.5700**	**20.0000**	10.3615	10.3615	12.2500	12.2500
Yongnizhen	**20.0300**	**20.0300**	**20.1400**	**20.2700**	10.4300	10.4300	12.2830	12.2830
Linchang	**20.1300**	**22.4700**	**20.4000**	**22.5400**	10.4800	10.4800	12.3500	12.5000
Nanjing	**23.0230**	**23.0230**	**23.1000**	**23.1000**	10.5800	11.0600	12.5800	13.0000
Train No.	T54		K8471		K374/1		T118/5	
Station	Arrival	Departure	Arrival	Departure	Arrival	Departure	Arrival	Departure
Fuliji	8.0547	8.0547						
Suzhou	8.1207	8.1207						
SuzhouNan	8.1342	8.1342	8.0300	8.0300				
Xisipo	8.1933	8.1933	8.0945	8.0945				
Luling	8.2300	8.2300	8.1345	8.1345				
Tangnanji	8.3012	8.3012	8.2400	8.3700	8.0500	8.0700		
Guzhen	8.3237	8.3237	8.4000	8.4000	8.1200	8.1200		
Liancheng	8.3607	8.3607	8.4415	8.4415	8.1800	8.2000		
Xinmaqiao	8.3932	8.3932	8.4815	8.4815	8.2600	8.2600		
Caolaoji	8.4847	8.4847	8.5900	10.1200	8.3630	8.3630		
Bengbu	9.0000	9.0200	9.3300	9.3500	8.5000	9.0800		
Bengbu Dong	9.0745	9.0745	9.4230	9.4230	9.1500	9.1500	8.0345	8.0345
Fengyang	9.1530	9.1530	9.5200	9.5200	9.2400	9.2400	8.1130	8.1130

(continued)

Table B.4 (continued)

Train No.	T54		K8471		K374/1		T118/5	
Station	Arrival	Departure	Arrival	Departure	Arrival	Departure	Arrival	Departure
Bangqiao	9.2020	9.2020	9.5730	9.5730	9.2930	9.2930	8.1620	8.1620
Xiaoxihe	9.2400	9.2400	10.0200	10.0200	9.3400	9.3400	8.2000	8.2000
Minggang	9.3250	9.3250	10.1030	10.1030	9.4330	10.1630	8.2850	8.2850
Bianzhuang	9.3550	9.3550	10.1430	10.1430	10.2030	10.3400	8.3150	8.3150
Guandian	9.4030	9.4030	10.2030	10.2030	10.4000	10.4000	8.3630	8.3630
Sanjie	9.4515	9.4515	10.2630	10.2630	10.4600	10.4600	8.4115	8.4115
Zhangbaling	9.5115	9.5115	10.3500	10.4730	10.5430	10.5430	8.4715	8.4715
Shaheji	9.5545	9.5545	10.3630	10.3630	11.0030	11.0030	8.5145	8.5145
Chuzhou	10.0245	10.0245	11.0100	11.0400	11.1000	11.2430	8.5845	8.5845
Danzhi	10.0915	10.0915	11.1300	11.1300	11.3200	11.3200	9.0515	9.0515
Dongge	10.1715	10.1715	11.2300	11.3500	11.4200	11.5400	9.1315	9.1315
Yongnizhen	10.2400	10.2400	11.4230	11.4230	12.0300	12.0400	9.2000	9.2000
Linchang	10.2830	10.2830	11.4800	11.4800	12.1100	12.2300	9.2430	9.2430
Nanjing	10.3530	10.4730	11.5600	−1.0000	12.3600	12.3900	9.3300	9.4600
Train No.	T284/1		T65		T131/4/1		K559/8	
Station	Arrival	Departure	Arrival	Departure	Arrival	Departure	Arrival	Departure
Fengyang	8.0330	8.0330						
Bangqiao	8.0820	8.0820	8.0020	8.0020				
Xiaoxihe	8.1200	8.1200	8.0400	8.0400				

(continued)

Table B.4 (continued)

Train No.	T284/1		T65		T131/4/1		K559/8	
Station	Arrival	Departure	Arrival	Departure	Arrival	Departure	Arrival	Departure
Minggang	8.2050	8.2050	8.1250	8.1250	8.0650	8.0650		
Bianzhuang	8.2350	8.2350	8.1550	8.1550	8.0950	8.0950		
Guandian	8.2830	8.2830	8.2030	8.2030	8.1430	8.1430		
Sanjie	8.3315	8.3315	8.2515	8.2515	8.1915	8.1915		
Zhangbaling	8.3915	8.3915	8.3115	8.3115	8.2515	8.2515	8.0430	8.0430
Shaheji	8.4345	8.4345	8.3545	8.3545	8.2945	8.2945	8.1200	8.1500
Chuzhou	8.5045	8.5045	8.4245	8.4245	8.3645	8.3645	8.2800	8.2800
Danzhi	8.5715	8.5715	8.4915	8.4915	8.4315	8.4315	8.3600	9.2200
Dongge	9.0515	9.0515	8.5715	8.5715	8.5115	8.5115	9.3330	9.3330
Yongnizhen	9.1200	9.1200	9.0400	9.0400	8.5800	8.5800	9.4100	9.4100
Linchang	9.1630	9.1630	9.0830	9.0830	9.0230	9.0230	9.4800	9.4800
Nanjing	9.2500	9.3300	9.1700	−1.0000	9.1100	9.1400	9.5800	10.0500
Train No.	T166/3		T140/37					
Station	Arrival	Departure	Arrival	Departure				
Danzhi	8.0430	8.0430						
Dongge	8.1130	8.1130						
Yongnizhen	8.1800	8.1800						
Linchang	8.2205	8.2205						
Nanjing	8.3000	8.3300	8.0800	8.1100				

Note XX.YY.ZZ denotes XX:YY:ZZ

Appendix C
The Timetables

See Figs. C.1, C.2, and C.3.

L. Jia et al., *High-Speed Railway Operation Under Emergent Conditions*,
Advances in High-speed Rail Technology,
https://doi.org/10.1007/978-3-662-63033-4

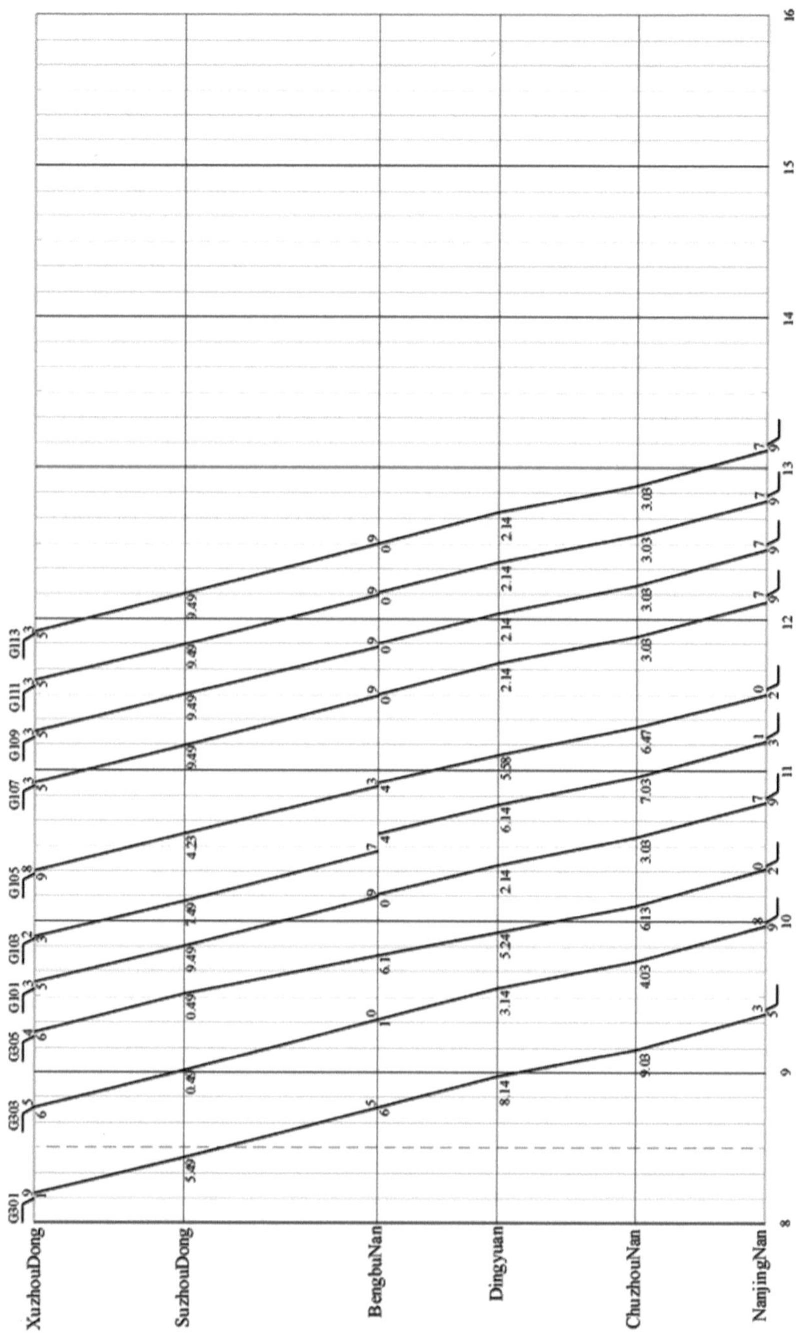

Fig. C.1 Scheduled timetable of high-speed railway. *Note* AA.3 denotes AA minutes after 30 s

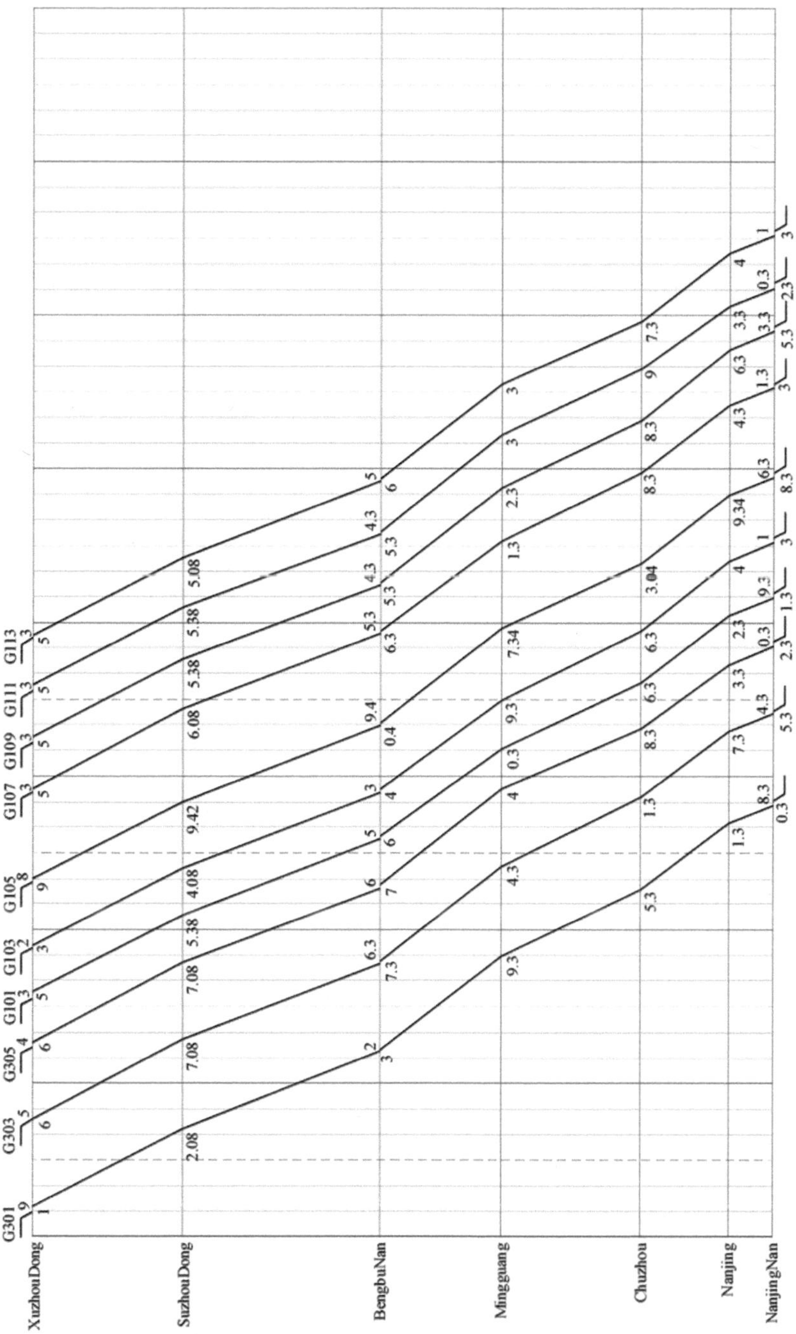

Fig. C.2 Rescheduled timetable of route 1. *Note* AA.3 denotes AA minutes 30 s

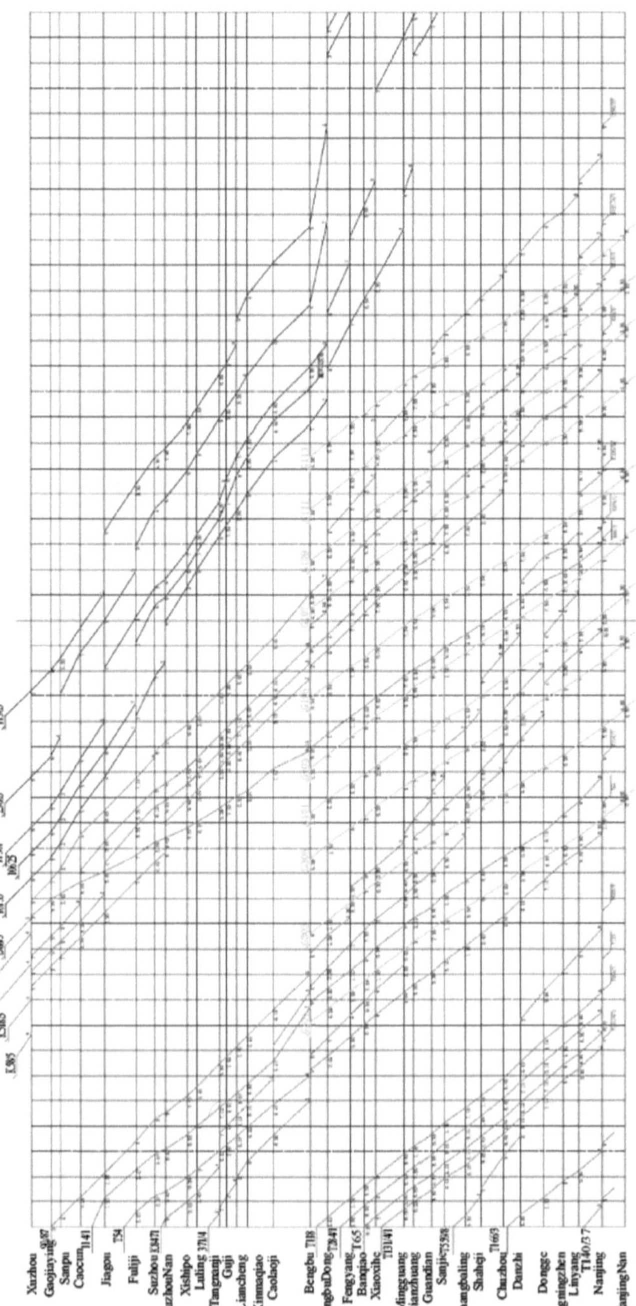

Fig. C.3 Rescheduled timetable of route 2 (route 1 and route 2 having the same Bengbu-Nanjing section)

Appendix D
The Free Flow Running Time of Sections

See Tables D.1 and D.2.

L. Jia et al., *High-Speed Railway Operation Under Emergent Conditions*,
Advances in High-speed Rail Technology,
https://doi.org/10.1007/978-3-662-63033-4

Table D.1 Minimal running time of Xuzhou-Nanjing section of common speed trains

Station	Distance/km	The free flow running time				
		High-speed train	Express train	Fast train	Common slow train	Freight train
Xuzhou	9	2.42	3.23	4.30	5.24	6.45
Gaojiaying	7	2.06	2.38	3.30	4.12	5.15
Sanpu	15	4.30	5.38	7.30	9.00	11.15
Caocun	14	4.12	5.15	7.00	8.24	10.30
Jiagou	16	4.48	6.00	8.00	9.36	12.00
Fuliji	14	4.12	5.15	7.00	8.24	10.30
Suzhou	2	0.36	0.45	1.00	1.12	1.30
SuzhouNan	15	4.30	5.38	7.30	9.00	11.15
Xisipo	8	2.24	3.00	4.00	4.48	6.00
Luling	15	4.30	5.38	7.30	9.00	11.15
Tangnanji	7	2.06	2.38	3.30	4.12	5.15
Guzhen	7	2.06	2.38	3.30	4.12	5.15
Liancheng	8	2.24	3.00	4.00	4.48	6.00
Xinmaqiao	14	4.12	5.15	7.00	8.24	10.30
Caolaoji	14	4.12	5.15	7.00	8.24	10.30
Bengbu	8	2.24	3.00	4.00	4.48	6.00
Bengbu Dong	16	4.48	6.00	8.00	9.36	12.00
Fengyang	11	3.18	4.08	5.30	6.36	8.15
Bangqiao	8	2.24	3.00	4.00	4.48	6.00
Xiaoxihe	18	5.24	6.45	9.00	10.48	13.30
Minggang	8	2.24	3.00	4.00	4.48	6.00
Bianzhuang	9	2.42	3.23	4.30	5.24	6.45
Guandian	11	3.18	4.08	5.30	6.36	8.15
Sanjie	13	3.54	4.53	6.30	7.48	9.45
Zhangbaling	10	3.00	3.45	5.00	6.00	7.30
Shaheji	13	3.54	4.53	6.30	7.48	9.45
Chuzhou	9	2.42	3.23	4.30	5.24	6.45
Danzhi	18	5.24	6.45	9.00	10.48	13.30
Dongge	6	1.48	2.15	3.00	3.36	4.30
Yongnizhen	9	2.42	3.23	4.30	5.24	6.45
Linchang	14	4.12	5.15	7.00	8.24	10.30
Nanjing						

Note AA.BB denotes AA min BB s

Table D.2 Minimal running time of Xuzhou Dong-Nanjing Nan section under different speed limitation

Station	Distance /km	The minimal running time		
		350 km/h	160 km/h	70 km/h
Xuzhou Dong	67	11.29	25.08	57.26
Suzhou Dong	88	15.05	33.00	75.26
Bengbu Nan	54	9.15	20.15	46.17
Ding yuan	62	10.38	23.15	53.09
Chuzhou Nan	59	10.07	22.08	50.34
Nanjing Nan				

Note AA.BB denotes AA min BB s

Appendix E
The Set of Sections and Stations in Routes

See Tables E.1 and E.2.

L. Jia et al., *High-Speed Railway Operation Under Emergent Conditions*,
Advances in High-speed Rail Technology,
https://doi.org/10.1007/978-3-662-63033-4

Table E.1 Weight of sections in different routes

Route 1			Route 2		
Station	Section length/km	Weight	Station	Section length/km	Weight
Xuzhou Dong	67	0.1671	Xuzhou	9	0.0260
Suzhou Dong	88	0.2195	Gaojiaying	7	0.0202
Bengbu Nan	54	0.1347	Sanpu	15	0.0434
Bengbu	8	0.0200	Caocun	14	0.0405
Bengbu Dong	16	0.0399	Jiagou	16	0.0462
Fengyang	11	0.0274	Fuliji	14	0.0405
Bangqiao	8	0.0200	Suzhou	2	0.0058
Xiaoxihe	18	0.0449	SuzhouNan	15	0.0434
Minggang	8	0.0200	Xisipo	8	0.0231
Bianzhuang	9	0.0224	Luling	15	0.0434
Guandian	11	0.0274	Tangnanji	7	0.0202
Sanjie	13	0.0324	Guzhen	7	0.0202
Zhangbaling	10	0.0249	Liancheng	8	0.0231
Shaheji	13	0.0324	Xinmaqiao	14	0.0405
Chuzhou	9	0.0224	Caolaoji	14	0.0405
Danzhi	18	0.0449	Bengbu	8	0.0231
Dongge	6	0.0150	Bengbu Dong	16	0.0462
Yongnizhen	9	0.0224	Fengyang	11	0.0318
Linchang	14	0.0349	Bangqiao	8	0.0231
Nanjing	11	0.0274	Xiaoxihe	18	0.0520
Nanjing Nan			Minggang	8	0.0231
			Bianzhuang	9	0.0260
			Guandian	11	0.0318
			Sanjie	13	0.0376
			Zhangbaling	10	0.0289
			Shaheji	13	0.0376
			Chuzhou	9	0.0260
			Danzhi	18	0.0520
			Dongge	6	0.0173
			Yongnizhen	9	0.0260
			Linchang	14	0.0405
			Nanjing		

Table E.2 Weight of stations in routes

Route 1			Route 2		
Station	Grade	Weight	Station	Grade	Weight
Xuzhou Dong	High-speed intermediate station	1.50	Xuzhou	Special station	1.00
Suzhou Dong	High-speed intermediate station	1.50	Gaojiaying	Fourth class station	0.25
Bengbu Nan	High-speed intermediate station	1.50	Sanpu	Fourth class station	0.25
Bengbu	First class station	1.00	Caocun	Fourth class station	0.25
Bengbu Dong	First class station	1.00	Jiagou	Fourth class station	0.25
Fengyang	Third class station	0.33	Fuliji	Third class station	0.33
Bangqiao	Fourth class station	0.25	Suzhou	Second class station	0.50
Xiaoxihe	Fourth class station	0.25	SuzhouNan	Fifth class station	0.20
Minggang	Third class station	0.33	Xisipo	Fourth class station	0.25
Bianzhuang	Fourth class station	0.25	Luling	Third class station	0.33
Guandian	Fourth class station	0.25	Tangnanji	Fourth class station	0.25
Sanjie	Fourth class station	0.25	Guzhen	Third class station	0.33
Zhangbaling	Fourth class station	0.25	Liancheng	Fourth class station	0.25
Shaheji	Fourth class station	0.25	Xinmaqiao	Fourth class station	0.25
Chuzhou	Second class station	0.5	Caolaoji	Third class station	0.33
Danzhi	Fourth class station	0.25	Bengbu	First class station	1.00
Dongge	Fourth class station	0.25	Bengbu Dong	First class station	1.00
Yongnizhen	Fourth class station	0.25	Fengyang	Third class station	0.33
Linchang	Fourth class station	0.25	Bangqiao	Fourth class station	0.25
Nanjing	First class station	1.00	Xiaoxihe	Fourth class station	0.25
			Minggang	Third class station	0.33
			Bianzhuang	Fourth class station	0.25
			Guandian	Fourth class station	0.25
			Sanjie	Fourth class station	0.25
			Zhangbaling	Fourth class station	0.25
			Shaheji	Fourth class station	0.25
			Chuzhou	Second class station	0.5
			Danzhi	Fourth class station	0.25
			Dongge	Fourth class station	0.25
			Yongnizhen	Fourth class station	0.25
			Linchang	Fourth class station	0.25
			Nanjing	First class station	1.00